PARTICLES, STRINGS, AND COSMOLOGY

To learn more about the AIP Conference Proceedings, including the Conference Proceedings Series, please visit the webpage
http://proceedings.aip.org/proceedings

PARTICLES, STRINGS, AND COSMOLOGY

13th International Symposium on Particles, Strings, and Cosmology

PASCOS 2007

London, United Kingdom 2 – 7 July 2007

EDITORS

Arttu Rajantie
Paul Dauncey
Carlo Contaldi
Horace Stoica

Imperial College London
London, United Kingdom

SPONSORING ORGANIZATIONS

Science and Technology Facilities Council
Centre for Theoretical Cosmology
Institute for Particle Physics Phenomenology
Institute of Physics
Imperial College London

AMERICAN INSTITUTE OF PHYSICS

Melville, New York, 2007
AIP CONFERENCE PROCEEDINGS ■ 957

Editors

Arttu Rajantie
Paul Dauncey
Carlo Contaldi
Horace Stoica

Department of Physics
Imperial College London
London SW7 2AZ
United Kingdom

E-mail: a.rajantie@imperial.ac.uk
p.dauncey@imperial.ac.uk
c.contaldi@imperial.ac.uk
f.stoica@imperial.ac.uk

L.C. Catalog Card No. 2007939146
ISBN 978-0-7354-0471-7
ISSN 0094-243X

Printed in the United States of America

Contents

PLENARY TALKS

PARTICLES

STRINGS

COSMOLOGY

PREFACE

The 13th International Symposium on Particles, Strings and Cosmology was held in the Physics Department of Imperial College London from 2 to 7 July 2007. It was the first PASCOS symposium organised in Europe, and a truly global event with 198 registered participants from 26 countries all around the world. The programme consisted of five days of scientific conference with more than a hundred talks and poster presentations, followed by a public event marking the 50th anniversary of the arrival of the late Nobel Laureate Abdus Salam at Imperial College. The symposium was part of Imperial's centenary celebrations.

Since the first PASCOS symposium at the Northeastern University in 1990, particle physics, cosmology and string theory have got ever closer together, and these symposia have been a driving force in this development. Today, string theory is the foundation on which most theoretical work in particle physics and cosmology is based. Discoveries in particle physics have immediate implications for cosmology, and vice versa, and one day they will hopefully be able to provide an empirical test of string theory. PASCOS symposia have thus become even more important as a way of keeping abreast of new developments in other fields.

The symposium was, by all accounts, a great success. It gave a comprehensive review of the status and the new developments in particle physics, cosmology and string theory. It also painted a picture of an exciting future, in an atmosphere of great anticipation, with LHC, Planck and many other ground-breaking experiments under way or being planned.

I would like to thank the members of the organising committees and the Imperial College staff and graduate students whose help was vital for the smooth running of the symposium. The symposium would not have been possible without generous financial support by the sponsors listed on the following page, whose help is gratefully acknowledged. And above all, I would like to thank all the speakers, poster presenters and other participants for contributing to the scientific discussion and exchange of ideas during the symposium.

I hope these proceedings will form a valuable resource not only for the participants but for all particle physicists, string theorists and cosmologists. They are supplemented by the original slides and video recordings of many of the talks, which are available on the conference website http://www.pascos07.org.

5 October 2007
Arttu Rajantie
Symposium Chair

ORGANISATION

International Advisory Committee

J. Bagger (Johns Hopkins)
P. Frampton (N. Carolina)
J. Gunion (UC Davis)
J.E. Kim (Seoul National)
P. Nath (Northeastern U)
D.P. Roy (TIFR, Mumbai)
S. Raby (OSU)
J. Silk (Oxford)
M.T. Vaughn (Northeastern U)
K.C. Wali (Syracuse)

Local Organising Committee

A. Rajantie (chair)
S.A. Abel (IPPP, Durham)
C.R. Contaldi
P.D. Dauncey
G.J. Davies
M.J. Duff
T.W.B. Kibble
R.J. Rivers
K.S. Stelle
H. Stoica
T. Wiseman

Parallel Session Conveners

Particles:
P.D. Dauncey (Imperial)
S.A. Abel (Durham)
B.C. Allanach (Cambridge)
G.J. Davies (Imperial)

Strings:
K.S. Stelle (Imperial)
H. Stoica (Imperial)
T. Wiseman (Imperial)

Cosmology:
C. Contaldi (Imperial)
C. Gordon (Oxford)
K. Koyama (Portsmouth)
H.V. Peiris (Chicago)

Symposium Sponsors

Science and Technology Facilities Council
Centre for Theoretical Cosmology
Institute for Particle Physics Phenomenology
Institute of Physics
Imperial College London

PLENARY TALKS

The case for an aggressive program of dark energy probes[1]

Andreas Albrecht

Department of Physics, One Shields Ave.; University of California; Davis, CA 95616

Abstract. The observed cosmic acceleration presents the physics and cosmology communities with amazing opportunities to make exciting, probably even radical advances in these fields. This topic is highly data driven and many of our opportunities depend on us undertaking an ambitious observational program. Here I outline the case for such a program based on both the exciting science related to the cosmic acceleration and the impressive impact that a strong observational program would have. Along the way, I challenge a number of arguments that skeptics use to question the value of a strong observational commitment to this field.

Keywords: cosmic acceleration, dark energy, dark energy experiments
PACS: 98.80.-k,95.35.+d

INTRODUCTION

This is truly remarkable time to be involved in cosmology research. There are many reasons for this, but none stand out quite as dramatically as the observed cosmic acceleration (attributed in current nomenclature to the "dark energy"). In the words of the Dark Energy Task Force (DETF) [1] "most experts believe that nothing short of a revolution in our understanding of fundamental physics will be required to achieve a full understanding of the cosmic acceleration". As things stand, this revolution is being motivated both by remarkable new data sets and by exciting theoretical developments.

Many ambitious researchers in cosmology and related fields have been galvanized by these extraordinary developments. There have been a number of excellent talks at this conference on the topic of cosmic acceleration which capture some of this climate. The DETF, charged with charting a way forward with dark energy observations, received 50 thoughtful and thorough whitepapers from leaders in the field (despite the lack of any specific commitment at that time to fund future dark energy experiments).

Most of us look at these developments and are astonished at our good fortune to be part of what will surely be viewed as one of the great moments in the history of science. Indeed, since the discovery of dark energy every group that has deliberated on future directions for the field has recognized the exciting opportunities and challenges presented by the cosmic acceleration[2]. But a small number of negative voices continue

---

[1] Invited talk at PASCOS 07, Imperial College, July 2007

[2] See for example [2, 3, 4]. Since the 2006 DETF report a number of panels have recommended pursuit of specific ground and space based dark energy projects, and just since PASCOS 07 the US National Research Council's *Committee on NASA's Beyond Einstein Program* named a "Joint Dark Energy Mission" the top priority for that program `http://nationalacademies.org/morenews/20070907b`

CP957, *Particles, Strings, and Cosmology, 13th International Symposium*
edited by A. Rajantie, C. Contaldi, P. Dauncey, and H. Stoica

to be heard alongside the building enthusiasm for further studies of the dark energy. To some degree this negativity may reflect the natural skepticism of scientists in the face of unbridled enthusiasm (such as the above). To the extent that that is the explanation, the skepticism is surely a good thing that will lead to a healthy debate and produce more rigorous research. On the other hand, I suspect some of the negativity is just the result of sloppy thinking and needs to be challenged and simply put to rest. Regardless of how one might attribute explanations and motivations, one purpose of this paper is to engage in this debate on a number of fronts.

Here are some illustrations of some of the negative views I am talking about. At lunch on the first day of the PASCOS 07 conference there was a lively discussion about the merits of dark energy experiments. One colleague commented

> *"Studies of dark energy are unlikely to be interesting because we already have a theory of dark energy."*

moments later, another cosmologist declared

> *"Studies of dark energy are unlikely to be interesting because we have no theory of dark energy."*

The two were united in their conclusion, and apparently not too bothered by subtle differences in their reasoning

One unavoidable feature of physics research is that at any given time the total amount of data is finite. That necessarily means there will be more than one theory that fits the data. In many fields, this universal fact is seen as a source of vitality. Curiosity about further resolving these degeneracies drives exciting new experiments and theoretical work. The momentous progress in physics over the last century can be seen in this light as the fundamental particles, the atomic theory of heat, quantum physics and general relativity emerged from "under the radar" of earlier data sets that were fit perfectly by more primitive theories. As was amply evident in the talks at the PASCOS 07 conference, we have good reason to look forward to similar advances as the LHC data starts coming in.

But for some reason, the fact that a given future dark energy experiment will not remove all uncertainties about the nature of dark energy seems to generate considerable angst among some physicists and astronomers[3] and is sometimes given as a reason to be discouraged from even doing more experiments. As I shall quantify below (and as was also shown by others at PASCOS 07), the proposed new experiments will have an impressive impact on our knowledge of dark energy. This is all one can ever ask of a new experiment.

This paper focuses on two key areas. In the next section I review some of the thriving theoretical work that has been stimulated by the cosmic acceleration. The striking theoretical issues raised by the cosmic acceleration and the remarkable directions we have been driven in our initial attempts to understand it are the key reasons I find this topic so deeply interesting. I also believe this is why so many excellent researchers are taking risks and changing direction in their careers in order to get involved.

---

[3] See the question section of my PASCOS 07 talk for an example[5]

The third section summarizes a number of results which demonstrate the tremendous impact future experiments can have on our understanding of dark energy, both on our understanding of the general properties of dark energy and in terms of constraining specific models of dark energy that are currently of interest. It is these results that demonstrate that an "aggressive program of dark energy probes" is indeed possible.

This paper is based on my talk at the PASCOS 07 meeting at Imperial College. My slides and a video of the talk are available online[5]. This online material as well as my "Origins of Dark Energy" talk[6] and related papers[7, 8] are a good source for the technical material on which this paper is based. My goal here is to assemble some key arguments in a concise form. Readers seeking more details should refer to this other material.

DARK ENERGY SCIENCE

Two types of acceleration

At the start of the era of observations of cosmic acceleration the cosmology community was familiar with two possible sources of cosmic acceleration. These two types of acceleration were sufficiently established that they could both be found in textbooks.

Cosmological constant: The cosmological constant arises as an "extra" constant term in Einstein's equations. A FRW universe with a cosmological constant with value Λ is equivalent to adding a matter component with density $\rho_\Lambda \equiv \frac{\Lambda}{8\pi G}$ and with equation of state $p = -\rho$. This equation of state insures that ρ_Λ remains constant throughout the evolution of the universe. Until the mid 1990's, a commonly held belief among particle physicists and cosmologists was that $\Lambda = 0$. For many this belief was partly motivated by the fact that naively quantum fields give contributions to Λ of order 10^{60} or 10^{120} times larger than observationally acceptable values. It was widely believed that finding some symmetry or dynamical process that sets Λ (including contributions from quantum fields) precisely to zero was the best hope to resolve this apparent discrepancy.

Dynamical acceleration: Another widely held belief that has had growing support over the last few decades is that the universe underwent a period of **cosmic inflation** in the distant past. A period of cosmic inflation appears to explain many features observed in the universe today, some of which seems puzzling before the idea of inflation came along.

Today inflation is quite well understood in terms of its phenomenology, but it still has a number of unresolved foundational questions. Despite these, it is certainly clear that cosmic inflation requires a period of cosmic acceleration that *cannot* be described by a cosmological constant. Cosmic inflation is understood to be driven by some matter field typically called the "inflaton" which exhibits an equation of state $p = w\rho$. During inflation w takes values that approach $w = -1$ but it is a requirement of inflation that strict equality does not hold. This is because inflation is fundamentally a dynamical process. Not only must the universe enter and exit inflation (leading to large variations in w) but small deviations from $w = -1$ are also necessary throughout the inflationary period in order for the mechanisms of inflation to work properly.

Today, some people (one quoted in the introduction) like to push the point of view that there is one superior theory of cosmic acceleration, namely the cosmological constant. It is very likely that before the late 1990's, these same people believed in cosmic inflation, and also believed $\Lambda = 0$. In fact, the pioneers who started contemplating $\Lambda \neq 0$ in the 1990's [9, 10] did so precisely to prop up inflation (and the associated "CDM" cosmology) in light of data which would have otherwise been problematic. In fact [9] notes the possibility of a dynamical source of acceleration and even seems to regard it somewhat favorably over a strict cosmological constant.

So the field really has two different ideas of how cosmic acceleration can come about. One is dynamical (inflation-like) and one is not dynamical (the cosmological constant). Because of this interesting dichotomy the DETF (and many others) see the discrimination between dynamical and non-dynamical sources of cosmic acceleration as the best starting point for analyzing dark energy data (both present and future).

Note that *if* the data continues to be consistent with a cosmological constant it will only ever *bound* the possibility that there is a dynamical aspect to the cosmic acceleration. In that case one can never prove that the acceleration is absolutely constant. This should be no more troubling to a physicist than the fact that we only have *bounds* on the photon mass, deviations from the equivalence principle and other aspects of the physical world that many presently regard as absolute truths. The reason massless photons and the equivalence principle are seen as "truth" is that we have sufficiently high quality data to know that any corrections will be very small. The impossibility of getting an absolute result in a cosmological constant driven universe should certainly not make us timid about seeking high quality data in this area as well.

Some remarkable features of accelerating cosmologies

A key part of the science case for further study of dark energy is the very exciting nature of the theoretical developments that surround this this topic. I will highlight a few of these in this section.

Strange numbers: I've already mentioned the fact that naive quantum field theory arguments predict a discrepancy in the value of Λ by a factor of 10^{60} or even 10^{120}. In addition, most dynamical models of dark energy end up requiring a new particle associated with a so called "quintessence field" that has a mass of $10^{-31}eV$ (many orders of magnitude smaller than current bounds of the photon mass). With a few noteworthy exceptions, theories of quintessence do not consider the problem of how to protect such a tiny mass from quantum corrections. Also, in general such a light particle will lead to long range forces, and most quintessence models do not have a mechanism for evading current bounds on such forces. Some dynamical models require additional tuning of parameters so that the onset of cosmic acceleration can occur in the right epoch of the universe (the "why now problem"). Theorists tend to take such features of a theory as a sign that something important is missing in our understanding.

Λ and equilibrium: If the universe has a true (non dynamical) cosmological constant with a positive value (sufficient to account for the current acceleration, or even smaller) the future of the universe will have some striking features. The effective "cosmological constant density" ρ_Λ will come to completely dominate the universe as the other types of matter dilute with the cosmic expansion. Under these conditions the universe will approach something called "de Sitter space", which has features which fit a (somewhat generalized) notion of equilibrium: The de Sitter space is approached asymptotically by huge fraction of all possible initial states, it has a (Hawking) temperature and an entropy. The entropy of a de Sitter space has been shown to be larger than the entropy of all other cosmological states with a non-zero cosmological constant[11].

This situation implies the following picture of cosmology in the presence of a fundamental cosmological constant: The universe is eternal, and spends most of its time in the equilibrium state described above. Cosmology as we know it must emerge from very rare large fluctuations from equilibrium. I have argued elsewhere[12] that this general picture of cosmology is much more powerful than more traditional approaches to cosmological "initial conditions". This is because the probabilities are in principle assigned to different cosmologies (or fluctuations from equilibrium) by concrete calculations based on fundamental physical laws rather than ad hoc proposals for "wavefunctions of the universe" which, far from being compelling, are notorious for generating controversy.

However, the first authors to calculate probabilities in this equilibrium picture concluded that that picture is a spectacular failure because cosmologies inconsistent with our observations were exponentially favored by their calculations[13]. Sorbo and I[12] have proposed an alternative approach to these calculations which exponentially favors the standard inflationary picture of cosmology. Deeper insights into the nature of quantum gravity are needed to learn which (if either) of the two approaches is correct.

My main point here is that considering the cosmic acceleration has driven theorists to contemplate radical new ideas about how to think about cosmology, equilibrium and initial conditions.

The string theory landscape: Andrei Linde and Renata Kallosh have given excellent talks at PASCOS 07 on the string theory landscape. I will refer you to those for details. I will just make a few brief remarks here on topics that link the string theory landscape to the main focus of my discussion.

The string theory landscape marks a radical change in thinking about fundamental physics. For many, it has replaced the old idea that the correct "theory of everything" will state the fundamental particles and their parameters and couplings in one tidy package. Instead, the string theory landscape tells us that when it comes to the physical observables that are important to us, they are chosen form an enormous collection (around 10^{1000}) of string theory "vacuua" that fundamentally appear to be equally valid choices.

As Linde and Kallosh have emphasized, this dramatically different approach to fundamental physics was driven in large part by attempts to reconcile string theory with the observed cosmic acceleration. In fact, many people believe that this picture leads to a successful (anthropic[14]) account of how the cosmological constant (or more correctly, the vacuum energy) should have more or less the value needed to explain the current cosmic acceleration[15]. Some of the most enthusiastic proponents of this view have even argued informally that this success is so compelling as to undermine the need to collect further data on the cosmic acceleration.

However, as you heard in other talks at this meeting, there is more to the landscape vacuua than stating that they all fundamentally appear to be equally valid. The dynamics of a state evolving in the landscape will tend to prefer some vacuua over others. Perhaps even an equilibrium picture could emerge. Basically, one has to do full cosmology in the landscape before real predictions can be made. I find the work along these lines extremely interesting, but the landscape is a complicated place and we appear to be very far from a final answer. It is entirely possible that once we understand the full dynamics of the landscape the results will be inconsistent with our observations (just as was found in [13] for the Λ case), causing a complete failure of the landscape picture. Also, the acceleration in the landscape picture is expected to be observationally indistinguishable from a cosmological constant, so any observation signal of $w \neq -1$ will rule out the landscape explanation.

The string landscape is another example where attempts to understand the cosmic acceleration have driven theory in dramatic new directions.

Other explanations: There are a variety of other approaches to explain cosmic acceleration. These include modifying Einstein gravity (see [6] for numerous talks on this subject), introducing new non-accelerating physics to explain the observations[16, 17], or "even" considering the possibility that more conventional physics produce acceleration [18] or lead to observations that we are currently misinterpreting as acceleration[19, 20].

THE IMPACT OF FUTURE EXPERIMENTS

The DETF

Based on 50 whitepapers, as well as internal expertise, the DETF developed mock data (or "data models") to represent a number of future dark energy observations. They divided these data models into stages. "Stage 2" represents data from projects that are already underway, "Stage 3" projects are medium term medium cost projects, and "Stage 4" are the larger more expensive projects currently under consideration.

The DETF considered the impact of the various projected data sets on cosmological parameters in a standard cosmological model. They parameterized the dark energy by its density today as well as a two parameter description of the equation of state parameter as it evolves with the cosmic scale factor a given by $w(a) = w_0 + w_a(1-a)$ (the "$w_0 - w_a$" parameterization).

The DETF considered cosmologies in which the cosmic curvature is allowed to be nonzero ($\Omega_k \neq 0$). While some are happy to take $\Omega_k = 0$ as a prior assumption, I don't believe there is a good case for doing so. For one, the main reason people like the $\Omega_k = 0$ prior is that that value is seen as a universal prediction of cosmic inflation. However, the only reason the $\Omega_k = 0$ universe has any chance of fitting the data is due to the presence of dark energy. Since the dark energy is so poorly understood, I feel it is a mistake to assume we know how to include it correctly, and how well it may or may not fit the $\Omega_k = 0$ picture when things are understood more completely. Secondly, there is a very interesting subclass of inflation models that allow a small deviation from $\Omega_k = 0$ (see for example [21]). Discriminatory power between these pictures is a valuable capability to expect from future observations.

The DETF used a figure of merit given by the inverse area in $w_o - w_a$ space constrained by a given experiment. The DETF showed that good Stage 3 data will improve on Stage 2 constraints in $w_0 - w_a$ space by a factor of order 3, and good Stage 4 experiments will result in an order of magnitude improvement. We note that in the parameter space considered by the DETF the current data (Stage 1) does not result in significant constraints, which is why current data is typically analyzed in terms of simpler cosmological models[22].

Beyond the DETF

The DETF report gives a useful perspective on the power of future experiments, but it left a number of interesting answered questions. One question is "did the DETF miss anything important by using their particular parameterization of the dark energy?" Some researchers are also concerned that the DETF dark energy parameterization is too abstract, and would like to see estimates of the impact of future experiments on specific existing models of dark energy. In the next two sections I will report on new work (since the DETF) that answers both these questions, and helps give a more complete picture of the full power of proposed dark energy experiments. Another key "beyond DETF" question involves the subtleties of combining data sets. This is an area where improvements on the DETF analysis will certainly be important and new work[23, 24] suggests such improvements will demonstrate an even greater impact for future experiments. Those improvements are not included in the analysis below, so these results should be seen as underestimates of the true potential of future experiments.

Beyond two parameters: The $w_0 - w_a$ parameterization used by the DETF allows a variety of linear functions $w(a)$. With Bernstein, I modeled $w(a)$ with a much larger number of parameters[7], resulting in stepwise constant functions $w(a)$. We used a Fisher matrix analysis that allowed us to consider an orthonormal basis of independently measured mode functions $w_j(a)$. An advantage of this approach is that when one increases the total number of parameters (by decreasing the step width) one simply gets an ever improving approximation to the finite number of continuous modes that are well-measured by the given experiment (most other parameterizations of $w(a)$ are highly unstable against changing the number of parameters). We argue in [7] that in the small bin limit our approach allows each experiment to show exactly what it is able to measure. The well-measured modes are chosen by the properties of the experiment and thus there is no information lost due to prior assumptions about the form of $w(a)$.

Figure 1 shows the figure of merit derived using DETF parameters (dark bars) and a suitably large number of step parameters to achieve convergence onto the well-measured modes (light bars). The different bars and panels correspond to different choices of DETF data models[7]. The upshot of [7] is that the best future data sets will measure many more than two parameters, resulting in massively higher figures of merit vs. the DETF estimates. These huge figures of merit translate directly into greater discovery power through greatly improved constraints in parameter space. However, aside from an overall rescaling of the figure of merit, the other conclusions of the DETF regarding the importance of combining techniques and the relative rankings of the different data models are unchanged.

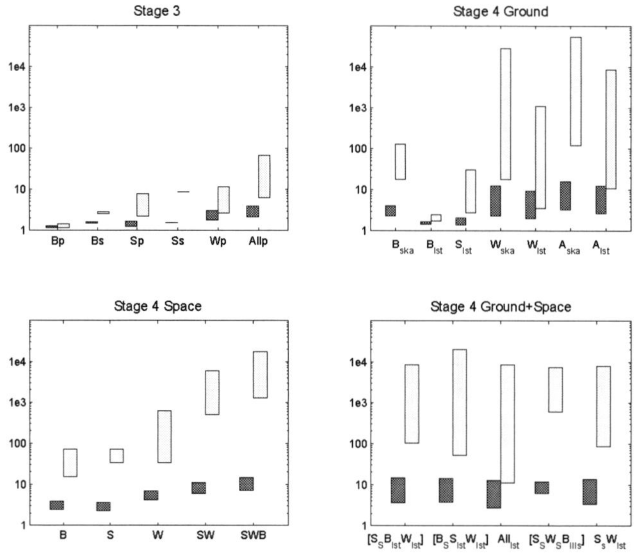

FIGURE 1. Figure 3 from [7] showing the DETF figure of merit (dark bars) and the figure of merit given by a complete set of well measured modes $w_j(a)$ (light bars). As detailed in [7], the higher dimensional parameter space gives a more complete picture of the ability of a given experiment to constrain dark energy properties. Good stage 4 experiments have a figure of merit many orders of magnitude higher than the DETF estimates.

Beyond abstract parameters: Theories of dark energy appear to be in a very primitive state, and are likely to change greatly before we have a deep understanding of the cosmic acceleration. To some the use of abstract parameters such as I've used above is the best approach. Basic characteristics of the dark energy can be discussed (such as whether or not it is dynamical) without dependence on a particular model that is unlikely to survive progress in this field.

Others would prefer to know if future experiments will have an impact on existing proposed explanations of dark energy. However flawed the existing explanations may turn out to be, they are what we have to work with for now and one certainly should expect proposed experiments to have an impact on these models.

To assess this point, my students and I have undertaken a number of studies of the impact of DETF data models on specific models of dark energy. You can learn more about this work in [5, 6, 8]. Here I will highlight a couple of main results.

Firstly, we found that the parameters of the specific models (i.e. parameters in the quintessence potential) were constrained to a similar degree as the DETF parameters. Typically a given model had two parameters that were constrained by cosmology, so the comparison to the two parameter DETF scheme made sense. In this picture, the higher number of parameters found in [7] shows up as discriminating power between a large variety of dark energy models (even though each tends to have a smaller number of parameters, parameterizing a limited set of functions $w(a)$ that are specific to that

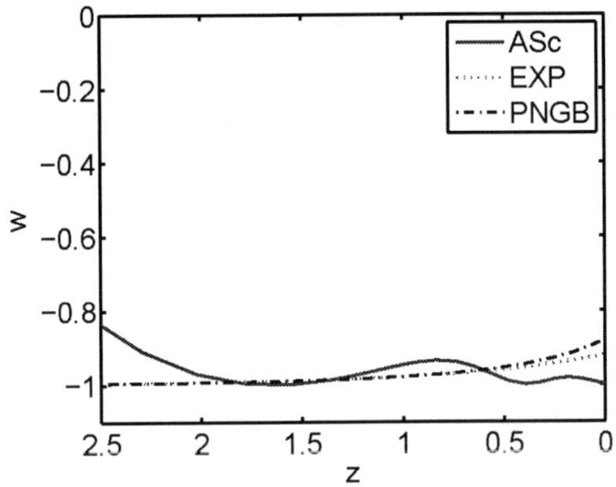

FIGURE 2. The three curves give $w(a)$ for three different models of dark energy. If the universe has chosen any of these models Stage 4 data would exclude a cosmological constant at at least *four sigma* .

model). t

Figure 2 gives another illustration of the impact of future experiments. The function $w(a)$ is shown for three different quintessence models with parameters fixed at specific values. If the universe has chosen any of these three models Stage 4 data will exclude a cosmological constant to at least four sigma. For these, the discriminatory power of even the best Stage 3 experiments would be below two sigma.

CONCLUSIONS AND SUMMARY

The case for aggressive pursuit of new data on dark energy is twofold. Firstly, I have outlined how the subject of dark energy has generated very exciting and often radical new theoretical ideas. These include dramatic proposals that change how we think about equilibrium and initial conditions in cosmology and even how we formulate fundamental theories. Second, impressive new experiments are within reach that could have a tremendous impact on our understanding of dark energy. The best experiments will constrain dark energy properties orders of magnitude better than the current or medium sized future experiments, and will have the ability to strongly discriminate among and even fully eliminate popular dark energy models based on subtle variations in the equation of state.

I have outlined and challenged some of the skeptical perspectives I have heard regarding future dark energy studies. The discovery of the cosmic acceleration has caused a great upheaval in our thinking about fundamental physics and cosmology. I often sense that the skeptics are hoping this upheaval will end quickly and are grasping for arguments that will allow things to rapidly return to normal. I feel this outcome is very unlikely, and this is exactly why I find the topic so exciting.

Nature has handed us an amazing opportunity. I hope that the physics and cosmology communities have the strength to face the challenge of the cosmic acceleration head on and give a response that we can be proud of when people write the history of this era.

ACKNOWLEDGMENTS

I would like to thank A. Abrahamse, M. Barnard, G. Bernstein, B. Bozek, L. Sorbo, M. Yashar and the DETF members who collaborated with me on some of the work reviewed here, and B. Bozek for helpful comments on the manuscript. Also, I thank the organizers, especially Arttu Rajantie, for a really excellent conference. This work was supported in part by DOE grant DE-FG03-91ER40674 and NSF grant AST-0632901.

REFERENCES

1. A. Albrecht, et al. (2006), astro-ph/0609591.
2. M. Turner, et al. (2003), washington, USA: Nat. Acad. Pr. (2003) 206 p.
3. A. Albrecht, et al. (2005), www.interactions.org/quantumuniverse/.
4. J. A. Peacock, et al. (2006), astro-ph/0610906.
5. A. Rajantie, Pascos (2007), imperial College, July 2007. Slides and video available at http://www.pascos07.org/.
6. C. Burgess, Origins of dark energy (2007), mcMaster University, May 2007. Slides available at http://origins.mcmaster.ca/darkenergy/.
7. A. Albrecht, and G. Bernstein, *Phys. Rev.* **D75**, 103003 (2007), astro-ph/0608269.
8. A. Abrahamse, A. Albrecht, M. Barnard, B. Bozek, and M. Yashar (2007), three papers in preparation involving various combinations of these authors.
9. G. Efstathiou, W. J. Sutherland, and S. J. Maddox, *Nature* **348**, 705–707 (1990).
10. L. M. Krauss, and M. S. Turner, *Gen. Rel. Grav.* **27**, 1137–1144 (1995), astro-ph/9504003.
11. G. W. Gibbons, and S. W. Hawking, *Phys. Rev.* **D15**, 2738–2751 (1977).
12. A. Albrecht, and L. Sorbo, *Phys. Rev.* **D70**, 063528 (2004), hep-th/0405270.
13. L. Dyson, M. Kleban, and L. Susskind, *JHEP* **10**, 011 (2002), hep-th/0208013.
14. S. Weinberg, *Rev. Mod. Phys.* **61**, 1–23 (1989).
15. R. Bousso, and J. Polchinski, *JHEP* **06**, 006 (2000), hep-th/0004134.
16. C. Csaki, N. Kaloper, and J. Terning, *Phys. Rev. Lett.* **88**, 161302 (2002), hep-ph/0111311.
17. C. Csaki, N. Kaloper, and J. Terning, *Ann. Phys.* **317**, 410–422 (2005), astro-ph/0409596.
18. S. Deser, and R. P. Woodard (2007), arXiv:0706.2151[astro-ph].
19. E. W. Kolb, S. Matarrese, and A. Riotto, *New J. Phys.* **8**, 322 (2006), astro-ph/0506534.
20. P. Martineau, and R. Brandenberger (2005), astro-ph/0510523.
21. B. Freivogel, M. Kleban, M. Rodriguez Martinez, and L. Susskind, *JHEP* **03**, 039 (2006), hep-th/0505232.
22. L. Knox, *Phys. Rev.* **D73**, 023503 (2006), astro-ph/0503405.
23. H. Zhan, *JCAP* **0608**, 008 (2006), astro-ph/0605696.
24. M. Schneider, L. Knox, H. Zhan, and A. Connolly (2006), astro-ph/0606098.

Search for Higgs Bosons and Supersymmetry at the Tevatron

Volker Büscher

Physikalisches Institut, Universität Bonn, 53115 Bonn, Germany

Abstract. Since the start of Run II in March 2001 the Tevatron collider has delivered $p\bar{p}$ collisions corresponding to about 3 fb^{-1} to the CDF and DØ experiments. This report presents a brief summary of recent results based on this dataset from the searches for Higgs bosons and Supersymmetry.

Keywords: Higgs, Supersymmetry
PACS: 12.60.Fr, 12.60.Jv, 13.85.Rm, 14.80.Bn, 14.80.Cp, 14.80.Ly

INTRODUCTION

At a centre-of-mass energy of 1.96 TeV, the datasets collected by the Tevatron experiments CDF and DØ provide the opportunity to search for particles that have not been observable previously. This includes Higgs bosons as predicted by the Standard Model and its supersymmetric extensions, as well as a large variety of new particles predicted in models of physics beyond the Standard Model. In this talk a selection of recent[1] Tevatron results from these areas is discussed. For a complete and uptodate set of results, the reader is referred to [1].

HIGGS BOSONS

Electroweak precision measurements strongly indicate that the Standard Model Higgs boson must be relatively light. The upper limit on the mass of the Higgs boson of 144 GeV at 95% C.L. [2] leaves only little room above the mass limit of 114.4 GeV set by direct searches at LEP. Similarly, at least one light Higgs boson is expected within the Minimal Supersymmetric Standard Model. This Higgs boson mass range is potentially accessible to the Tevatron experiments once enough integrated luminosity has been accumulated. The current status of searches for Standard Model and SUSY Higgs bosons is reviewed in the following sections.

Heavy Standard Model Higgs Boson

Within the mass range of interest at the Tevatron, the Standard Model Higgs boson is expected to decay dominantly either into a pair of b-quarks for a Higgs boson mass m_H

---

[1] as of July 2007

CP957, *Particles, Strings, and Cosmology, 13th International Symposium*
edited by A. Rajantie, C. Contaldi, P. Dauncey, and H. Stoica

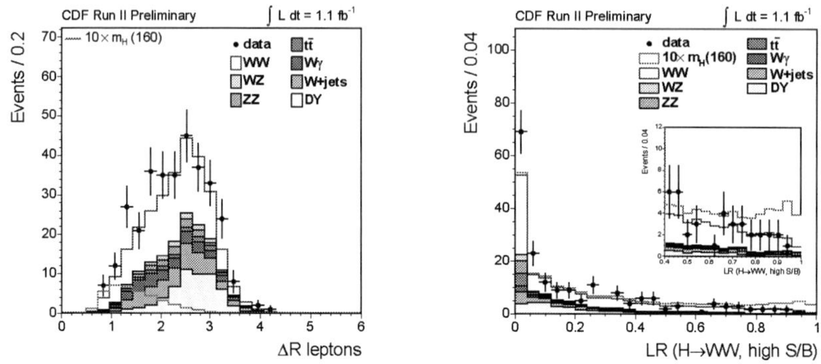

FIGURE 1. Lepton opening angle ΔR (left) and likelihood ratio LR (right) in the search for H$\rightarrow$WW$\rightarrow$ $\ell\nu\ell\nu$. Data (points) are compared to backgrounds (histograms) and signal (open red histogram) with a cross section enhanced by a factor of 10 (from [3]).

below about 135 GeV, or into a pair of W bosons for $m_H \gtrsim 135$ GeV. Exploiting the leptonic decay mode of the W boson, it is possible to search for heavy Higgs bosons produced via gluon fusion, with a relatively high cross section of the order of 1 pb.

Both CDF and DØ collaborations have searched for H$\rightarrow$WW$\rightarrow$ $\ell\nu\ell\nu$ (ℓ =e,μ) in datasets corresponding to roughly 1 fb^{-1}. Requiring two isolated electrons or muons with high transverse momentum and well-measured large missing transverse energy, backgrounds are dominated by W+jet production and irreducible background from WW production. After more cuts on event kinematics, the signal can be isolated further by exploiting spin correlations between the leptons: since for signal both W bosons originate from the decay of a scalar Higgs boson, the two leptons are preferentially emitted in the same direction. As visible in Fig. 1(left), signal events therefore accumulate at small opening angle ΔR between the leptons. After all cuts, 1.8 (3.9) events from H$\rightarrow$WW are expected for a Higgs boson mass of 160 GeV in the DØ (CDF) analysis [3], on top of a background of 45$\pm$8 (286$\pm$23) events. 37 (323) data events have been observed, consistent with the Standard Model expectation. To ensure an optimal test for the presence of a high-mass Higgs signal, the CDF analysis is exploiting the full kinematical information contained in each event to calculate a likelihood ratio for signal and background hypotheses, as shown in Fig. 1(right). An interpretation in terms of an upper limit on the Higgs boson production cross section is discussed later in this document.

Light Standard Model Higgs Boson

For a Standard Model Higgs boson with mass below about 135 GeV, the decay into a pair of b-quarks is dominant. To be able to separate events with H$\rightarrow$bb from the massive multijet backgrounds, the search relies on Higgs production in association with a W or Z boson. For these production channels, the leptonic decays of the vector bosons allow to trigger efficiently and to suppress the multijet background to an acceptable level.

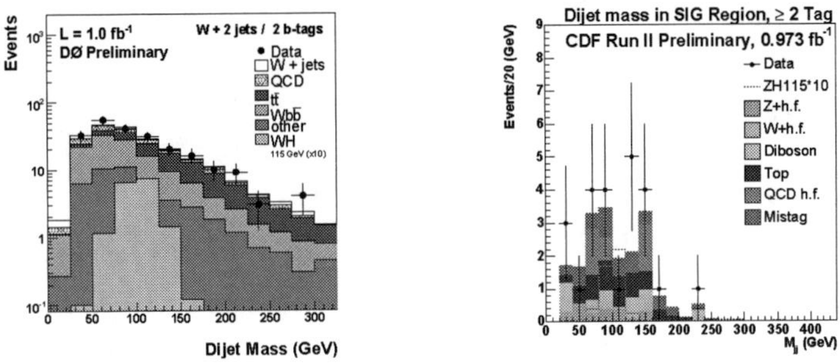

FIGURE 2. Dijet mass distribution after all cuts in the search for WH→ $\ell\nu$bb (left) and ZH→ $\nu\nu$bb (right) with two b-tagged jets. Data (points) are compared to backgrounds (shaded histograms) and signal (lightly shaded or open histogram) with a cross section enhanced by a factor of 10 (from [4]).

After selecting events with W→ $\ell\nu$, Z→ $\ell\ell$ or Z→ $\nu\nu$ plus two jets using requirements for isolated leptons and/or missing transverse energy, the sample can be further enhanced in signal by requiring one or both jets to be b-tagged. Irreducible background from vector boson production in association with b-quarks can be distinguished from Higgs events by reconstructing the dijet invariant mass to test for the presence of a Higgs resonance. Fig. 2 shows the invariant mass of two b-tagged jets in the DØ WH and CDF ZH searches, based on datasets corresponding to 1 fb^{-1}[4].

A variety of channels have been analyzed by both collaborations, so far without any evidence for a significant excess compared to the expected backgrounds. Limits on production cross sections of light Higgs bosons have been set as presented in the following section.

SM Higgs Combination

Given the amount of integrated luminosity that will be available at the Tevatron, sensitivity to the Standard Model Higgs boson can only be achieved by combining the information from all search channels and both experiments. Currently, results from the channels listed above based on datasets of 0.26 fb^{-1} up to 1.0 fb^{-1} have been combined to set upper limits on Higgs production at the Tevatron [5]. The combined 95%. C.L. limit is shown as a function of Higgs boson mass in Fig. 3. The limit is expressed as multiple of the cross section expected for the Standard Model Higgs boson. Clearly, more luminosity and additional analysis improvements will be needed before sensitivity can be reached.

Higgs bosons in Supersymmetry

Supersymmetric extensions of the Standard Model require the introduction of five physical Higgs bosons: two CP-even Higgs bosons h and H, one CP-odd Higgs boson A, and two charged Higgs bosons $H^{\pm}$. Within the Minimal Supersymmetric Standard

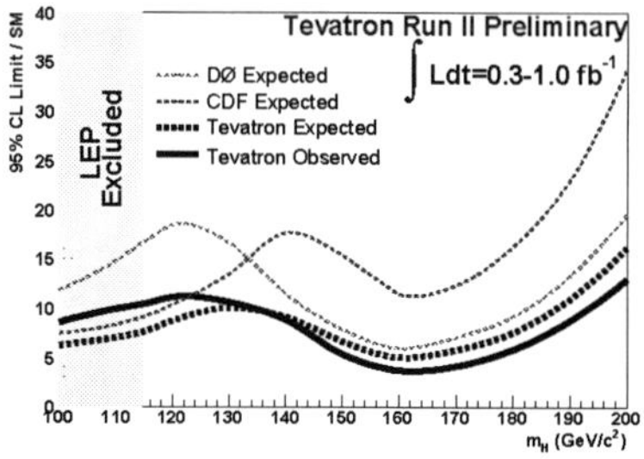

FIGURE 3. Ratio of expected (dashed lines) and observed (solid line) cross section limits to the Standard Model Higgs boson production cross section for the individual and combined CDF and DØ analyses as a function of Higgs boson mass (from [5]).

Model (MSSM), all Higgs boson masses and couplings are specified at tree level by two parameters, typically chosen as the mass of the A boson m_A and the ratio of vacuum expectation values $\tan\beta$.

The searches for a light Standard Model Higgs boson discussed in the previous sections can be re-interpreted as a search for the MSSM Higgs boson h, which is predicted to have a mass of less than $\approx$135 GeV. However, since the Wh and Zh production cross sections are suppressed with respect to the Standard Model, there is no sensitivity with the limited amount of integrated luminosity currently available.

On the other hand, the cross section of MSSM Higgs boson production via gluon fusion is enhanced by a factor of $\tan^2\beta$ (at tree level) with respect to the Standard Model. For large enough $\tan\beta$, the enhancement is sufficient to achieve sensitivity with the data currently available to the inclusive production of Higgs bosons decaying to two tau leptons. Both CDF and DØ collaborations have searched for Higgs bosons in this channel in datasets corresponding to about 1 fb^{-1}.

The analyses require at least one leptonic tau decay to trigger the events and to suppress multijet backgrounds. Hadronic tau decays are separated from the jet background by relying on differences in shower shape, track multiplicity, isolation and a number of kinematic quantitities. Significant irreducible background from Z$\rightarrow\tau\tau$ production remains, and can only be distinguished from the signal by reconstructing the invariant mass of the visible ditau system. As illustrated in Fig. 4(left), this allows a separation of massive Higgs bosons from Z decays, even though the neutrinos from tau decays cannot be fully reconstructed. The slight excess observed by CDF at ditau masses around 140 GeV (significance less than 2σ) has not been confirmed by the DØ analysis. Limits on $\tan\beta$ can be derived as a function of m_A as shown in Fig. 4(right) [6].

SUPERSYMMETRY

In addition to an enlarged Higgs sector, supersymmetry (SUSY) predicts a new particle for each Standard Model particle. The partner particles have the same quantum numbers except for the spin, i.e. each fermionic (bosonic) degree of freedom in the Standard

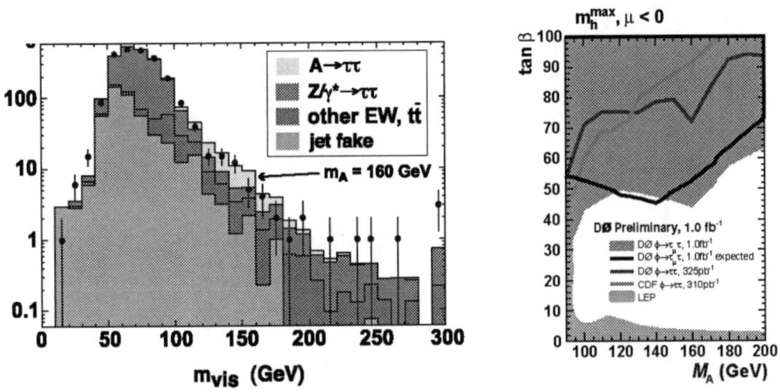

FIGURE 4. Left: Visible ditau mass distribution in the CDF search for Higgs bosons decaying into two tau leptons for data (points), sum of all backgrounds (blue histogram) and signal (yellow histogram); Right: Regions in the $(\tan\beta, m_A)$-plane excluded by Tevatron and LEP searches for the m_h-max scenario with $\mu < 0$ (from [6]).

Model is associated with a new bosonic (fermionic) superpartner. Since no superpartners have been observed experimentally, supersymmetry is assumed to be broken by an unknown mechanism that results in higher masses for the superpartners. Searches for supersymmetry at the Tevatron consider a large variety of SUSY breaking mechanisms, each leading to substantially different mass hierarchies which result in different final state signatures.

Squarks/Gluinos

Squarks and gluinos, the superpartners of quarks and gluons, are generally expected to be heavier than most SUSY particles. Therefore they can decay via cascades into the lightest supersymmetric particle (LSP), which in this context is assumed to be the lightest neutralino, one of the partners of the neutral gauge and Higgs bosons. This leads to final states with at least two jets, missing transverse energy from the LSPs escaping undetected, and potentially additional leptons from the cascade decays.

Both CDF and DØ have designed inclusive searches for multijet final states with missing transverse energy $\not{E}_t$, with results available based on datasets corresponding to about 1 fb^{-1}. To suppress the massive background from Standard Model multijet production, the selections have to rely on a combination of requirements of high jet multiplicities or large jet transverse momenta as well as high $\not{E}_t$. As shown in Fig. 5, a SUSY signal would then manifest itself as an excess on top of irreducible background from production of vector bosons in association with jets, where W$\rightarrow \ell\nu$ or Z$\rightarrow \nu\nu$ decays give rise to high missing transverse energy. The modeling of these backgrounds is a challenge, since the differential cross sections directly depend on higher-order QCD corrections. The implementation of these corrections in Monte Carlo generators has to be validated with data, as demonstrated in Fig. 6: Tevatron data for e.g. the transverse momentum spectrum of the third-leading jet in Z+jets events is insufficiently modelled

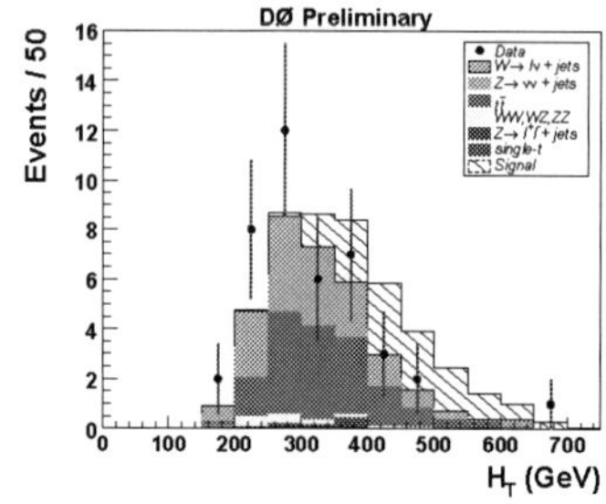

FIGURE 5. Distribution of the scalar sum H_T of all jet transverse energies in the search for squarks and gluinos for data (points), Standard Model background (shaded histogram) and SUSY signal (hatched histogram) at the end of the selection (from [8]).

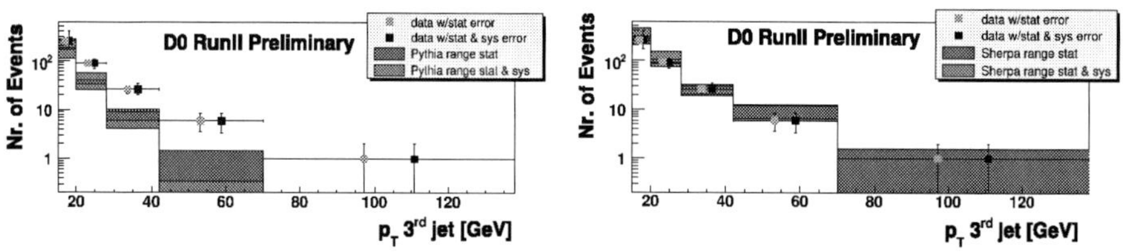

FIGURE 6. Distribution of transverse momentum of the third-leading jet in Z+jets events for data (points) in comparison with the predictions by the Pythia (left) and Sherpa (right) generators (from [7]).

by the pythia generator (leading-order matrix element), whereas the inclusion of higher orders as implemented in the sherpa generator allows to model the data within statistical and systematic uncertainties [7].

Neither CDF nor DØ have observed any significant excess in their searches for squarks and gluinos [8][8], and therefore have placed limits on squark/gluino masses as displayed in Fig. 7. In addition to these inclusive searches, dedicated analyses exist that have been optimized to detect the decays of bottom- or top-squarks, since these particles are potentially significantly lighter than all other squarks. Improved mass limits are available for the decays $\tilde{t} \to c\tilde{\chi}_1^0$, $\tilde{t} \to b\ell\tilde{\nu}$ and $\tilde{b} \to b\tilde{\chi}_1^0$ [1].

Charginos, Neutralinos

Another promising avenue for the discovery of supersymmetry at the Tevatron is the search for production of charginos and neutralinos, the fermionic partners of the gauge and Higgs bosons. The lightest chargino $\tilde{\chi}_1^{\pm}$ can be as light as 100 GeV, the mass limit set by the LEP experiments. Searches at the Tevatron therefore focus on charginos and neutralinos decaying directly to the LSP via the exchange of virtual W/Z bosons or sfermions. A particularly clean SUSY signature is provided by the associated production of the lightest chargino with the second-lightest neutralino $\tilde{\chi}_2^0$, with both decaying leptonically into a total of three charged leptons plus missing transverse energy

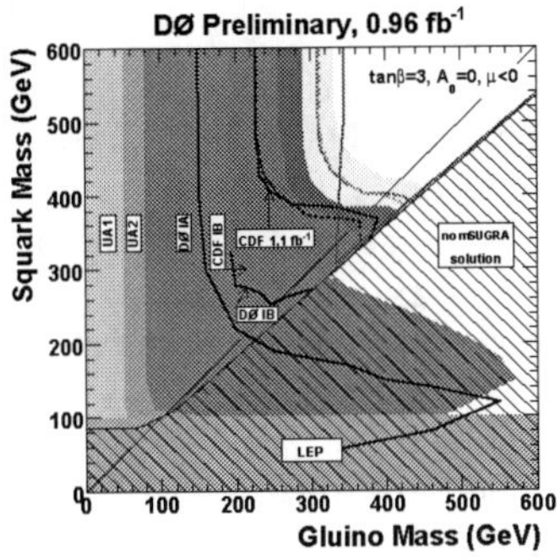

FIGURE 7. Regions in the plane of squark versus gluino mass excluded by searches at UA1, UA2, CDF, DØ and LEP (from [9]).

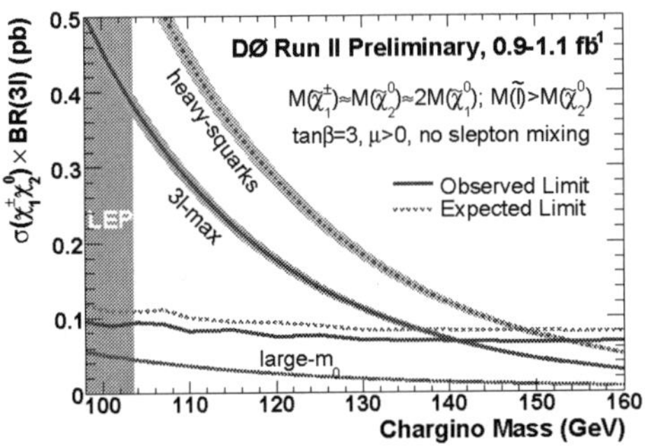

FIGURE 8. Limit on the product of production cross section and branching fraction for $\tilde\chi_1^\pm\tilde\chi_2^0 \to 3\ell + \not{E}_t$ as a function of the chargino mass in comparison with the prediction from a selection of MSSM model choices (from [10]).

carried away by neutrinos and the two LSPs.

Both CDF and DØ collaborations have searched for this signal in a number of multi-lepton analyses, based on datasets of up to 1.1 fb^{-1}. The combination of requirements for three relatively loosely identified leptons and significant missing transverse energy is sufficient to reduce the background to the small but irreducible component from WZ$\to \ell\nu\ell\ell$ production. No hint for a signal from chargino/neutralino production has been observed by either experiment, resulting in limits on the product of cross section and leptonic branching fraction as shown in Fig. 8. Currently, chargino masses of up to 140 GeV are probed for favourable SUSY scenarios with large leptonic branching fractions [10].

SUMMARY

With an integrated luminosity of more than 3 fb^{-1} already delivered, and bright prospects for more data to be accumulated over the next few years, the Tevatron collider continues to provide unique opportunities to search for Higgs bosons and new particles.

Combined Tevatron results from the search for the Standard Model Higgs boson are available based on all major channels, but so far do not provide sensitivity. With more data and further analysis improvements, it is hoped to reach sensitivity over the course of the next year. Searches for SUSY Higgs bosons are already probing the high $\tan\beta$ region, but have not uncovered any significant excess.

A large variety of signatures is considered in searches for new particles, including signals expected within models of supersymmetry. No hint of any deviation with respect to expectations from the Standard Model has been observed, and limits well beyond the reach of previous experiments have been set.

ACKNOWLEDGMENTS

The author would like to thank the organizers for the invitation to speak at this excellent conference, as well as the CDF and DØ contact persons for providing the material for this talk.

REFERENCES

1. CDF Collaboration: http://www-cdf.fnal.gov/physics/physics.html; DØ Collaboration: http://www-d0.fnal.gov/Run2Physics/WWW/results.htm.
2. LEP Electronweak Working Group, http://lepewwg.web.cern.ch/LEPEWWG/.
3. CDF Collaboration, "Search for H→WW Production with Matrix Element Methods in $p\bar{p}$ collisions at $\sqrt{s}$=1.96 TeV", CDF Note 8774; DØ Collaboration, "Search for the Higgs boson in H→WW decays with 950 pb^{-1} at DØ in Run II", DØ Note 5194-CONF and 5063-CONF.
4. DØ Collaboration, "A Search for WH Production at $\sqrt{s}$=1.96 Tev with 1 fb^{-1} of Data", DØ Note 5357-CONF; CDF Collaboration, "Search for the Standard Model Higgs boson in the MET and b-jet Signature", CDF Note 8442.
5. The TEVNPH Working Group, "Combined DØ and CDF Upper Limits on Standard-Model Higgs-Boson Production", hep-ex/0612044, CDF Note 8384, DØ Note 5227.
6. CDF Collaboration, "Search for Neutral MSSM Higgs Bosons Decaying to Tau Pairs", CDF Note 8676; DØ Collaboration, "Search for Neutral Higgs Boson Production in the Decay h→ $\tau_\mu \tau$ with the DØ Detector at $\sqrt{s}$=1.96 TeV", DØ Note 5331-CONF.
7. DØ Collaboration, "Z+jet production in the DØ experiment: A comparison between data and the PYTHIA and SHERPA Monte Carlos", DØ Note 5066-CONF.
8. DØ Collaboration, "Search for squarks and gluinos in events with jets and missing transverse energy with the DØ detector using 1 fb^{-1} of Run IIa data", DØ Note 5312-CONF; CDF Collaboration, "Search for Gluino and Squark Production in Multijets plus missing E_T Final States", CDF Note 8405.
9. P. Verdier, private communication.
10. DØ Collaboration, "Search for the Associated Production of Chargino and Neutralino in Final States with three Leptons with the DØ detector", DØ Note 5348-CONF; CDF Collaboration, "Combined limit for the trileptons analyses Summer 2006", CDF Note 8653.

Dynamics of Dark Energy

Edmund J. Copeland

School of Physics and Astronomy
University of Nottingham
University Park
Nottingham NG7 2RD

Abstract. I briefly review attempts that have been made to model dark energy. These include models of a cosmological constant, dynamical models where a scalar field may be responsible for the observed late time acceleration through to the possibility that we are not fully in control of the gravity sector and the acceleration may be some manifestation of modified gravity. In all cases we will see some degree of fine tuning is required with the current models.

Keywords: Dark Energy, Cosmology
PACS: 95.36+x, 98.80-k

INTRODUCTION

The universe is expanding, we have known that for many years. The conventional wisdom was that the presence of non relativistic matter as a dominant source of energy density led to a Universe which was decelerating in an understandable way. It all made sense, but unfortunately through the 1980's and early 1990's a number of observations of the large scale structure in the universe began to question the accepted paradigm of a cold dark matter (CDM) dominated $\Omega_{\rm matter} = 1$ universe. For example in a study of the large scale clustering of IRAS galaxies performed in [1] it was realised that it was difficult to reconcile the observed clustering of these galaxies with that predicted in the standard paradigm at the time. In fact it led a number of authors to begin questioning the paradigm and in 1990 Efstathiou et al, [3] actually proposed that a better fit to the data available at that time would be a scenario with a "spatially flat cosmology in which as much as 80 percent of the critical density is provided by a positive cosmological constant." This remarkable conclusion was reached well before the observations of distant Type Ia supernova hit the headlines in 1998. Their reasoning was reinforced in the analysis of the APM Galaxy Survey data in which there was strong evidence for more large-scale structure in the survey than predicted by the standard scale-invariant CDM model [2]. Perhaps we should not be too surprised that there were people back in the 80's and 90's who were turning their attention to the prospect of a cosmological constant. After all, there were many groups of astronomers trying to reconcile the fact that the large scale structure data was looking so inconsistent with a standard CDM scenario. However, I think it is fair to say that not too many people (especially from the particle physics camp) were prepared to face the prospect that we had to have a Λ present today if our cosmology was to make sense. Two exceptions to this were Weinberg [4] and Linde [5], as they argued for a Λ but from an anthropic standpoint as we will shortly see The real catalyst though which sparked theorists into action was the reported observations

CP957, *Particles, Strings, and Cosmology, 13th International Symposium*
edited by A. Rajantie, C. Contaldi, P. Dauncey, and H. Stoica

by the Supernova groups in 1998 [6, 7] of deviations in the Hubble diagram at large redshifts providing evidence of an accelerating Universe. It seemed to galvanise people into action and has led to an unprecedented amount of attention being lavished on the question of the origin of the dark energy as the source of the accelerating universe. In this brief review I will try and give a flavour of the many attempts that have been made to access the nature of dark energy. Given that there have been over 1200 papers written on the subject since 1998, I apologise from the outset. I can not cover the field properly, nor give proper references. There are a number of reviews on the subject and I urge the interested reader to look there for details. Here are a number of them [8, 9, 10, 11, 12]. Neither will I go into great details over any particular approach, this is going to be a schematic review, aimed at giving the reader a flavour of the attempts that have been made to understand the nature of the dark energy.

In what follows I will describe attempts to model the dark energy as a pure cosmological constant Λ. This remains the best fit to the data, although from a particle physics perspective it is fraught with problems. I will then discuss a range of dynamical dark energy models in which we assume there is some as yet unknown mechanism which has forced the real cosmological constant to vanish to day, and the dark energy is associated with say the potential energy of some slowly rolling scalar field. This too has problems, not least being the fact such a field generally has to have a very small mass, which means that it is subject to the severe constraints that exist in the form of fifth force experiments. Finally, we will look at attempts to modify the left hand side of Einstein's equations, the geometry part. Perhaps it is not the matter that needs looking at, it is the spacetime to which the matter is coupled. However, we will see that even in this case, it is pretty hard to modify Einstein's theory too much to accommodate the late time accelerating universe. All in all, the dark energy issue remains one of the most interesting and perplexing problems in physics and astrophysics.

A LITTLE BACKGROUND

A brief reminder of the background equations. We have the Friedmann and fluid equations describing the evolution of the scale factor $a(t)$ in an isotropic and homogeneous universe of spatial curvature K containing matter of density ρ and pressure p, related by an equation of state $p = w\rho$.

$$H^2 \equiv \left(\frac{\dot{a}}{a}\right)^2 = \frac{8\pi G\rho}{3} - \frac{K}{a^2}, \tag{1}$$

$$\dot{H} = -4\pi G(p+\rho) + \frac{K}{a^2}, \tag{2}$$

$$\dot{\rho} + 3H(\rho + p) = 0. \tag{3}$$

where H is the Hubble parameter.

Equation (3) can be derived from Eqs. (1) and (2), which means that two of Eqs. (1), (2) and (3) are independent. Eliminating the K/a^2 term from Eqs. (1) and (2), we obtain

$$\frac{\ddot{a}}{a} = -\frac{4\pi G}{3}(\rho + 3p) \,. \tag{4}$$

Hence the accelerated expansion occurs for $\rho + 3p < 0$.

Let us consider the evolution of the universe filled with a barotropic perfect fluid with an equation of state

$$w = p/\rho \,, \tag{5}$$

where w is assumed to be constant. Then by solving the Einstein equations given in Eqs. (1) and (2) with $K = 0$, we obtain

$$H = \frac{2}{3(1+w)(t-t_0)} \,, \tag{6}$$

$$a(t) \propto (t - t_0)^{\frac{2}{3(1+w)}} \,, \tag{7}$$

$$\rho \propto a^{-3(1+w)} \,, \tag{8}$$

where t_0 is constant. We note that the above solution is valid for $w \neq -1$. The radiation dominated universe corresponds to $w = 1/3$, whereas the dust dominated universe to $w = 0$. In these cases we have

$$\text{Radiation}: \quad a(t) \propto (t - t_0)^{1/2} \,, \quad \rho \propto a^{-4} \,, \tag{9}$$

$$\text{Dust}: \quad a(t) \propto (t - t_0)^{2/3} \,, \quad \rho \propto a^{-3} \,. \tag{10}$$

Both cases correspond to a decelerated expansion of the universe.

From Eq. (4) an accelerated expansion ($\ddot{a}(t) > 0$) occurs for the equation of state given by

$$w < -1/3 \,. \tag{11}$$

In order to explain the current acceleration of the universe, we require an exotic energy dubbed "dark energy" with equation of state satisfying Eq. (11). From Eq. (3) the energy density ρ is constant for $w = -1$. This is the famous cosmological constant case and it follows that the Hubble rate is also constant from Eq. (1), giving the evolution of the scale factor:

$$a \propto e^{Ht} \,, \tag{12}$$

which is the de-Sitter universe.

Now introducing the dimensionless density parameter $\Omega(t) = \rho(t)/\rho_c(t)$ where $\rho_c(t) = 3H^2(t)/8\pi G$ is the critical density (gives a spatially flat universe), we can rewrite the Friedmann equation as an elegant algebraic expression $\Omega_m + \Omega_\Lambda + \Omega_k = 1$ where I have divided the density up into three contributions arising from matter, a cosmological constant type term and spatial curvature. Turning briefly to the observational constraints on these three quantities. First Ω_m the contribution from matter.

Well a number of surveys of large scale structure, coupled to weak lensing tests, cluster baryon abundance measurements using X-ray measurements of intracluster gas, or SZ measurements and numerical simulations of structure growth all appear to indicate that $\Omega_m \sim 0.23$. For example the most recent 2df result, coupled with WMAP indicates $\Omega_m = 0.231 \pm 0.021$ [13]. whereas a similar analysis of the Sloan data set gives $\Omega_m = 0.24 \pm 0.02$ [14]. The bottom line is that it is much less than unity. Analysing the peak structure of the CMB anisotropies provides evidence for the contribution of Ω_k. In fact to a good approximation, the position of the first peak in ℓ space is given by $\ell_{\rm peak} \sim \frac{220}{\sqrt{\Omega_m+\Omega_\Lambda}}$. The most recent beautiful WMAP3 data then leads to the constraint $|1-(\Omega_m+\Omega_\Lambda)| = 0.03^{+0.026}_{-0.025}$, hence the universe is perfectly consistent with being spatially flat $\Omega_k \simeq 0$ [15]. However a slight word of caution, always be aware that these numbers come with certain prior assumptions made about some of the other parameters. There could well be some residual dependence on these parameters that has not been properly taken into account. Finally, the observations of the Type Ia supernova indicates that the Hubble diagram is no longer well fitted by a CDM dominated universe, and prefers a universe that is dominated with a cosmological constant today – well not actually today. We don't know what the universe is doing right now, we can only know what it was doing around a redshift of $z \sim 0.2-1$ as far as the supernova are concerned. For all we know it has actually stopped accelerating today and we just haven't received the information yet! Coupled with large scale structure surveys and WMAP3 data the favourite model appears to have $\Omega_\Lambda \sim 0.71, \Omega_m \sim 0.24, \Omega_{\rm baryons} \sim 0.05, \Omega_k \sim 0$, with $w = -0.97 \pm 0.07$ perfectly consistent with the cosmological constant result $w = -1$.

MODELS OF Λ

Why is the observed value of the cosmological constant a problem? For many it isn't, it is just another parameter that makes up the cosmological parameter set, and its observed contribution should just be taken as an empirical fact. However, for those who believe it has something to do with the world of particle physics then it is a real problem because its energy scale is so small compared to the value we would naively expect it to have. This cosmological constant problem has been with us for a long time. It was assumed the cosmological constant (or Λ) was zero today, but we did not have any real reason to set it to that value, rather we relied on some as yet unknown symmetry to protect it and maintain it at zero. To be a bit more precise, as we have seen today it is of order the critical density inferring an energy density scale of $10^{-47}(GeV)^4$. Now in the world of quantum field theory we expect that the zero point fluctuations of any scalar fields (for example) present will contribute towards an effective cosmological constant. Moreover the scale of that constant will be governed by the cut-off scale associated with the theory. So if I associated that with the Planck scale, I would expect a contribution of vacuum fluctuations to be at a scale of $10^{71}(GeV)^4$ a mere 10^{118} orders of magnitude larger than is observed. If you didn't trust Planck scale physics (and who could blame you!) then how about the electroweak scale at around a TeV. Then we would expect a contribution of around $1(TeV)^4$ which is a mismatch of 10^{61} or so. Thus even invoking supersymmetry (which brings us down to a TeV scale) we still have an incredible mismatch of scales. This is the cosmological constant problem, why has the universe not been completely dominated by these energy sources before now? Why are they effectively decoupling from the cosmological dynamics ? Needless to say there has been a great deal of interest in the possibility that a cosmological constant may pop out of string theory and in the braneworld. This has met with mixed success but has opened up an area of research which has led people into the myriad of possibilities known as the String Landscape. It was known for some time that in conventional supergravity models there were 'No go' theorems which forbid cosmic acceleration in cosmological solutions arising from compactification of pure supergravity models where the internal space is a time-independent, non-singular compact manifold without boundary [16]. Now one way of avoiding the theorem is to relax the conditions under which it applies such as allowing the internal space to be time-dependent, the analogue of a time-dependent scalar fields (radion). However, doing this, whilst leading to some solutions which have an accelerating late time evolution, they generally had mixed fortune in that they were preceded by a kinetic energy dominated evolution rather than a more useful matter dominated era. An alternative approach was developed by a number of authors who were interested in the pressing issue of stabilising the extra dimensions (moduli fields) in string theory [17, 18]. They realised that by using antisymmetric tensor fields to stabilise the extra dimensions (basically by wrapping them in fluxes), the net effect was that the resulting effective cosmological constant was described as a combination of the bare cosmological constant and the energy density of the fluxes (see also earlier work by Linde [5]). Given suitably quantised fluxes they argued that it was possible to have the required precision to account for the observed cosmological constant with of order 100 fluxes. This was easy to formulate in string theory. In fact given that there are believed to be of order 10^{500} possible vaccua in string theory, it opened up the possibility that this new Landscape would inevitably have regions (possibly many of them) where the vacuum energy matched our observed one [19]. This Landscape is good and bad. The good aspect of it, is that should it exist, it seems quite plausible that we can expect there to be many many Λ's and a large number of these will be in the range of values we require. Indeed in a pioneering paper back in the late 1980's Weinberg effectively predicted we would have to find a Λ term today with the kind of scale it appears to have [4]. Anything larger would have meant the universe would expand too rapidly for structures to collapse to form galaxies. The downside is that it is proving almost impossible to calculate anything properly in this framework. We dont yet have exact solutions we can use and so tend to end up adopting Anthropic arguments to justify our claims. The Landscape is clearly an interesting feature of string theory, and the hope for many of us is that one day we will be able to really test the notion that this is the origin of our observed universe. Other string related approaches to Λ include: using the extra dimensions to remove the gravitational effect of the vacuum energy, leaving us with an effectively flat four dimensional theory [20**?**]; of using a mechanism whereby Λ relaxes through the nucleation of branes coupled to a gauge potential [21] and by using the Cyclic universe as a means to give the cosmological constant plenty of time to relax to its present day value [22].

ROLLING SCALAR FIELDS – QUINTESSENCE AND K-ESSENCE

An alternative approach to explain the nature of the dark energy is to say that we don't know why Λ is zero, but accept it is and try to find an alternative explanation for the dark energy in the form of an evolving scalar field – a late time inflaton field if you like. Many of us have had a go at this, far too many to mention, but I will briefly describe some of the neat features associated with it. Many Quintessence scenarios have the attractive feature that there are attractor solutions where the system evolves so as to not dominate the dynamics at early times but come to dominate at late times. This means that the system can evolve from a wide range of initial conditions in the safe knowledge that it will inevitably dominate the evolution at late times. The idea first developed by Ratra and Peebles [23] as well as Wetterich [24] in the mid 1980's was later used for this purpose in [25, 26]. I am not going to go into the details of any of the scalar field models – there are so many of them, but I feel it should be made clear that they all have the same basic fine tuning issues. First of all the parameters in the potential need to be generally adjusted so that today the scalar field potential matches that required for it to be dark energy. The potentials have to be flat enough to have slow roll inflation and this needs to be protected against supergravity corrections. Secondly, it has to come to dominate at the correct time, namely in the recent past. In almost all cases of Quintessence the net affect is that the mass of the Quintessence field has to be very small, $m_Q \simeq 10^{-33}$eV, with an associated energy density of the field being $V_0 \simeq (10^{-3}eV)^4$. Such small mass scale should cause a few alarm bells to go off. A light field implies long range interactions, hence fifth force tests. There are strong limits on the allowed couplings of such fields to matter, and basically any such field needs to be protected from having too strong a coupling to matter otherwise we would have seen it as a fifth force. One possible route therefore is to consider models that have protected masses such as through using Nambu-Goto Goldstone bosons [27, 28, 29]. Following the equation of state for the scalar field during its evolution we see how it tracks that of the background radiation, then matter before coming to dominate like a cosmological constant today. However, it is not a constant, it is evolving. One of the biggest questions we face is how can we test whether dark energy is evolving and moreover can we see that it might be different from a true cosmological constant?

An alternative approach to having a suitably flat potential, is to consider the case of a non-canonical kinetic sector to the theory driving the late time acceleration[30, 31]. Such an approach is called K-essence and has generated a lot of interest because of the fact that non-canonical terms appear often in string models. One of the nice features of the approach is that a long period of perfect tracking is followed by domination of dark energy triggered by the transition to matter domination – an epoch during which structures can form. This solution of the 'why now' problem is therefore addressed because rather than being put in as an arbitrary time scale, the transition occurs around the epoch of radiation-matter equality an epoch we know happened. However, a word of warning. It is not clear that K-essence solves the coincidence problem. The basin of attraction into the regime of tracker solutions is small compared to those where it immediately goes into K-essence domination, and so there may well be a degree of fine tuning required in the scenario that is not obvious at first sight [32].

Before leaving the idea of scalar fields let me just mention that there have been a number of dark energy papers written making use of tachyon fields, phantom fields, chameleon fields, acceleron fields (scalar fields coupled to massive neutrinos) and chaplygin gases amongst others. The idea of the chameleon field is worth describing in a bit more detail [33]. The key idea is that in order to avoid fifth force type constraints on Quintessence models, why not have a situation where the mass of the field depends on the local matter density? It can then be massive in high density regions (i.e. solar system) and light ($m \sim H$) in low density regions (i.e. cosmological scales). In that way it can explain dark energy without violating solar system bounds. There is plenty of work currently going on to investigate the nature and consistency of chameleon fields.

MODIFYING GRAVITY

Up to now we have worked on the assumption that the dark energy is to be explained as a new form of matter, such as the cosmological constant. However, it could well be that it has nothing to do with the matter, but rather the gravity of the universe that we have got wrong. In other words perhaps we need to modify Einstein's gravity. It is worth mentioning one approach in a bit more detail, but it gives a flavour of the many attempts that have been made over the past few years. In the same spirit of the pioneering work of Starobinsky [34], Carroll et al looked at a modification of the usual Einstein action [35] in which late time acceleration can be realized by terms containing inverse powers of the Ricci scalar added to the Einstein Hilbert action. Unfortunately the original model ($\mathscr{L} \propto R + 1/R$) is not compatible with solar system experiments and this remains the real test of modified gravity models today. A large number of people have generalised the model of Carroll et al to more general $R + f(R)$ type models. However usually $f(R)$ struggles to satisfy both solar system bounds on deviations from General Relativity and late time acceleration. It brings in an extra light degree of freedom which once again implies fifth force constraints come into play. A possible get out clause that has been investigated is to make the scalar degree of freedom massive in the high density solar vicinity and therefore hidden from solar system tests by a chameleon type mechanism. In general it appears to require a form for $f(R)$ where the mass squared of the scalar is large and positive at high curvature. In fact very recently Amendola and Tsujikawa have analysed these models and concluded that the $f(R)$ models satisfying cosmological and local gravity constraints are practically indistinguishable from the Lambda CDM model, at least at the background level [36]. Other modified gravity models include ideas based on Braneworlds, where the extra dimensions can lead to modifications of the four dimensional gravity we are used to. Of particular popularity has been the model of Dvali-Gabadadze-Porrati (DGP) [37] in which the brane is embedded in a *Minkowski* bulk. In a DGP braneworld, gravity remains four dimensional at short distances but can leak into the bulk at large distances leading to infrared modifications to Einstein gravity. There is a cross-over scale around which gravity manifests these higher-dimensional properties. This scenario is a simple one parameter model which can account for the current acceleration of the universe provided the cross-over scale is fine tuned to match observations. However there are problems with the model, perhaps the most significant and potentially lethal one being the fact that in the Self accelerating branch there are ghosts in the spectrum for any value of the brane tension. Such an instability is serious and suggests that all is not well with the model [38].

I will finish on the controversial note – perhaps there is no such thing as dark energy and we are just misinterpreting the data. Although this is not the view of the majority of cosmologists, it has been suggested by a small number of authors and the result does not appear to be easy to shoot down. For example Hunt and Sarkar [39] have recently suggested that we can do away with the need for dark energy if we allow for the possibility that we live in an inhomogeneous universe, where inflation proceeded leading to features (bumps) in the primordial spectrum so that it is not scale free. We could be living in a local void where the Hubble flow is 30% faster than the global rate. They acknowledge a possible problem making their data consistent with the observed baryon oscillations in the large scale structure power spectra but otherwise they fit the data very well. Another inhomogeneous model has also been proposed recently by Wiltshire [40], and alternatives to Dark Matter and Dark Energy have been proposed by Mannheim [41]. Basically, we should always be prepared for the unexpected in this field.

CONCLUSIONS

This is a subject in which observations are transforming the field, especially the CMBR and LSS. The data to date suggests that the universe is accelerating and is being driven by a pure cosmological constant. However we don't really know why it is accelerating and what if anything has generated the constant to be the value it is. Moreover we don't even know if it is a cosmological constant and determining that must be one of the most important questions we would like to know the answer to. In particular, is $w = -1$, the cosmological constant and if not, then what value has it? On the other hand is w(z) dynamical? Perhaps we are actually seeing some new Gravitational Physics. There are lots of models of dark energy but it may yet prove too difficult to separate one from another such as cosmological constant. However we need to try though! Rather frustratingly for some of us, perhaps we will only be able to determine Λ from anthropic arguments and not from fundamental theory. And finally, perhaps we are wrong and we do not need a lambda term at all. The next decade is going to be exciting in this field – fingers crossed we get to determine the nature of our accelerating universe !

ACKNOWLEDGMENTS

I am very grateful to the organisors of Pascos 2007 in particular to Arttu Rajantie for inviting me to such an interesting meeting.

REFERENCES

1. G. Efstathiou, N. Kaiser, W. Saunders, A. Lawrence, M. Rowan-Robinson, R. S. Ellis and C. S. Frenk, Mon. Not. Roy. Astron. Soc. **247** (1990) 10.
2. S. J. Maddox, G. Efstathiou and W. J. Sutherland, M.N.R.A.S **283**, 1227 (1996) arXiv:astro-ph/9601103.

3. G. Efstathiou, W. J. Sutherland and S. J. Maddox, Nature **348**, 705 (1990).
4. S. Weinberg, Rev. Mod. Phys. **61**, 1 (1989).
5. A. D. Linde, Rept. Prog. Phys. **47**, 925 (1984).
6. S. Perlmutter *et al.*, Astrophys. J. **517**, 565 (1999).
7. A. G. Riess *et al.*, Astron. J. **116**, 1009 (1998); Astron. J. **117**, 707 (1999).
8. V. Sahni and A. A. Starobinsky, Int. J. Mod. Phys. D **9**, 373 (2000); V. Sahni, Lect. Notes Phys. **653**, 141 (2004) [arXiv:astro-ph/0403324].
9. S. M. Carroll, Living Rev. Rel. **4**, 1 (2001).
10. T. Padmanabhan, Phys. Rept. **380**, 235 (2003); T. Padmanabhan, Current Science, **88**, 1057 (2005) [arXiv:astro-ph/0510492].
11. P. J. E. Peebles and B. Ratra, Rev. Mod. Phys. **75**, 559 (2003).
12. E. J. Copeland, M. Sami and S. Tsujikawa, Int. J. Mod. Phys. D **15**, 1753 (2006) [arXiv:hep-th/0603057].
13. S. Cole *et al.* [The 2dFGRS Collaboration], Mon. Not. Roy. Astron. Soc. **362** (2005) 505 [arXiv:astro-ph/0501174].
14. M. Tegmark *et al.*, Phys. Rev. D **74**, 123507 (2006) [arXiv:astro-ph/0608632].
15. D. N. Spergel *et al.* [WMAP Collaboration], Astrophys. J. Suppl. **170**, 377 (2007) [arXiv:astro-ph/0603449].
16. G.W. Gibbons, *Aspects of Supergravity Theories*, in *Supersymmetry, Supergravity, and Related Topics*, eds. F. del Aguila, J.A. de Azcárraga and L.E. Ibañez (World Scientific 1985) pp. 346-351.
17. R. Bousso and J. Polchinski, JHEP **0006**, 006 (2000).
18. S. Giddings, S. Kachru and J. Polchinski, Phys. Rev. D **66**, 106006 (2002).
19. L. Susskind, arXiv:hep-th/0302219.
20. S. Kachru, M. Schulz, and E. Silverstein, Phys. Rev. **D62**, 045021 (2000); N. Arkani-Hamed, S. Dimopoulos, N. Kaloper, and R. Sundrum, Phys. Lett. B **480**, 193 (2000);
21. J. L. Feng, J. March-Russell, S. Sethi and F. Wilczek, Nucl. Phys. B **602**, 307 (2001).
22. P. J. Steinhardt and N. Turok, Science **312**, 1180 (2006) [arXiv:astro-ph/0605173].
23. B. Ratra and J. Peebles, Phys. Rev D **37**, 321 (1988).
24. C. Wetterich, Nucl. Phys B. **302**, 668 (1988).
25. R. R. Caldwell, R. Dave and P. J. Steinhardt, Phys. Rev. Lett. **80**, 1582 (1998).
26. I. Zlatev, L. M. Wang and P. J. Steinhardt, Phys. Rev. Lett. **82**, 896 (1999); P. J. Steinhardt, L. M. Wang and I. Zlatev, Phys. Rev. D **59**, 123504 (1999).
27. J. A. Frieman, C. T. Hill, A. Stebbins and I. Waga, Phys. Rev. Lett. **75**, 2077 (1995).
28. K. Choi, Phys. Rev. D**62**, 043509 (2000).
29. J. E. Kim and H. P. Nilles, Phys. Lett. B **553**, 1 (2003).
30. C. Armendáriz-Picón, V. Mukhanov, and P. J. Steinhardt, Phys. Rev. Lett. **85**, 4438 (2000).
31. C. Armendáriz-Picón, V. Mukhanov, and P. J. Steinhardt, Phys. Rev. D **63**, 103510 (2001).
32. M. Malquarti, E. J. Copeland, A. R. Liddle and M. Trodden, Phys. Rev. D **67**, 123503 (2003); M. Malquarti, E. J. Copeland and A. R. Liddle, Phys. Rev. D **68**, 023512 (2003).
33. J. Khoury and A. Weltman, Phys. Rev. Lett. **93**, 171104 (2004).
34. A. A. Starobinsky, Phys. Lett. B **91**, 99 (1980).
35. S. M. Carroll, V. Duvvuri, M. Trodden and M. S. Turner, Phys. Rev. D 70, 043528 (2004).
36. L. Amendola and S. Tsujikawa, arXiv:0705.0396 [astro-ph].
37. G. R. Dvali, G. Gabadadze and M. Porrati, Phys. Lett. B **485**, 208 (2000).
38. C. Charmousis, R. Gregory, N. Kaloper and A. Padilla, JHEP **0610**, 066 (2006) [arXiv:hep-th/0604086].
39. P. Hunt and S. Sarkar, arXiv:0706.2443 [astro-ph].
40. D. L. Wiltshire, arXiv:gr-qc/0702082.
41. P. D. Mannheim, Prog. Part. Nucl. Phys. **56**, 340 (2006) [arXiv:astro-ph/0505266]

New Stringy Instanton Effects[1]

M. Cvetič, R. Richter and T. Weigand

Department of Physics and Astronomy, University of Pennsylvania, Philadelphia, PA 19104-6396, USA

Abstract. D-brane instantons can generate open string couplings in the superpotential which violate global abelian symmetries and are therefore perturbatively forbidden. After discussing the main ingredients, focussing for concretenes on Type IIA orientifold compactifications, we exemplify the generation of perturbatively forbidden Yukawa couplings as well as Majorana mass terms for right-handed neutrinos in SU(5) GUT-like models.

Keywords: Document processing, Class file writing, LaTeX 2_ε
PACS: 43.35.Ei, 78.60.Mq

INTRODUCTION

During the last year there has been some progress towards a more complete understanding of non-perturbative effects in supersymmetric four-dimensional string compactifications on Calabi-Yau orientifolds. As was realized in [1, 2, 3, 4], D-brane instantons can induce couplings between open string fields which are perturbatively forbidden because they violate global $U(1)$ selection rules. These effects are intrinsically stringy in that they cannot be described by conventional gauge instantons. For Type IIA orientifolds with intersecting D6-branes the relevant class of instantons is given by Euclidean D2-brane instantons, short E2-instantons, wrapping special Lagrangian three-cycles of the internal Calabi-Yau space [1, 3].

In general the gauge group $U(N_a)$, carried by N_a coincident D-branes, contains an anomalous $U(1)_a$ which becomes massive via the generalized Green-Schwarz mechanism and survives as a global perturbative symmetry. It is due to this $U(1)$ selection rule that particular phenomenologically important couplings are absent in intersecting brane worlds. These include Majorana masses for the right-handed neutrino, μ-terms for the MSSM Higgs sector or Yukawa-couplings of type $\mathbf{10} \cdot \mathbf{10} \cdot \mathbf{5}_H$ in $SU(5)$-like GUT models. On the other hand, in the dual strongly coupled description in terms of M-theory compactified on G_2-manifolds the $U(1)_a$ decouples completely. There is therefore no associated selection rule and no obstruction for the above mentioned couplings to exist. The resolution to this puzzle is given by $U(1)_a$ breaking non-perturbative terms in the type IIA picture.

Indeed, from the axionic shift symmetries under the abelian symmetries induced by the Chern-Simons couplings of the N_a $D6_a$-branes one finds the $U(1)_a$ transformation

---

[1] Talk given by M. Cvetič.

CP957, *Particles, Strings, and Cosmology, 13th International Symposium*
edited by A. Rajantie, C. Contaldi, P. Dauncey, and H. Stoica

of the instanton [1, 3]

$$e^{-S_{E2}} = \exp\left[\frac{2\pi}{\ell_s^3}\left(-\frac{1}{g_s}\mathrm{Vol}_\Xi + i\int_\Xi C^{(3)}\right)\right] \longrightarrow e^{iQ_a(E2)\Lambda_a}\, e^{-S_{E2}} \tag{1}$$

with

$$Q_a(E2) = N_a\, \Xi \circ (\Pi_a - \Pi_a'). \tag{2}$$

Thus the exponential suppression factor $e^{-S_{E2}}$ characteristic for instantonic couplings transforms under the global $U(1)$ symmetries in such a way that the full coupling

$$W_{np} = \prod_i \Phi_{a_i b_i}\, e^{-S_{E2}} \tag{3}$$

is invariant again. In this way an instanton with the appropriate zero mode structure has the potential to generate perturbatively forbidden couplings.

Various applications of the associated effects in different branches of the string landscape have recently appeared in [5, 6, 7, 8, 9, 10, 11, 12, 13, 14, 15, 16, 17, 18, 19, 20], with directly related earlier work including [21, 22].

INSTANTON ZERO-MODES

In order for the instanton to contribute to the superpotential it has to carry the correct zero-mode structure. The subsequent description of E2-instanton zero modes follows mainly [8] (see also [11]).

Consider a Type IIA orientifold with spacetime filling D6-branes in the presence of a single Euclidean D2-brane wrapping the three-cycle Ξ of the internal manifold. In general, if Ξ is not invariant under the orientifold action we have to consider also the image D2-brane wrapping Ξ'. For this topological sector to correspond to a local minimum of the full string action, Ξ has to be volume minimizing in its homology class, i.e. special Lagrangian. Standard arguments taking into account the localization of the E2-brane in the external dimensions show that if the cycle Ξ is supersymmetric with respect to the orientifold plane, the instanton preserves the supercharges $Q_\alpha, \overline{Q}'_{\dot\alpha}$ and breaks $\overline{Q}_{\dot\alpha}, Q'_\alpha$, where the unprimed and primed quantities $Q_\alpha, \overline{Q}_{\dot\alpha}$ and $Q'_\alpha, \overline{Q}'_{\dot\alpha}$ generate the $\mathcal{N}=1$ subalgebra preserved by the orientifold and the one orthogonal to it, respectively.

To restore supersymmetry in the topological sector containing the E2-brane we have to integrate all amplitudes over the corresponding Goldstone fermions $\overline{\theta}_{\dot\alpha}$ and θ'_α associated with the violation of $\overline{Q}_{\dot\alpha}$ and Q'_α. Depending on the details of the orientifold projection, the θ'_α can be projected out provided the cycle Ξ is invariant, $\Xi = \Xi'$. In this case, the associated topological sector contributes to the anti-holomorphic superpotential involving the anti-chiral superfields. We identify it as the anti-instanton sector. By contrast, the instanton, wrapping a cycle which is anti-supersymmetric to the orientifold plane, preserves the supercharges $\overline{Q}_{\dot\alpha}, Q'_\alpha$ and violates $Q_\alpha, \overline{Q}'_{\dot\alpha}$, thus contributing to the holomorphic superpotential provided the $\overline{\theta}'_{\dot\alpha}$ are projected out. This is the situation we are interested in when computing corrections to the superpotential.

In the sequel it will be useful to consider only 'aligned' cycles Ξ wrapped by the (anti-)instanton. At the CFT level, the fact that the instanton is actually 'anti-aligned' w.r.t. the D6-branes internally is taken into account by projecting the spectrum between the E2 and the D6-branes of the model onto its GSO-odd part. From the worldvolume perspective, the two objects clearly carry opposite charge under the RR three-form C_3 coupling to the worldvolume. The classical part of the Euclidean (anti-)instanton action appearing as $e^{-S_{E2}}$ in corresponding F-term couplings reads

$$S_{E2} = \frac{2\pi}{\ell_s^3}\left(\frac{1}{g_s}\mathrm{Vol}_\Xi \mp i\int_\Xi C_3\right), \tag{4}$$

with the (lower) upper sign corresponding to (anti-)instantons[2].

The zero modes of the (anti-)instanton can be computed in setups where the $\mathcal{N} = (2,2)$ CFT describing the internal sector is known exactly. The general form of the various vertex operators can be found in [1]. For the sake of concreteness and as preparation for our explicit computations, we specialize here to the case that the D6- and the E2-branes wrap factorizable three-cycles of toroidal orientifolds.

One distinguishes two different kinds of zero-modes, namely the uncharged ones associated with strings starting and ending on the instanton and the charged ones which arise at intersections between the instanton and D-branes.

• Uncharged zero modes:
This sector contains the usual four bosonic zero modes x_E^μ corresponding to the Goldstone bosons associated with the breakdown of four-dimensional Poincaré invariance due to the localization of the instanton in spacetime. Their vertex operators (in the (-1) picture) are simply given by

$$V_{x_E^\mu}(z) = \Omega_{E2E2}\, x_E^\mu \frac{1}{\sqrt{2}}\psi_\mu(z)\, e^{-\varphi(z)}. \tag{5}$$

Here Ω_{E2E2} denotes the Chan Paton factor. The polarization x_E^μ carries no mass dimension, corresponding to a field in $D = 0$ dimensions. It is related to the position x_0^μ of the instanton in external spacetime via $x_E^\mu = x_0^\mu/\ell_s$.

The fermionic superpartners are given by four Weyl spinors $\theta_\alpha, \overline{\theta}_{\dot\alpha}$ with vertex operators (in the (-1/2) picture)

$$\begin{aligned} V_\theta(z) &= \Omega_{E2E2}\,\theta_\alpha\, S^\alpha(z)\prod_{I=1}^{3} e^{\frac{i}{2}H_I(z)} e^{-\varphi(z)/2}, & q &= 3/2, \\ V_{\overline{\theta}}(z) &= \Omega_{E2E2}\,\overline{\theta}_{\dot\alpha}\, S^{\dot\alpha}(z)\prod_{I=1}^{3} e^{-\frac{i}{2}H_I(z)} e^{-\varphi(z)/2}, & q &= -3/2. \end{aligned} \tag{6}$$

[2] We define the string length as $\ell_s = \sqrt{2\pi\alpha'}$.

Here $H_I(z)$ denotes the bosonization of the complexified internal fermions Ψ^I and S^α ($S^{\dot\alpha}$) are the left-(right-)handed four-dimensional spin fields. We have also included the worldsheet charge q.

In general the E2-E2 sector also comprises $b_1(\Xi)$ chiral superfields corresponding to the Wilson line and position moduli of the E2-brane. On toroidal backgrounds, they are associated with the moduli along those two-tori in which the E2-brane is not fixed. There appear additional zero modes at the intersection of the E2-brane and its orientifold image. Due to the Dirichlet-Dirichlet boundary conditions, the orientifold projection picks up an additional minus sign as compared to the D6-D6' sector. For single instantons, we therefore find $\frac{1}{2}(I_{E2E2'}+I_{O6E2})$ bosonic and/or fermionic zero modes carrying charge 2 under $U(1)_{E2}$. Their vertex operators are identical to the more familiar massless states between D6-branes and their images. For further details see [16].

- Charged zero modes :

Additional zero modes, called charged zero modes, arise at the intersection of the E2-brane with the various D6-branes from open strings localized at the intersection point. Open strings in this sector are subject to Dirichlet-Neumann boundary conditions in the extended four dimensions and to mixed DN boundary conditions internally, depending on the concrete intersection angles. The external DN conditions shift the oscillator moding in these directions by $1/2$. In the Ramond sector, the zero point energy is still vanishing and we find massless fermions. The novelty as compared to the case of spacetime filling branes at angles is that the degeneracy of states is lifted in that the four-dimensional spin fields S_α or $S_{\dot\alpha}$ are no longer present. This also affects the details of the GSO-projection. In the NS sector, the vacuum energy is zero only for completely parallel branes, which is the only situation with bosonic zero modes in the non-singular geometric phase.

In case the instanton wraps the cycle Ξ such that all intersection angles θ^I_{E2a} are positive for some cycle Π_a wrapped by a D6-brane the vertex operators for the fermionic zero mode λ_a at the intersection $E2-a$ is given by

$$V_{\lambda_a} = \Omega_{aE2}\,\lambda_a\,\Sigma(z)\prod_{I=1}^{3}\sigma_{1-\theta^I_{E2a}}(z)\,e^{-i(\theta^I_{E2a}-\frac{1}{2})H_I(z)}\,e^{-\varphi(z)/2}. \tag{7}$$

Here $\Sigma(z)$ denotes the bosonic twist field ensuring Dirichlet-Neumann boundary conditions in spacetime. The λ_a are Grassmannian variables and represent the polarization of the fermionic zero mode, normalized again as a field in $D=0$ dimensions. Note that the GSO-projection forces us to keep only the state in the sector starting from the D6-brane and ending on the E2 and projects out the state with the reversed orientation. The relevant intersection angles are therefore negative and lead to the above form of the vertex carrying worldsheet charge $q=-1/2$. As indicated by the CP indices, it transforms as $(\mathbf{N_a},-1_E)$. The various remaining cases are dealt with analogously and the result is displayed in table 1.

In conclusion a generic single E2-instanton wrapping a rigid cycle with no intersection with its orientifold image carries the uncharged zero modes x^μ, θ^α and $\theta_{\dot\alpha}$. Apart from the latter it also carries the charged zero modes λ depending on the geometric intersection number with the D-branes present in the setup.

TABLE 1. zero modes on E2, D6 intersections

zero modes	Reps.	number
$\lambda_{a,I}$	$(-1_E,\Box_a)$	$I=1,\ldots,[\Xi\cap\Pi_a]^+$
$\overline{\lambda}_{a,I}$	$(1_E,\overline{\Box}_a)$	$I=1,\ldots,[\Xi\cap\Pi_a]^-$
$\lambda_{a',I}$	$(-1_E,\overline{\Box}_a)$	$I=1,\ldots,[\Xi\cap\Pi'_a]^+$
$\overline{\lambda}_{a',I}$	$(1_E,\Box_a)$	$I=1,\ldots,[\Xi\cap\Pi'_a]^-$

PHENOMENOLOGICAL APPLICATIONS

In the following we discuss two examples of instanton generated perturbatively forbidden superpotential terms in Type IIA orientifolds. In this case, space-time instantons are given by E2-branes wrapping three-cycles Ξ in the Calabi–Yau, so that they are pointlike in four-dimensional Minkowski space.

Generically such instantons provide all in the previous section described zero modes. All these zero modes appear in the instanton measure and in order to to give contributions to the holomorphic superpotential, whose measure is $d^4x\,d^2\theta$, the instanton has to meet several constraints. Rigidity of the instanton ($b_1(\Xi)=0$) avoids the appearance of additional zero modes in the $E2$-$E2$ sector apart from the universal ones. If one chooses $\Xi=\Xi'$ the universal zero modes are subject to the Orientifold action $\Omega\overline{\sigma}$. Depending on the orientifold action one obtains $SO(N)$ or $USp(N)$ gauge group. In the latter case the zero modes x_μ,θ are anti-symmetrised and the mode $\overline{\theta}$ gets symmetrised, while for the $SO(N)$ instanton x_μ,θ are symmetrised and $\overline{\theta}$ gets anti-symmetrised. So in case of a single $O(1)$ instanton we get the desired measure $d^4x\,d^2\theta$ [9, 10, 11]. In the following we show that such $O(1)$ instantons can generate perturbatively forbidden couplings.

$SU(5)$ Yukawa couplings

Grand Unified $SU(5)$-like models based on intersecting $D6$-branes[3] generically suffer under the absence of the important Yukawa coupling $\mathbf{10}\cdot\mathbf{10}\cdot\mathbf{5}_H$ and are therefore ruled out from being considered realistic.

To construct such a model one only needs two stacks a and b of branes giving rise to $U(5)_a\times U(1)_b$ gauge symmetry. The $U(5)_a$ splits into $SU(5)_a\times U(1)_a$, where the anomalous $U(1)_a$ gets massive via the generalized Green-Schwarz mechanism and appears as a global symmetry in the effective action. While matter transforming as $\mathbf{10}$ under $SU(5)_a$ arises at intersections of stack a with its image a' the matter fields transforming as $\mathbf{5}$ as well as Higgs fields $\mathbf{5_H}$ and $\overline{\mathbf{5}}_\mathbf{H}$ are located at intersections of stack a with b and b'. Assuming that after applying the generalized Green Schwarz mechanism only the combination $U(1)_X=\frac{1}{4}U_a-\frac{5}{4}U(1)_b$ remains massless one obtains a flipped $SU(5)\times U(1)_X$ model. Table 2 displays the intersection numbers for such a model as

[3] See e.g. [23, 24] for global constructions of this type.

well as their charges under the particular gauge group.

TABLE 2. GUT $SU(5)$ intersecting D6-brane model

sector	number	$U(5)_a \times U(1)_b$ reps.	$U(1)_X$
(d',a)	3	$\mathbf{10}_{(2,0)}$	$\frac{1}{2}$
(a,b)	3	$\overline{\mathbf{5}}_{(-1,1)}$	$-\frac{3}{2}$
(b',b)	3	$\mathbf{1}_{(0,-2)}$	$\frac{5}{2}$
(d',b)	1	$\mathbf{5}^H_{(1,1)} + \overline{\mathbf{5}}^H_{(-1,-1)}$	$(-1)+(1)$

Looking at the $U(1)_{a,b}$ charges it is clear that perturbatively the two Yukawa couplings $\mathbf{10}\,\overline{\mathbf{5}}\,\overline{\mathbf{5}}^{\mathrm{H}}$ and $\overline{\mathbf{5}}\,\mathbf{5}^{\mathrm{H}}\,\mathbf{1}$ are present. For flipped $SU(5)$ these gives masses to the heavy (u,c,t)-Quarks and the leptons. However, the Yukawa couplings for the light (d,s,b)-quarks

$$\mathbf{10}_{(2,0)}\,\mathbf{10}_{(2,0)}\,\mathbf{5}^H_{(1,1)} \tag{8}$$

is not invariant under the two $U(1)$s.

Now, if the model contains an $O(1)$-instanton with intersections [14]

$$[\Xi\cap\Pi_a]^+ = 0, \quad [\Xi\cap\Pi_a]^- = 1, \quad [\Xi\cap\Pi_b]^+ = 0, \quad [\Xi\cap\Pi_b]^- = 1, \tag{9}$$

then we get five zero modes $\overline{\lambda}^i_{[5]}$ from the intersections of the instanton with $D6_a$ and one zero mode $\overline{\nu}_{[-1]}$ from the intersection with $D6_b$. Since the instanton lies in an $\Omega\overline{\sigma}$ invariant position, one can absorb these six matter zero modes with the three disc diagrams depicted in figure 1.

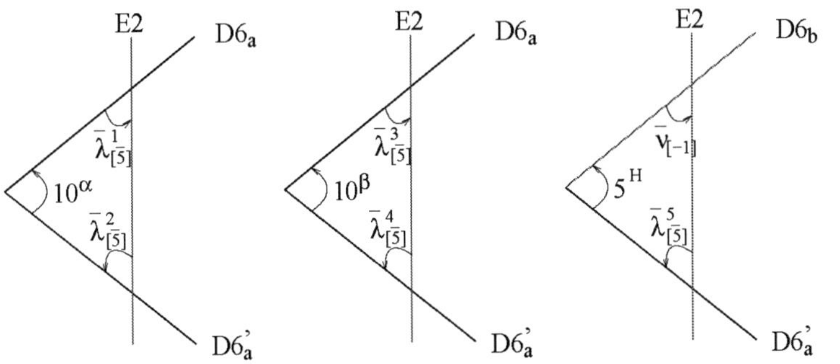

FIGURE 1. Absorption of zero modes

All charge selection rules are satisfied. Due to the Grassmanian nature of the fermionic zero modes $\overline{\lambda}^i$ the index structure of the Yukawa coupling is

$$W_Y = x^{\alpha\beta}\, e^{-\frac{2\pi}{\alpha_{GUT}}\frac{\mathrm{Vol}_{E2}}{\mathrm{Vol}_{D6}}}\, \varepsilon_{ijklm}\, \mathbf{10}^{\alpha}_{ij}\, \mathbf{10}^{\beta}_{kl}\, \mathbf{5}^H_m \tag{10}$$

where $\alpha\beta = 1,2,3$ are generation indices. Here we have assumed that both D6-branes have the same volume and gauge coupling unification at the GUT scale. The parameter $x^{\alpha\beta}$ contains the contribution from the one-loop determinants and additional prefactors. The location of the matter fields depend on the flavour index and therefore also the Yukawa couplings do.

To summarize, we find that D-brane instantons can generate the $\mathbf{10}\,\mathbf{10}\,\mathbf{5}_H$ Yukawa coupling. For the case of flipped SU(5) the hierarchy between the (u,c,t) quarks and the (d,s,b) quarks is explained by the E2-instanton suppression $\exp(-S_{E2}(U))$, whereas the hierarchy between the first, second and third family is due to the different suppression of world-sheet disc instantons $\exp(-S_{ws}(T))$, where U denote the complex structure and T the Kähler superfields.

Majorana masses

Another interesting perturbatively forbidden coupling is the Majorana mass term for right-handed Neutrino. These Majorana masses together with non-perturbative Dirac masses give rise to the seesaw mechanism. Generically the right-handed Neutrino is located at intersections between to abelian gauge groups $U(1)_a$ and $U(1)_b$ which get massive due to the generalized Green-Schwarz mechanism. In order to get the other matter content of the MSSM we have to introduce additional branes. For simplicity let us focus on a supersymmetric $SU(5)$ model, where we need only one additional brane providing a $SU(5)$ gauge group. Table 3 displays the matter content as well as the transformation under the gauge groups.

TABLE 3. GUT $SU(5)$ intersecting D6-brane model

sector	number	$U(5) \times U(1)_a \times U(1)_b$
(c',c)	3	$\mathbf{10}_{(0,0)}$
(c,b)	3	$\overline{\mathbf{5}}_{(1,0)}$
(a,c)	1	$\mathbf{5}^H_{(1,0)} + \overline{\mathbf{5}}^H_{(-1,0)}$
(a,b)	3	$\mathbf{1}_{(-1,1)}$

We note that the coupling generating the Dirac mass is realized $\overline{\mathbf{5}}\,\mathbf{5}^H\,\mathbf{1}$ while a Majorana mass term for the righthanded Neutrino

$$M_{N_R^c}\,\mathbf{1}_{(-1,1)}\,\mathbf{1}_{(-1,1)} \tag{11}$$

is forbidden due to global $U(1)_a$ and $U(1)_b$. If the model contains an $O(1)$-instanton with intersections [1, 3, 8, 11, 13]

$$[\Xi\cap\Pi_a]^+ = 0, \quad [\Xi\cap\Pi_a]^- = 2, \quad [\Xi\cap\Pi_b]^+ = 2, \quad [\Xi\cap\Pi_b]^- = 0, \tag{12}$$

then we get the right number of charged zero modes which can be absorbed via two disc diagrams. Performing the integration over the instantonic zero modes gives [1, 8]

$$M_{N^c_{N_R}}^{\alpha\beta} = x^{\alpha\beta} M_s e^{-\frac{2\pi}{\alpha_{GUT}} \frac{\mathrm{Vol_{E2}}}{\mathrm{Vol_{D6c}}}}, \tag{13}$$

where M_s is the string mass and α_{GUT} denotes the gauge coupling at the GUT scale. The prefactor x contains contribution from the one loop determinant as well as from the disc computation and is of order 1. Again the result is flavor depending due to world sheet instantons. For a ratio of the volumes of $0.06 < \frac{\mathrm{Vol_{E2}}}{\mathrm{Vol_{D6c}}} < 0.02$ we obtain Majorana masses in the range of $10^{11} - 10^{15} GeV$ which gives with the Dirac Neutrino masses the desired see-saw masses. For a local supersymmetric realization as well as the computation of the instanton induced Majorana mass term for right-handed Neutrino see [8].

These examples illustrate the importance of taking into account non-perturbative effects in exploring the string landscape. It remains as a challenge for future work to discover globally defined realistic string vacua in which such and similar genuinely stringy effects naturally realize the hierarchies of scale which seem so puzzling from a purely four-dimensional effective field theoretical point of view.

REFERENCES

1. R. Blumenhagen, M. Cvetič, and T. Weigand, *Nucl. Phys.* **B771**, 113–142 (2007), hep-th/0609191.
2. M. Haack, D. Krefl, D. Lüst, A. Van Proeyen, and M. Zagermann, *JHEP* **01**, 078 (2007), hep-th/0609211.
3. L. E. Ibáñez, and A. M. Uranga, *JHEP* **03**, 052 (2007), hep-th/0609213.
4. B. Florea, S. Kachru, J. McGreevy, and N. Saulina (2006), hep-th/0610003.
5. S. A. Abel, and M. D. Goodsell (2006), hep-th/0612110.
6. N. Akerblom, R. Blumenhagen, D. Lüst, E. Plauschinn, and M. Schmidt-Sommerfeld (2006), hep-th/0612132.
7. M. Bianchi, and E. Kiritsis (2007), hep-th/0702015.
8. M. Cvetič, R. Richter, and T. Weigand (2007), hep-th/0703028.
9. R. Argurio, M. Bertolini, G. Ferretti, A. Lerda, and C. Petersson (2007), arXiv:0704.0262 [hep-th].
10. M. Bianchi, F. Fucito, and J. F. Morales (2007), arXiv:0704.0784 [hep-th].
11. L. E. Ibáñez, A. N. Schellekens, and A. M. Uranga (2007), arXiv:0704.1079 [hep-th].
12. N. Akerblom, R. Blumenhagen, D. Lüst, and M. Schmidt-Sommerfeld (2007), arXiv:0705.2366 [hep-th].
13. S. Antusch, L. E. Ibáñez, and T. Macri (2007), arXiv:0706.2132 [hep-ph].
14. R. Blumenhagen, M. Cvetič, D. Lüst, R. Richter, and T. Weigand (2007), arXiv:0707.1871 [hep-th].
15. O. Aharony, and S. Kachru (2007), arXiv:0707.3126 [hep-th].
16. R. Blumenhagen, M. Cvetič, R. Richter, and T. Weigand (2007), arXiv:0708.0403 [hep-th].
17. O. Aharony, S. Kachru, and E. Silverstein (2007), arXiv:0708.0493 [hep-th].
18. M. Billo, et al. (2007), arXiv:0708.3806 [hep-th].
19. M. Billo, et al. (2007), arXiv:0709.0245 [hep-th].
20. M. Aganagic, C. Beem, and S. Kachru (2007), arXiv:0709.4277 [hep-th].
21. E. Witten, *Nucl. Phys.* **B474**, 343–360 (1996), hep-th/9604030.
22. O. J. Ganor, *Nucl. Phys.* **B499**, 55–66 (1997), hep-th/9612077.
23. M. Cvetič, I. Papadimitriou, and G. Shiu, *Nucl. Phys.* **B659**, 193–223 (2003), hep-th/0212177.
24. C.-M. Chen, V. E. Mayes, and D. V. Nanopoulos (2006), hep-th/0612087.

Beyond the Standard Model at the LHC and Beyond

John Ellis

Theoretical Physics Division, Physics Department, CERN, CH 1211 Geneva 23, Switzerland

Abstract. Many of the open questions beyond the Standard Model will be addressed by the LHC, including the origin of mass, supersymmetry, dark matter and the possibility of large extra dimensions. A linear e^+e^- collider (LC) with sufficient centre-of-mass energy would add considerable value to the capabilities of the LHC.

Keywords: Beyond the Standard Model, Higgs, Supersymmetry, LHC, ILC
PACS: 11.15.Ex,11.30.Pb,11.30.Qc,12.10.-g,12.60.Jv,14.80.Bn

OPEN QUESTIONS BEYOND THE STANDARD MODEL

There is a standard list of fundamental open questions beyond the Standard Model:
• What is the origin of particle masses?
Are they indeed due to a Higgs boson, as hypothesized within the Standard Model? If so, is the Higgs boson accompanied by some other physics such as supersymmetry? If not, what replaces the Higgs boson?
• Why are there so many types of matter particles?
Related to this question is the mixing of the different flavours of quarks and leptons, and the mechanism for CP violation. This matter-antimatter difference is thought to be linked to cosmological baryogenesis. However, the Standard Model cannot, by itself, explain the cosmological matter-antimatter asymmetry, adding urgency to the search for flavour and CP violation beyond the Standard Model.
• Are the fundamental forces unified?
If so, in the simplest models this unification occurs only at some very high energy $\sim 10^{16}$ GeV. Physics at this scale may be probed via neutrino physics, or possibly less directly at accelerators via measurements of particle masses and couplings, and looking for unification relations between them.
• What is the quantum theory of gravity?
The best candidate for such a theory may be (super)string theory, which generously predicts extra space-time dimensions as well as supersymmetry, but at what energy scale?

The good news is that all of these fundamental open questions will be addressed by the LHC: its energy should be ample for resolving the problem of mass, including the questions whether there is a Higgs boson and/or supersymmetry, a dedicated experiment will be examining matter-antimatter differences, models of unification could be probed

CP957, *Particles, Strings, and Cosmology, 13th International Symposium*
edited by A. Rajantie, C. Contaldi, P. Dauncey, and H. Stoica

via measurements of sparticle masses and couplings, and string theory might be probed via supersymmetry breaking, extra dimensions or even black hole production and decay. Accordingly, most of this talk will be concerned with LHC physics, accompanied by some asides about linear collider (LC) physics.

Supersymmetry may play a role in answering most of the open questions, for example by stabilizing the scale of electroweak symmetry breaking or by aiding the unification of the gauge coupling, and it also seems to play an essential role in string theory. Therefore, I have invested many or my personal efforts in supersymmetry, and apologize for giving pride of place to it in the rest of this talk.

HIGGS PHYSICS

Several different Higgs production mechanisms will be important at the LHC, including gluon-gluon and WW fusion, and production in association with gauge bosons or heavy quarks. Several decay modes will also be important, including $\gamma\gamma$, four leptons, $\tau^+\tau^-$, WW, $\bar{b}b$, etc. [1, 2]. Accordingly the search for the Higgs boson at the LHC will require the combination of many different signatures and hence excellent understanding of all components of ATLAS and CMS. As seen in Fig. 1, once both detectors have accumulated and analyzed a couple of hundred pb^{-1} of data, they may be able to start excluding certain ranges of Higgs masses at the 95 % confidence level [3]. With $\sim 1\ \mathrm{fb}^{-1}$ of analyzed data, they might be able to exclude a Standard Model Higgs boson over the entire mass range up to 1 TeV, or they might be able to establish a five-σ signal if the Higgs boson weighs between $\sim$ 150 and 500 GeV. Several fb^{-1}, corresponding to several months at one tenth of the design luminosity, would be needed to discover the Higgs boson over the entire mass range.

With some luck, it may also be possible to determine the spin of the Higgs boson [4, 5, 6]. For example, if it is light and is seen to decay into $\gamma\gamma$, it cannot have spin one. If it has higher mass and is seen to decay into four leptons, their angular correlations will enable J^P states other than 0^+ to be excluded with high significance. Analogous studies could be made at a LC, also using angular distibutions and the threshold behaviour for $e^+e^- \to ZH$ [7].

As seen in Fig. 2, the LHC may also make a first analysis of the proportionality of the Higgs couplings to particle masses [3]. If $m_H \sim 120$ GeV, it may be able to measure the couplings to $\tau^+\tau^-$, $\bar{b}b$, WW, ZZ and $\bar{t}t$ each with an accuracy ~ 20 %. Much more precise measurements could be made subsequently with a LC [8].

As well as its importance for particle physics, the discovery of a Higgs boson would also be very important for cosmology. An elementary Higgs boson would have caused a phase transition in the early Universe when it was $\sim 10^{-12}$ s old, and might have generated the matter in the Universe via electroweak barygenesis. Further back in the history of the Universe, a related inflaton might have expanded the Universe exponentially when it was $\sim 10^{-35}$ s old. Coming back to the present, naively the Higgs boson of the Standard Model would contribute a factor $\sim 10^{56}$ too much to the present-day dark energy, apparently requiring some 'miraculously' fine-tuned cancellation. Cosmologists should be as interested as particle physicists in the dénouement of the Higgs saga. For the time being, the LHC and a possible subsequent LC will be our only direct windows on this

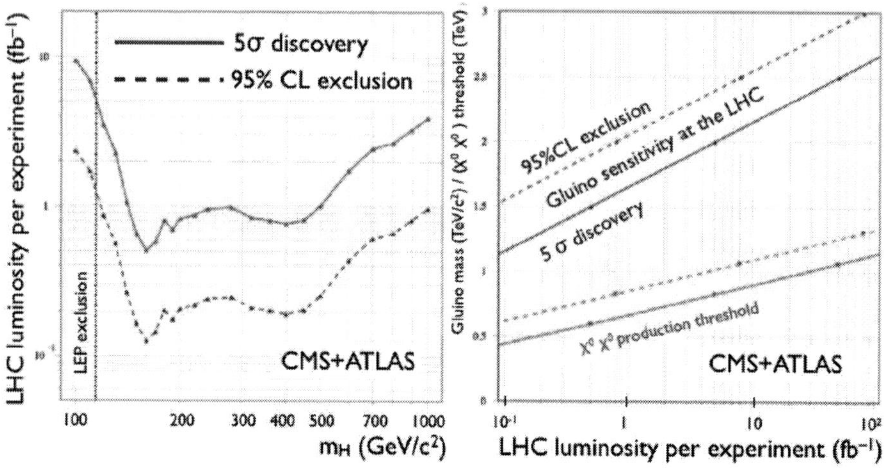

FIGURE 1. The combined sensitivities of ATLAS and CMS to a Standard Model Higgs boson (left), and the gluino (right), as a function of the analyzed LHC luminosity. The right panel also shows the threshold for sparticle pair production at a LC for the corresponding gluino mass, calculated within the CMSSM [3].

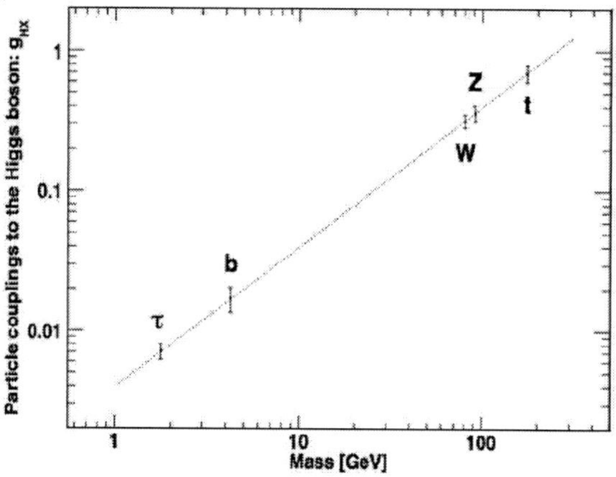

FIGURE 2. Estimates of the accuracy with which experiments at the LHC could measure the couplings of the Higgs boson to various particles [3].

physics.

SUPERSYMMETRY?

My personal favourite candidate for new physics beyond the Higgs boson is supersymmetry, for several reasons. True, it is intrinsically beautiful and (almost) an essential ingredient in string theory, etc., but these are not the reasons that motivate me to expect that it may appear at the LHC. There are four specific reasons why one might expect supersymmetry to appear around the TeV scale, and hence be accessible to the LHC. One is the naturalness or hierarchy problem [9], another is the unification of the gauge couplings [10], another is the supersymmetric prediction of a light Higgs boson as preferred by the precision electroweak data [11], and another is that many supersymmetric models predict the existence of cold dark matter with a density comparable to that required by astrophysics and cosmology [12].

In the following, I concentrate on the minimal supersymmetric extension of the Standard Model (MSSM). In addition to the coupling μ between its two gauge multiplets, and the ratio $\tan\beta$ of their vacuum expectation values, the MSSM has many apparently arbitrary soft supersymmetry-breaking parameters, including scalar masses m_0, gaugino masses $m_{1/2}$, trilinear soft couplings A_0 and a bilinear soft coupling B_0. It is commonly assumed that the $m_0, m_{1/2}$ and A_0 are each universal at some input GUT scale, a framework known as the constrained MSSM (CMSSM). This is not the same as minimal supergravity (mSUGRA), which fixes in addition the gravitino mass: $m_{3/2} = m_0$, and imposes $B_0 = A_0 - m_0$. We will see later some potential implications of these extra conditions.

There are direct limits on sparticle masses from their absences at LEP and the Tevatron, and indirect constraints from the LEP lower limit $m_h > 114$ GeV and from B physics, including in particular measurements of $b \to s\gamma$ decay. One possible indication of new physics at the TeV scale may be provided by the BNL measurement of the anomalous magnetic moment of the muon [13], that seems to exhibit a three-σ discrepancy with the Standard Model, though this is still somewhat controversial. The strongest constraint on (one combination of) supersymmetric model parameters is provided by the density of cold dark matter: $0.094 < \Omega_\chi h^2 < 0.124$, assuming that it is mainly composed of the lightest neutralino χ.

This not the only possibility: presumably the LSP should have neither strong nor electromagnetic interactions, but there are other candidates that also have these properties. The supersymmetric partners of the neutrinos have been excluded by a combination of LEP and direct dark matter searches, but the LSP might be the spartner of some particle beyond the Standard Model, such as the gravitino.

In a minimal supersymmetric model with universal soft supersymmetry-breaking parameters, one may consider these constraints in the $(m_{1/2}, m_0)$ plane. As shown in the left panel of Fig. 3, where the LSP is assumed to be the lightest neutralino, the resulting allowed regions include a narrow (pale turquoise) strip near the boundary beyond which $m_\chi > m_{\tilde{\tau}_1}$ (dark brown shading) where $m_0 \sim 100$ GeV, and another strip (not shown) where $m_0 > 1$ TeV near the boundary beyond which electroweak symmetry breaking is no longer possible (this region is disfavoured by $g_\mu - 2$, indicated by pink shading).

The fact that these allowed strips extend to relatively large values of $m_{1/2}$ and m_0 implies that even the lightest visible sparticles may not be very light, as seen in the right panel of Fig. 3. We generated a large sample of possible CMSSM scenarios (the red symbols), of which a subsample (shown in darker blue) provide the right amount of cold dark matter [14]. Also shown (in paler green) are the scenarios that would be detectable at the LHC. We see that most (but not all) of the dark matter scenarios should be visible at the LHC: one may be hopeful, but there is no guarantee that the LHC will discover

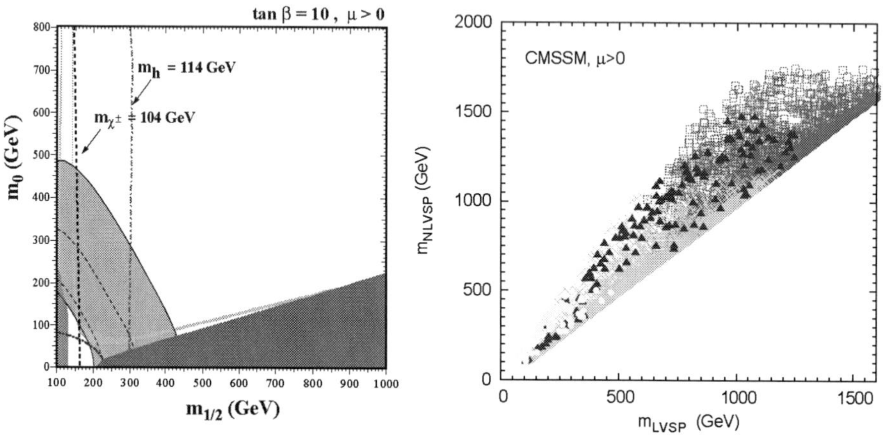

FIGURE 3. The $(m_{1/2}, m_0)$ plane in the CMSSM for $\tan\beta = 10$, $\mu > 0$ and $A_0 = 0$, incorporating the theoretical, experimental and cosmological constraints described in the text (left), and the masses of the lightest and next-to-lightest visible supersymmetric particles in a sampling of CMSSM scenarios (right) [14]. The right panel also indicates the scenarios providing a suitable amount of cold dark matter (blue), those detectable at the LHC (green) and those where the astrophysical dark matter might be detected directly (yellow).

supersymmetry. On the other hand, direct astrophysical dark matter search experiments (very pale yellow) may have less chances in the foreseeable future. An LC with centre-of-mass energy ~ 3 TeV would be needed to see sparticles in all the CMSSM dark matter scenarios we sampled.

Can one estimate the scale of supersymmetry on the basis of present data [15]? The precision electroweak measurements of m_W and $\sin^2\theta_W$ both have some sensitivity to $m_{1/2}$ through radiative corrections, and m_W slightly prefers smaller values of $m_{1/2}$, though this trend is hardly significant. The agreement of $b \to s\gamma$ with the Standard Model offers no encouragement to enthusiasts for light supersymmetry, and other B-decay observables such as $B_s \to \mu^+\mu^-$ and $B_u \to \tau\nu$ do not yet provide much information about the possible scale of supersymmetry breaking. On the other hand, the disagreement between the experimental value of $g_\mu - 2$ with the theoretical value calculated in the Standard Model using low-energy e^+e^- data could be explained by light supersymmetry, as shown already in the left panel of Fig. 3.

As shown in the left panel of Fig. 4 [15], a global fit to precision electroweak and B-decay observables indicates a preference for relatively small values of $m_{1/2}$. This is due predominantly to $g_\mu - 2$, but there is some support from the measurements of m_W. Correspondingly, the most likely value for the mass of the lightest supersymmetric Higgs boson is only slightly above the LEP lower limit, as seen in the right panel of Fig. 4.

The classic signature of sparticle production is the appearance of events with missing energy-momentum carried away by invisible dark matter particles. Studies indicate that such a signature should be observable at the LHC above instrumental and Standard Model backgrounds [1, 2]. In many supersymmetric scenarios, the LHC would produce a pair of squarks and/or gluinos primarily, and these would subsequently decay into a

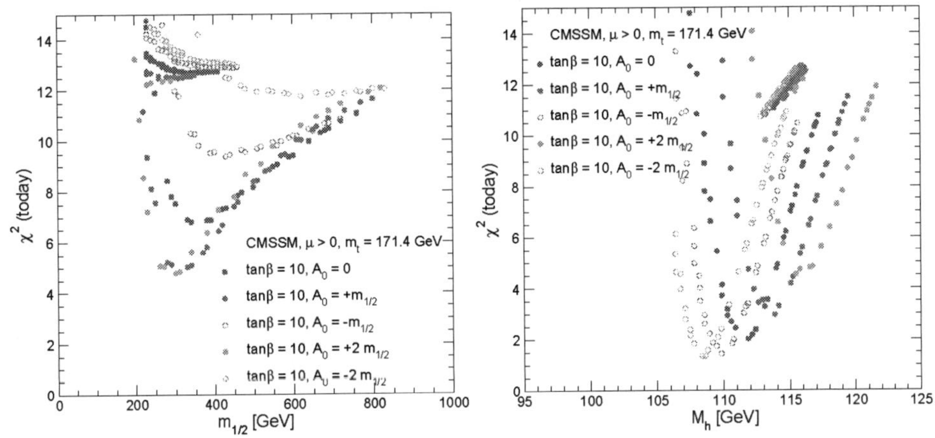

FIGURE 4. Evaluation of χ^2 for a global fit to electroweak and B-decay observables as a function of $m_{1/2}$ (left) and m_h (right) in the CMSSM for $\tan\beta = 10$, $\mu > 0$ and different values of A_0 [15].

cascade of secondary lighter sparticles, At the end of each decay chain would appear the LSP, namely the lightest neutralino χ in the CMSSM framework considered so far. Studies indicate that in many benchmark scenarios a number of different sparticle species could be observed at the LHC and their masses measured quite accurately [16, 17]. A LC would produce democratically all the sparticle species that are kinematically accessible, and therefore would be particularly interesting for producing sparticles without strong interactions, such as sleptons, charginos and neutralinos, which are expected to be lighter than squarks and gluinos in CMSSM scenarios.

Even a limited amount of LHC luminosity would enable the LHC to see squarks or gluinos weighing a TeV or more. As shown in the right panel of Fig. 1, it is estimated [1, 2] that 0.1 fb^{-1} of LHC luminosity would be sufficient to observe a gluino with a mass of 1.2 GeV at the five-σ level, or to exclude a gluino weighing < 1.5 TeV. The discovery and exclusion reaches would extend to about 2.2 and 2.5 TeV, respectively with 10 fb^{-1} of LHC luminosity.

As also shown in the right panel of Fig. 1, assuming universal input sparticle masses at the GUT scale, the corresponding thresholds for sparticle pair production at a linear e^+e^- collider would be 0.5 (0.6) or 0.8 (1.0) TeV [3, 18]. Hence, for example, if the LHC discovers the gluino with 0.1 fb^{-1}, one may expect that sparticle pair production would be accessible to a linear collider with centre-of-mass energy of 0.5 TeV, whereas if the LHC does not discover the gluino even with 10 fb^{-1}, the e^+e^- sparticle pair production threshold may be above 1 TeV. At least in such a simple model, the LHC will tell us how much energy a linear collider would need to find supersymmetry [18].

If Nature is kind, and sparticles not only exist but also are quite light, it will be possible to test directly unification of the gauge couplings and universality of the soft supersymmetry-breaking scalar masses with high precision, in particular by comparing measurements at the LHC and the ILC [19]. However, there is no guarantee that any sparticles will be light enough to be produced at the ILC, even with a centre-of-mass

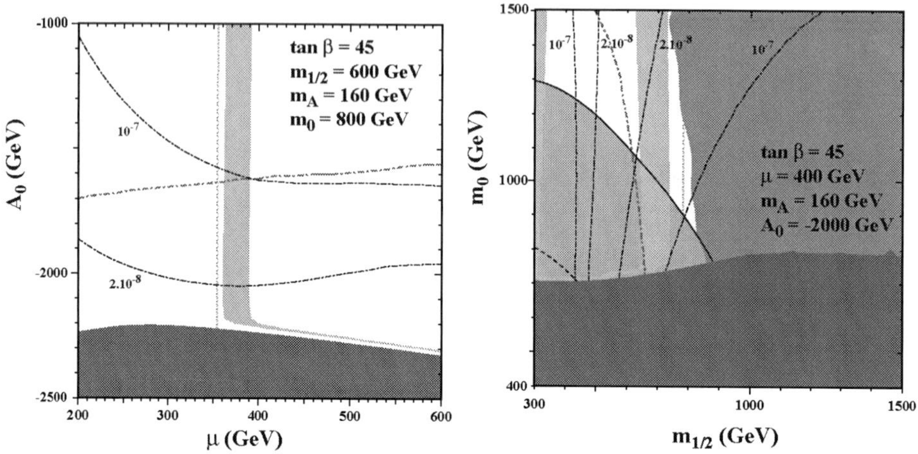

FIGURE 5. A light $m_A \sim 160$ GeV could be accommodated within the NUHM, if $\tan\beta \sim 50, m_{1/2} \sim 600$ GeV, $m_0 \sim 800$ GeV, $\mu \sim 400$ GeV and $A_0 \sim -2$ TeV, in which case $m_h, b \to s\gamma, B_s \to \mu^+\mu^-, B_u \to \tau\nu$ and the cold dark matter scattering rates would all be very close to the present experimental limits [22].

energy of 1 TeV. For this reason, and also because even if there are some light sparticles the heavier ones will be produced only at higher energies, a high-energy LC would be advantageous, which is why CERN and its collaborators are developing the CLIC technology that should be capable of reaching 3 TeV in the centre of mass [20].

So far, I have concentrated on the CMSSM, assuming that all the soft supersymmetry-breaking scalar masses are universal at the GUT scale. However, this may not be the case: in particular, there is no good theoretical or phenomenological reason why the Higgs scalar masses should be universal. In models with non-universal Higgs masses (NUHM), there are two additional degrees of freedom, and the Higgs mixing parameter μ and the pseudoscalar Higgs mass m_A may be treated as free parameters [21].

We have studied [22] whether the NUHM framework could accommodate a pseudoscalar supersymmetric Higgs boson as light as the value not yet excluded by direct searches at the Tevatron. Recall that the CDF experiment saw a two-σ excess in the $\tau^+\tau^-$ spectrum that could be explained by supersymmetric Higgs bosons weighing ~ 160 GeV [23], whereas the D0 experiment saw no such excess [24]. We found [22] that $m_A \sim 160$ GeV could indeed be accommodated within the NUHM, if $\tan\beta \sim 50, m_{1/2} \sim 600$ GeV, $m_0 \sim 800$ GeV, $\mu \sim 400$ GeV and $A_0 \sim -2$ TeV, as seen in Fig. 5. This would be quite an extreme scenario, but it would have the merit of being testable in the near future. In this corner of the NUHM parameter space, $m_h, b \to s\gamma, B_s \to \mu^+\mu^-, B_u \to \tau\nu$ and the cold dark matter scattering rates would all be very close to the present experimental limits.

GRAVITINO DARK MATTER?

The LSP should have no strong or electromagnetic interactions, otherwise it would bind to conventional matter and be detectable as some anomalous heavy nucleus. Apart from the lightest neutralino χ, possible weakly-interacting scandidates include the sneutrino, which is excluded by LEP and direct searches in simple models, and the gravitino $\tilde{G}$, which has attracted relatively little detailed attention until recently. It would be a nightmare for astrophysical detection, but a bonanza for the LHC [25, 26, 27], as we now discuss.

The next-to-lightest sparticle (NLSP) would have a very long lifetime in models with gravity-mediated supersymmetry breaking, due to the weak gravitational strength of the interactions responsible for its decays. For example, if the NLSP is the lighter stau slepton $\tilde{\tau}_1$ one has

$$\Gamma_{\tilde{\tau}_1 \to \tau \tilde{G}} = \frac{1}{48\pi} \frac{1}{M_P^2} \frac{m_{\tilde{\tau}_1}^5}{m_{\tilde{G}}^2} \left(1 - \frac{m_{\tilde{G}}^2}{m_{\tilde{\tau}_1}^2}\right)^4 , \tag{1}$$

leading to a lifetime that could be measured in hours, days, weeks, months or even years! The stau is not the only candidate for the NLSP. Other generic possibilities include the lightest neutralino χ, a sneutrino [28], or even the lighter stop squark [29]. In the case of minimal supergravity (mSUGRA), one finds that a gravitino LSP combined with a stau NLSP is as generic as the conventional neutralino LSP scenario, as seen by the light (yellow) shaded region in the left panel of Fig. 6. All of these possibilities are constrained by astrophysics and cosmology, particularly through limits on the decays of the metastable NLSP [30], and its bound states if it is charged [31]. These effects may even improve the agreement between cosmological nucleosynthesis calculations and the observed Lithium abundances, as shown in the pink (darker) shaded region in the right panel of Fig. 6 for one particular non-mSUGRA scenario [32]. This scenario could be probed only with a higher-energy LC [33].

We recently examined how a stau NLSP scenario could be explored at the LHC [25, 26, 27]. We found that the normal experimental triggers on jets and energetic muons and electrons would select the events containing staus [26, 27]. It would then be possible to identify the stau tracks in the events with quite high efficiency, and measure the stau mass very accurately via a combination of momentum and time-of-flight measurements, as seen in Fig. 7. The stau could be combined with jets in the event to reconstruct the masses of heavier sparticles in the supersymmetric decay cascades, as in neutralino LSP scenarios, as shown in Fig 8.

Very slow-moving staus might be stopped either in the detector material or in the rock surrounding the cavern [34, 35]. One could use the muon system to locate the stau's impact point on the cavern wall with an uncertainty ~ 1 cm, and its impact angle with an accuracy $\sim 10^{-3}$. One might then be able to bore into the cavern wall and remove a core from the rock, which one could then store while waiting for the stau to decay. Before boring, one would have to wait for a shut-down of the LHC, since the radioactivity in the cavern is quite high while it is operating. However, it is planned to stop the accelerator for a couple of days every month, so this strategy might work if the lifetime exceeds about 10^6 s!

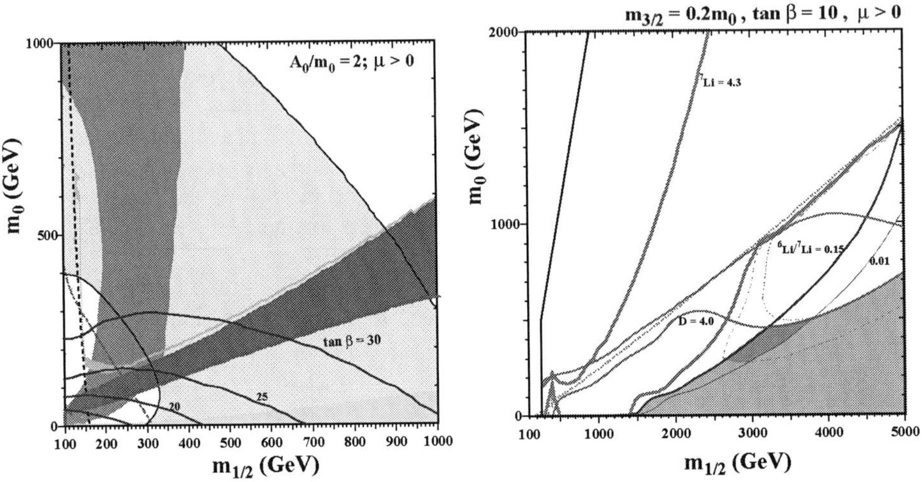

FIGURE 6. In the mSUGRA scenario with $A_0 = 2m_0$ and $\mu > 0$ (left panel), in addition to the allowed χ dark matter strip (light turquoise) and the forbidden $\tilde{\tau}_1$ LSP wedge (dark brown), there is a wedge (pale yellow) with a metastable $\tilde{\tau}_1$ whose decays do not disturb the cosmological light element abundances. In the non-SUGRA scenario with $m_{\tilde{G}} = 0.2m_0$ (right panel), there is also a region (darker pink) where $\tilde{\tau}_1$ bound-state effects [31] may improve the Lithium abundances [32].

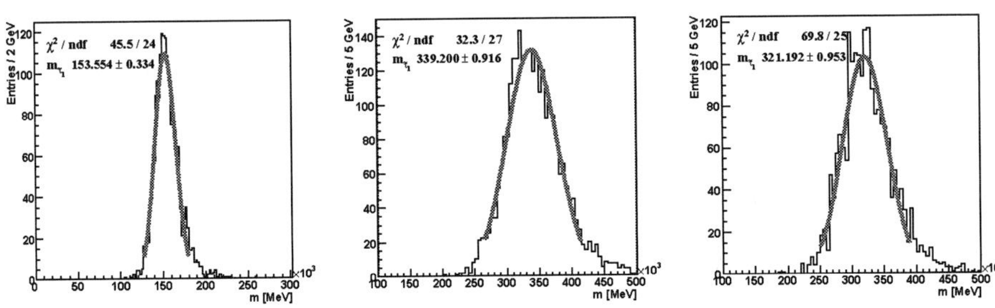

FIGURE 7. The mass of a metastable stau could be measured quite accurately at the LHC [27], as exemplified in three benchmark scenarios [25].

EXTRA DIMENSIONS?

Supersymmetry is certainly not the only possibility for new physics at the LHC. Another possibility offered by string theory is that there might be large extra dimensions. When string theory was originally proposed as a 'Theory of Everything', it was imagined that all the extra dimensions would be curled up on length scales comparable to the Planck length $\sim 10^{-33}$ cm. However, then it was realized that string unification could be achieved more easily if one of these dimensions was somewhat smaller than the GUT scale [36], and a number of scenarios with much larger extra dimensions have

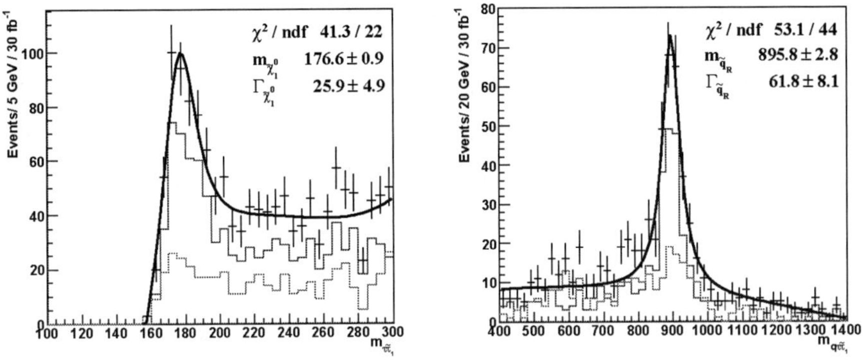

FIGURE 8. The reconstruction of heavier sparticles decaying into the $\tilde{\tau}$ in a scenario where it is the NLSP and the gravitino is the LSP [26].

been considered. For example, an extra dimension of size ~ 1 TeV^{-1} could help break supersymmetry [37] and/or the electroweak gauge symmetry, an extra dimension of micron size could help rewrite the hierarchy problem [38], and even infinite extra dimensions are allowed if they are warped appropriately [39].

In many of these scenarios, there are potential signals to be found at the LHC, such as Kaluza-Klein excitations of gravitons, or missing energy 'leaking' into an extra dimension. The most spectacular possibility would occur if gravity becomes strong at the TeV energy scale, in which case microscopic black holes might be produced at the LHC. These would be very unstable, decaying rapidly via Hawking radiation into multiple jets, leptons and photons, that would be easily detectable [40] and distinguishable from supersymmetry and other extra-dimensional scenarios [41].

SUMMARY

The origin of mass is the most pressing problem in particle physics, and requires a solution within the LHC energy range. Will it be a simple Higgs boson? and/or supersymmetry? The LHC will tell! there are many speculative ideas for other possible new physics beyond the Standard Model, such as grand unification, strings, and extra dimensions. The LHC is also capable of probing many of these speculations including, as the last two examples show, novel ideas that were undreamt of when the LHC and its experiments were designed. We do not know what the LHC will find, but we can be sure that its discoveries will set the agenda for possible future projects, such as a linear e^+e^- collider.

REFERENCES

1. ATLAS Collaboration, *Detector and Physics Performance Technical Design Report*, LHCC 99-14/15 (1999).
2. CMS Collaboration, *CMS Physics Technical Design Report. Volume II: Physics Performance*, CERN/LHCC 2006-021, CMS TDR 8.2 (2006).
3. A. Blondel, L. Camilleri, A. Ceccucci, J. Ellis, M. Lindroos, M. Mangano and G. Rolandi, *Physics opportunities with future proton accelerators at CERN*, arXiv:hep-ph/0609102.
4. S. Y. Choi, D. J. Miller, M. M. Muhlleitner and P. M. Zerwas, Phys. Lett. B **553** (2003) 61 [arXiv:hep-ph/0210077].
5. C. P. Buszello, I. Fleck, P. Marquard and J. J. van der Bij, Eur. Phys. J. C **32** (2004) 209 [arXiv:hep-ph/0212396].
6. E. Accomando *et al.*, *Workshop on CP studies and non-standard Higgs physics*, arXiv:hep-ph/0608079.
7. D. J. Miller, S. Y. Choi, B. Eberle, M. M. Muhlleitner and P. M. Zerwas, Phys. Lett. B **505** (2001) 149 [arXiv:hep-ph/0102023].
8. G. Weiglein *et al.* [LHC/LC Study Group], Phys. Rept. **426** (2006) 47 [arXiv:hep-ph/0410364]; A. Djouadi, J. Lykken, K. Monig, Y. Okada, M. J. Oreglia and S. Yamashita *et al.*, *International Linear Collider Reference Design Report Volume 2: Physics at the ILC*, arXiv:0709.1893 [hep-ph].
9. L. Maiani, *All You Need To Know About The Higgs Boson*, Proceedings of the Gif-sur-Yvette Summer School On Particle Physics, 1979, pp.1-52; G. 't Hooft, in *Recent developments in Gauge Theories*, Proceedings of the NATO Advanced Study Institute, Cargèse, 1979, eds. G. 't Hooft et al. (Plenum Press, NY, 1980); E. Witten, Phys. Lett. B **105** (1981) 267.
10. J. R. Ellis, S. Kelley and D. V. Nanopoulos, Phys. Lett. B **249** (1990) 441 and Phys. Lett. B **260** (1991) 131; U. Amaldi, W. de Boer and H. Furstenau, Phys. Lett. B **260** (1991) 447; C. Giunti, C. W. Kim and U. W. Lee, Mod. Phys. Lett. A **6** (1991) 1745.
11. M. W. Grunewald, arXiv:0709.3744 [hep-ex].
12. J. R. Ellis, J. S. Hagelin, D. V. Nanopoulos, K. A. Olive and M. Srednicki, Nucl. Phys. B **238** (1984) 453.
13. H. N. Brown *et al.* [Muon g-2 Collaboration], Phys. Rev. Lett. **86** (2001) 2227 [arXiv:hep-ex/0102017].
14. J. R. Ellis, K. A. Olive, Y. Santoso and V. C. Spanos, Phys. Lett. B **603** (2004) 51 [arXiv:hep-ph/0408118].
15. J. R. Ellis, S. Heinemeyer, K. A. Olive, A. M. Weber and G. Weiglein, arXiv:0706.0652 [hep-ph].
16. I. Hinchliffe, F. E. Paige, M. D. Shapiro, J. Soderqvist and W. Yao, Phys. Rev. D **55** (1997) 5520 [arXiv:hep-ph/9610544].
17. M. Battaglia, A. De Roeck, J. R. Ellis, F. Gianotti, K. A. Olive and L. Pape, Eur. Phys. J. C **33** (2004) 273 [arXiv:hep-ph/0306219]; A. De Roeck, J. R. Ellis, F. Gianotti, F. Moortgat, K. A. Olive and L. Pape, arXiv:hep-ph/0508198.
18. This discussion is being extended in J. R. Ellis, K. A. Olive and P. Sandick, *in preparation.*
19. P. Bechtle, K. Desch, W. Porod and P. Wienemann, Eur. Phys. J. C **46** (2006) 533 [arXiv:hep-ph/0511006].
20. E. Accomando *et al.* [CLIC Physics Working Group], *Physics at the CLIC multi-TeV linear collider*, arXiv:hep-ph/0412251.
21. J. Ellis, T. Falk, K. Olive and Y. Santoso, *Nucl. Phys.* **B 652** (2003) 259, hep-ph/0210205.
22. J. R. Ellis, S. Heinemeyer, K. A. Olive and G. Weiglein, Phys. Lett. B **653** (2007) 292 [arXiv:0706.0977 [hep-ph].
23. CDF Collaboration, CDF note 8676, see: `http://www-cdf.fnal.gov/~aa/mssm_htt_1fb/note/cdf8676.pdf`.
24. D0 Collaboration, D0 Note 5331-CONF, see: `http://www-d0.fnal.gov/cgi-bin/d0note?5331`.
25. A. De Roeck, J. R. Ellis, F. Gianotti, F. Moortgat, K. A. Olive and L. Pape, Eur. Phys. J. C **49** (2007) 1041 [arXiv:hep-ph/0508198].
26. J. R. Ellis, A. R. Raklev and O. K. Oye, JHEP **0610** (2006) 061 [arXiv:hep-ph/0607261].
27. J. R. Ellis, A. R. Raklev and O. K. Oye, *in preparation.*
28. L. Covi and S. Kraml, arXiv:hep-ph/0703130.

29. J. L. Diaz-Cruz, J. R. Ellis, K. A. Olive and Y. Santoso, JHEP **0705** (2007) 003 [arXiv:hep-ph/0701229].
30. R. H. Cyburt, J. R. Ellis, B. D. Fields and K. A. Olive, Phys. Rev. D **67** (2003) 103521 [arXiv:astro-ph/0211258].
31. M. Pospelov, Phys. Rev. Lett. **98** (2007) 231301 [arXiv:hep-ph/0605215].
32. R. H. Cyburt, J. R. Ellis, B. D. Fields, K. A. Olive and V. C. Spanos, JCAP **0611** (2006) 014 [arXiv:astro-ph/0608562].
33. O. Cakir, I. T. Cakir, J. R. Ellis and Z. Kirca, arXiv:hep-ph/0703121.
34. J. L. Feng and B. T. Smith, Phys. Rev. D **71** (2005) 015004 [Erratum-ibid. D **71** (2005) 0109904] [arXiv:hep-ph/0409278].
35. K. Hamaguchi, Y. Kuno, T. Nakaya and M. M. Nojiri, Phys. Rev. D **70** (2004) 115007 [arXiv:hep-ph/0409248].
36. E. Witten, Nucl. Phys. B **471** (1996) 135 [arXiv:hep-th/9602070].
37. I. Antoniadis, Phys. Lett. B **246** (1990) 377.
38. I. Antoniadis, N. Arkani-Hamed, S. Dimopoulos and G. R. Dvali, Phys. Lett. B **436** (1998) 257 [arXiv:hep-ph/9804398].
39. L. Randall and R. Sundrum, Phys. Rev. Lett. **83** (1999) 3370 [arXiv:hep-ph/9905221].
40. C. M. Harris, M. J. Palmer, M. A. Parker, P. Richardson, A. Sabetfakhri and B. R. Webber, JHEP **0505** (2005) 053 [arXiv:hep-ph/0411022].
41. J. M. Smillie and B. R. Webber, JHEP **0510** (2005) 069 [arXiv:hep-ph/0507170].

Amplitudes of Field Theories

Bo Feng

Blackett Laboratory & The Institute for Mathematical Sciences, Imperial College, London, SW7 2AZ, UK and Center of Mathematical Science, Zhejiang University, Hangzhou, China

Abstract.
In this talk, we will present recent progresses in perturbative calculations of scattering amplitudes at tree and one-loop levels. At tree level, we will discuss MHV-diagram method and on-shell recursion relation. At one-loop level, we will establish the framework of Unitarity cut method.

Keywords: On-shell recursion relation; Unitarity cut method
PACS: 11.15.Bt; 11.55.Bq; 12.38.Bx

INTRODUCTION

With the coming of LHC (Large Hadron Collide) experiments, one big challenge we are facing is data analysis. The data collected by experimenters, presumedly include both backgrounds (contributions from standard model) and signals (contributions from "new physics"). With current technology, the measurement can be very accurate with only a few percentages of uncertainties. To be able to diagnose the existence of new physics, we must be able to calculate scattering amplitudes of backgrounds and signals *theoretically* with the same level of accuracy. In another word, one-loop amplitudes are required by most of interaction channels.

In principal, we can use Feynman diagram method to calculate all amplitudes. However, with following reasons, it is not a very efficient method, especially when we deal with QCD processes. First, if the process is a little bit complicated, we will end up with a lot of diagrams. Second, each diagram will have complicated tensor and group structures. Thirdly, each diagram is not gauge invariant and the result would to be long and complicated. Only when we sum up some subsets of diagrams together we achieve gauge invariance and huge cancelation will happen in middle steps.

Because these, calculations using Feynman diagrams are not very fast numerically for each phase point. The situation becomes worse when we do the final higher dimension phase space integration. It will be hard to accept if one process will take one year of computer time.

It is the aim of this talk to present some very promising progresses toward speeding up the calculation of scattering amplitudes needed by LHC experiments.

SOME BACKGROUNDS

In this talk we will use spinor notations extensively, so we will explain them first[1]. We all know that the lie algebra of Lorentz group can (locally) be written as the sum of

CP957, *Particles, Strings, and Cosmology, 13th International Symposium*
edited by A. Rajantie, C. Contaldi, P. Dauncey, and H. Stoica

two sl_2, i.e, we will have two fundamental representations: *negative chirality spinor* λ_α with representation $(1/2,0)$ and *positive chirality spinor* $\widetilde{\lambda}_{\dot{\alpha}}$ with representation $(0,1/2)$. Any other representations of Lorentz group can always be constructed as proper tensor of these fundamental two. For example, vector representation can be written as bi-spinor notation through Pauli-sigma matrix $V_{\alpha\dot{\alpha}} = V_\mu \sigma^\mu_{\alpha\dot{\alpha}}$. One important consequence is that $V^\mu \cdot V_\mu = det(V_{\alpha\dot{\alpha}})$. Applying this to on-shell momentum of massless particle we will have $p^2 = 0 = det(p_{\alpha\dot{\alpha}})$. In another word, as two-by-two matrix, $p_{\alpha\dot{\alpha}}$ is degenerated, so we can write it as $p_{\alpha\dot{\alpha}} = \lambda_\alpha \widetilde{\lambda}_{\dot{\alpha}}$.

The last equation is, in fact, the key for all simplifications involving massless particles. The lesson we learn from this equation is that for on-shell massless particles, the right variables to describe them are not momentum vector and polarization vector, but *spinors*. The reason is following. First from spinors we can reconstruct momentum vector. Second, the polarization of fermions is given exactly by spinors while for vector bosons, it is given by

$$\varepsilon^+(p) = \frac{\lambda_k \widetilde{\lambda}_p}{\langle k\ p\rangle}, \quad \varepsilon^-(p) = \frac{\lambda_p \widetilde{\lambda}_k}{[p\ k]} \tag{1}$$

where k is an arbitrary auxiliary light-cone momentum which is related to gauge freedom. For simplification, we have defined following notations:

$$|p\rangle \equiv \lambda_p, \quad |p] \equiv \widetilde{\lambda}_p; \quad \langle k\ p\rangle \equiv \varepsilon_{\alpha\beta} \lambda_k^\alpha \lambda_p^\beta, \quad [p\ k] \equiv \varepsilon_{\dot{\alpha}\dot{\beta}} \widetilde{\lambda}_p^{\dot{\alpha}} \widetilde{\lambda}_k^{\dot{\beta}}. \tag{2}$$

Using these, if $P = \sum_a p_a$, $p_a^2 = 0$, we can define $\langle k_1|P|k_2] = \sum_a \langle k_1\ p_a\rangle [p_a\ k_2]$ and $\langle k_1|PQ|k_2\rangle = \sum_a \sum_b \langle k_1\ p_a\rangle [p_a\ q_b] \langle q_b\ k_2\rangle$ etc..

TREE LEVEL AMPLITUDES

Now we discuss tree-level amplitudes, especially for theories with gauge symmetry. Gauge invariance makes final results much simpler than naively expected from Feynman diagrams since many cancelations happen in middle steps. However, gauge invariance does much more than what we thought. There is another not-so-obvious simple result if we use right variables, the famous Parke-Taylor formula[2, 3] for MHV (maximum helicity violation) amplitudes at tree-level[1]:

$$A(1^+, 2^+, \ldots, i^-, \ldots, j^-, \ldots, (n-1)^+, n^+) = \frac{\langle i\ j\rangle^4}{\prod_{k=1}^n \langle k\ (k+1)\rangle}. \tag{3}$$

To understand this result, Witten proposed twistor string theory in [4] and showed that MHV amplitude in twistor space corresponds to straight line, thus gave a geometric explanation of MHV results. In some sense, all recent developments are intrigued by this work. However, in this talk we will not mention too much about the twistor string theory.

---

[1] Here and later we will discuss *color-ordered amplitudes*. For explanation, see [1].

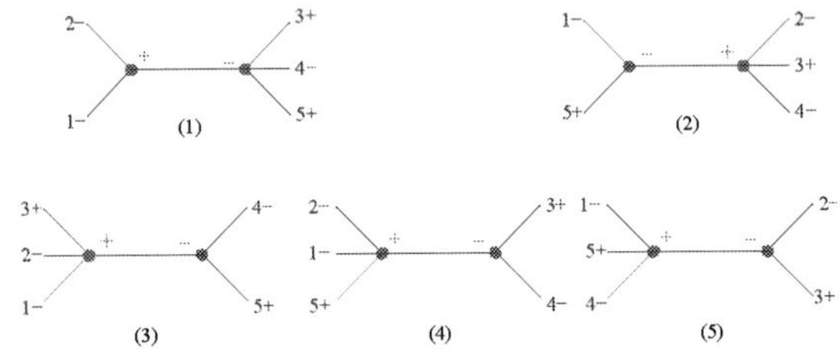

FIGURE 1. MHV diagram for five gluons

MHV-diagram method

The first method at tree-level we want review is the MHV-diagram method which was proposed by Cachazo, Svrcek, Witten in [5] and later was proved in [6, 7]. The basic idea is that we can use *MHV as vertex to construct amplitudes* just like using Feynman vertex to construct Feynman diagrams.

Let us give one example $A(1^-,2^+,3^+,4^-,5^+)$ to demonstrate the idea. From the Figure 1 we can see that there are five contributions. To find the contribution for the first one, we need just to write down

$$\left(\frac{\langle 1\, 2\rangle^3}{\langle 2\, P\rangle \langle P\, 1\rangle}\right) \frac{1}{P^2} \left(\frac{\langle P\, 4\rangle^4}{\langle P\, 3\rangle \langle 3\, 4\rangle \langle 4\, 5\rangle \langle 5\, P\rangle}\right), \quad P = p_1 + p_2\,. \tag{4}$$

However, there is a technical point: in general $P = p_1 + p_2$ is not along light-cone, so we need to find a way to define its chiral spinor. The prescription is given by $|P\rangle \equiv |P|\eta]$ where $|\eta]$ is an arbitrary anti-chiral spinor which will drop out at final result (however we need to take same $|\eta]$ for all diagrams).

One good thing about the MHV-method is that gauge invariance is built in from start. However, when the number of negative helicity increases, the number of MHV diagrams will increase a lot. This constrains the practical use of MHV method.

On-shell recursion relation

In this subsection we will introduce another method to calculation tree level amplitudes: the on-shell recursion relation (or BCFW recursion relation) [8, 6]. However, before doing that, let us recall some general properties of tree level amplitudes in field theories.

The first important property of amplitudes is the *analyticity*: *amplitudes are analytic function of (complex) momenta.* In reality, momentum is a real vector. However, if we analytically extend each real component into complex number, amplitudes, as a function of momentum, will become analytic function. Because this, a lot of very general results of amplitudes can be derived just by assuming analyticity like S-matrix theory.

One consequence of analyticity is that at tree level, amplitudes as functions of momenta have only pole structure, i.e., there is no branch cut. Furthermore, these poles are single poles with locations at these points $P^2 - m^2 = 0$ where P is the sum of some subset of external momenta. This conclusion can be easily seen from Feynman diagrams using Feynman rules since all denominators are given by propagators $P^2 - m^2$.

The second important property of amplitudes is the *factorization form* when $P^2 \to m^2$. In another word, this property tells us that in the on-shell limit $P^2 \to m^2$, which terms are dominant in whole amplitudes. At tree level, the result is very simple and can be written as

$$A(1,2,...,n) \sim \sum_h A_L(1,2,..,k,P_k^h)\frac{1}{P_k^2 - m^2}A_R(P_k^{-h},k+1,...,n), \quad P_k = \sum_{i=1}^k p_i \tag{5}$$

where summing h is over all possible helicities along propagator. There is one important thing about this form we want to remark: in each term A_L, A_R are on-shell tree level amplitudes since $P_k^2 \to m^2$, so it is gauge invariant.

With there preparations, we will start to derive the on-shell recursion relation. Let us consider the tree amplitude with n massless particles, for example, gluons, $A(1,2,...,n)$. As we have mentioned before, instead of using momenta and polarization vectors, we can use $\lambda_i, \widetilde{\lambda}_i$, thus the amplitudes will be function of these $\lambda_i, \widetilde{\lambda}_i$. Now let us define following analytic function

$$A(z) = A(\widetilde{\lambda}_i(z), \lambda_j(z)) \tag{6}$$

where we have selected two particles i^-, j^+ and made following shifting

$$\lambda_j(z) = \lambda_j - z\lambda_i, \quad \widetilde{\lambda}_i(z) = \widetilde{\lambda}_i + z\widetilde{\lambda}_j\ . \tag{7}$$

The wanted amplitude is just $A(z=0)$. The function $A(z)$ defined above has following special properties. First, with this shifting of $\lambda_j, \widetilde{\lambda}_i$, momenta $p_i(z), p_j(z)$ become complex number. Using our general properties discussed at beginning of this subsection, we know that $A(z)$ will be analytic function of z with only single poles. The positions of poles are where $P(z)^2 = 0$ where $P(z)$ is the sum of ordered momenta including $p_i(z)$, but not including $p_j(z)$. Secondly, momentum conservation is still respected because $p_i(z) + p_j(z) = p_i(z=0) + p_j(z=0)$ by the special form of shifting we have chosen.

To find $A(z=0)$ we do following contour integration $I = \oint dz\frac{A(z)}{z}$ where the contour is chosen to be big enough such that all single poles of $A(z)$ are inside the contour. For this integration we have two ways to do it. First we evaluate it using poles inside and get

$$I = A(z=0) + \sum_{poles\ z_i} \text{Residue}\left(\frac{A(z)}{z}\right)_{z_i} \tag{8}$$

where the first term is exactly the one we are looking for. The second way is to evaluate around poles at infinity. It is given by $I = -A(z=\infty)$, which is the boundary contribution of $A(z)$. Identifying both we get

$$A(z=0) = -\sum_{poles\ z_k} \text{Residue}\left(\frac{A(z)}{z}\right)_{z_k} - A(z=\infty), \tag{9}$$

How to evaluate residues of single poles? Here the factorization form tells us that the residue is nothing, but given by

$$Residue_{z_k} = \sum_h A_L(\widetilde{\lambda}_i(z_k), P_k(z_k)^h)\frac{1}{P_k^2}A_R(\lambda_j(z_k), P_k(z_k)^{-h}) \ . \tag{10}$$

Since both A_L, A_R are on-shell amplitudes with less number of particles, equation (9) gives the wanted on-shell recursion relation. The last piece is the boundary term $A(z = \infty)$. It can be shown that it is zero [6] with our special choice of shifting related to the helicities of particles i, j.

Putting all together we can write recursion relation as following

$$A_n = \sum_{i=1}^{n-3}\sum_{h=\pm} A_{i+2}^h \frac{1}{P_{n,i}^2} A_{n-i}^{-h} \ .$$

Before we finish this subsection, we have two remarks. First properties we have used to derive above recursion relation are very general properties of field theories. Using similar idea, about recursion relation has been generalized to include massive particles, gravitons, fermions etc. Second, this method has one tricky point, i.e., the boundary contribution. If the shifting is chosen improperly, the boundary contribution is not zero and the recursion relation is not so easy to use.

ONE-LOOP AMPLITUDES

The calculation of one-loop amplitudes is obviously much more difficult than the one of tree-level amplitudes. However, there is a general framework, so called PV-reduction (Passarino-Veltman reduction) [9], which has been used extensively in practices.

The general idea is following. By some algebraic manipulations, any one-loop amplitude with complex tensor and group structures can be written as the sum of some bases with rational coefficients. These bases are universal, so the whole calculation can be divided into two parts: studying bases and finding coefficients.

The choice of bases is not unique, but for simplicity, scalar bases are chosen in most practices and we know how to deal with them: some of them have analytic expressions and some of them, numerical packages. More concretely, in our FDH-scheme (four dimensional helicity scheme), the scalar bases are following five topologies: tadpoles, bubbles, triangles, boxes and pentagons [10] (see Figure 2)[2] and the prototype of integration is given by

$$I_n = \int d^{4-2\varepsilon}\ell \frac{1}{\ell^2(\ell-K_1)^2....(\ell-\sum_{j=1}^{n-1}K_j)^2} \ . \tag{11}$$

After knowing bases, the left task is to find rational coefficients. There are various ways to do it and here we will discuss the unitarity-cut-method proposed in [11]. The

[2] In a lot of situations, the tadpole contribution is zero.

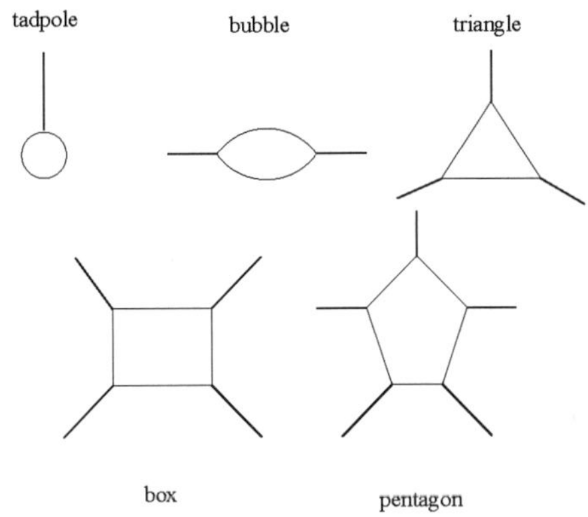

FIGURE 2. The bases

general idea is following. We know that $A_{1-loop} = \sum_i c_i I_i$ where I_i are bases. Now we take imaginary part at both sides $\Delta A = \sum_i c_i \Delta I_i$ where since c_i are rational coefficients, they do not have imaginary part. Furthermore, each ΔI_i is unique function, so we can distinguish them without any ambiguity. Thus by comparing both sides, we can determine all c_i.

From above discussion, we can see that the key point of this method is to calculate ΔA (as well as ΔI_i). The expression is given by

$$\Delta A = \int d^{4-2\varepsilon} \ell A_L A_R \delta(\ell^2) \delta((\ell - K)^2) \,. \tag{12}$$

The advantages of unitarity-cut method is following. First the input is the product of left and right *on-shell* tree level amplitudes, thus gauge invariance is built in and no huge cancelation coming from individual Feynman diagrams. Second, in each calculation we find only subset of coefficients related to the given cut, so we efficiently separate one big problem into several small pieces which are easier.

Now we discuss how to evaluate ΔA. To do so, first we rewrite integral momentum $\ell = \widetilde{\ell}^{4D} + \vec{\mu}^{-2\varepsilon}$ where $\widetilde{\ell}$ is in pure four-dimension and $\vec{\mu}^{-2\varepsilon}$, in the extra -2ε-dimension. For the $\widetilde{\ell}$ we rewrite it as $\widetilde{\ell} = \ell + zK$ with $\ell^2 = 0$ and K is the cut-momentum. Under this new variables, phase space integration is given by [12, 13]

$$\Delta A = \int_0^1 du\, u^{-1-\varepsilon} \int d^4 \ell \delta^+(\ell^2) \delta((1-2z)K^2 - 2K \cdot \ell) A_L A_R \tag{13}$$

where $z = \frac{1-\sqrt{1-u}}{2}$ and $u = \frac{4\mu^2}{K^2}$.

The expression (13) hints that whole integration can be done in two stages. At stage one, we do pure 4D space integration define by $C[u] = \int d^4 \ell \delta^+(\ell^2) \delta((1-2z)K^2 - 2K \cdot \ell) A_L A_R$ and at stage two, we do $\int_0^1 du\, u^{-1-\varepsilon} C[u]$. The reason we have made this division is that the calculation of first stage has been solved already in [14, 15] using *holomorphic anomoly*. One important point about the stage one is that the final result will be the form $C[u] = \sum_i c_i(u) \Delta I_i(u)$ where $c_i(u)$ is *polynomial* of u. To demonstrate, we list $C[\Delta I_i]$ for bubbles, triangles and boxes as following

$$\Delta I_2(K) = \int_0^1 du\, u^{-1-\varepsilon}\sqrt{1-u}\,, \tag{14}$$

$$\Delta I_3(K_1;K_3) = -\int_0^1 du\, u^{-1-\varepsilon}\frac{1}{\sqrt{\Delta_3}}\ln\left(\frac{Z+\sqrt{1-u}}{Z-\sqrt{1-u}}\right), \tag{15}$$

$$\Delta I_4(K;P_1,P_2) = \frac{1}{2K^2}\int_0^1 du\, u^{-1-\varepsilon}\frac{1}{\sqrt{B-Au}}\ln\left(\frac{D-Cu+\sqrt{1-u}\sqrt{B-Au}}{D-Cu-\sqrt{1-u}\sqrt{B-Au}}\right) \tag{16}$$

For second stage, although it seems that we have integration to do, in fact, we can do it algebraically by recursion relation. To see that, let us use following example:

$$\mathrm{Tri}^{(n)}(Z) \equiv \int_0^1 du\, u^{-1-\varepsilon}u^n \ln\left(\frac{Z+\sqrt{1-u}}{Z-\sqrt{1-u}}\right).$$

We can derive following recursion relation by partial integration

$$\mathrm{Tri}^{(n)}(Z) = -\frac{(Z^2-1)(n-1-\varepsilon)}{(n-\varepsilon)}\mathrm{Tri}^{(n-1)}(Z) + \frac{2Z(n-1-\varepsilon)}{(n-\varepsilon)}\mathrm{Bub}^{(n-1)}.$$

Similar recursion relations can be derived for all other bases.

Combining these two stages, we see that we have solved the problem of finding coefficients of bases using unitarity-cut method. We want to emphasize that coefficients will be function of ε through recursion relation at the second stage, so we get analytic result for all order of ε.

Algebraic results

Although we have established the whole framework for the phase space integration, the actual calculation still need some efforts. In this subsection, we will show that in fact we can do it once for all [16].

The idea is that after t-integration, the general form of remaining integration at stage one will be

$$\int \langle\lambda\, d\lambda\rangle[\widetilde{\lambda}\, d\widetilde{\lambda}]\frac{G(\lambda)\prod_{j=1}^{n+k-2}[a_j\,\ell]}{\langle\ell|K|\ell]^n\prod_{i=1}^{k}\langle\ell|Q_i|\ell]}$$

Using Schouten identity, we can split them into following sum

$$\sum_{k=2}^{n} G_k(\lambda)\frac{\prod_{j=1}^{k-2}[b_j\,\ell]}{\langle\ell|K|\ell]^k} + \sum_{i=1}^{k} F_i(\lambda)\frac{1}{\langle\ell|K|\ell]\,\langle\ell|Q_i|\ell]}.$$

The point about this splitting is that now we can identify which term in above splitting gives contribution to which bases. More concretely, coefficient of box given by K,Q_i,Q_j

is given by the residue of pole $\langle \ell | Q_i Q_j | \ell \rangle$ in the $F_i(\lambda)$. For triangle defined by K, Q_i, it is given by the residue of pole $\langle \ell | Q_i K | \ell \rangle$ in the $F_i(\lambda)$. Finally for bubble it is given by residues from G_k. Calculating these residues we arrive some algebraic functions which give the wanted coefficients of bases. For example, the box coefficients is given by

$$C_{box;ij} = \frac{K^2}{(1-2z)} \left(\frac{F_{i,j}(P_1^{(ij)}) + F_{i,j}(P_2^{(ij)})}{2} \right),$$

For others, it can be found in [16].

ACKNOWLEDGMENTS

I would like to thank the organizers of PASCOS, 2007 for inviting me to present this result in the conference. The work is supported by the Marie-Curie Research Training Network under contact MRTN-CT-2004-005104.

REFERENCES

1. L. J. Dixon, arXiv:hep-ph/9601359.
2. S. J. Parke and T. R. Taylor, Phys. Rev. Lett. **56**, 2459 (1986).
3. F. A. Berends and W. T. Giele, Nucl. Phys. B **306**, 759 (1988).
4. E. Witten, Commun. Math. Phys. **252**, 189 (2004) [arXiv:hep-th/0312171].
5. F. Cachazo, P. Svrcek and E. Witten, JHEP **0409**, 006 (2004) [arXiv:hep-th/0403047].
6. R. Britto, F. Cachazo, B. Feng and E. Witten, Phys. Rev. Lett. **94**, 181602 (2005) [arXiv:hep-th/0501052].
7. K. Risager, JHEP **0512**, 003 (2005) [arXiv:hep-th/0508206].
8. R. Britto, F. Cachazo and B. Feng, Nucl. Phys. B **715**, 499 (2005) [arXiv:hep-th/0412308].
9. G. Passarino and M. J. G. Veltman, Nucl. Phys. B **160**, 151 (1979).
10. Z. Bern, L. J. Dixon and D. A. Kosower, Nucl. Phys. B **412**, 751 (1994) [arXiv:hep-ph/9306240].
11. Z. Bern, L. J. Dixon, D. C. Dunbar and D. A. Kosower, Nucl. Phys. B **425**, 217 (1994) [arXiv:hep-ph/9403226].
12. C. Anastasiou, R. Britto, B. Feng, Z. Kunszt and P. Mastrolia, Phys. Lett. B **645**, 213 (2007) [arXiv:hep-ph/0609191].
13. C. Anastasiou, R. Britto, B. Feng, Z. Kunszt and P. Mastrolia, JHEP **0703**, 111 (2007) [arXiv:hep-ph/0612277].
14. F. Cachazo, P. Svrcek and E. Witten, JHEP **0410**, 077 (2004) [arXiv:hep-th/0409245].
15. R. Britto, E. Buchbinder, F. Cachazo and B. Feng, Phys. Rev. D **72**, 065012 (2005) [arXiv:hep-ph/0503132].
16. R. Britto and B. Feng, Phys. Rev. D **75**, 105006 (2007) [arXiv:hep-ph/0612089].

Black Hole Attractors in Extended Supergravity

Sergio Ferrara* and Alessio Marrani†

*Physics Department, Theory Unit, CERN, CH 1211, Geneva 23, Switzerland;
INFN - Laboratori Nazionali di Frascati, Via Enrico Fermi 40,00044 Frascati, Italy;
Department of Physics and Astronomy, University of California, Los Angeles, CA USA.
†Museo Storico della Fisica e Centro Studi e Ricerche "Enrico Fermi", 00184 Roma, Italy;
INFN - Laboratori Nazionali di Frascati, Via Enrico Fermi 40,00044 Frascati, Italy.

Abstract. We review some aspects of the attractor mechanism for extremal black holes of (not necessarily supersymmetric) theories coupling Einstein gravity to scalars and Maxwell vector fields. Thence, we consider $\mathcal{N}=2$ and $\mathcal{N}=8$, $d=4$ supergravities, reporting some recent advances on the moduli spaces associated to BPS and non-BPS attractor solutions supported by charge orbits with non-compact stabilizers.

Keywords: Black Holes; Attractors; Moduli Spaces; Supergravity
PACS: 04.70.Bw; 04.65.+e.

The so-called *attractor mechanism* was first considered in the framework of $\mathcal{N}=2$, $d=4$ ungauged supergravity coupled to n_V vector multiplets [1]-[5]. It concerns the stabilization of the scalar fields ϕ^i ($i=1,...,n_V$) of the theory near the event horizon of an extremal, static, spherically symmetric and asymptotically flat black hole (BH) [6]. An extremal BH can be defined to have vanishing temperature ($T=0$), and thus it is thermodynamically stable. The asymptotical behavior of the scalars ϕ^i is defined by the limits

$$\begin{aligned} lim_{r\rightarrow\infty}\phi^i(r) &= \phi^i_\infty \in \mathcal{M}; & (1)\\ lim_{r\rightarrow r_H}\phi^i(r) &= \phi^i_H(q,p), & (2)\end{aligned}$$

where $\mathcal{M}$ is the scalar manifold, r_H is the radial coordinate of the event horizon, and (q,p) denotes the set $\{q_\Lambda, p^\Lambda\}$ of the electric and magnetic charges of the BH ($\Lambda = 0,1,...,n_V$), which are conserved due to the overall $(U(1))^{n_V+1}$ gauge-invariance of the considered theory. The dynamical flow determining the radial evolution of the scalars $\phi^i(r)$ between the above two asymptotical limits is non-singular near the horizon, provided that

$$\left.\frac{\partial V_{BH}(\phi,q,p)}{\partial\phi^i}\right|_{\phi^j=\phi^j_H} = 0, \tag{3}$$

where V_{BH} is a certain positive definite, charge-dependent function in $\mathcal{M}$, named *BH effective potential* [5]. The condition (3) determines the so-called *attractor equations*, whose solutions are the purely charge-dependent, stabilized horizon configurations $\phi^i_H(q,p)$ in the r.h.s. of Eq. (2). By using the Bekenstein-Hawking entropy-area formula [7, 5], the classical BH entropy reads

$$\mathcal{S}(q,p) = \frac{A_H}{4} = \pi V_{BH}(\phi_H(q,p),q,p), \tag{4}$$

CP957, *Particles, Strings, and Cosmology, 13th International Symposium*
edited by A. Rajantie, C. Contaldi, P. Dauncey, and H. Stoica

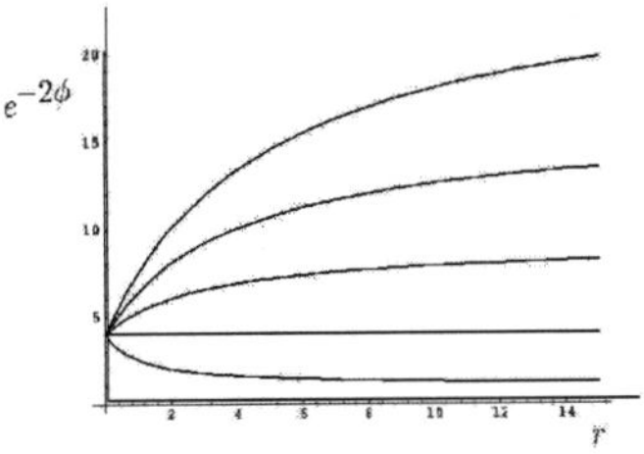

FIGURE 1. Realization of the attractor mechanism in the $\frac{1}{2}$-BPS dilatonic BH [3, 4, 6]. Independently on the set of asymptotical ($r \to \infty$) scalar configurations, the near-horizon evolution of the dilatonic function $e^{-2\phi}$ converges towards a fixed *attractor* value, which is purely dependent on the (ratio of the) quantized conserved charges of the BH.

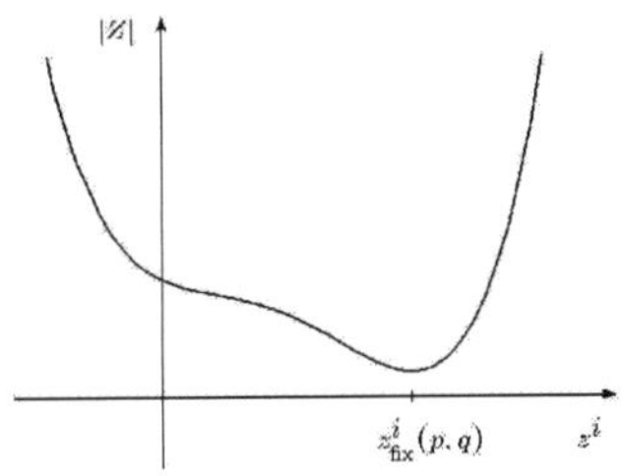

FIGURE 2. Minimization of the absolute value of the "central charge" function Z in $\mathscr{M}$. In the picture $z^i_{fix}(p,q)$ stands for the *attractor*, purely charge-dependent value of the scalars at the event horizon of the considered $\frac{1}{2}$-BPS extremal BH. The attractor mechanism fixes the extrema of the central charge to correspond to the discrete *fixed* points of the *attractor variety* [8] $\mathscr{M}$. Of course, the dependence of the central charge on scalars is shown for a given supporting BH charge configuration.

where A_H is the area of the BH event horizon.

The horizon geometry of extremal, asymptotically flat BHs in $\mathscr{N} = 2$, $d = 4$ supergravity is a maximally supersymmetric $\mathscr{N} = 2$ background, namely the Bertotti-Robinson (BR) $AdS_2 \times S^2$ BH metric [9, 10], which in turn is a particular case of the extremal p-brane horizon geometry $AdS_{p+2} \times S^{d-p-2}$ [11].

The first class of attractors to be studied was the $\frac{1}{2}$-BPS one, which preserves 4 supersymmetries out of the 8 pertaining to the asymptotical $\mathscr{N} = 2$, $d = 4$ Poincaré superalgebra. Examples of such attractors are given by Figures 1 and 2. Recently, many important advances have been performed in the study of extremal BH attractors, mainly concerning new classes of attractor configurations, corresponding to non-BPS, non-supersymmetric horizon geometries [12]–[46].

For asymptotically flat extremal BHs V_{BH} is given in terms of the scalar-dependent, complex symmetric matrix $\mathscr{N}_{\Lambda\Sigma}(\phi)$ (with $Im\mathscr{N}_{\Lambda\Sigma}$ negative definite), determining the couplings of the Maxwell field strength terms $\mathscr{F}^2$ and $\mathscr{F}\widetilde{\mathscr{F}}$ in the Lagrangian density, and of the electric and magnetic BH charges [5]:

$$V_{BH}(\phi,q,p) = -\frac{1}{2}\left(q_\Lambda - \mathcal{N}_{\Lambda\Sigma}p^\Sigma\right)(Im\mathcal{N})^{-1|\Lambda\Delta}\left(q_\Delta - \overline{\mathcal{N}}_{\Delta\Gamma}p^\Gamma\right). \tag{5}$$

Such a formula is valid for any (not necessarily supersymmetric) theory coupling Einstein gravity to scalars and Maxwell vector fields, whose Lagrangian density in general has the form

$$\begin{aligned}\frac{\mathcal{L}}{\sqrt{-g}} &= -\tfrac{1}{2}R - g_{ab}\left(\partial_\mu\phi^a\right)\left(\partial_\nu\phi^b\right)G^{\mu\nu} + \\ &+ (Im\mathcal{N}_{\Lambda\Sigma})\mathcal{F}^\Lambda_{\mu\nu}\mathcal{F}^{\Sigma|\mu\nu} + \frac{1}{2\sqrt{-g}}(Re\mathcal{N}_{\Lambda\Sigma})\varepsilon^{\mu\nu\rho\lambda}\mathcal{F}^\Lambda_{\mu\nu}\mathcal{F}^\Sigma_{\rho\lambda} + \ldots,\end{aligned} \tag{6}$$

where g_{ab} is the metric of the scalar manifold and $G_{\mu\nu}$ is the space-time metric.

An equivalent (but manifestly duality-covariant) expression reads [5]

$$V_{BH}(\phi,q,p) = -\frac{1}{2}Q^T M(\mathcal{N})Q, \tag{7}$$

where Q^T is the $1\times(2n_V+2)$ vector $\left(p^\Lambda,q_\Lambda\right)$ of the BH charges, and $M(\mathcal{N})$ is the symplectic $(2n_V+2)\times(2n_V+2)$ real, negative definite symmetric matrix

$$\begin{aligned}&M = R^T M_D R,\ R\equiv\begin{pmatrix} I & 0\\ -Re\mathcal{N} & I\end{pmatrix},\ M_D\equiv\begin{pmatrix} Im\mathcal{N} & 0\\ 0 & (Im\mathcal{N})^{-1}\end{pmatrix},\\ &M\Omega M = \Omega,\ \Omega\equiv\begin{pmatrix} 0 & -I\\ I & 0\end{pmatrix}.\end{aligned} \tag{8}$$

In $\mathcal{N}=2$, $d=4$ supergravity the scalar manifold is endowed with the so-called special Kähler geometry (see *e.g.* [47]) ; its metric $g_{i\bar{j}} = \partial_i\overline{\partial}_{\bar{j}}K$ (K being the Kähler potential) and $\mathcal{N}_{\Lambda\Sigma}$ are respectively given by the formulæ:

$$\begin{aligned}&g_{i\bar{j}} = -ie^K\left[\left(\overline{D}_{\bar{j}}\overline{X}^\Lambda\right)D_iF_\Lambda - \left(D_iX^\Lambda\right)\overline{D}_{\bar{j}}\overline{F}_\Lambda\right];\\ &\mathcal{N}_{\Lambda\Sigma} = h_{I\Lambda}\left(f^{-1}\right)^I_\Sigma,\ f^\Lambda_I \equiv e^{K/2}\left(X^\Lambda,\overline{D}_{\bar{i}}\overline{X}^\Lambda\right),\ h_{I\Lambda}\equiv e^{K/2}\left(F_\Lambda,\overline{D}_{\bar{i}}\overline{F}_\Lambda\right),\end{aligned} \tag{9}$$

where D_i denotes the Kähler-covariant derivative, and $\left(X^\Lambda,F_\Lambda\right)$ are the holomorphic sections of the Hodge bundle over $\mathcal{M}$ ($F_\Lambda = \frac{\partial F(X)}{\partial X^\Lambda}$, whenever the holomorphic prepotential function $F(X)$ exists) (see *e.g.* [47] and Refs. therein).

The symplectic-covariant formulation of the $\mathcal{N}=2$ special Kähler geometry can be actually generalized to all extended ($\mathcal{N}=3,\ldots,8$) $d=4$ supergravities [48, 25, 49]. In such theories, V_{BH} can be expressed as

$$V_{BH} = \frac{1}{2}\left|Z_{AB}\right|^2 + \left|Z^I\right|^2, \tag{10}$$

where Z_{AB} $(A,B=1,...,\mathcal{N})$ is the antisymmetric *central charge matrix* and Z^I are the so-called *dressed* (or *matter*) *charges*, respectively appearing in the supersymmetry transformations of the gravitinos $\psi_{\mu A}$ and of the other fermions λ_A^I of the theory in the considered BH background:

$$\begin{aligned}\delta_\varepsilon \psi_{\mu A}\big|_{BH} &\sim Z_{AB}\gamma_\mu\varepsilon^B;\\ \delta_\varepsilon\lambda_A^I\big|_{BH} &\sim Z^I\varepsilon_A,\end{aligned} \tag{11}$$

where γ_μ are the γ-matrices and ε is the parameter of the supersymmetric transformation.

Let us consider the maximal $d=4$ supergravity, *i.e.* $\mathcal{N}=8$ supergravity, based on the real 70-dim. symmetric manifold $\frac{E_{7(7)}}{SU(8)}$ [50]. In this case no matter multiplets are coupled to the gravity one, thus Eq. (10) simplifies to $(A,B=1,...,8)$

$$V_{BH}=\frac{1}{2}\left|Z_{AB}\right|^2, \tag{12}$$

with $Z_{AB}=L_{AB}^{\Lambda}(\phi)Q_\Lambda$, where $L(\phi)\in E_{7(7)}$ and Q is the charge vector. Under a transformation h of the stabilizer $SU(8)$, the matrix Z transforms as [42]

$$Z(\phi,Q)\longmapsto Z(\phi_g,Q)=hZ\left(\phi_g,g^{-1}Q\right)\Longrightarrow V_{BH}(\phi,Q)=V_{BH}\left(\phi_g,g^{-1}Q\right). \tag{13}$$

By computing V_{BH} at one of its critical points, one obtains a completely charge-dependent expression:

$$V_{BH}\big|_{\frac{\partial V_{BH}}{\partial\phi}=0}\equiv V_{BH,cr.}(Q)=V_{BH,cr.}\left(g^{-1}Q\right)\sim\sqrt{|\mathcal{I}_4|}, \tag{14}$$

$\mathcal{I}_4$ being the quartic Cartan-Cremmer-Julia invariant of the fundamental representation **56** of $E_{7(7)}$ [50, 51].

The local $SU(8)$ symmetry allows one to go to the so-called "normal frame" [52]. In such a frame, Z_{AB} and $\mathcal{I}_4$ respectively read as follows:

$$\begin{aligned}Z_{AB,normal} &= skew-diag(\rho_1,\rho_2,\rho_3,\rho_4)\,e^{i\varphi/4};\\ \mathcal{I}_{4,normal} &= \left[(\rho_1+\rho_2)^2-(\rho_3+\rho_4)^2\right]\left[(\rho_1-\rho_2)^2-(\rho_3-\rho_4)^2\right]+8\rho_1\rho_2\rho_3\rho_4(cos\varphi-1),\end{aligned} \tag{15}$$

with $\rho_i\in R^+\ \forall i=1,...,4$. Note that $Z_{AB,normal}$ has an $(SU(2))^4$ symmetry.

From the analysis performed in [53, 54, 25], the $\mathcal{N}=8$ attractor equations yield only 2 distinct classes of solutions with non-vanishing entropy ($\frac{1}{8}$-BPS for $\mathcal{I}_4>0$, non-BPS for $\mathcal{I}_4<0$):

1. $\frac{1}{8}$-BPS: $\rho_1=\rho_{\frac{1}{8}-BPS}\in R_0^+$ and all the others vanish, $\mathcal{I}_{4,normal,\frac{1}{8}-BPS}>0$, and

$$\mathcal{S}_{\frac{1}{8}-BPS}=\pi\sqrt{\mathcal{I}_{4,normal,\frac{1}{8}-BPS}}=\pi\rho_1^2. \tag{16}$$

The corresponding orbit of supporting BH charges in the **56** of $E_{7(7)}$ is $\mathcal{O}_{\frac{1}{8}-BPS} = \frac{E_{7(7)}}{E_{6(2)}}$. Moreover, $Z_{AB,normal,\frac{1}{8}-BPS}$ has symmetry enhancement (*m.c.s.* stands for *maximal compact subgroup*)

$$(SU(2))^4 \longrightarrow SU(6) \otimes SU(2) = m.c.s.\left(E_{6(2)}\right). \tag{17}$$

Notice that $\varphi_{\frac{1}{8}-BPS}$ is actually undetermined.

2. non-BPS: all ρs are equal to $\rho_{non-BPS} \in R_0^+$, $\varphi_{non-BPS} = \pi$, $\mathcal{I}_{4,normal,non-BPS} < 0$, and

$$\mathcal{S}_{non-BPS} = \pi\sqrt{-\mathcal{I}_{4,normal,non-BPS}} = 4\pi\rho^2. \tag{18}$$

The corresponding orbit of supporting BH charges in the **56** of $E_{7(7)}$ is $\mathcal{O}_{non-BPS} = \frac{E_{7(7)}}{E_{6(6)}}$. Furthermore, $Z_{AB,normal,non-BPS}$ has symmetry enhancement

$$(SU(2))^4 \longrightarrow USp(8) = m.c.s.\left(E_{6(6)}\right). \tag{19}$$

Thus, the symmetry of $Z_{AB,normal}$ gets enhanced at the particular points of $\frac{E_{7(7)}}{SU(8)}$ given by the solutions of $\mathcal{N} = 8$, $d = 4$ attractor equations with non-vanishing $\mathcal{I}_4$. In general, *the invariance properties of the solutions to attractor Eqs. with $\mathcal{I}_4 \neq 0$ are given by the m.c.s. of the stabilizer of the corresponding supporting BH charge orbit.*

The 70×70 Hessian matrix of V_{BH} at the $\frac{1}{8}$-BPS critical points has rank 30; its 30 strictly positive and 40 vanishing eigenvalues respectively correspond to the 15 vector multiplets and to the 10 hypermultiplets of the $\mathcal{N} = 2$, $d = 4$ spectrum obtained by reducing $\mathcal{N} = 8$ supergravity according to the following branching of the **70** (four-fold antisymmetric) of $SU(8)$ [55]:

$$\begin{aligned} &SU(8) \longrightarrow SU(6) \otimes SU(2); \\ &\mathbf{70} \longrightarrow \left[(\mathbf{15},\mathbf{1}) \oplus (\overline{\mathbf{15}},\mathbf{1})\right]_{m\neq 0} \oplus (\mathbf{20},\mathbf{2})_{m=0}. \end{aligned} \tag{20}$$

On the other hand, at the non-BPS critical points the Hessian matrix has rank 28; such a splitting of the mass spectrum can be interpreted according to the following branching of the **70** of $SU(8)$ [39]:

$$\begin{aligned} &SU(8) \longrightarrow USp(8); \\ &\mathbf{70} \longrightarrow (\mathbf{1} \oplus \mathbf{27})_{m\neq 0} \oplus (\mathbf{42})_{m=0}. \end{aligned} \tag{21}$$

As shown in [42], the massless modes of the critical Hessian matrix actually correspond to flat directions of V_{BH} itself. This can be easily realized by noticing that the stabilizers of the charge orbits are non-compact, so that

$$\begin{aligned} &g_Q Q^{BPS} = Q^{BPS},\ \forall g_Q \in E_{6(2)}; \\ &g_Q Q^{non-BPS} = Q^{non-BPS},\ \forall g_Q \in E_{6(6)}, \end{aligned} \tag{22}$$

and thus at the critical points (recall Eq. (13))

$$V_{BH}\left(\phi_{g_Q}, g_Q^{-1}Q\right) = V_{BH}\left(\phi_{g_Q}, Q\right) = V_{BH}\left(\phi, Q\right). \tag{23}$$

This implies that each of the two classes of $\mathcal{N} = 8$, $d = 4$ extremal BH attractors with non-vanishing entropy has an associated moduli space:

$$\begin{aligned} &BPS: \frac{E_{6(2)}}{SU(6)\otimes SU(2)}, \ quaternionic\ manifold\ with\ dim_R = 40; \\ &non-BPS: \frac{E_{6(6)}}{USp(8)}, \ \mathcal{N} = 8,\ d = 5\ scalar\ manifold\ with\ dim_R = 42. \end{aligned} \tag{24}$$

The same reasoning, which is actually independent on the number d of space-time dimensions and on $\mathcal{N}$, will apply to *all* theories of the kind considered above, whose scalar manifold is an homogeneous (not necessarily symmetric) space, when the stabilizer of the orbit of the attractor-supporting charge vector Q is non-compact [42]. For $\mathcal{N} > 2$ this will apply to both BPS and non-BPS critical points (as shown above for $\mathcal{N} = 8$, $d = 4$). However, for $\mathcal{N} = 2$, $d = 4$ the stabilizer of the ($\frac{1}{2}$-)BPS orbit is compact, and no flat directions will occur (apart from hypermultiplets). This is strictly true as far as the metric of the scalar manifold is strictly positive definite at the considered BPS critical points. Indeed, by using special Kähler geometry one can prove the following result, holding for any $\mathcal{N} = 2$, $d = 4$ supergravity [5] (such a result, *mutatis mutandis*, holds also for $d = 5$ [56]):

$$\left(D_i\overline{D}_{\overline{j}}V_{BH}\right)_{BPS} = 2\left(g_{i\overline{j}}V_{BH}\right)_{BPS}. \tag{25}$$

Reconsidering $\mathcal{N} = 2$, $d = 4$ supergravity, the Riemann tensor of the special Kähler scalar manifold satisfies the following relation (see *e.g.* [47] and Refs. therein)

$$R_{i\overline{j}l\overline{k}} = -g_{i\overline{j}}g_{l\overline{k}} - g_{i\overline{k}}g_{l\overline{j}} + C_{ilp}\overline{C}_{\overline{j}\overline{k}\overline{p}}g^{p\overline{p}}, \tag{26}$$

where the rank-3 completely symmetric tensor C_{ijk} has the properties

$$\overline{D}_{\overline{l}}C_{ijk} = 0,\ D_{[l}C_{i]jk} = 0. \tag{27}$$

In particular, for homogeneous symmetric cubic special Kähler geometries another set of relations holds [57, 58] (see also [49], [34] and Refs. therein; here and below z^i denote the complex scalars):

$$\begin{aligned} &D_lC_{ijk} = 0; \\ &C_{ijk} = e^K\partial_i\partial_j\partial_k f(z),\ \ f(z) \equiv \tfrac{1}{3!}d_{ijk}z^iz^jz^k; \\ &\overline{E}_{\overline{i}ijpq} \equiv g^{k\overline{k}}g^{r\overline{j}}C_{r(pq}C_{ij)k}\overline{C}_{\overline{k}\overline{i}\overline{j}} - \tfrac{4}{3}g_{(q|\overline{i}}C_{|ijp)} = 0; \\ &d_{ABC}d^{B(PQ}d^{LM)C} = \tfrac{4}{3}\delta_A^{(P}d^{QLM)}. \end{aligned} \tag{28}$$

The $\mathcal{N}=2$, $d=4$ attractor equations read [5]

$$2\overline{Z}D_iZ+iC_{ijk}g^{j\bar{j}}g^{k\bar{k}}\left(\overline{D}_{\bar{j}}\overline{Z}\right)\overline{D}_{\bar{k}}\overline{Z}=0, \tag{29}$$

Z denoting the $\mathcal{N}=2$ covariantly holomorphic *central charge function* [47]

$$Z\equiv e^{K/2}\left(X^{\Lambda}q_{\Lambda}-F_{\Lambda}p^{\Lambda}\right). \tag{30}$$

Eqs. (29) yield three classes of solutions [23, 27]:

i) $\frac{1}{2}$-BPS solutions [5], with $Z\neq 0$ and $D_iZ=0\ \forall i$. They saturate the BPS bound [59]:

$$M^2_{ADM,BPS}=|Z|^2_{BPS}, \tag{31}$$

M_{ADM} being the Arnowitt-Deser-Misner BH mass [60].

ii) non-BPS solutions with $Z\neq 0$ and $D_iZ\neq 0$ for at least some i [5, 13, 15, 16, 23]. They do not preserve any supersymmetry and do not saturate the BPS bound; indeed, for symmetric spaces it holds that [27]:

$$M^2_{ADM,non-BPS,Z\neq 0}=4|Z|^2_{non-BPS,Z\neq 0}>|Z|^2_{non-BPS,Z\neq 0}. \tag{32}$$

Such a result actually holds for homogeneous non-symmetric [34] and also for generic cubic (at least within some particular assumptions [16]) special Kähler geometries.

iii) non-BPS solutions with $Z=0$ and $D_iZ\neq 0$ for at least some i [24, 27, 34, 46]. They do not preserve any supersymmetry and do not saturate the BPS bound:

$$M^2_{ADM,non-BPS,Z=0}=\left[g^{i\bar{j}}\left(D_iZ\right)\overline{D}_{\bar{j}}\overline{Z}\right]_{non-BPS,Z=0}>0. \tag{33}$$

As mentioned above, $\frac{1}{2}$-BPS critical points are stable; they have no massless Hessian modes at all, and thus they do not have any associated moduli space. The moduli spaces associated to the $\mathcal{N}=2$, $d=4$ non-BPS solutions with $Z\neq 0$ and $Z=0$ and to the $\mathcal{N}=2$, $d=5$ non-BPS solutions have been recently determined in [42] (see also [45]); they are respectively given by Tables 2, 3 and 4 of [42].

As obtained in [16], the $2n_V\times 2n_V$ (real form of the) Hessian matrix of V_{BH} at its non-BPS $Z\neq 0$ critical points in a generic cubic special Kähler geometry of complex dimension $dim_{\mathbb{C}}=n_V$ has n_V+1 strictly positive and n_V-1 vanishing eigenvalues. As pointed out above, in the homogeneous (not necessarily symmetric) case, these latter n_V-1 massless Hessian modes actually correspond to n_V-1 flat directions of $V_{BH,non-BPS,Z\neq 0}$ [42].

The same result holds also for generic cubic special Kähler geometries, at least for some particular BH charge configurations [45]. This is simply seen *e.g.* by splitting the complex scalars as $z^i=x^i-i\lambda^i$, and considering the peculiar non-BPS $Z\neq 0$-supporting BH charge configuration $Q_0^T=\left(p^0,0,q_0,0\right)$, for which the criticality conditions $\frac{\partial V_{BH}}{\partial x^i}=0$ can be solved by putting $x^i=0\ \forall i$. For such a case, in [45] V_{BH} was shown to acquire the following simple form:

$$V_{BH}|_{x^i=0\ \forall i,\ Q=Q_0}=\frac{1}{2}\left[\left(p^0\right)^2\mathcal{V}+\left(q_0\right)^2\mathcal{V}^{-1}\right]\equiv V^*_{BH}\left(\mathcal{V},p^0,q_0\right), \tag{34}$$

where $\mathscr{V} \equiv \frac{1}{3!}d_{ijk}\lambda^i\lambda^j\lambda^k$. By rescaling $\lambda^i \equiv \mathscr{V}^{1/3}\widehat{\lambda}^i$, it is immediate to realize that $V^*_{BH}\left(\mathscr{V}, p^0, q_0\right)$ *does not depend on any of the* $\widehat{\lambda}^i$. By definition, the $\widehat{\lambda}^i$s belong to the geometrical *locus* $\frac{1}{3!}d_{ijk}\widehat{\lambda}^i\widehat{\lambda}^j\widehat{\lambda}^k = 1$; thus, they parameterize $n_V - 1$ "flat" directions of $V_{BH}|_{x^i=0\ \forall i}$ at its non-BPS $Z \neq 0$ critical points supported by the charge configuration Q_0. Such $n_V - 1$ "flat" directions turn out to span nothing but the $(n_V - 1)$-dim. real special scalar manifold of the corresponding $\mathscr{N} = 2$, $d = 5$ parent supergravity theory [45].

Let us now consider an explicit example, namely the *magic* $\mathscr{N} = 2, d = 4$ supergravity theory based on the exceptional Jordan algebra J_3^O over the octonions (see *e.g.* [27, 61, 39, 42] and Refs. therein). It is based on the rank-3 homogeneous symmetric special Kähler manifold $\frac{E_{7(-25)}}{E_{6(-78)}\otimes U(1)}$ with $dim_C = n_V = 27$; the charge vector Q sits in the fundamental representation **56** of $E_{7(-25)}$.

The $\frac{1}{2}$-BPS attractors are supported by a Q belonging to the BPS orbit $\frac{E_{7(-25)}}{E_{6(-78)}}$ ($dim_R = 55$); due to the compactness of $E_{6(-78)}$, there is no BPS moduli space at all.

On the other hand, the non-BPS $Z \neq 0$ attractors are supported by a Q belonging to the 55-dim. non-BPS orbit $\frac{E_{7(-25)}}{E_{6(-26)}}$, $E_{6(-26)}$ being a non-compact real form of the exceptional group E_6; the corresponding non-BPS $Z \neq 0$ moduli space reads $\frac{E_{6(-26)}}{F_{4(-52)}}$ ($dim_R = 26$), where $F_{4(-52)} = m.c.s.\left(E_{6(-26)}\right)$ [62]. It is nothing but the rank-2, real special scalar manifold of the corresponding $\mathscr{N} = 2$, $d = 5$ parent supergravity theory [45].

The non-BPS $Z = 0$ attractors are supported by a Q belonging to the 55-dim. non-BPS orbit $\frac{E_{7(-25)}}{E_{6(-14)}}$, $E_{6(-14)}$ being the only other non-compact real form of E_6 contained in $E_{7(-25)}$ [62]; the corresponding non-BPS $Z = 0$ moduli space is the rank-2, homogeneous symmetric (not special) Kähler manifold $\frac{E_{6(-14)}}{SO(10)\otimes U(1)}$ ($dim_C = 16$) [45], where $SO(10) \otimes U(1) = m.c.s.\left(E_{6(-14)}\right)$ [62].

The corresponding parent theory in $d = 5$ is the *magic* $\mathscr{N} = 2$, $d = 5$ supergravity over J_3^O (see *e.g.* [57, 54, 42] and Refs. therein). For such a theory, the BPS charge orbit coincides with $\frac{E_{6(-26)}}{F_{4(-52)}}$ itself [54], and there are no BPS massless Hessian modes [56]. The unique class of non-BPS attractors with non-vanishing cubic invariant I_3 (see *e.g.* [56] and Refs. therein) is supported by the 26-dim. BH charge orbit $\frac{E_{6(-26)}}{F_{4(-20)}}$, $F_{4(-20)}$ being the only non-compact real form of the exceptional group F_4 contained in $E_{6(-26)}$ [62]. The corresponding non-BPS moduli space is the rank-1, homogeneous symmetric manifold $\frac{F_{4(-20)}}{SO(9)}$ ($dim_R = 16$) [42], where $SO(9) = m.c.s.\left(F_{4(-20)}\right)$ [62].

Finally, it is worth remarking that the non-BPS $d = 5$ attractors can give rise to both $Z \neq 0$ and $Z = 0$ non-BPS $d = 4$ critical points, depending on the sign of an extra Kaluza-Klein charge [45]. This implies that the moduli space of non-BPS $d = 5$ attractors is contained in the moduli spaces of both species ($Z \neq 0$ and $Z = 0$) of non-BPS $d = 4$ attractors, as pointed out in [45] (and as given by the Tables 2, 3 and 4 of [42]).

ACKNOWLEDGMENTS

The topics covered in this review are mainly based on collaborations with: L. Andrianopoli, S. Bellucci, A. Ceresole, R. D'Auria, G. Gibbons, E. Gimon, M. Günaydin, R. Kallosh, A. Strominger, M. Trigiante.

The work of S.F. has been supported in part by European Community Human Potential Program under contract MRTN-CT-2004-005104 *"Constituents, fundamental forces and symmetries of the universe"* and the contract MRTN-CT-2004-503369 *"The quest for unification: Theory Confronts Experiments"*, in association with INFN Frascati National Laboratories and by D.O.E. grant DE-FG03-91ER40662, Task C.

The work of A.M. has been supported by a Junior Grant of the *"Enrico Fermi"* Center, Rome, in association with INFN Frascati National Laboratories.

REFERENCES

1. S. Ferrara, R. Kallosh and A. Strominger, *$\mathcal{N}=2$ Extremal Black Holes*, Phys. Rev. **D52**, 5412 (1995), hep-th/9508072.
2. A. Strominger, *Macroscopic Entropy of $\mathcal{N}=2$ Extremal Black Holes*, Phys. Lett. **B383**, 39 (1996), hep-th/9602111.
3. S. Ferrara and R. Kallosh, *Supersymmetry and Attractors*, Phys. Rev. **D54**, 1514 (1996), hep-th/9602136.
4. S. Ferrara and R. Kallosh, *Universality of Supersymmetric Attractors*, Phys. Rev. **D54**, 1525 (1996), hep-th/9603090.
5. S. Ferrara, G. W. Gibbons and R. Kallosh, *Black Holes and Critical Points in Moduli Space*, Nucl. Phys. **B500**, 75 (1997), hep-th/9702103.
6. G. Gibbons, in *Unified theories of Elementary Particles. Critical Assessment and Prospects*, Proceedings of the Heisemberg Symposium, München, West Germany, 1981, ed. by P. Breitenlohner and H. P. Dürr, Lecture Notes in Physics Vol. 160 (Springer-Verlag, Berlin, 1982); G. W. Gibbons, in *Supersymmetry, Supergravity and Related Topics*, Proceedings of the XVth GIFT International Physics, Girona, Spain, 1984, ed. by F. del Aguila, J. de Azcárraga and L. Ibáñez, (World Scientific, 1995), pag. 147; P. Breitenlohner, D. Maison and G. W. Gibbons, *Four Dimensional Black Holes from Kaluza–klein theories*, Commun. Math. Phys. **120**, 295 (1988); R. Kallosh, A. D. Linde, T. Ortin, A. W. Peet and A. Van Proeyen, *Supersymmetry as a cosmic censor*, Phys. Rev. D **46**, 5278 (1992), hep-th/9205027; R. Kallosh, T. Ortin and A. W. Peet, *Entropy and action of dilaton black holes*, Phys. Rev. D **47**, 5400 (1993), hep-th/9211015; R. Kallosh, *Supersymmetric black holes*, Phys. Lett. B **282**, 80 (1992), hep-th/9201029; R. Kallosh and A. W. Peet, *Dilaton black holes near the horizon*, Phys. Rev. D **46**, 5223 (1992), hep-th/9209116; R. R. Khuri and T. Ortin, *Supersymmetric Black Holes in $\mathcal{N}=8$ Supergravity*, Nucl. Phys. B **467**, 355 (1996), hep-th/9512177; A. Sen, *Black Hole Solutions In Heterotic String Theory On A Torus*, Nucl. Phys. B **440** (1995) 421, hep-th/9411187; A. Sen, *Quantization of dyon charge and electric magnetic duality in string theory*, Phys. Lett. B **303** (1993) 22, hep-th/9209016; A. Sen, *Extremal black holes and elementary string states*, Mod. Phys. Lett. A **10** (1995) 2081, hep-th/9504147; M. J. Duff, J. T. Liu and J. Rahmfeld, *Four-dimensional string/string/string triality*, Nucl. Phys. **B459**, 125 (1996), hep-th/9508094; M. Cvetic and C. M. Hull, *Black Holes and U-Duality* , Nucl. Phys. **B 480** (1996) 296, hep-th/9606193; M. Cvetic and I. Gaida, *Duality Invariant Non-Extreme Black Holes in Toroidally Compactified String Theory*, Nucl. Phys. **B 505** (1997) 291, hep-th/9703134; M. Cvetic and D. Youm, *All the Static Spherically Symmetric Black Holes of Heterotic String on a Six Torus*, hep-th/9512127; M. Cvetic and A. A. Tseytlin, *Solitonic Strings and BPS Saturated Dyonic Black Holes*, Phys.Rev. **D 53** (1996) 5619.
7. J. D. Bekenstein, Phys. Rev. **D7**, 2333 (1973); S. W. Hawking, Phys. Rev. Lett. **26**, 1344 (1971), in C. DeWitt, B. S. DeWitt, *Black Holes (Les Houches 1972)* (Gordon and Breach, New York, 1973); S. W. Hawking, Nature **248**, 30 (1974); S. W. Hawking, Comm. Math. Phys. **43**, 199 (1975).

8. G. Moore, *Attractors and Arithmetic*, hep-th/9807056; G. Moore, *Arithmetic and Attractors*, hep-th/9807087; G. Moore, *Les Houches Lectures on Strings and Arithmetic*, hep-th/0401049.
9. T. Levi-Civita, R.C. Acad. Lincei **26**, 519 (1917); B. Bertotti, *Uniform Electromagnetic Field in the Theory of General Relativity*, Phys. Rev. **116**, 1331 (1959); I. Robinson, Bull. Acad. Polon. **7**, 351 (1959).
10. D. A. Lowe and A. Strominger, *Exact Four-Dimensional Dyonic Black Holes and Bertotti-Robinson Space-Times in String Theory*, Phys. Rev. Lett. **73**, 1468 (1994), hep-th/9403186.
11. G. W. Gibbons and P. Townsend, *Vacuum Interpolation in Supergravity via Super p-Branes*, Phys. Rev. Lett. **71**, 3754 (1993), hep-th/9307049.
12. A. Sen, *Black Hole Entropy Function and the Attractor Mechanism in Higher Derivative Gravity*, JHEP **09**, 038 (2005), hep-th/0506177.
13. K. Goldstein, N. Iizuka, R. P. Jena and S. P. Trivedi, *Non-Supersymmetric Attractors*, Phys. Rev. **D72**, 124021 (2005), hep-th/0507096.
14. A. Sen, *Entropy Function for Heterotic Black Holes*, JHEP **03**, 008 (2006), hep-th/0508042.
15. R. Kallosh, *New Attractors*, JHEP **0512**, 022 (2005), hep-th/0510024.
16. P. K. Tripathy and S. P. Trivedi, *Non-Supersymmetric Attractors in String Theory,* JHEP **0603**, 022 (2006), hep-th/0511117.
17. A. Giryavets, *New Attractors and Area Codes*, JHEP **0603**, 020 (2006), hep-th/0511215.
18. K. Goldstein, R. P. Jena, G. Mandal and S. P. Trivedi, *A C-Function for Non-Supersymmetric Attractors,* JHEP **0602**, 053 (2006), hep-th/0512138.
19. M. Alishahiha and H. Ebrahim, *Non-supersymmetric attractors and entropy function*, JHEP **0603**, 003 (2006), hep-th/0601016.
20. R. Kallosh, N. Sivanandam and M. Soroush, *The Non-BPS Black Hole Attractor Equation*, JHEP **0603**, 060 (2006), hep-th/0602005.
21. B. Chandrasekhar, S. Parvizi, A. Tavanfar and H. Yavartanoo, *Non-supersymmetric attractors in R^2 gravities*, JHEP **0608**, 004 (2006), hep-th/0602022.
22. J. P. Hsu, A. Maloney and A. Tomasiello, *Black Hole Attractors and Pure Spinors*, JHEP **0609**, 048 (2006), hep-th/0602142.
23. S. Bellucci, S. Ferrara and A. Marrani, *On some properties of the Attractor Equations*, Phys. Lett. **B635**, 172 (2006), hep-th/0602161.
24. S. Bellucci, S. Ferrara and A. Marrani, *Supersymmetric Mechanics. Vol.2: The Attractor Mechanism and Space-Time Singularities* (LNP **701**, Springer-Verlag, Heidelberg, 2006).
25. S. Ferrara and R. Kallosh, *On $\mathcal{N}=8$ attractors*, Phys. Rev. D **73**, 125005 (2006), hep-th/0603247.
26. M. Alishahiha and H. Ebrahim, *New attractor, Entropy Function and Black Hole Partition Function*, JHEP **0611**, 017 (2006), hep-th/0605279.
27. S. Bellucci, S. Ferrara, M. Günaydin and A. Marrani, *Charge Orbits of Symmetric Special Geometries and Attractors*, Int. J. Mod. Phys. **A21**, 5043 (2006), hep-th/0606209.
28. D. Astefanesei, K. Goldstein, R. P. Jena, A. Sen and S. P. Trivedi, *Rotating Attractors*, JHEP **0610**, 058 (2006), hep-th/0606244.
29. R. Kallosh, N. Sivanandam and M. Soroush, *Exact Attractive non-BPS STU Black Holes*, Phys. Rev. **D74**, 065008 (2006), hep-th/0606263.
30. P. Kaura and A. Misra, *On the Existence of Non-Supersymmetric Black Hole Attractors for Two-Parameter Calabi-Yau's and Attractor Equations*, hep-th/0607132.
31. G. L. Cardoso, V. Grass, D. Lüst and J. Perz, *Extremal non-BPS Black Holes and Entropy Extremization*, JHEP **0609**, 078 (2006), hep-th/0607202.
32. S. Bellucci, S. Ferrara, A. Marrani and A. Yeranyan, *Mirror Fermat Calabi-Yau Threefolds and Landau-Ginzburg Black Hole Attractors*, hep-th/0608091.
33. G.L. Cardoso, B. de Wit and S. Mahapatra, *Black hole entropy functions and attractor equations*, hep-th/0612225.
34. R. D'Auria, S. Ferrara and M. Trigiante, *Critical points of the Black-Hole potential for homogeneous special geometries*, hep-th/0701090.
35. S. Bellucci, S. Ferrara and A. Marrani, *Attractor Horizon Geometries of Extremal Black Holes*, Contribution to the Proceedings of the XVII SIGRAV Conference,4–7 September 2006, Turin, Italy, hep-th/0702019.

36. A. Ceresole and G. Dall'Agata, *Flow Equations for Non-BPS Extremal Black Holes*, JHEP **0703**, 110 (2007), hep-th/0702088.
37. L. Andrianopoli, R. D'Auria, S. Ferrara and M. Trigiante, *Black Hole Attractors in $\mathcal{N} = 1$ Supergravity*, hep-th/0703178.
38. K. Saraikin and C. Vafa, *Non-supersymmetric Black Holes and Topological Strings*, hep-th/0703214.
39. S. Ferrara and A. Marrani, *$\mathcal{N} = 8$ non-BPS Attractors, Fixed Scalars and Magic Supergravities*, Nucl. Phys. **B** 2007, in press, ArXiV:0705.3866.
40. S. Nampuri, P. K. Tripathy and S. P. Trivedi, *On The Stability of Non-Supersymmetric Attractors in String Theory*, ArXiV:0705.4554.
41. L. Andrianopoli, R. D'Auria, E. Orazi, M. Trigiante, *First Order Description of Black Holes in Moduli Space*, ArXiV:0706.0712.
42. S. Ferrara and A. Marrani, *On the Moduli Space of non-BPS Attractors for $\mathcal{N} = 2$ Symmetric Manifolds*, Phys. Lett. **B** 2007, in press, ArXiV:0706.1667.
43. G. L. Cardoso, A. Ceresole, G. Dall'Agata, J. M. Oberreuter, J. Perz, *First-order flow equations for extremal black holes in very special geometry*, ArXiV:0706.3373.
44. A. Misra and P. Shukla, *'Area codes', large volume (non-)perturbative alpha-prime and instanton: Corrected non-supersymmetric (A)dS minimum, the 'inverse problem' and 'fake superpotentials' for multiple-singular-loci-two-parameter Calabi-Yau's*, ArXiV:0707.0105.
45. A. Ceresole, S. Ferrara and A. Marrani, *4d/5d Correspondence for the Black Hole Potential and its Critical Points*, ArXiV:0707.0964.
46. S. Bellucci, A. Marrani, E. Orazi and A. Shcherbakov, *Attractors with Vanishing Central Charge*, ArXiV:0707.2730.
47. A. Ceresole, R. D'Auria and S. Ferrara, *The Symplectic Structure of $\mathcal{N} = 2$ Supergravity and Its Central Extension*, Talk given at ICTP Trieste Conference on Physical and Mathematical Implications of Mirror Symmetry in String Theory, Trieste, Italy, 5-9 June 1995, Nucl. Phys. Proc. Suppl. **46** (1996), hep-th/9509160.
48. L. Andrianopoli, R. D'Auria and S. Ferrara, *U-duality and central charges in various dimensions revisited*, Int. J. Mod. Phys. **A13**, 431 (1998), hep-th/9612105.
49. L. Andrianopoli, R. D'Auria, S. Ferrara and M. Trigiante, *Extremal Black Holes in Supergravity*, in : *"String Theory and Fundamental Interactions"*, M. Gasperini and J. Maharana eds. (LNP, Springer, Berlin-Heidelberg, 2007), hep-th/0611345.
50. E. Cremmer and B. Julia, *The $SO(8)$ Supergravity*, Nucl. Phys. **B159**, 141 (1979).
51. E. Cartan, *Oeuvres complètes* (Editions du Centre National de la Recherche Scientifique, Paris, 1984).
52. S. Ferrara, C. A. Savoy and B. Zumino, *General Massive Multiplets In Extended Supersymmetry*, Phys. Lett. **B100**, 393 (1981).
53. S. Ferrara and J. M. Maldacena, *Branes, central charges and U-duality invariant BPS conditions*, Class. Quant. Grav. **15**, 749 (1998), hep-th/9706097.
54. S. Ferrara and M. Günaydin, *Orbits of Exceptional Groups, Duality and BPS States in String Theory*, Int. J. Mod. Phys. **A13**, 2075 (1998), hep-th/9708025.
55. L. Andrianopoli, R. D'Auria and S. Ferrara, *U-Invariants, Black-Hole Entropy and Fixed Scalars*, Phys. Lett. **B403**, 12 (1997), hep-th/9703156.
56. S. Ferrara and M. Günaydin, *Orbits and attractors for $\mathcal{N} = 2$ Maxwell-Einstein supergravity theories in five dimensions*, Nucl. Phys. **B759**, 1 (2006), hep-th/0606108.
57. M. Günaydin, G. Sierra and P. K. Townsend, *The Geometry of $\mathcal{N} = 2$ Maxwell-Einstein Supergravity and Jordan Algebras*, Nucl. Phys. **B242**, 244 (1984).
58. E. Cremmer and A. Van Proeyen, *Classification of Kähler Manifolds in $\mathcal{N} = 2$ Vector Multiplet Supergravity Couplings*, Class. Quant. Grav. **2**, 445 (1985).
59. G. W. Gibbons and C. M. Hull, *A Bogomol'ny Bound for General Relativity and Solitons in $\mathcal{N} = 2$ Supergravity*, Phys. Lett. **B109**, 190 (1982).
60. R. Arnowitt, S. Deser and C. W. Misner, *The Dynamics of General Relativity*, in : *"Gravitation: an Introduction to Current Research"*, L. Witten ed. (Wiley, New York, 1962).
61. S. Ferrara, E. G. Gimon and R. Kallosh, *Magic supergravities, $\mathcal{N} = 8$ and black hole composites*, Phys. Rev. **D74**, 125018 (2006), hep-th/0606211.
62. R. Gilmore, *Lie Groups, Lie Algebras, and Some of Their Applications* (Dover Publications, 2006).

High-energy black hole production

Steven B. Giddings[1]

Department of Physics, University of California, Santa Barbara, CA 93106

Abstract. Black hole production in high-energy collisions is briefly surveyed. Included is a summary of recent developments and open problems relevant to collider (LHC) production, as well as of some theoretical issues pointing towards fundamental principles of quantum gravity.

Keywords: Black holes, LHC, quantum gravity
PACS: 04.50.Gh, 11.25.Wx, 04.60.-m, 04.70.Dy

INTRODUCTION

Black hole production may be the most spectacular physics at future colliders, perhaps even beginning with the 2008 startup of the LHC. And even if it is not phenomenologically accessible in the near future, it raises some very profound theoretical issues. This talk will present a summary and update on some of the physics of black hole creation at high energies.[2]

The basic idea is that once we reach collision energies exceeding the Planck mass, which we denote M_D, collisions of particles can form black holes. The Planck mass is traditionally expected to be of order 10^{19} GeV, but in TeV-scale gravity scenarios the Planck mass could be lowered to the accessible energy ~ 1 TeV. Possible realizations of TeV-scale gravity include large extra dimensions[4, 5] and large warping[6, 7, 8]. If the D-dimensional metric is of the general warped form,

$$ds^2 = e^{2A(y)}dx_4^2 + g_{mn}(y)dy^m dy^n \ , \tag{1}$$

where y^m are coordinates for the extra dimensions and e^{2A} is the warp factor, the basic relation between the four-dimensional Planck scale and the D-dimensional Planck scale is

$$M_4^2 = M_D^{D-2} \int \frac{d^{D-4}y}{(2\pi)^{D-4}} \sqrt{g_{D-4}} e^{2A} \ . \tag{2}$$

If the integral on the RHS of the equation, called the "warped volume," is large, this can yield the measured four-dimensional gravitational constant with an M_D significantly below $10^{19} GeV$. In the large extra dimensions case, this is readily understood as due to dilution of the field lines in the extra dimensions. One must supplement this picture with a brane-world scenario in order that the ordinary gauge forces behave four-dimensionally. Many such scenarios are being investigated, in string theory and elsewhere.

---

[1] Email: giddings@physics.ucsb.edu.

[2] Due to space limitations only some representative references can be given. Other reviews include [1, 2, 3].

CP957, *Particles, Strings, and Cosmology, 13th International Symposium*
edited by A. Rajantie, C. Contaldi, P. Dauncey, and H. Stoica

Once it became clear that the true Planck scale could be as low as a TeV, it was obvious that black hole production could take place at this scale. Early discussions of this in large extra dimension scenarios and warped scenarios are [9, 10] respectively.

Some features of such black hole production – and its precursor physics at the Planck scale – are highly dependent on the detailed model of the extra dimensional spacetime and brane-world realization of gauge forces and matter. However, there are certain features which are fairly model independent and "generic." Typically these dominate in an expansion in M_D/E, where E is the collision energy. This discussion will focus on such generic features.

To describe the collider phenomenology of such black holes, one must understand their production and decay. Many details of both were investigated in the papers [11, 12], where it was in particular found that black hole production could be both *prolific* and have *outstanding signatures*. After formation, a black hole decays in several phases[11], termed *balding, spindown, Schwarzschild*, and *Planck*. This paper will summarize these basic stages, and indicate improvements in understanding since these original papers. Some parts of the story that are still missing are also indicated.

FORMATION

In a collision at energies $E \gg M_D$, one believes that many features of quantum gravity are adequately summarized by classical physics, with subleading effects suppressed by powers of M_D/E. (For more discussion see [13, 14, 15, 16] and references therein.) Classical production of black holes at high-energies was first studied by Penrose, who in unpublished work[17] showed that a collision at zero impact parameter forms a closed-trapped surface. The "area theorem" of general relativity then implies that this corresponds to formation of a true black hole. A more general construction for non-zero impact parameter and general dimension was later found in [18], which confirmed the expectation [19] that black holes would form out to impact parameters comparable to the Schwarzschild radius,

$$R_S(E) = \left[\frac{2(2\pi)^{D-4}E}{(D-2)\Omega_{D-2}M_D^{D-2}}\right]^{1/(D-3)} \sim \quad M_D^{-1}(E/M_D)^{1/(D-3)} \,. \tag{3}$$

(Here $\Omega_{D-2} = 2\pi^{(D-1)/2}/\Gamma[(D-1)/2)]$ is the volume of the unit $(D-2)$-sphere.) An important observation is that the trapped surface forms before the particles actually collide. This is because the incident particles travel at essentially the speed of light, so that they can enter into a region smaller than R_S before any signal of their arrival deforms the geometry in this region. Thus, the trapped surface construction of [17, 18] can be made entirely in the well-understood pre-collision geometry, which consists of two Aichelburg-Sexl solutions.

This construction then yields important features of production. Existence of a maximum impact parameter $b \sim R_S$ implies a cross section

$$\sigma \sim \pi R_S^2 \,. \tag{4}$$

Moreover, the area theorem states that classically the horizon area of the ultimate black hole must be greater than that of the original trapped surface, giving a lower bound on the mass of the black hole. The construction of [18] explicitly yields the trapped surface in $D = 4$, but numerical calculation of its shape had to be performed in the important work [20], which then yielded more precise results for maximum impact parameters and lower bounds on black hole masses. Another advance was the realization [21] that pushing the trapped surface construction as far into the future of the collision as possible before losing control of the dynamics yields modest improvements on the estimate of the cross section and a slight improvement on the lower bound on the mass.

In the context of hadron colliders, the cross section (4) is the parton-level cross section, and parton distribution functions must be folded in to get production rates. This question will be further discussed shortly, in the section on experimental expectations. Some remaining uncertainties in production include the claim that charge effects can reduce cross sections[22], as well as the fact that at LHC one doesn't really expect to be fully in the asymptotic regime $E \gg M_D$, so many other effects dependent on the detailed brane/standard model construction could make important corrections. Improved understanding of these questions is needed.

DECAY

Balding phase. When first formed, the black hole is very asymmetrical – both its gravitational and other fields have very high multipole moments. However, a stationary black hole has no "hair" – that is, to become stationary, the black hole must shed its higher multipole moments through a process termed *balding*. In the $E \gg M_D$ limit, this is an essentially classical process, which takes place on a characteristic time scale $t \sim R_S$. At the same time, the black hole should shed any charges carried by particles with large charge to mass ratios, through an analog of the Schwinger process; this is expected to include color and electric charge.

Thus, during balding, the black hole will emit gravitational and gauge radiation, and its charge. The result will be a spinning, uncharged Kerr black hole. The essentially classical balding process is very poorly understood at any level of detail – this is a place for future improvement, perhaps via numerical methods. An important question for the rest of the decay is the mass and spin of the resulting black hole, accounting for this "inelasticity." Since balding (minus the charge loss) is a classical process, a lower bound on mass is given by the trapped surface area as described above. Improved understanding of balding would also be useful to determine its contribution to experimental signatures.

Spindown phase. After balding the black hole is classically stable, but as predicted by Hawking, will continue to decay quantum-mechanically. A rotating black hole first decays by preferentially shedding its angular momentum, in *spindown*. The four-dimensional version of the process was studied long ago[23]. To calculate details of this process, one must calculate higher-dimensional Hawking emission rates, which depend not just on the well known thermality of the Hawking radiation, but also on detailed "gray-body" factors, that parametrize the effects of potential barriers for various modes near the black hole. This is a hard problem.

A basic parameter for higher dimensional black holes is the evaporation time scale,

$$t_H \sim M_D^{-1}(E/M_D)^{(D-1)/(D-3)} . \tag{5}$$

Initial estimates for other parameters of spindown were made in [11], based on extrapolations from the D=4 results[23]. (This phase was not considered in [12].) There has been much subsequent work to actually calculate gray-body factors and provide more precise estimates ([24, 25] and references therein), and we are nearing a more complete understanding of this phase. The rate of spin loss vs. mass loss is qualitatively similar to that in D=4, though with the more refined calculations, [25] claims that a black hole will lose over 50% of its mass during spindown. This is somewhat dependent on the rather arbitrary definition of the end of the spindown phase, but stresses the relevance of this phase. It is clearly important to better understand the details of spindown, and experimental features of black hole decay will depend on these details (uncertainties over bulk graviton emission[26, 27] also remain). One potentially important signature was pointed out in [11]: one might look for dipolar patterns characteristic of the black hole shedding its spin.

Detailed study of spindown has also begun to reveal other important parameters, like improved estimates of the relative rates of emission of vector, spinor, and scalar particles. Apparently vector and spinor emission dominate. The resulting ratios can yield detailed predictions for hadron to lepton ratios, etc., improving rough estimates of [11, 12]. It is important for this program to be brought to completion, working out these and other specific signatures.

Schwarzschild phase. There has been greater focus on the details of this theoretically-simplest stage, where the black hole is non-rotating and decays via Hawking radiation. Power spectra, relative emission rates of different particle species, etc. are largely determined by the fact that such a black hole emits essentially thermally, at an instantaneous temperature determined by its mass:

$$T_H = \frac{D-3}{4\pi R_S} \propto M^{-1/(D-3)} , \tag{6}$$

and this yields estimates of relative multiplicities of decay products[11, 12]. However, as with spindown, there can be important corrections due to gray-body factors, which can now be calculated[24, 25] as opposed to estimated. Modifications to the thermal multiplicities include suppression of low-energy gauge boson emission and other corrections.

Future improvements, which seem close to being within reach, are a full study of the evolution of the black hole through *both* the spindown and Schwarzschild phases, properly incorporating gray-body factors, and integrating over the evolution, in order to give detailed predictions for critical features such as the energy spectrum, total and relative multiplicities, event shape parameters (such as angular distributions), etc.

Planck phase. When the black hole reaches the mass scale $M \sim M_D$, all known physics breaks down and predictions cannot be made: this is the fully quantum-gravitational realm. Conversely, to the extent to which we can see experimental signatures from this phase, we will learn about the dynamics of quantum gravity. This is thus the most interesting phase. One might in general expect emission of a few particles (or, if string theory is right, excited string states) with energies $E \sim M_D$ and high transverse momenta. But who knows what surprises this phase might yield, if indeed it can be seen.

EXPERIMENTAL EXPECTATIONS

Production rates. The above discussion can now be assembled into a description of experimental expectations. The first question is how low the Planck scale could be. Different normalization conventions exist for this scale; for further discussion see [28]. The current treatment has adopted the convention of [29] (also used by the Particle Data Book), which differs from the definition of M_p used in [11] by a factor:

$$M_p = 2^{1/(D-2)} M_D\ ; \tag{7}$$

this formula can be used to convert formulas in that paper. Current bounds, quoted for example at SUSY 2007[30] are dimension-dependent but lie around

$$M_D \gtrsim 1\text{TeV}\ , \tag{8}$$

with similar expectations for warped compactifications such as the toy model of [6] or in string theory[7, 8]. Of course there are numerous theoretical difficulties in finding a complete model with such a scale, but we will take the viewpoint that the true Planck scale could lie in this vicinity and explore the consequences.

For a given Planck mass, the next question is for what collision energies the M_D/E expansion begins to yield reliable results. There are a number of possible criteria, many discussed in [11]; this reference advocated that one particularly useful criterion is that the entropy of the black hole be large, so that a thermal approximation begins to make sense. A non-rotating hole of mass M has entropy

$$S_{BH} = [2M/(D-2)M_D]^{(D-2)/(D-3)}\, (2\pi)^{(2D-7)/(D-3)} \Omega_{D-2}^{-1/(D-3)}\ . \tag{9}$$

For example, for the representative values $D=6$ and $D=10$, a black hole with mass

$$M = 5M_D \tag{10}$$

has entropy $S_{BH} \simeq 24$. Thus, a plausible threshold for semiclassical black hole effects could lie around $5M_D$.

Next consider production rates. The rough cross section (4) can be improved by the results of [18, 20, 21]. One must account not only for the modification of the cross-section, but also for the fact that the inelasticity grows at increasing impact parameter, so one can drop below a threshold such as (10) at sufficiently large impact parameter. The trapped-surface calculations of [18, 20, 21] provide *lower* bounds on the mass of the black hole. In four dimensions and with zero impact parameter, such a trapped-surface bound gives $M \geq E/\sqrt{2} \simeq .71E$. However, subsequent estimates of [31] raise the estimated mass to $M \approx .84E$. While inelasticity was considered in [11], its effects were underestimated based on these then state-of-the-art four-dimensional estimates – [18] revealed that inelasticity increases with dimension. Indeed, in the higher-dimensional situation, inspection of the figures in [21] show that the trapped-surface bounds are lower in higher dimensions. Analogs of the estimates [31] don't yet exist.[3]

The trapped surface lower bounds of [18, 20, 21] can be roughly parametrized by a curve of the form

$$M = .5E\ , b < .5R_S \quad ; \quad M = 0\ , b > .5R_S\ . \tag{11}$$

This parametrization follows from fig. 10 of [21], for $D = 10$, with the cutoff taking into account a threshold of the form (10). For lower D this parametrization represents a significant *underestimate*. Moreover, as exemplified by the four-dimensional estimates [31], one expects that the actual mass should be higher. Thus, for the purposes of making estimates, we will consider the following two "production efficiency curves:"

$$(I): \qquad M = .6E\ , b < .5R_S \quad ; \quad M = 0\ , b > .5R_S \tag{12}$$

$$(II): \qquad M = .7E\ , b < .5R_S \quad ; \quad M = 0\ , b > .5R_S\ . \tag{13}$$

These very rough estimates have corrections of both signs; more precise calculations should be part of future work.

With a black hole threshold of 5 TeV, the curves (12) correspond to a minimum parton CM energy of 8.3 and 7.1 TeV, respectively. By these energies, the parton distribution functions have fallen significantly. Total cross sections are readily estimated; the parameterizations (I) and (II) give us a parton cross section $\sigma = \pi R_S^2/4$ for partons above the critical energy. From these cross sections, calculations using the CTEQ6M parton distribution functions[4] yield:

$$(I): \sigma = 1.8 \times 10^2 fb \quad ; \quad (II): \sigma = 1.8 \times 10^3 fb\ . \tag{14}$$

These (still rough) estimates are significantly lower than the original estimates of [11, 12], due to the substantial inelasticity revealed in the subsequent work [18, 20, 21]. This illustrates the great sensitivity of cross sections to the production efficiency curves, motivating their better understanding. Still, the estimates (I), (II) yield respectable rates: at the nominal LHC luminosity of $10^{34}/cm^2 s$, one black hole every ten minutes for (I), and one every minute for (II). However, such considerations also illustrate a narrowed parameter window where black hole production is possible, as has been stressed in [33, 32].

Signatures. To determine explicit and quantitative signatures, one needs to finalize the full integrated description of the evolution through the spindown and the Schwarzschild phases, including the important effects of gray-body factors, as described above. These results could for example be used in conjunction with event generators. Three such generators exist: TRUENOIR, CHARYBDIS, and Catfish. However, none of

---

[3] Recent work[32] pessimistically takes such lower bounds to represent the actual mass.

[4] I gratefully acknowledge Tom Rizzo's calculation of these cross sections.

them properly account for spindown, which is an important feature of the decay[11, 25]. Thus significant work remains to be done.

Nonetheless, some rather striking qualitative signatures can be inferred[11, 12]. These include

- potentially large cross sections, approaching 10^3 fb or more (also an increase of cross section with energy, according to (4), which will be hard to see at LHC)
- relatively high sphericity;
- high multiplicity ($\propto S_{BH}$) of primary particles produced;
- hard transverse leptons and jets, in significant numbers;
- approximately thermally determined ratios of species;
- angular distributions characterizing the spindown phase;
- suppression of highest-energy jets[9, 11, 34];
- decrease of primary final state lepton/parton energy with total event transverse energy, resulting from decreasing Hawking temperature with mass.

Some of these signatures may be most visible through multi-event statistics, particularly near threshold, where individual event multiplicities may not yield good statistics. (Indeed, [33] estimates multiplicity $\langle N \rangle \approx S_{BH}/3$ – which should now be checked with improved gray-body calculations.)

Cosmic ray production. Immediately after [11, 12], it was realized that black holes should be produced in collisions of hadronic ultrahigh-energy cosmic rays with the atmosphere, whose CM energies range upwards of 100 TeV, but that the rate would be overwhelmed by QCD processes and thus be too low to be observable[28]. However, one expects such cosmic rays to interact with the microwave background and produce a flux of ultrahigh-energy GZK cosmic ray neutrinos. These could produce an observable rate of black hole events[35, 36, 37, 28]. In particular, the possibility of seeing such events at cosmic ray facilities such as AMANDA, Auger, IceCube, or others has now been widely investigated.

In particular, [33] argued that non-observation of such events with five years of Auger data could yield a bound $M_D \gtrsim 2$ TeV, and rule out production of semiclassical black holes at LHC. It will be interesting to keep an eye on this data. There are possible uncertainties in such a discussion; examples include arguments that neutrino fluxes can be suppressed in large extra dimension scenarios[38] and a proposal[39] that $\nu p \to BH$ events could be suppressed by the physics preserving baryon number, and uncertainties over primary composition[40]. So, while we may be lucky enough to see hints from Auger, ultimately we should wait to see what LHC brings.

THEORETICAL ISSUES

Whether or not black hole production lies within reach depends on whether nature is able to realize a TeV-scale gravity scenario. Construction of such scenarios has been very challenging, but there has been a lot of progress and ongoing developments too numerous to review here. Ultimately, experiment will be the jury.

However, even if black hole production is not experimentally accessible, it is an extremely important theoretical problem, as it forces confrontation with our most profound theoretical issues. Notable among these is the black hole information paradox.[5] The basic statement of this paradox is that consideration of the fate of quantum information in the context of evaporating black holes apparently forces us to abandon a cherished principle of physics. The possibilities include abandoning unitary quantum-mechanical evolution, as originally suggested by Hawking[43], with the apparent consequent disastrous abandonment of energy conservation[44]; abandoning stability, as implied by a black hole remnant scenario, or abandoning macroscopic locality, in order that information can escape a black hole in Hawking radiation.

Nonlocality. There is not space here for a full discussion of the situation, but it is worth summarizing some highlights. First, an increasingly widespread belief, particularly in the string theory community, is that locality is not correct, and information escapes during evaporation of a black hole. There has been a great deal of discussion of this in this in the context of ideas such as the "holographic principle." However, if nonlocality is the answer to the paradox, it is very important to answer some basic questions, such as: 1) what is the mechanism for locality violation; 2) how is Hawking's original argument for information loss avoided and 3) in what contexts do semiclassical general relativity(GR) together with local quantum field theory(QFT) fail? One thing that would be very interesting is to properly establish the boundaries of a *correspondence limit* for the new physics, whether string theory or something else, that replaces GR+QFT.

An obvious possibility, widely considered by string theorists, is that such nonlocality has as its origin the nonlocal extendedness of strings. This possibility can be investigated precisely in the context of high-energy collisions. The result is recent arguments[15, 16] that high-energy scattering evidences nonlocal behavior, but *there is no evidence of a role for string extendedness.* Instead, it has been proposed[45] that nonlocality could emerge as an effect intrinsic to the non-perturbative dynamics of quantum gravity.

An important question is how precisely one can test the notion of locality. In QFT, locality is encoded in commutativity of local observables at spacelike separations. Diffeomorphism invariance, the gauge symmetry of gravity, implies that there are no such local observables. One can instead formulate certain "proto-local" observables[46, 47], which are diffeomorphism-invariant objects that approximately reduce to local observables in appropriate limits. However, limits on when this reduction is possible include those from situations where gravity becomes strong, such as in formation of a black hole. Another test of locality is in the asymptotic behavior of scattering amplitudes at high energies. Behavior such as (4), and Hawking (thermal) decay amplitudes, violate the known useful criteria for locality such as the Froissart[48] and possibly Cerulus-Martin[49] bounds. Together, these statements suggest that there is no precise notion of locality in quantum gravity, and support the above statement that its breakdown is associated with non-perturbative gravitational dynamics.

This guides us towards parametrizing the correspondence limit where GR+QFT are expected to break down; one parametrization is in the form of the "locality bounds"

[5] For reviews see [41, 42].

of [45, 50, 51]. For example, one expects a Fock space description of a two-particle state consisting of minimum-uncertainty wavepackets with approximate positions x, y and momenta p, q to only be a good approximation to a complete quantum description when

$$|x-y|^{D-3} > G|p+q| , \tag{15}$$

where G is a constant proportional to the D-dimensional Newton constant.

Another important question is how these ideas relate to the notion that there should be a good effective field theory description of the interior of a large black hole, which lies at the heart of Hawking's argument for information loss. It may be that new non-local effects in black holes are as invisible to the framework of GR+QFT as quantum effects are to a classical description of the atom. However, there are some initial indications otherwise[52]. In particular, a careful attempt to justify Hawking's derivation of information loss in an expansion in M_D/E reveals an apparent breakdown of the calculation, at the long time scales where we would expect information to begin to escape a black hole; this is a proposed resolution to the information paradox. Whether or not this breakdown of a perturbative derivation of information loss is the ultimate rationale, a proposal is that *information escapes a black hole through nonperturbative gravitational effects that only become important at long times*[45, 52, 53]. These would not respect usual notions of locality.

These ideas should also be relevant in the cosmological context[53, 52, 51], and may alter the picture that has led to conundrums of modern cosmology, such as the landscape and Boltzman brains. Such issues are under active investigation.

Without prejudicing the discussion in favor of string theory, an ultimate question in this direction is what are the basic principles of such a "nonlocal mechanics," providing the complete nonperturbative description of quantum gravity.

ACKNOWLEDGMENTS

I thank the organizers of PASCOS for the invitation to a very stimulating conference. I am particularly grateful to T. Rizzo for conversations and cross-section computations, and to A. Shapere for conversations. This work was supported in part by the DOE under Contract DE-FG02-91ER40618, and by grant RFPI-06-18 from the Foundational Questions Institute (fqxi.org).

REFERENCES

1. P. Kanti, Int. J. Mod. Phys. A **19**, 4899 (2004) [arXiv:hep-ph/0402168].
2. B. Webber, SLAC Summer Institute on Particle Physics (SSI 2005) [arXiv:hep-ph/0511128].
3. B. Koch, M. Bleicher and H. Stoecker, J. Phys. G **34**, S535 (2007) [arXiv:hep-ph/0702187].
4. N. Arkani-Hamed, S. Dimopoulos and G. R. Dvali, Phys. Lett. B **429**, 263 (1998) [arXiv:hep-ph/9803315].
5. I. Antoniadis, N. Arkani-Hamed, S. Dimopoulos and G. R. Dvali, Phys. Lett. B **436**, 257 (1998) [arXiv:hep-ph/9804398].
6. L. Randall and R. Sundrum, Phys. Rev. Lett. **83**, 3370 (1999) [arXiv:hep-ph/9905221].
7. S. B. Giddings, S. Kachru and J. Polchinski, Phys. Rev. D **66**, 106006 (2002) [arXiv:hep-th/0105097].
8. M. R. Douglas and S. Kachru, Rev. Mod. Phys. **79**, 733 (2007) [arXiv:hep-th/0610102].

9. T. Banks and W. Fischler, arXiv:hep-th/9906038.
10. S. B. Giddings and E. Katz, J. Math. Phys. **42**, 3082 (2001) [arXiv:hep-th/0009176].
11. S. B. Giddings and S. D. Thomas, Phys. Rev. D **65**, 056010 (2002) [arXiv:hep-ph/0106219].
12. S. Dimopoulos and G. L. Landsberg, Phys. Rev. Lett. **87**, 161602 (2001) [arXiv:hep-ph/0106295].
13. S. D. H. Hsu, Phys. Lett. B **555**, 92 (2003) [arXiv:hep-ph/0203154].
14. S. B. Giddings and V. S. Rychkov, Phys. Rev. D **70**, 104026 (2004) [arXiv:hep-th/0409131].
15. S. B. Giddings, Phys. Rev. D **74**, 106006 (2006) [arXiv:hep-th/0604072].
16. S. B. Giddings, D. J. Gross and A. Maharana, arXiv:0705.1816 [hep-th].
17. R. Penrose, unpublished (1974).
18. D. M. Eardley and S. B. Giddings, Phys. Rev. D **66**, 044011 (2002) [arXiv:gr-qc/0201034].
19. K. S. Thorne, *In *J R Klauder, Magic Without Magic*, San Francisco 1972, 231-258*
20. H. Yoshino and Y. Nambu, Phys. Rev. D **67**, 024009 (2003) [arXiv:gr-qc/0209003].
21. H. Yoshino and V. S. Rychkov, Phys. Rev. D **71**, 104028 (2005) [arXiv:hep-th/0503171].
22. H. Yoshino and R. B. Mann, Phys. Rev. D **74**, 044003 (2006) [arXiv:gr-qc/0605131].
23. D. N. Page, Phys. Rev. D **14**, 3260 (1976).
24. G. Duffy, C. Harris, P. Kanti and E. Winstanley, JHEP **0509**, 049 (2005) [arXiv:hep-th/0507274]; M. Casals, P. Kanti and E. Winstanley, JHEP **0602**, 051 (2006) [arXiv:hep-th/0511163]; M. Casals, S. R. Dolan, P. Kanti and E. Winstanley, JHEP **0703**, 019 (2007) [arXiv:hep-th/0608193].
25. D. Ida, K. y. Oda and S. C. Park, Phys. Rev. D **67**, 064025 (2003) [Erratum-ibid. D **69**, 049901 (2004)] [arXiv:hep-th/0212108]; Phys. Rev. D **71**, 124039 (2005) [arXiv:hep-th/0503052]; Phys. Rev. D **73**, 124022 (2006) [arXiv:hep-th/0602188].
26. A. S. Cornell, W. Naylor and M. Sasaki, JHEP **0602**, 012 (2006) [arXiv:hep-th/0510009].
27. V. Cardoso, M. Cavaglia and L. Gualtieri, JHEP **0602**, 021 (2006) [arXiv:hep-th/0512116].
28. S. B. Giddings, *In the Proceedings of APS/DPF/DPB Summer Study on the Future of Particle Physics - Snowmass 2001, Snowmass, Colorado, 30 Jun - 21 Jul 2001, pp P328*, ed. N. Graf. [arXiv:hep-ph/0110127].
29. G. F. Giudice, R. Rattazzi and J. D. Wells, Nucl. Phys. B **544**, 3 (1999) [arXiv:hep-ph/9811291].
30. A. Duperrin, talk at SUSY 2007, Karlsruhe, Germany.
31. P. D. D'Eath, Class. Quant. Grav. **10**, S207 (1993).
32. P. Meade and L. Randall, arXiv:0708.3017 [hep-ph].
33. L. A. Anchordoqui, J. L. Feng, H. Goldberg and A. D. Shapere, Phys. Lett. B **594**, 363 (2004) [arXiv:hep-ph/0311365].
34. S. Hofmann et. al, arXiv:hep-ph/0111052.
35. D. Dorfan, S.B. Giddings, T. Rizzo, and S. Thomas, unpublished (2001).
36. J. L. Feng and A. D. Shapere, Phys. Rev. Lett. **88**, 021303 (2002) [arXiv:hep-ph/0109106].
37. L. Anchordoqui and H. Goldberg, Phys. Rev. D **65**, 047502 (2002) [arXiv:hep-ph/0109242].
38. J. Lykken, O. Mena and S. Razzaque, arXiv:0705.2029 [hep-ph].
39. D. C. Dai, G. D. Starkman and D. Stojkovic, Phys. Rev. D **73**, 104037 (2006) [arXiv:hep-ph/0605085].
40. L. A. Anchordoqui, H. Goldberg, D. Hooper, S. Sarkar and A. M. Taylor, arXiv:0709.0734 [astro-ph].
41. S. B. Giddings, arXiv:hep-th/9412138; arXiv:hep-th/9508151.
42. A. Strominger, arXiv:hep-th/9501071.
43. S. W. Hawking, Phys. Rev. D **14**, 2460 (1976).
44. T. Banks, L. Susskind and M. E. Peskin, Nucl. Phys. B **244**, 125 (1984).
45. S. B. Giddings, Phys. Rev. D **74**, 106005 (2006) [arXiv:hep-th/0605196]; Phys. Rev. D **74**, 106009 (2006) [arXiv:hep-th/0606146].
46. S. B. Giddings, D. Marolf and J. B. Hartle, Phys. Rev. D **74**, 064018 (2006) [arXiv:hep-th/0512200].
47. M. Gary and S. B. Giddings, Phys. Rev. D **75**, 104007 (2007) [arXiv:hep-th/0612191].
48. M. Froissart, Phys. Rev. **123** (1961) 1053.
49. F. Cerulus and A. Martin, Phys. Lett. **8** (1964) 80.
50. S. B. Giddings and M. Lippert, Phys. Rev. D **65**, 024006 (2002) [arXiv:hep-th/0103231]; Phys. Rev. D **69**, 124019 (2004) [arXiv:hep-th/0402073].
51. S. B. Giddings and D. Marolf, arXiv:0705.1178 [hep-th].
52. S. B. Giddings, arXiv:hep-th/0703116, to appear in Phys. Rev. D.
53. N. Arkani-Hamed, S. Dubovsky, A. Nicolis, E. Trincherini and G. Villadoro, JHEP **0705**, 055 (2007) [arXiv:0704.1814 [hep-th]].

The Measure of the Universe

S. W. Hawking

DAMTP, University of Cambridge, Wilberforce Road, Cambridge CB3 0WA, UK

Cosmology has no predictive power without a theory of initial conditions. Because of the singularity theorems of Penrose and myself, many people assume that the initial state is necessarily of trans-Planckian curvature. We have no ideas of how to formulate initial conditions in such situations. String theory, at least in the form we know it, is based on perturbations about flat space, and so would break down along with classical general relativity. However, the singularity theorem relevant to cosmology, though not that for black holes, depends on the strong energy condition:

$$T_{ab}V^aV^b \geq \frac{1}{2}TV^aV_a$$

for any timelike or null vector V^a. This is always satisfied by gauge fields, but can be violated locally by scalar fields. It is therefore possible for the universe either to bounce or to approach a de Sitter state in the past. Such non-singular solutions form only a small subset of the space of all scalar field gravity solutions, but I shall show that the no boundary condition implies that they provide the dominant contribution to the present state. The curvature of the universe need never have been at the Planck level, and the birth of the universe can have been entirely within the semi classical domain. String theory is not necessary for cosmology.

The no boundary condition gives a measure on solutions for the universe, which seems to be heavily biased towards little or no inflation. However, I shall argue that the true measure of the universe is given by the amplitude, times a volume factor. This volume weighting restores the probability of high inflation, starting at a saddle point of the potential.

Our best guess for the structure of spacetime at the present time, is that it is approximately of the form,

$$M = X \otimes Y$$

where X is four dimensional Minkowski space, and Y is a six or seven dimensional internal space. The geometry of the internal space will determine the effective particle physics theory at low energy. The metric of Y will be Ricci flat at tree level, and will depend on a finite number of parameters, or moduli. However, one would expect quantum corrections and super symmetry breaking to remove the degeneracy, and introduce an effective potential, V, which was a function on the moduli space of Y. If ϕ are local coordinates on Y, they can be regarded as scalar fields on X. The potential, V, could have a large number of local minima, corresponding to a landscape of possible vacuum states of M theory. So what is it that determines that we are in the standard model state, and not one of the possible alternative vacuum states?

CP957, *Particles, Strings, and Cosmology, 13th International Symposium*
edited by A. Rajantie, C. Contaldi, P. Dauncey, and H. Stoica

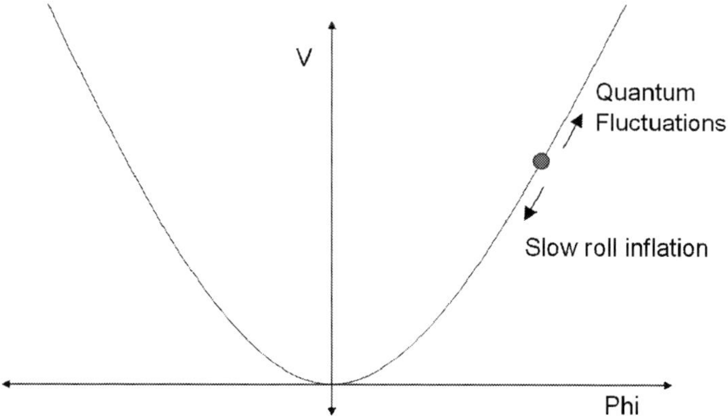

FIGURE 1. Eternal Inflation Scenario. Widely separated regions were supposed to fall into different local minima of the potential which would give the universe a mosaic structure.

To answer this, one has to turn to cosmology. One idea that has been advanced is the Eternal Inflation Scenario. In this, the scalar fields, ϕ, are supposed to fluctuate up in some regions, and down in others. There will be as many regions in which it fluctuates down the potential hill as there are in which it fluctuates up, but the regions that fluctuate up, will expand faster. The upwards fluctuating regions will dominate later surfaces of constant time if a certain condition is met: the condition for eternal inflation. Widely separated regions that fluctuate down the potential hill would fall into different local minima of the potential, which would give the universe a mosaic structure, with different parts of the universe in different vacuum states. This derivation of eternal inflation is not gauge invariant, and violates energy conservation and the Hamiltonian constraint. However I will derive a similar condition from a very different argument.

Eternal inflation is an essentially classical picture, which assumes there is a single metric for the universe. That is why its advocates feel it is necessary to suppose the universe has a mosaic structure, to accommodate the possibility that the universe could be in any of the vacuum states, in the same universe. However in a fully quantum theory, the universe can have any metric with suitable boundary conditions, which I shall take to be the no boundary condition.

$$\Psi[h_{ij},\phi] = \int Dg e^{-S[g]}$$

The amplitude for a state with metric g, and matter fields ϕ, on a spacelike surface S, is given by the path integral over all no boundary metrics, with those values on the surface S. One can also calculate the amplitude for inhomogeneous final states which are a mosaic of different vacuum states. They will in general be lower than the amplitudes for homogeneous final states.

The amplitude, Ψ, is the wave function of the universe. It will obey the Wheeler DeWitt equation:

$$\left[-G_{ijkl}\frac{\delta^2}{\delta h_{ij}\delta h_{kl}} - {}^3R(h)h^{\frac{1}{2}} + 2\Lambda h^{\frac{1}{2}}\right]\Psi[h_{ij}] = 0$$

where G_{ijkl} is the metric on superspace,

$$G_{ijkl} = \frac{1}{2}h^{-\frac{1}{2}}(h_{ik}h_{jl} + h_{il}h_{jk} + h_{ij}h_{kl})$$

and ${}^3R$ is the scalar curvature of the intrinsic geometry of the three-surface.

In the case that the surfaces, S, are three-spheres of radius a, and the matter is a single scalar field ϕ, this is a wave equation in the (a,ϕ) plane, with a playing the role of time.

$$\frac{1}{2}\left[\frac{\partial^2}{\partial a^2} - a^2 - \frac{1}{a^2}\frac{\partial}{\partial\phi^2} + a^4V\right]\Psi(a,\phi) = 0$$

In the Euclidean region, $a^2V < 1$, there will be a real Euclidean solution of the field equations, and the wave function will be exponential. Outside this region, however, there will only be complex solutions, and the wave function will oscillate rapidly. One can represent the wave function as the product of a rapidly varying phase, C, and slowly varying amplitude, B. Plugging this in the Wheeler DeWitt equation, one finds that C obeys the Hamilton Jacobi equation.

$$\Psi = Be^{iC}$$
$$\nabla C \cdot \nabla C - J = 0$$
$$\nabla B \cdot \nabla C = 0$$

One can therefore interpret the wave function by WKB, as corresponding to a family of Lorentzian solutions of the field equations. The trajectories of the solutions are given by the gradient of C, raised by the Wheeler DeWitt metric. The amplitude, B, obeys a conservation equation, which implies that the amplitudes of individual solutions are constant over the evolution of the solutions.

$$a = H^{-1}\cosh(Ht), \text{ where } H^2 = V_1.$$

The wave function of the universe given by the no boundary proposal, corresponds to solutions that bounce at the boundary of the Euclidean region. The potential will be a maximum at the minimal surface. The amplitude of the solution will be the amplitude of a $K = 0$ surface, at the potential at the bounce. There will be a mismatch in the derivatives of the scalar field, but this will be small if the potential satisfies the slow roll condition, ∇V small compared to V.

The amplitude of a solution will be e^I, where $I = -\frac{3}{4}\pi V$, is the Euclidean action of half a four-sphere with curvature scalar, $R = 8\pi V$, and V is the value of the potential at the bounce. The amplitude will be a maximum for solutions that bounce at the minimum of the potential. However, such solutions will just be empty deSitter space, and so not candidates for the universe we observe. To obtain a matter filled universe with structure, like galaxies and stars, it seems necessary for the universe to have a large number, N, of e-foldings of slow roll inflation. If one weights solutions with their amplitude, the probability distribution would be strongly biased to low values of N, the number of e-foldings of inflation. This would predict that our universe would have the least value of N compatible with our existence, which would not produce the universe we observe.

This has been recognized to be a problem with the no boundary proposal for some time. I think the answer is that there are two different probabilities involved. The amplitude gives the probability for the entire universe. However, one does not observe the entire universe, but only a Hubble volume around oneself. The number of such Hubble volumes at a given matter density, is proportional to the volume of the universe at that time, which in turn is proportional to e^{3N}. Thus on a frequency definition of probability, the probability of observing a Hubble volume of a given history, is proportional to the probability of that history, times e^{3N}.

$$\begin{aligned} P(\text{Entire universe}) &= |\Psi^2| \\ P(\text{Hubble volume}) &= |\Psi^2| e^{3N} \end{aligned}$$

The volume weighting transforms the probability distribution over N, the amount of inflation. It can more than compensate for the reduction in amplitude, due to a higher value of the potential at the bounce, if the slow roll parameter, ε, is less than the potential, V, in Planck units,

$$\varepsilon = \frac{\nabla V \cdot \nabla V}{V^2} < V$$

This is the same as the condition for eternal inflation but derived in a gauge invariant manner.

When the microwave fluctuations we observe left the horizon, this condition was not satisfied, because $\frac{V}{\varepsilon}$ was about 10^{-5} at the time. But inflation may have started before that, at higher V and lower ε. For solutions that start at the Planck density, in a polynomial potential, this condition will be satisfied only near the Planck density. The probability distribution would still be overwhelmingly in favor of low N. On the other hand, for a solution that starts at a maximum or a saddle point, the probability distribution would favour very large N.

The dominant contribution is likely to come from broad saddle points well below the Planck density.

$$\frac{V''}{V} > -2$$

$$\text{Amplitude} = \exp(V_1{}^{-\frac{1}{2}})$$

The metrics will be well within the semi classical regime. They would start out with a Hawking Moss instanton, a de Sitter like state which is unstable, and begins to run down the potential hill. The origin of the universe, is in the low energy regime of M theory, in which four dimensional general relativity is a good approximation. This is supported by the fact that calculations based on four dimensional general relativity, are in excellent agreement with observations of the microwave background. One would not expect this, if the four dimensional approximation, $X \otimes Y$, broke down before one gets back to the time of inflation. This would indicate that the internal space, Y, is smaller than 10^6 times the Planck length. The only situation in which 4D general relativity would break down, and in which one would need string theory, or some other approach, would be the final stages of evaporation of a black hole.

The only vacuum states that will have significant amplitudes to be matter filled, will be those where the minimum of the potential, lies on the line of descent from a broad saddle point of index one. By this I mean that there is only one direction in which the action decreases. This is in accord with the general principle, that the instanton that describes the decay of an unstable state, should have one, and only one, negative mode. In this case, the instanton would be the Hawking Moss or de Sitter instanton, with the value of the vacuum energy at the saddle point. The negative mode, would be the homogeneous mode in which the scalar fields everywhere move along the line of steepest descent from the saddle point. If the saddle point is broad, that is, if $\frac{V''}{V}$, is small and negative, the lowest inhomogeneous mode will be positive. This will give the Hawking Moss instanton one, and only one, negative mode, as required.

In the usual, bottom up, approach, one assumes that the universe started in a state of high symmetry, which then evolved to the present broken symmetry state. The symmetry breaking would happen in different directions in different places, leading to topological defects, such as domain walls, cosmic strings, and monopoles. On the other hand, according to the top down approach I have described, the solution that gives the dominant contribution to the amplitude, could have the same broken symmetry all the way back. In this case, there would be no production of topological defects.

The no boundary condition, enables us in principle to calculate volume weighted quantum amplitudes, for the whole universe to be in different states in the landscape. It would be a mistake to assume that we should be in the state with the highest volume weighted amplitude. That would be like saying I should be Chinese, because there are many more Chinese than Brits. If the volume weighted amplitude for the standard model vacuum is non zero, it is irrelevant what the volume weighted amplitudes for other vacuum states are. The theory can not predict a unique vacuum state. Instead, we have to input that we live in the standard model vacuum.

The bouncing universes that the no boundary proposal predicts, might seem at first sight, similar to the Ekpyrotic or cyclic universes. However, there is an important difference.In the Ekpyrotic universe, it is implicitly assumed that the state in the infinite past, is one of minimum excitation, although this is not clearly stated, or well defined. This means perturbations would be small in the infinite past, and grow during collapse, and subsequent expansion. In other words, the thermodynamic arrow of time, will point forward in both the contraction, and the expansion. By contrast, the no boundary solutions will have minimum excitations at the bounce, where the no boundary condition is imposed. This means the arrow of time will point forward in the expanding phase, and backward in the contracting phase.

In fact, the physical relevance of the contracting phase, is questionable. It is like the analytical continuation of the semi-classical solution that describes pair creation in an electric field.. This consists of an electron and a positron that come in from infinity at $t < 0$, are brought to rest at $t = 0$, and accelerate away from each other for $t > 0$. However, physically there are no incoming particles. Instead one says that the electron positron pair was quantum created at $t = 0$. In a similar way, one should not attach any physical significance to the early contracting phase of the universe, but say the universe was quantum created at the bounce.

The no boundary proposal, and the top down approach, allow us to calculate amplitudes for different states at the present time. These overwhelmingly favor low or zero inflation, which would lead to an almost empty deSitter like universe. However, I argue that the probability of the entire universe, given by the amplitude, should be multiplied by the volume, to get the probability of observing a Hubble volume. This volume weighting favors inflation starting at a saddle point of the potential, and leads to the prediction that the universe should be flat, within the limits of the fluctuations in the microwave background.

The dominant histories in the path integral for these amplitudes, are bouncing solutions of the field equations, which lie entirely in the semi classical regime of four dimensional general relativity. However, these should probably not be interpreted as describing bouncing universes, but rather as the quantum creation of universes in de-Sitter like states.

The amplitudes will be highest for states in which the whole universe is in a single state, rather than a mosaic of different states, as predicted by eternal inflation. There will be no primordial production of topological defects, such as monopoles, and cosmic strings. Not all states in the landscape will have significant amplitudes, but there will be more than one that do, so M theory does not predict a unique low energy particle physics theory. It is implausible that life is possible only in one of these states, so we might have chosen a better location.

A Brief Review of Aspects of Lattice QCD

R.R. Horgan

University of Cambridge, DAMTP-CMS, Wilberforce Road, Cambridge CB3 0WA

Abstract. A brief introduction and overview of some current results and methods is given for Lattice QuantunChromodynamics (LQCD).

Keywords: Lattice, QCD, quark, renormalization, fermion
PACS: 11.15Ha, 12.38Gc

LATTICE FORMALISM

In the lattice formulation of quantum gauge field theory the lattice spacing a provides a non-perturbative gauge invariant regulator. In Figure 1 typical gauge invariant contributions to the gauge action are shown.

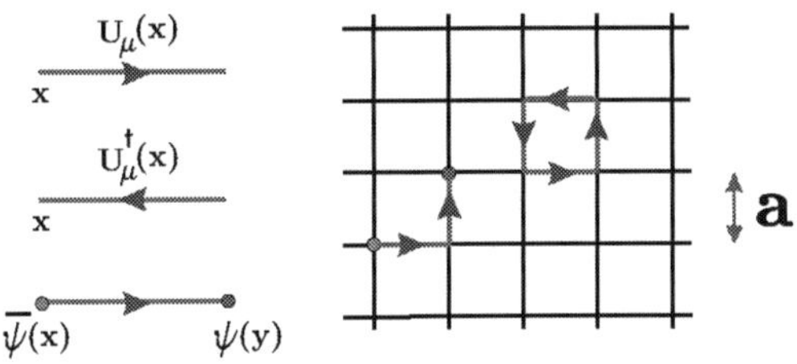

FIGURE 1. A sketch of typical gauge invariant contributions to a lattice regularized gauge theory

In general the path integral from which all observable quantities are derived is

$$Z = \int \mathcal{D}\psi \mathcal{D}\bar{\psi} \mathcal{D}U e^{\bar{\psi}M(U)\psi - \beta S_g(U)}$$

where

$$U_\mu(\mathbf{x}) = e^{-iag_0 A_\mu(\mathbf{x})} \in SU(N_c), \quad A_\mu(\mathbf{x}) \text{ is gauge field}$$

The lattice coupling constant is $g_0(a)$, with $\beta = \frac{6}{g_0(a)^2}$, and $\alpha_L(a) = \frac{g_0(a)^2}{4\pi}$. The quark contribution is $M(U) \sim \not{D} + m, \quad \not{D} = \gamma_\mu \Delta_\mu$, where Δ_μ is a lattice covariant derivative. The action is invariant under the gauge transformation $g(x) \in SU(N_c)$:

$$U_\mu(\mathbf{x}) \to g(\mathbf{x}) U_\mu(\mathbf{x}) g^\dagger(\mathbf{x}+e_\mu), \quad \psi(\mathbf{x}) \to g(\mathbf{x})\psi(\mathbf{x}), \quad \bar{\psi}(\mathbf{x}) \to \bar{\psi}(\mathbf{x}) g^\dagger(\mathbf{x}).$$

We view the 4D Euclidean Lattice Field Theory as a 4D Statistical Field Theory (SFT) on lattice of spacing a and the continuum limit is $a \to 0$ in physical units: $ma \to 0$, where m is a physical mass scale. From the SFT viewpoint the correlation length $\xi = 1/m$ diverges in lattice units in the continuum limit: $\xi/a \to \infty$. This signifies a continuous

CP957, *Particles, Strings, and Cosmology, 13th International Symposium*
edited by A. Rajantie, C. Contaldi, P. Dauncey, and H. Stoica

phase transition which is controlled by a fixed point. We require any formulation of LQCD to have the same continuum limit controlled by the correct, asymptotically free, fixed point. All such formulations are equivalent: This is called UNIVERSALITY

OPERATORS AND OBSERVABLES

The expectation value of an operator is

$$\langle \hat{O} \rangle = \frac{1}{Z} \int \mathscr{D}\psi \mathscr{D}\bar{\psi} \mathscr{D}U O[\psi, \bar{\psi}, U] e^{\bar{\psi} M(U) \psi - \beta S_g(U)}$$

An example of a fermion operator expectation value is

$$\langle \bar{\psi}\Gamma\psi \rangle = -\int \mathscr{D}U \frac{\partial}{\partial \eta} \Gamma \frac{\partial}{\partial \bar{\eta}} e^{-\bar{\eta} M^{-1}(U)\eta} \underbrace{\frac{1}{Z} \mathrm{det} M(U) e^{-\beta S_g(U)}}_{P(U)} |_{\eta,\bar{\eta}=0} = \langle \mathrm{Tr}(\Gamma M^{-1}) \rangle_P \, .$$

where $\eta, \bar{\eta}$ are Grassmann sources for $\bar{\psi}, \psi$ which have been integrated over using Grassmann integration. If $\mathrm{det}M(U)$ is real and non-negative then $P(U)$ is a probability distribution over U and is realizable by Monte-Carlo procedures. However, $\mathrm{det}M(U)$ is computationally expensive but gives a realistic, unquenched theory of dynamical fermions. If we omit $\mathrm{det}M(U)$ we get an unrealistic, quenched theory but one that is cheap to simulate.

We use LQCD to non-perturbatively measure hadron masses and the hadronic matrix elements which determine predictions for experimental observables. The two-point function is

$$G(t) = \langle \hat{A}(t) \hat{B}(0) \rangle_c \underbrace{\longrightarrow}_{t \to \infty} c_A c_B e^{-M_{AB} t} \, .$$

and we compute the lowest mass M_{AB} in the channel coupling to both $\hat{A}$ and $\hat{B}$.

A typical hadronic matrix element governing a physical process is defined by

$$\langle H_1 | \hat{O} | H_2 \rangle = \lim_{\substack{t_1 \to -\infty \\ t_2 \to \infty}} e^{M_1 |t_1| + M_2 |t_2|} \frac{\langle \hat{H}_1(t_1) \hat{O}(0) \hat{H}_2(t_2) \rangle}{c_{H_1} c_{H_2}} \, .$$

where $\hat{H}_1$ and $\hat{H}_2$ are respective interpolating operators for hadrons H_1 and H_2.

THE STANDARD MODEL

A major objective of LQCD is to test the Standard Model. The charged quark current in the Standard Model is

$$J_\mu = \bar{u}_{L,a} \gamma_\mu V_{ab} d_{L,b}, \quad \begin{pmatrix} u \\ d \end{pmatrix}_{a=1,2,3} = \begin{pmatrix} u \\ d \end{pmatrix} \begin{pmatrix} c \\ s \end{pmatrix} \begin{pmatrix} t \\ b \end{pmatrix} \, ,$$

where $q_L = \frac{1}{2}(1 - \gamma_5) q$ and V_{ab} is the unitary CKM matrix.

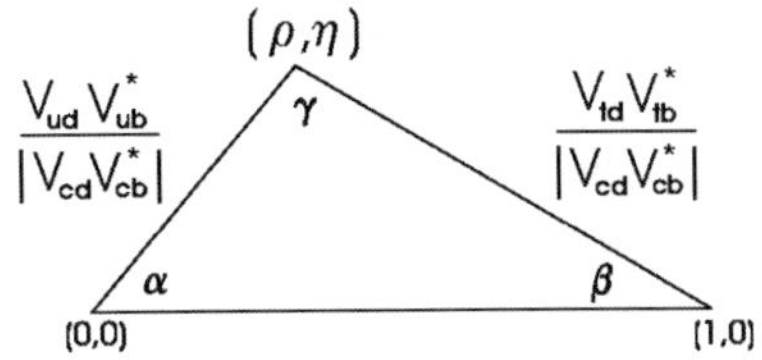

FIGURE 2. One of the standard unitarity triangles

LQCD gives predictions for exclusive processes involving "gold-plated" hadrons. E.g., $B \to l\nu$, $B \to \pi l\nu$, $B-\bar{B}$ mixing. Non-lattice methods also deal with inclusive processes. These include sum-rules, RG improvement and gluon resummation techniques. E.g., $B \to X_u l\nu$, $B \to X_s\gamma$. We use experiment and theoretical prediction to over-constrain the elements of V_{ab} and look for inconsistency in the unitarity constraint as a signal for New Physics (NP) Beyond the Standard Model (BSM). A major property is that Flavour-Changing Neutral Current (FCNC) processes are forbidden at tree level by the GIM mechanism. Thus FCNC processes are sensitive to NP since they are mediated by loop diagrams which can contain new particles. E.g., $B \to K^*\gamma, B \to X_s\gamma$, and $B-\bar{B}$ mixing.

The unitarity constraints are encoded in the Unitarity Triangle (UT) such as in Figure 2. The area of the UT is measure of CP-violation $\Longrightarrow \quad \sin(2\beta) > 0$. The angles α, β, γ can be determined from experiment free from uncertainties of hadronic matrix elements: UTangles. However, the sides of the UT require knowledge of hadronic matrix elements from lattice measurements: UTlattice. The comparison with data from the UTfit collaboration [1] is shown in Figure 3. The consistency of allowed regions within and between the plots gives a precision test of SM. Also require that different analyses of processes mediated by tree diagrams in which NP contributions are negligible should agree. This observation is relevant to the determination of $|V_{ub}|$ discussed later.

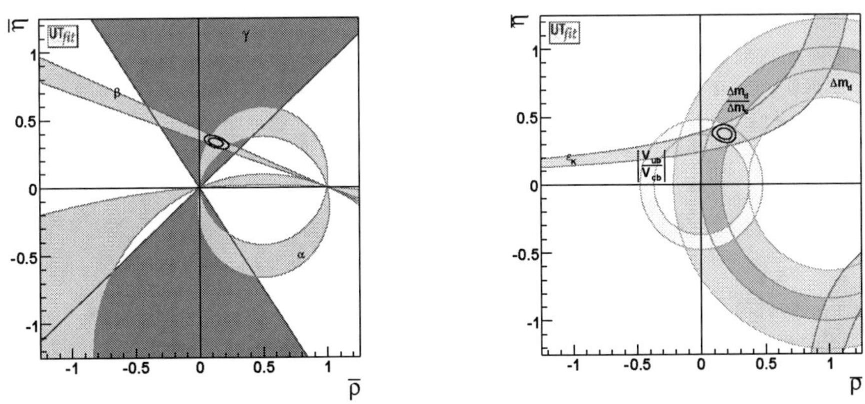

FIGURE 3. UTangles on the left and UTlattice on the right. From UTfit [1].

ERROR BUDGET

The goal is to reduce errors on target observables to $2-3\%$ or less.

Operator matching in SM: Integrate out internal heavy particles: W, Z, top.

$$H_{weak} \xrightarrow[\text{dim}-\text{reg}]{\text{OPE}} \underset{4-\text{fermi ops etc}}{\text{continuum effective QFT}} \xrightarrow[\text{matching}]{} \text{lattice EQFT} .$$

Require that the LQFT operator has same matrix elements as continuum EQFT operator. Use perturbative matching at scale $\Lambda \sim \pi/a$ exploiting automated methods [2]. Currently 1-loop in most cases but 2-loops for $\overline{MS}$ quark mass [3] and 3-loops for strong coupling constant [4]. The error due to 2-loops is $3-5\%$ but it largely cancels in ratios. Alternatively, use non-perturbative matching methods such as RG/MOM and Ward identity methods. These need matching scale $\mu \ll 1/a$ but μ must still be sufficiently large to be perturbative $\rightarrow 2-5\%$ errors. Step scaling techniques use matching of HQET to QCD at small (a,V) and then use RG step scaling to determine HQET couplings for lattices with (a,V) sufficiently large to allow simulation. Initial, small, $a \Longrightarrow$ unquenched calculation could be expensive. See lectures by Sommer [5] and references therein.

Chiral extrapolation:

$$LQCD \xrightarrow[E<1\text{GeV}]{} \underbrace{SU(3)_L \otimes SU(3)_R}_{\text{chiral theory of hadrons}}$$

The physical values of pseudoscalar masses are $m_\pi^2/m_K^2 \sim 2m_l/m_s \Longrightarrow m_l/m_s \sim 1/27$. However, computing $\det[\not{D}+m_q]$ is problematic for $m_q a$ too small since it becomes too expensive. In best current simulations $m_l/m_s = 0.1 \Longrightarrow m_\pi \sim 240\text{MeV}$ [6].

$$\text{typical cost} \propto \left(\frac{L}{a}\right)^4 \left(\frac{1}{a}\right)^{1-2} \left(\frac{1}{m_q}\right)^{2-3} .$$

For $m_l/m_s > 1/27$ extrapolate to $m_\pi = 140\text{MeV}$ using chiral loop perturbation theory for which m_π must be sufficiently small for reliable extrapolation. E.g.,

$$\begin{aligned} f_\pi &= f\left\{1-2I_\mu - I_K + \frac{4}{f^2}\mu(m_u+m_d)L_5 + \frac{8}{f^2}\mu(m_u+m_d+m_s)L_4\right\}, \\ I_X &= \frac{m_X^2}{2(4\pi f)^2}\ln\left(\frac{m_X^2}{(4\pi f)^2}\right). \end{aligned}$$

Improved algorithms: New algorithms improve the condition number of matrix operations and allow smaller m_l/m_s and smaller lattice spacing for the same CPU time. The condition number of $MM^\dagger$ is $\kappa(M) \sim 1/m_q^2$. The Rational Hybrid Monte-Carlo (RHMC) approach reduces the effective m_q dependence of κ by mass preconditioning, giving an improvement by factor $\sim 6-10$ for $m_l/m_s = 0.1$ [7, 8]. Other methods include the Domain Decomposition HMC [9].

Lattice artifacts: Errors $\sim (\Lambda_{QCD}a)^m, (m_q a)^n, \quad m,n = 1,2\ldots$. Symanzik improvement techniques for actions and operators reduce artifacts in on-shell observables. Use perturbative and non-perturbative methods. The gluon action is improved to $O(a^2)$, and additionally at $O(\alpha_s a^2)$ for dynamical staggered fermions [10]. Artifacts are reduced by operator improvement $\Longleftrightarrow$ matching. The ASQTAD staggered fermion action, which is tree-level $O(a^2)$ and tadpole improved, is widely used for current results. Its successor is HISQ, a relativistic action for fermions up to the charm mass [11]. It has no tree-level $O(a^2), O((am)^4)$ artifact errors and is effective for $am_q \leq 1$. For $a = 0.13$fm (coarse lattice) find $m_q \leq 1.6$GeV. It also has much reduced 1-loop "taste"-changing $O(a^2)$ errors.

Statistics: Faster and bigger computers but they cost money!

MEASUREMENTS AND RESULTS

Must accurately pin down predictions in the next few years. Flavour-physics tests of SM from theory, B-factories and LHCb will complement direct LHC searches for NP.

The fermions defining $P(U)$ are called sea quarks: $P(U) = \frac{1}{Z}\det M_{sea}(U)e^{-\beta S_g(U)}$ and the fermions used in operators are called valence quarks: $\langle \bar{\psi}\Gamma\psi \rangle = \langle \mathrm{Tr}(\Gamma M_{val}^{-1}) \rangle_P$. Can use physical masses for valence quarks even though m_l/m_s for sea quarks is too large for them to be physical $\Rightarrow$ chiral extrapolation in sea quark masses. Valence and sea quarks may be from different formalisms; question of violation of unitarity – expect restoration as $a \to 0$. Excellent progress with ASQTAD staggered fermions by MILC, HPQCD, FNAL collaborations.

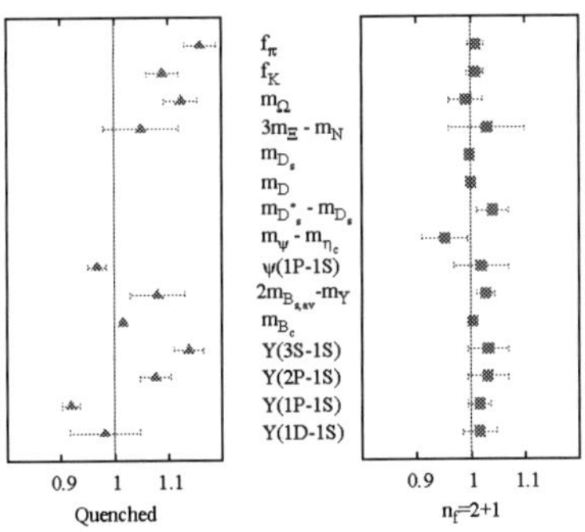

FIGURE 4. Summary of results for LQCD/Expt. from MILC,HPQCD,FNAL collaborations

Expect agreement for "Gold-plated" observables which means for stable particles with no nearby thresholds. Quenched results are ambiguous. E.g., conflicting determination of a from observables at different physical scales and the incorrect running of $\alpha_0(a)$. Unquenched results are unambiguous and the excellent results are shown in Figure 4.

The strong coupling constant: Set the scale for a from $M_{\Upsilon'} - M_\Upsilon$, measure various Wilson plaquettes for various a, deduce $\alpha_{\overline{MS}}(a^{-1})$ from 3-loop lattice and continuum pert. theory, and use 3-loop β-function to run to scale M_Z [4]. The smallest error is on the LQCD calculation:

LQCD: $\alpha^{(5)}_{\overline{MS}}(M_Z) = 0.1170(12)$ World average [12]: $\alpha^{(5)}_{\overline{MS}}(M_Z) = 0.1176(20)$

Light quark masses: For $q = u,d,s$ tune lattice masses $m_q(a)$ for various a to fit physical π, K, η masses using staggered chiral P.Th. to extrapolate to physical m_l/m_s. Have $m_q^{\overline{MS}}(\mu) = Z(\mu a)\, m_q(a)$ with $Z(\mu a)$ computed from matching 2-loop continuum and lattice perturbation theories for staggered quarks [3, 13] and 1-loop for Wilson quarks [14]. In Table 1 we show $m^{\overline{MS}}(\mu = 2\text{GeV})$ for staggered fermions $n_f = 2+1$ flavour QCD:

TABLE 1. quark masses $m^{\overline{MS}}(\mu = 2\text{GeV})$ from LQCD.

	Q Mason et al. [3]	C. Bernard et al. [13]	JLQCD [14]
m_u	1.9(0)(1)(1)(2)MeV	2.0(0)(1)(2)(1)MeV	3.54(+64)(-35)MeV
m_d	4.4(0)(2)(2)(2)MeV	4.6(0)(2)(2)(1)MeV	3.54(+64)(-35)MeV
m_s	87(0)(4)(4)(0)MeV	90(0)(5)(4)(0)MeV	91.1(+14.6)(-6.2)MeV

Göckeler et al. [15] use $n_f = 2$, $O(a)$ improved, Wilson fermions and non-perturbative matching to find $m_s^{\overline{MS}}(\mu = 2\text{GeV}) = 100 - 130\text{MeV}$

Decay constants: f_P is defined for pseudoscalar mesons by the matrix element $\langle 0|A_\mu|P_{q\bar{q}'}\rangle = q_\mu f_P$, and determines the meson decay rate $\Gamma(P_{q\bar{q}'} \to l\bar{\nu}) \propto f_P^2 |V_{qq'}|^2$. There are accurate new results from HISQ valence on ASQTAD sea [11]:

$$f_K/f_\pi = 1.189(7) \Longrightarrow \ V_{us} = 0.2262(14)\,.$$

This is to be compared with the current world average $V_{us} = 0.2257(10)$ from reactions including $K \to (\pi) l\nu, \tau \to$ hadrons [16] and which includes the preliminary Domain Wall Fermion (DWF) result for K_{l3} form-factor [17]. Unitarity requires that $|V_{ud}|^2 + |V_{us}|^2 + |V_{ub}|^2 = 1 - \delta$ with $\delta = 0$. After including new LQCD results have $\delta \sim (0.8 \pm 0.7)\,10^{-3}$.

For leptonic D-decay the results from HISQ are $4-5$ times more accurate than previous LQCD results:

$$(f_{D_s}/f_D)/(f_K/f_\pi) = 0.977(10) \quad \chi\text{PTh} \ \Rightarrow \ \sim 1\,.$$

Alternatively, deduce $|V_{cs}/V_{cd}| = 4.42(4)_{th}(41)_{expt}$. For leptonic B-decay find [11, 18, 19]

$$\text{ASQTAD:} \quad f_B = 216(9)(19)(6)\text{MeV} \qquad f_{B_s} = 269(29)\text{MeV}$$

$$\text{Expt.:} \ f_B = \begin{cases} 229\binom{+36}{-31}\binom{+30}{-34}\text{MeV} & \text{(Belle 2006)} \\ 167\binom{+48}{-66}\text{MeV} & \text{(BaBar 2006)} \end{cases}$$

To deduce f_B, experiments measure $f_B|V_{ub}|$ and currently use the value for $|V_{ub}|$ from inclusive semi-leptonic B-decay which differs from the exclusive value: see below.

Semi-leptonic exclusive decay: LQCD measures $\langle \pi|V_\mu|B\rangle$ as function of q^2. The form factors $f_+(q^2), f_0(q^2)$ are computed and a fit to the data determines $|V_{ub}|$. However, the lattice data extend only over fraction of q^2 range of physical $B \to \pi\ l\nu$ decay. Typically $q^2 > 16\text{GeV}^2$. A fit to the q^2 dependence of the form factor, deduced from analyticity and unitarity, allows extrapolation to small q^2 for experimental comparison.

Semi-leptonic inclusive decay: Alternatively, can determination $|V_{ub}|$ from inclusive $B \to X_u\ l\nu$. Use effective theory, scale separation, RG improvement and perturbative

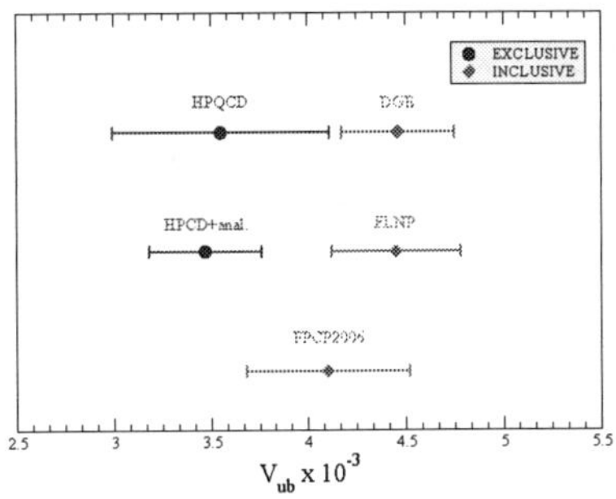

FIGURE 5. A comparison of $|V_{ub}|$ determined from exclusive (LQCD) and inclusive (DGE,BLNP) decays

resummation. Two approaches: The Dressed Gluon Exponentiation (DGE) [20] and the BLNP method [21]. The validity of these different approaches imposes different kinematic cuts on data and averaging of results. The results for $|V_{ub}|$ from both exclusive and inclusive decays are shown in Figure 5. There is clear disagreement which, because it is at tree-level, cannot easily be explained by NP, although there could be a right-handed current from NP. The unitarity plot is consistent with the exclusive, lattice, result but there is currently a discrepancy with the inclusive measurement [22, 23, 20, 24, 21].

$B_q - \bar{B}_q$ **mixing:** Mediated by a Flavour-Changing Neutral Current, for example $O_L = (\bar{q}b)_{V-A}(\bar{q}b)_{V-A}$. Calculate Δm_q, the mass difference between the heavy and light eigenstates, and measure Δm_q experimentally by observing $B_q - \bar{B}_q$ oscillations which were observed by the CDF and D0 collaborations in 2006. We have

$$\langle \bar{B}^0_q | H_{eff} | B^0_q \rangle = \frac{G_F^2}{6\pi^2} M_W^2 (V_{tb} V_{tq}^*)^2 \eta_{B_q} S_0(x_t) m_{B_q}^2 f_{B_q}^2 \hat{B}_{B_q}, \quad x_t = m_t/M_W$$

$$\frac{\Delta m_d}{\Delta m_s} = \frac{m_{B_d} f_{B_d}^2 \hat{B}_{B_d} |V_{td}|^2}{m_{B_s} f_{B_s}^2 \hat{B}_{B_s} |V_{ts}|^2} \propto (1-\bar{\rho})^2 + \bar{\eta}^2 .$$

Must match the continuum O_L with its lattice counterpart. Currently use 1-loop lattice P.Th. Note that errors in $\Delta m_d/\Delta m_s$ are smaller than in $\Delta m_d, \Delta m_s$ separately; see Figure 3.

$$\langle B_q | O_L | \bar{B}_q \rangle_\mu^{\overline{MS}} = \frac{8}{3} f_{B_q}^2 B(\mu) M_{M_q}^2, \qquad \hat{B} = 1.534\, B(\mu = m_b).$$

$$\xi = f_{B_d}^2 \hat{B}_{B_d} / f_{B_s}^2 \hat{B}_{B_s}$$

$\xi = 1.23(6)$ from LQCD [25] which is consistent with the UT value of $\xi = 1.24(8)$ [1] [26]. Both values need errors to be improved but for LQCD the lag in error control is due to complexity of numerical and matching calculations.

Rare decays: Consider important example of $B \to K^*\gamma$. It is related to inclusive $B \to X_s\gamma$ analysed by resummation methods. The dominant operator in effective theory is the magnetic penguin:

$$O_7 = \frac{e}{8\pi^2} m_b \bar{s}_\alpha \sigma^{\mu\nu} (1+\gamma_5) b_\alpha F_{\mu\nu}$$

TABLE 2. Typical resource timings for physical or near physical quark masses

a fm	m_l/m_s	size	L fm	MC traj	TF Years	Formalism
0.09	1/20	$56^3 \times 96$	5.0	4200	0.71	ASQTAD/MILC/RHMD
0.06	1/10	$64^3 \times 144$	3.8	4500	1.41	ASQTAD/MILC/RHMD
0.06	1/27	$100^3 \times 144$	6.0	7450	32	ASQTAD/MILC/RHMD
0.09	1/27	$64^3 \times 128$	5.8	10000	230	DWF/RHMC $L_5 = 24$
0.06	0.084	$64^3 \times 128$	3.8	10000	130	DWF/RHMC $L_5 = 16$
0.06	1/27	$96^3 \times 128$	5.8	10000	680	DWF/RHMC $L_5 = 16$

The amplitude $\propto V_{ts}^* V_{tb}$ determines $|V_{ts}|$ since $|V_{tb}| \sim 1$. Δm_s determines the same CKM combination and any mismatch signals NP. This decay has been seen, ~ 1000 events, at CLEO, Belle, BaBar. Expect ~ 68000 events at LHCb. We use NRQCD for the b-quark to match $O_7^{\overline{MS}}(\mu)$ to $O_7^L(a)$ using 1-loop lattice perturbation theory. The alternative non-perturbative matching must ensure correct m_b dependence. Must calculate $\langle K^*|O_7^L|B\rangle$ at $q^2 = 0$, which is difficult if the B-meson is in rest-frame since the K^* is too energetic. Remedy this by using moving NRQCD where b-quark has optimal frame velocity v.

Timings and costs: For m_l/m_s close to or at physical ($m_l/m_s = 1/27$) values, CPU costs for typical future lattices using ASQTAD and DWF formalisms are shown in Table 2. For analysis assume additional 3 times ASQTAD simulation CPU requirements.

REFERENCES

1. V. Lubicz, UTfit collaboration. "Heavy Quarks and Leptons", HQL06, Munich. hep-ph/0702204.
2. A. Hart, et al., *J. Comp. Phys.* **209**, 340–353 (2005), hep-lat/0411009.
3. Q. Mason, et al., *Phys. Rev.* **D73**, 114501 (2006), hep-ph:0511160.
4. Q. Mason, et al., *Phys. Rev. Lett.* **95**, 052002 (2005), hep-lat/0503005.
5. R. Sommer, "Workshop on Perspectives in Lattice QCD", 2005, Nara, Japan. hep-lat/0611020.
6. C. Bernard, et al., *PoS LAT2006* p. 163 (2006), MILC collaboration. hep-lat/0609053.
7. M. Hasenbusch, *PoS LAT2005* p. 116 (2006), hep-lat/0509080.
8. M. Clark, and K. A.D., *Phys. Rev.* **D75**, 011502 (2007), hep-lat/0610047.
9. M. Lüscher, *Comp. Phys. Comm.* **165**, 199–220 (2005), hep-lat/0409106.
10. Z. Hao, et al., HPQCD collboration. To appear in Phys. Rev. D. arXiv:0705.4660.
11. E. Follana, et al., HPQCD collaboration. arXiv:0706.1726.
12. Y. w. M., et al., *J. Phys.* **G33**, 1 (2006).
13. C. Bernard, et al., MILC collaboration. "Chiral Dynamics 2006", Duke University. hep-lat/0610031.
14. T. Ishikawa, et al., CP-PACS/JLQCD collaboration. arXiv:0704.1937.
15. M. Göckeler, et al., *PoS LAT2006* p. 160 (2006), hep-lat/0610071.
16. P. Ball, Cosener's Forum on Heavy Flavour Physics, 2007.
17. D. Antonio, UKQCD/RBC collaboration. CKM2006, Japan. hep-lat/0702026.
18. A. Gray, *Phys. Rev. Lett.* **95**, 212011 (2005), hep-lat/0507015.
19. M. Wingate, et al., *Phys. Rev. Lett.* **92**, 162001 (2004), hep-ph/0311130.
20. E. Gardi, "CKM workshop", CKM2006, Nagoya, Japan. hep-ph/0703036.
21. M. Neubert, "Flavour Physics and CP-violation Conference", FPCP2006, Vancouver.
22. E. Gulez, *Phys. Rev.* **D73**, 074502 (2006), hep-lat/060121.
23. J. Flynn, and J. Nieves, arXiv:0705.3553.
24. E. Barberio, "Flavor Physics and CP-Violation Conference", FPCP2006, Vancouver. hep-ex/0605098.
25. E. Dalgic, et al., *Phys. Rev.* **D76**, 011501 (2007), HPQCD collaboration. hep-lat/0610104.
26. M. Okamoto, *PoS LATT2005* p. 013 (2005), hep-lat/0510113.

The Search for Gravitational Waves

James Hough for the LIGO Scientific Collaboration

SUPA, Institute for Gravitational Research.
University of Glasgow.
Glasgow G12 8QQ, UK

Abstract. An extended data taking run involving the long baseline gravitational wave detectors – LIGO, Virgo and GEO 600 – is close to an end and results of astrophysical significance are being produced. In this short review the development of the detectors and plans for the future will be discussed.

Keywords: Laser interferometry, neutron stars, black holes.
PACS: 03.67.-a; 04.80.Nn; 07.60.Ly; 42.50.Dv; 95.55.Ym

INTRODUCTION

Ever since their prediction by Einstein in 1917, gravitational waves have been the subject of some controversy as to whether they can be directly detected. However with the commissioning, operation and potential upgrading of the long baseline detectors LIGO (USA), GEO 600 (Germany/UK), Virgo (Italy, France) and TAMA 300 (Japan) we are heading into an era where this controversy will be resolved. A full discussion may be found in [1].

GRAVITATIONAL WAVES

Gravitational wave signals are expected over a wide range of frequencies; from 10^{-17} Hz in the case of ripples in the cosmological background, through signals in the audio band from the formation of neutron stars in supernova explosions, to 10^{10} Hz from the cosmological background itself. The most predictable sources are binary star systems. However there are many sources of much greater astrophysical interest associated with black hole interactions and coalescences, neutron star coalescences, low-mass X-ray binaries such as Sco-X1, stellar collapses to neutron stars and black holes (supernova explosions), rotating asymmetric neutron stars such as pulsars, and the physics of the early Universe. The signals from all these sources are at a level where detectors of very high strain sensitivity – of the order of 10^{-22} to 10^{-23} over relevant timescales – will be eventually required to allow a full range of observations and such detectors may be on ground or in space. From an experimentalist's point of view gravitational waves are best thought of as fluctuating strains in space-time.

CP957, *Particles, Strings, and Cosmology, 13th International Symposium*
edited by A. Rajantie, C. Contaldi, P. Dauncey, and H. Stoica

LONG BASELINE INTERFEROMETRIC DETECTORS

There are currently in operation, or in final stages of commissioning, a network of gravitational wave detectors each of which is based on using laser interferometric sensing to detect the changes, induced by gravitational waves, in the relative arm lengths of large-scale variants of Michelson-type interferometers, where the arms are formed between freely-hung mirrors.

The American LIGO project [2] comprises two detector systems with arms of 4 km length, one in Hanford, Washington State, and one in Livingston, Louisiana. One half length, 2 km, interferometer has also been built inside the same evacuated enclosure at Hanford. Construction of LIGO began in 1996 and progress has been outstanding with operation at instrumental design sensitivity now having been achieved as is indicated in Fig.1.

The French/Italian Virgo detector [3] of 3 km arm length at Cascina near Pisa, as shown in Fig.2, is close to completing commissioning and the Japanese TAMA 300 detector [4], which has arms of length 300 m, is operating at the Tokyo Astronomical Observatory.

All the systems mentioned above are designed to use resonant cavities in the arms of the detectors and use standard wire sling techniques for suspending the test masses.

However the German/British detector, GEO 600 [5], is somewhat different. It makes use of a four pass delay line system with an advanced optical signal enhancement technique known as signal recycling and utilises very low loss fused silica suspensions for the test masses.

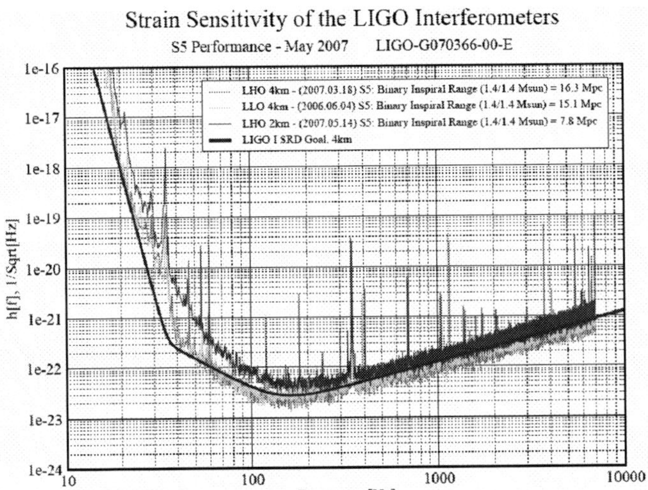

FIGURE 1. Amplitude spectral density of the noise from the LIGO detectors compared with the design goal (Courtesy of LIGO).

FIGURE 2. Bird's eye view a) of the LIGO detector site at Hanford (courtesy of the LIGO project) and b) of the Virgo detector site at Cascina in Italy (courtesy of the Virgo project).

GEO is now fully built and its sensitivity is being continuously improved. Currently it is within a factor of 3 of design sensitivity over a significant fraction of its frequency range.

Four science runs have so far been completed with these new interferometric detectors. All have involved the LIGO detectors, three have involved the GEO detector and two had TAMA involvement. The TAMA collaboration has completed 9 runs, 2 jointly with LIGO.

New 'upper limits' have been set on the strength of gravitational waves from a range of sources: coalescing compact binaries, pulsars, burst sources and a stochastic background of gravitational waves [6]. The fifth science run of the LIGO detectors started on 4th Nov 2005 with GEO having joined for data taking periods from January 2006 and Virgo from May 2007. This run is expected to terminate at 0000 UTC 1st October 2007 when the LIGO system will have one complete year of 'triple coincidence' data from all three of its detectors and will be the longest stretch of data taking with interferometric detectors to date.

During the next few years following intermediate upgrades to the LIGO and Virgo detectors we can expect to see searches for gravitational wave signals at a sensitivity level where a detection could be made. Recent discoveries of additional compact binary systems have improved the statistics for the expected rate of binary coalescences detectable by the LIGO system by a significant factor, giving some possibility of a first detection over the next few years [7]. An upper limit on the rate of BH-BH coalescences could be around 2 events per year.

However detection at the current level of sensitivity of the initial detectors is in no way guaranteed; thus improvement of the order of a factor of 10 to 15 in sensitivity of the current interferometric detectors is needed to reach sensitivity levels where many signals are expected.

Thus plans for an upgraded LIGO, 'Advanced LIGO', are already well formed. Advanced LIGO will incorporate 40 kg fused silica test mass mirrors, suspended by fused silica fibers or ribbons, along with an improved seismic isolation system, and increased laser power, close to 200 W. The upgrade is expected to commence in 2010, with full installation and initial operation of the upgraded system by 2014. On approximately the same timescale we can expect to see a similar upgrade to VIRGO, the rebuilding of GEO as a detector aiming at high sensitivity in the kHz frequency region and the building of a long-baseline underground detector, LCGT [8], in Japan. To go beyond this point, however, a number of challenges involving mechanical losses in coatings, thermal loading effects, and the use of non-classical light to bypass the standard quantum limit will have to be met. Cryogenic test mirrors and non-transmissive optics [9] are likely to be adopted, using materials of high thermal conductivity such as silicon.

Thus research groups in the field are already looking towards the next generation of ground-based detectors and a European design study for such a system funded under the EC Framework 7 programme is expected to commence early in 2008. Meanwhile there is a high level of activity in research towards the space-borne detector LISA, the Laser Interferometer Space Antenna [10] illustrated in Fig.3.

LONG BASELINE DETECTORS IN SPACE

Perhaps the most interesting gravitational wave signals (those resulting from the formation and coalescence of black holes in the range 10^3 to 10^6 solar masses) will lie in the region of 10^{-4} Hz to 10^{-1} Hz, and a detector whose strain sensitivity is approximately 10^{-23} over relevant timescales is required to search for these. The most promising way of looking for such signals is to fly a laser interferometer in space. LISA is being proposed by an American/European team; it consists of an array of three drag-free spacecraft at the vertices of an equilateral triangle of length of side 5 x 10^6 km. This cluster is placed in an Earth-like orbit at a distance of 1 AU from the Sun, and 20 degrees behind the Earth. Proof masses inside the spacecraft (two in each spacecraft) form the end points of three separate but not independent interferometers. LISA is a project in NASA's Beyond Einstein Program and the advisory committee reviewing the program (BEPAC) has very recently recommended that LISA be the Flagship mission for the program. They have recommended that LISA be preceded by the Joint Dark Energy Mission (JDEM) and thus LISA is expected to be launched as a joint NASA/ESA mission in 2018 and to be producing data for up to ten years. A demonstrator mission, LISA Pathfinder, is in phase C/D of development and hardware is being built for a launch in late 2010.

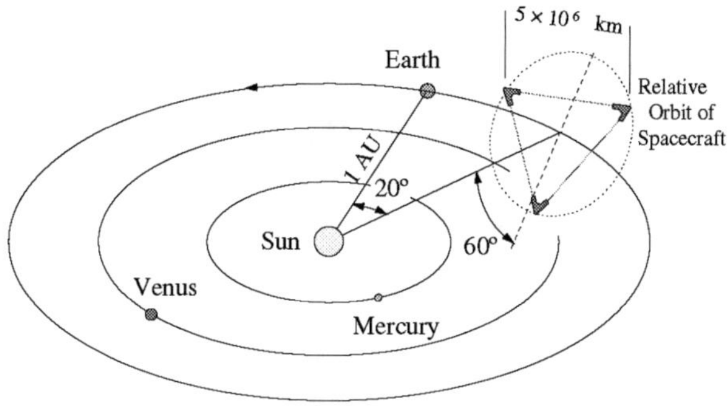

FIGURE 3. Schematic diagram of LISA and its orbit around the sun

ACKNOWLEDGMENTS

JH gratefully acknowledges the support of the United States National Science Foundation for the construction and operation of theLIGO Laboratory and the Science and Technology Facilities Council of the United Kingdom, the Max-Planck-Society and the State of Niedersachsen/Germany for support of the construction and operation of the GEO600 detector. The authors also gratefully acknowledge the support of the research by these agencies and by the Australian Research Council, the Council of Scientific and Industrial Research of India, the Istituto Nazionale di Fisica Nucleare of Italy, the Spanish Ministerio de Educacion y Ciencia, the Conselleria d'Economia Hisenda i Innovacio of the Govern de les Illes Balears, the Scottish Funding Council, the Scottish Universities Physics Alliance, The National Aeronautics and Space Administration, the Carnegie Trust, the Leverhulme Trust, the David and Lucile Packard Foundation, the Research Corporation, and the Alfred P. Sloan Foundation.
J.H. is also grateful to colleagues in Virgo, and the LISA International Science Team for help, and to the University of Glasgow for support.

REFERENCES

1. J. Hough, S. Rowan and B.S. Sathyaprakash, J. Phys. B: At. Mol. Opt. Phys. 38 S497-S519 2005
2. http://www.ligo.caltech.edu/

3. http://wwwcascina.virgo.infn.it/
4. http://tamago.mtk.nao.ac.jp/
5. http://www.geo600.uni-hannover.de/
6. R.E. Frey *for the LIGO Scientific Collaboration* LIGO: Status and Recent Results LIGO-P070079-01-Z available at: http://www.ligo.caltech.edu/docs/P/P070079-01.pdf
7. V. Kalogera et al, Astrophys. J. 601 L179 2004
8. http://www.icrr.u-tokyo.ac.jp/gr/gre.html
9. R.L. Byer in *Gravitational Astronomy: Instrument Design and Astrophysical Prospects* Eds. McClelland D E and Bachor H-A (World Scientific, Singapore) 1990
10. http://lisa.jpl.nasa.gov/

Metastable Supersymmetry Breaking in a Cooling Universe[1]

Vadim S. Kaplunovsky

Physics Theory Group, University of Texas
1 University Station, C1608, Austin, TX 78712, USA

Abstract. I put metastable supersymmetry breaking in a cosmological context. I argue that under reasonable assumptions, the cooling down early Universe favors metastable SUSY-breaking vacua over the stable supersymmetric vacua. To illustrate the general argument, I analyze the early-Universe history of the Intriligator–Seiberg–Shih model.

Keywords: supersymmetry breaking, cosmology
PACS: 12.60.Jv

Let me start with a historical note. The metastable supersymmetry breaking (MSB) is actually quite old — Michael Dine and Willy Fischler [1] constructed interesting models with both SUSY and non-SUSY vacua back in 1981. But later, when people searched for SUSY breaking driven by strong interactions (in a UV-free but IR-strong hidden sector) but didn't have techniques for analyzing effective potential in strongly interacting theories, they focused on models where SUSY *had to break* because there were no SUSY vacua at all, and no runaway directions [2]. Although many new techniques for analyzing IR-strong gauge theories emerged in mid-nineties, the search for SUSY breaking remained focused on *true vacua* (lowest-energy states) without SUSY. It took the Intriligator–Seiberg–Shih paper [3] to bring the MSB back into limelight.

Following Intriligator, Seiberg and Shih, there was a flood [8]–[24][2] of metastable SUSY–breaking (MSB) models, many of them string based. Such models have multiple vacua, some supersymmetric and some SUSY-breaking; sometimes there also supersymmetric runaway directions. The physically-interesting non-SUSY vacua are metastable but very long lived. Given infinite time, they would eventually tunnel to a SUSY vacuum or a runaway state, but this takes much longer then the present age of the Universe. So if a model somehow ended in the metastable state soon after the Big Bang, it would stay there until today and long afterwards.

Naturally, this raises **the Big Question:** *Given an MSB model with multiple vacua, which vacuum would be selected by the cosmological history of the Early Universe?* In this talk I am presenting **Our Answer:** *Under reasonable assumptions, the Early Universe favors the metastable SUSY–breaking vacua.*

---

[1] Plenary talk at Pascos–07 conference (Imperial College (London), July 2007), based on paper [4] by W. Fischler, myself, C. Krishnan, L. Manelli, and M. Torres, and on subsequent (not yet published) work by M. Torres and myself.

[2] As of this writing, the arXiv has over 130 papers on the subject, but I cannot quote them all because of space limitations. The papers [8]–[24] are just a small sample of this flood.

CP957, *Particles, Strings, and Cosmology, 13th International Symposium*
edited by A. Rajantie, C. Contaldi, P. Dauncey, and H. Stoica

Let me start by summarizing the common features of MSB models which can be used as SUSY-breaking hidden sectors of phenomenologically viable theories.

- For phenomenological reasons, the scale of SUSY breaking should be either 10^5–10^6 GeV (for the direct gauge mediation of SUSY breaking to the Standard Model), or 10^{10}–10^{11} GeV (for the indirect gauge mediation, or for the SUGRA+Kähler mediation). In any case, the SUSY breaking itself (as opposed to its mediation) does not depend on SUGRA effects and can be approximated by the rigid SUSY.
- The model must have an approximate $U(1)_{\mathrm{R}}$ symmetry to facilitate the spontaneous SUSY breakdown. I am not sure if this R-symmetry is quite as necessary as Seiberg *et al* claim [7], but it certainly helps, and thus far all known MSB models do have an approximate R–symmetry.
- In order to give masses to the Standard Model's gauginos, the R-symmetry must be broken. Usually, a small *explicit* breaking of the R-symmetry is amplified via spontaneous breaking. Alternatively, a small *explicit* breaking of the R-symmetry in the SUSY-breaking hidden sector is amplified in the mediator sector. But a purely spontaneous R-symmetry breaking would be bad because of exactly-massless Goldstone bosons.
- Explicit breaking of the R-symmetry leads to additional vacua with unbroken SUSY. For *small* R-symmetry breaking, those SUSY vacua are far away (in field space) from the non-SUSY vacua. That is, the scale σ of VEVs and masses in the SUSY vacuum is much bigger then the scale μ in the non-SUSY vacuum,

$$\frac{\sigma}{\mu} \sim \Big(\text{R-symmetry breaking}\Big)^{\text{some negative power}} \gg 1. \tag{1}$$

- The non-SUSY vacuum is metastable because it has higher energy density then the SUSY vacua. *But for $\sigma \gg \mu$, its lifetime is very long.* Indeed, the potential barrier between the SUSY and non-SUSY vacua is very wide, $\Delta\Phi = O(\sigma)$, while the potential difference is only $\Delta V = O(\mu^4)$. The tunneling action of a Euclidean bubble of the true vacuum inside the false vacuum is

$$S \sim \frac{V^2(\Delta\Phi)^4}{(\Delta V)^3} \gtrsim \frac{(\Delta\Phi)^4}{\Delta V} \sim \left(\frac{\sigma}{\mu}\right)^4 \gg 1. \tag{2}$$

  Thus, *for $\sigma \gtrsim 10\,\mu$, the metastable SUSY-breaking vacuum would easily survive until the present age of the Universe.*

To place an MSB model in a broader context, I make the following **assumptions**:

⋆ The SUSY-breaking hidden sector has nothing to do with inflating the Universe. The Inflation happens due to dynamics of a completely separate sector of the overall theory.

⋆ The overall theory has yet another sector, which cancels the cosmological constant due to SUSY breaking in the metastable vacuum.

⋆ After the Inflation, the reheating temperature is high enough for the high-temperature phase of the SUSY-breaking sector,

$$T_{\mathrm{reheat}} > O(\sigma) \gg \mu. \tag{3}$$

I claim that under these assumptions, the cosmological evolution of the MSB sector during the early Universe tends to end up in the metastable non-SUSY vacuum state. Here is the **basic argument**:

After the Inflation is over, the Hubble expansion of the Universe is slow ($H \ll T$) and the temperature decreases slowly enough for the quasistatic approximation: At any given time, the fields and particles are in thermal equilibrium for the appropriate temperature, and the free energy is minimized. Or rather, the free energy density $\mathscr{F}$ is always in a *local* minimum.

Multiple local minima of $\mathscr{F}$ correspond to multiple phases: one stable (the global minimum), the others metastable. Transitions between the phases require tunneling or thermal activation, and can be very slow. If they take longer then the Hubble time, they never happen, and the SUSY-breaking sector stays in a metastable phase.

The non-SUSY phase has higher potential energy then the "SUSY" phase[3] but also higher entropy (because it has lighter particles, $\mu \ll \sigma$). At higher temperatures, the entropy wins over the potential energy, which favors the non-SUSY phase. And at very high temperatures ($T > O(\sigma)$) the SUSY phase disappears altogether, because the slope of the entropy function overwhelms the minimum of the scalar potential.

I assume the Universe reheats to $T \gg \sigma$ and then slowly cools down. At first, the MSB sector has only the non-SUSY phase. As the temperature drops below $O(\sigma)$, other phases develop, but the non-SUSY phase has lowest free energy, and the sector remains in that phase.

Much later, for $T = O(\mu)$, the scalar potential wins over entropy, and the non-SUSY phase becomes metastable, while the SUSY phase becomes thermodynamically stable. But the first-order transition from the metastable to the stable phase requires either tunneling or thermal activation of a bubble, and both processes are very slow for $\sigma \gg \mu$:

$$\Gamma_{\text{tunneling}} \sim \exp(-S[\text{Eucl. 4D bubble}]) \ll \exp\left(-O(1) \times \left(\frac{\sigma}{\mu}\right)^4\right), \qquad (4)$$

$$\Gamma^{\text{thermal}}_{\text{activation}} \sim \exp\left(-\frac{E[\text{3D bubble}]}{T}\right) \ll \exp\left(-O(1) \times \left(\frac{\sigma}{\mu}\right)^3\right). \qquad (5)$$

Thus today, 13.5 gigayears since the temperature crossed the transition point, the theory remains in the metastable SUSY-breaking phase, and will stay there for many more gigayears.

To illustrate this general argument with a specific example, let us consider the **Intriligator–Seiberg–Shih model** [3]. In that model, the **UV** theory is simply SQCD with massive but light quarks, $m_q \ll \Lambda$, and $N_c < N_f < \frac{3}{2}N_c$. The **IR** theory at energies below Λ follows by Seiberg duality: it's SQCD with $N = N_f - N_c$ colors, $N_f > 3N$ flavors, and N_f^2 extra gauge singlets $\Phi_{ff'}$. The singlets originate from the mesons of the UV theory, $\Phi_{ff'} = \Lambda^{-1}\langle\tilde{q}_f q_{f'}\rangle$; their flavor quantum numbers are $\mathbf{Adj} + \mathbf{1}$. The IR quarks Q_f and antiquarks $\tilde{Q}_f$ and the IR gauge fields do not have clear UV origins. The

[3] By "SUSY" phase I mean the phase which for zero temperature reduces to the SUSY vacuum. At finite temperatures SUSY is broken, hence the quote marks.

superpotential is

$$W = h\,\mathrm{tr}(\Phi\tilde{Q}Q) - h\mu^2\,\mathrm{tr}(\Phi) + hNC\Big(\det(\Phi)\Big)^{1/N} \tag{6}$$

where $\mu^2 \simeq \Lambda m_q \ll \Lambda^2$, and $C \simeq \Lambda^{3-(N_f/N)}$. The Kähler function is approximately canonical (modulo perturbative renormalization), because the theory is IR free, $\beta_h, \beta_g > 0$.

Without the non-renormalizable third term, there is exact $U(1)_{\mathrm{R}}$ symmetry, and SUSY has to break:

$$F_\Phi \propto \tilde{Q}Q - \mu^2 \times \mathbf{1}_{N_f\times N_f} \neq 0 \quad \text{because } \mathrm{rank}(\tilde{Q}Q) \leq N < N_f. \tag{7}$$

In the non-SUSY vacuum,

$$\langle\Phi\rangle = 0, \qquad \langle Q\rangle = \langle\tilde{Q}\rangle^\top = \mu \times \begin{pmatrix} \mathbf{1}_{N\times N} \\ \mathbf{0}_{N\times(N_f-N)} \end{pmatrix}. \tag{8}$$

The determinant term in W breaks the R-symmetry, and leads to an additional SUSY vacuum (or rather $N_f - N = N_c$ vacua) with

$$\langle Q\rangle = \langle\tilde{Q}\rangle = 0, \qquad \langle\Phi\rangle = \sigma \times \mathbf{1}_{N_f\times N_f}, \tag{9}$$

$$\sigma = (\mu^2/C)^{N/(N_f-N)} \simeq \left(\mu^{2N}\Lambda^{N_f-3N}\right)^{1/(N_f-N)} \gg \mu. \tag{10}$$

To analyze and depict various phase transitions in this model, I am restricting its fieldscape to a two-parameter ansatz:

$$\Phi = \varphi \times \mathbf{1}_{N_f\times N_f}, \qquad Q = \tilde{Q}^\top = q \times \begin{pmatrix} \mathbf{1}_{N\times N} \\ \mathbf{0}_{N\times(N_f-N)} \end{pmatrix}, \tag{11}$$

and real φ and q (for real σ and μ). In this ansatz, the tunneling from the non-SUSY to the SUSY vacuum happens along the following path:

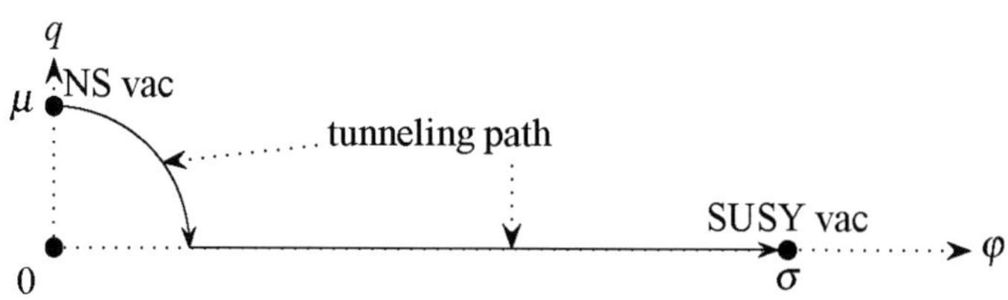

At zero temperature, the effective potential of the model is approximately

$$V(\varphi,q) \approx \frac{N_f h^2\mu^4}{Z_\Phi(\varphi)} \times \left(1 - (\varphi/\sigma)^{(N_f/N)-1}\right)^2 + Nh^2\left(q^4 - 2\mu^2q^2 + 2\varphi^2q^2\right) \tag{12}$$

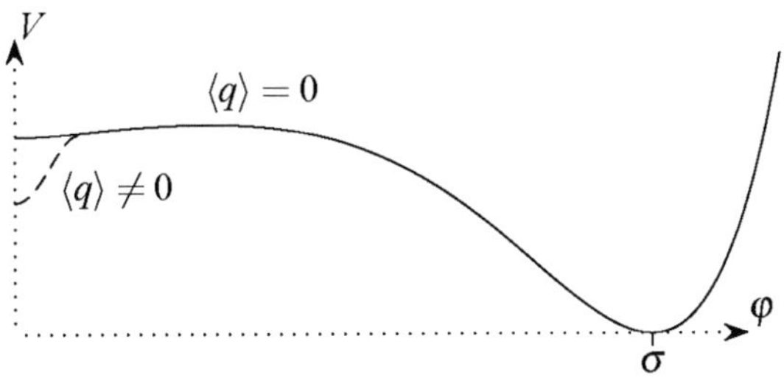

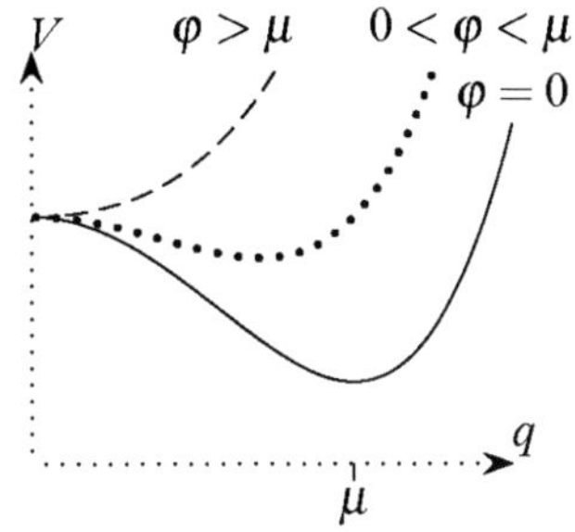

At finite temperatures, the effective potential — *i.e.*, the free energy density — comprises

$$\mathscr{F}(\varphi,q) = V(\varphi,q) + \mathscr{F}_T^{1\,\text{loop}}(\varphi,q) + \text{higher order corrections}, \tag{13}$$

$$\mathscr{F}_T^{1\,\text{loop}} = T \times \int \frac{d^3p}{8\pi^3}\,\text{Str}\left(1 - (-1)^F \exp\left(-\sqrt{p^2+M^2}/T\right)\right), \tag{14}$$

where the spectrum of M^2 depends on φ and q.

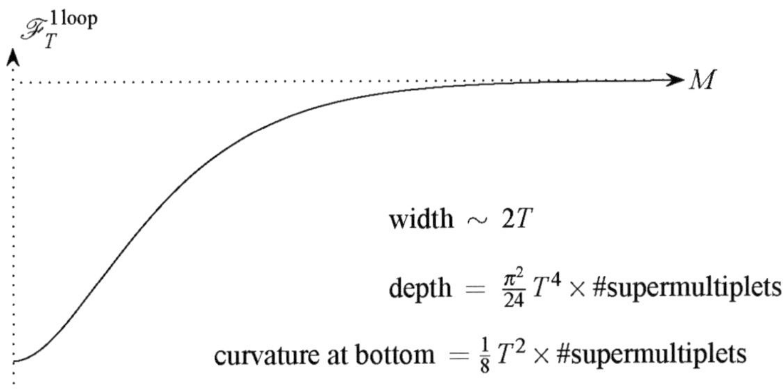

For high temperatures, $T \gg \sigma$, the thermal energy $\mathscr{F}_T^{1\,\text{loop}}$ completely overwhelms the scalar potential V. Consequently, the net free energy $\mathscr{F}(\varphi,q)$ has only one minimum at $\varphi = q = 0$, which means there is a unique high-temperature phase HT. Note that this phase is distinct from the non-SUSY phase at low temperatures because of different squark expectation values ($\langle q\rangle = 0$ in the HT phase versus $\langle q\rangle = \mu$ in the low-temperature NS phase).

For medium temperatures, $\mu \ll T \ll \sigma$, the thermal energy overwhelms the scalar potential for $q,\varphi \lesssim \mu \ll T$. But for $h\varphi \gg T$, the $\mathscr{F}_T^{1\,\text{loop}}(\varphi)$ flattens out (because all masses are either much larger or much smaller than T), so the minimum of V at $\varphi = \sigma$ becomes visible in the overall free energy:

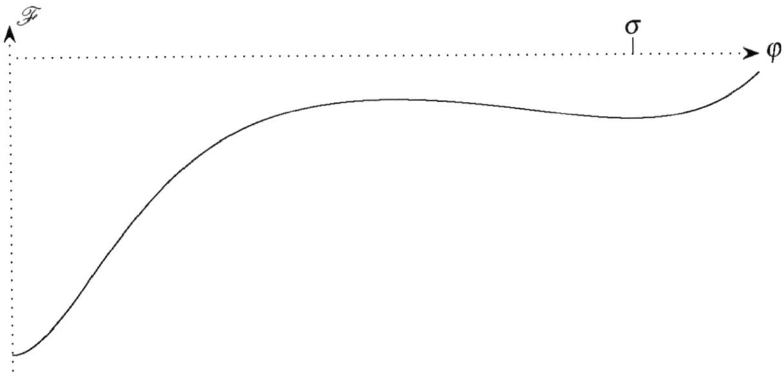

This gives us two phases: the stable HT phase with $\langle q\rangle = \langle\varphi\rangle = 0$, and the metastable NS (non-SUSY) phase with $\langle q\rangle = 0$ and $\langle\varphi\rangle \approx \sigma$. *In the cooling Universe, the system is in the HT phase before temperature drops below $O(\sigma)$, and afterwards it remains in the HT phase because it's stable.*

As the Universe cools down further, the energy difference between the HT and the SUSY phases becomes smaller, and eventually changes sign at the critical temperature

$$T_c^\phi \approx \frac{2\mu}{\sqrt{N}} \times \sqrt[4]{6\frac{Nh^2}{8\pi^2}} \tag{15}$$

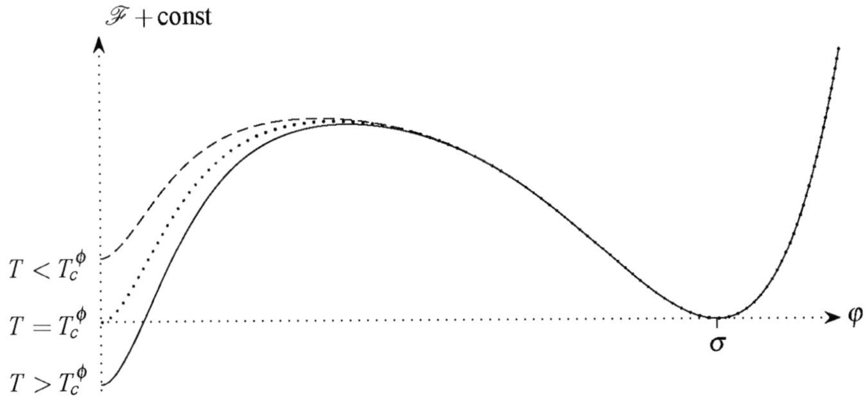

For temperatures below T_c^ϕ, the HT phase becomes metastable while the SUSY phase becomes thermodynamically stable. *Nevertheless, the model remain in the now-metastable HT phase because the first order transition between the two phases is extremely slow.*

At somewhat lower temperature

$$T_c^q \approx \frac{2\mu}{\sqrt{N_f + 2N}} \sim (0.4 \text{ to } 0.75) \times T_c^\phi \tag{16}$$

there is another phase transition in the squark direction:

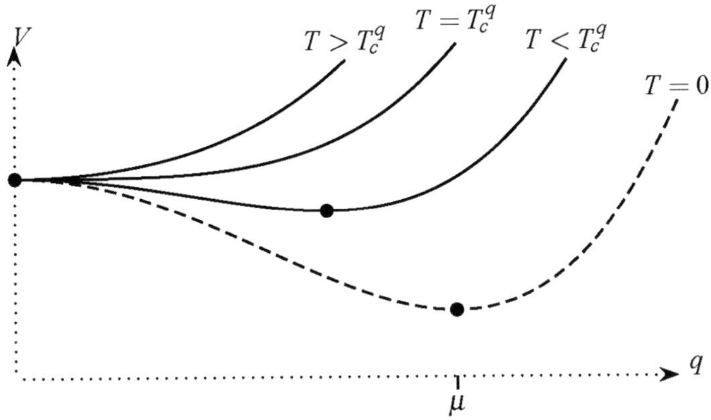

This transition is second order, and proceeds without delay. As soon as the Universe cools down to T_c^q, the HT phase with $\langle q \rangle = \langle \varphi \rangle = 0$ disappears, and the model enters the low-temperature non-SUSY phase NS with $\langle \varphi \rangle = 0$ but $\langle q \rangle \neq 0$.

Similar to the HT phase below T_c^ϕ, *the NS phase is metastable.* Given infinite time, it would eventually decay into the SUSY phase with $\langle q \rangle = 0$ and $\langle \varphi \rangle \approx \sigma$. But for $\sigma \gtrsim 20\mu$, the tunneling and the thermal activation are both very slow — *cf.* eqs. (4–5) — and the decay takes longer then the present age of the Universe.

Instead of decay, the model remains in the metastable NS phase. As the temperature drops, the squark VEV grows toward $\langle q \rangle = \mu$, and *the model cools down to the non-SUSY vacuum.*

Besides the Intriligator–Seiberg–Shih model, M. Torres and I have analyzed similar models with weakly gauged flavor symmetries (the whole $SU(N_f)_V$ or its subgroups). Such models have spontaneously broken R-symmetry at $T = 0$ and more complicated phase structures at $t > 0$. But of the end of the evolution, they too end up in metastable non-SUSY vacuum states.

To summarize our results, **Metastable SUSY breaking is OK**. In models with both non-SUSY and SUSY vacua where the latter have much larger VEVs and masses then the former, this little hierarchy not only keeps the metastable SUSY-breaking vacua very long lived, but also leads the cosmological evolution of the model toward those vacua. **But the devil is in details**:

- Above all, the model must work! And mediation of SUSY breaking to the SSM should also work.
- There should be no way *around* the potential barrier between the vacua. The pseudo-moduli directions are particularly dangerous.
- The phase diagram of the model should direct its thermal evolution toward the desired non-SUSY vacuum. In models with several distinct vacua, this could be quite a challenge.
- The mediators should not screw things up.
- *Etc., etc. ...*

Acknowledgements: Simultaneously with our paper [4] on which this talk is based, two other groups published independent papers [5, 6] with similar conclusions.
The research of our group was supported by the NSF under grant PHY–0455649.

REFERENCES

1. M. Dine and W. Fischler, *"A Phenomenological Model Of Particle Physics Based On Supersymmetry"*, Phys. Lett. B **110**, 227 (1982).
2. I. Affleck, M. Dine and N. Seiberg, *"Supersymmetry Breaking By Instantons"*, Phys. Rev. Lett. **51**, 1026 (1983).
   Also, five more papers by the same authors in 1983–84, and great many subsequent papers by others.
3. K. Intriligator, N. Seiberg and D. Shih, *"Dynamical SUSY breaking in meta-stable vacua"*, JHEP **0604**, 021 (2006) [arXiv:hep-th/0602239].
4. W. Fischler, V. Kaplunovsky, C. Krishnan, L. Mannelli and M. A. C. Torres, *"Meta-Stable Supersymmetry Breaking in a Cooling Universe"*, JHEP **0703** (2007) 107 [arXiv:hep-th/0611018].
5. N. J. Craig, P. J. Fox and J. G. Wacker, *"Reheating metastable O'Raifeartaigh models"*, Phys. Rev. D **75**, 085006 (2007) [arXiv:hep-th/0611006].
6. S. A. Abel, C. S. Chu, J. Jaeckel and V. V. Khoze, *"SUSY breaking by a metastable ground state: Why the early universe preferred the non-supersymmetric vacuum"*, JHEP **0701**, 089 (2007) [arXiv:hep-th/0610334].
7. See for example: A. E. Nelson and N. Seiberg, *"R symmetry breaking versus supersymmetry breaking"*, Nucl. Phys. B **416**, 46 (1994) [arXiv:hep-ph/9309299].
8. S. Franco and A. M. Uranga, *"Dynamical SUSY breaking at meta-stable minima from D-branes at obstructed geometries"*, JHEP **0606**, 031 (2006) [arXiv:hep-th/0604136].
9. I. Garcia-Etxebarria, F. Saad and A. M. Uranga, *"Local models of gauge mediated supersymmetry breaking in string theory"*, JHEP **0608**, 069 (2006) [arXiv:hep-th/0605166].
10. H. Ooguri and Y. Ookouchi, *"Landscape of supersymmetry breaking vacua in geometrically realized gauge theories"*, Nucl. Phys. B **755**, 239 (2006) [arXiv:hep-th/0606061].
11. M. Aganagic, C. Beem, J. Seo and C. Vafa, *"Geometrically induced metastability and holography"*, arXiv:hep-th/0610249.
12. M. Dine and J. Mason, *"Gauge mediation in metastable vacua"*, arXiv:hep-ph/0611312.
13. R. Kitano, H. Ooguri and Y. Ookouchi, Phys. Rev. D **75**, 045022 (2007) [arXiv:hep-ph/0612139].
14. H. Murayama and Y. Nomura, *"Gauge mediation simplified"*, Phys. Rev. Lett. **98**, 151803 (2007) [arXiv:hep-ph/0612186].
15. C. Csaki, Y. Shirman and J. Terning, *"A simple model of low-scale direct gauge mediation"*, JHEP **0705**, 099 (2007) [arXiv:hep-ph/0612241].
16. O. Aharony and N. Seiberg, *"Naturalized and simplified gauge mediation"*, JHEP **0702**, 054 (2007) [arXiv:hep-ph/0612308].
17. S. A. Abel and V. V. Khoze, *"Metastable SUSY breaking within the standard model"*, [arXiv:hep-ph/0701069].
18. S. A. Abel, J. Jaeckel and V. V. Khoze, *"Naturalised supersymmetric grand unification"*, [arXiv:hep-ph/0703086].
19. D. Shih, *"Spontaneous R-symmetry breaking in O'Raifeartaigh models"*, [arXiv:hep-th/0703196].
20. K. Intriligator, N. Seiberg and D. Shih, *"Supersymmetry Breaking, R-Symmetry Breaking and Metastable Vacua"*, JHEP **0707**, 017 (2007) [arXiv:hep-th/0703281].
21. H. Ooguri, Y. Ookouchi and C. S. Park, *"Metastable Vacua in Perturbed Seiberg-Witten Theories"*, [arXiv:0704.3613 (hep-th)].
22. K. van den Broek, *"Vscape V1.1.0: An interactive tool for metastable vacua"*, [arXiv:0705.2019 (hep-ph)].
23. E. Dudas, J. Mourad and F. Nitti, *"Metastable Vacua in Brane Worlds"*, [arXiv:0706.1269 (hep-th)].
24. H. Y. Cho and J. C. Park, *"Dynamical $U(1)_R$ Breaking in the Metastable Vacua"*, [arXiv:0707.0716 (hep-ph)].

High-Energy Cosmic Rays

Giorgio Matthiae
On behalf of the Pierre Auger Collaboration

Department of Physics of the University and Sezione INFN of Roma "Tor Vergata"

Abstract. After a brief description of the main experiments dedicated to the study of high-energy cosmic rays, the results of the Auger Observatory on anisotropy, mass composition and energy spectrum are presented and compared to those of the other experiments.

Keywords: Cosmic Rays, Astroparticle Physics.
PACS: 96.50.sb, 96.50.sd.

INTRODUCTION

The flux of cosmic rays, shown in Fig.1 as a function of energy, follows approximately a power law $E^{-\gamma}$ with spectral index γ roughly equal to 3.

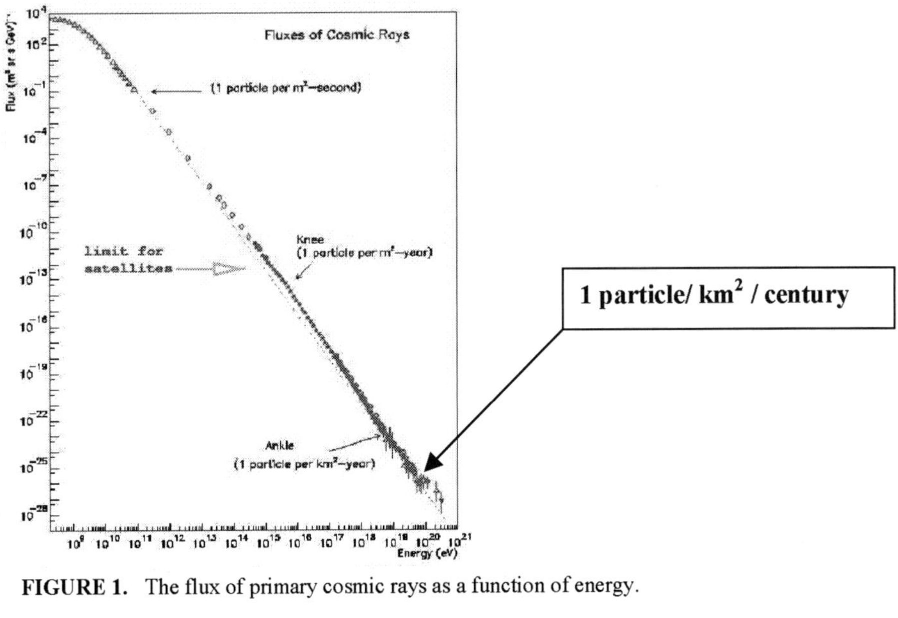

FIGURE 1. The flux of primary cosmic rays as a function of energy.

The spectrum exhibits interesting features, usually called the "knee" and the "ankle". At the energy of the "knee" (~ $3x10^{15}$ eV) the spectral index changes from approximately 2.7 to 3.1. The "ankle", a kind of undulation around a few 10^{18} eV,

CP957, *Particles, Strings, and Cosmology, 13th International Symposium*
edited by A. Rajantie, C. Contaldi, P. Dauncey, and H. Stoica

has been actively studied by recent experiments and will be discussed in the Section on the spectrum. In the region above 10^{19} eV the flux on earth is extremely low, of the order of 1 particle/ km^2/ century. Therefore the study of cosmic rays in this very high-energy region requires detectors with very large acceptance.
A comprehensive review of the experimental methods for studying very high-energy cosmic rays and of the earlier observations in this field can be found in ref.1.

HIGH-ENERGY COSMIC RAY EXPERIMENTS

In this Section recent experiments dedicated to the detection of cosmic rays in the very high-energy region, above 10^{18} eV, are briefly described.

AGASA. The Akeno Giant Air Shower Array [2], located in Japan at the latitude of about 35^0N and altitude of 900 m above sea level was in operation from 1990 until 2004. It was the first large surface array detector, designed to measure the front of the cosmic ray showers as they reach ground. The array consisted of 111 plastic scintillators with size of 2.2 m^2 deployed with separation of 1 km and covering an area of 100 km^2. The array was complemented by 27 muon detectors consisting of proportional counters placed below absorbers. The cumulative exposure was about 1600 km^2 sr yr.

HiRes. It is the new and sophisticated version of the pioneering experiment Fly's Eye of the Utah group based on the detection of the fluorescence light from the nitrogen molecules excited by the charged particles of the cosmic ray showers. It was in operation from 1997 until 2006. The HiRes experiment [3] consists of two sites (HiRes I and HiRes II) 12.6 km apart located at Dugway in Utah (USA) where fluorescence telescopes were installed (22 at HiRes I and 42 at HiRes II). The telescopes cover the full 360^0 azimuth and in elevation from 3^0 up to 17^0 (HiRes I) and from 3^0 up to 31^0 (HiRes II). The main components of each telescope are a spherical mirror of about 4 m^2 size and an array of 256 photomultipliers as sensitive element. UV filters to cut light outside the 300-400 nm interval of the nitrogen fluorescence were also used. The cumulative exposure was 4-5 times larger than AGASA.

Telescope Array (TA). It is being built in Utah. The basic configuration will combine a large scintillator ground array with HiRes type fluorescence telescopes.

AUGER. Two Observatories, one in the Northern and one in the Southern hemisphere are foreseen in the Auger project, to achieve a full exploration of the sky. The Southern Auger Observatory [4] is located near the small town of Malargüe in the province of Mendoza (Argentina) at the latitude of about 35^0 S and altitude of 1400 above see level. The region is flat, with very low population density and favorable atmospheric conditions. The Observatory is a hybrid system, a combination of a large surface array and a fluorescence detector.

The surface detector (SD) is a large array of 1600 water Cherenkov counters spaced at a distance of 1.5 km and covering a total area of 3000 km^2. Each counter is a plastic tank of cylindrical shape with size 10 m^2 x 1.2 m filled with purified water. Technical details of a tank are given in Fig.2. The surface detector measures the front of the shower as it reaches ground. The tanks activated by the event record the particle density and the time of arrival. The fluorescence detector (FD) consists of 24 telescopes located in four stations which are built on the top of small elevations on the perimeter of the site. The telescopes measure the shower development in the air by observing the fluorescence light. Each telescope has a 12 m^2 spherical mirror with curvature radius of 3.4 m and a camera with 440 photomultipliers. The field of view of each telescope is 30^0 x 30^0. UV filters were used as in HiRes.

The Southern Auger Observatory started to collect data in 2004. The exposure reached until present is about 5100 km^2 sr yr. The Observatory will be completed in 2007.

The Northern Auger Observatory which is now being designed will be located in Colorado (USA).

We notice that the present Auger Observatory is the only detector exploring the Southern hemisphere. The other experiments, either finished (AGASA, HiRes) or in preparation (Telescope Array) are in the Northern hemisphere. The Telescope Array will tend to work at energies lower than those of the other experiments.

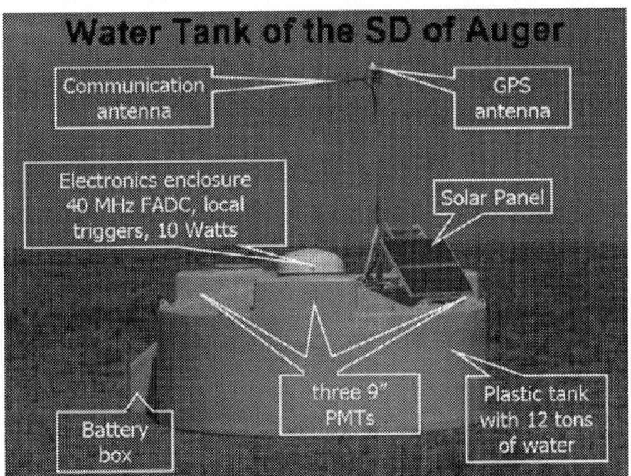

FIGURE 2. Picture of a water tank of the Surface Detector of the Auger Observatory. The insets give explanations for the various components of the system.

ANISOTROPY STUDIES

The identification of the sources of high-energy cosmic rays has been a great challenge. The region of the Galactic centre is an obvious and interesting place to look for localized excess of cosmic rays at EeV energies. Observation of an excess from the region of the Galactic centre at the level of 4.5 σ , in the energy region 1.0 – 2.5 EeV and with angular scale of 20^0 , was reported by AGASA [5].

The Auger Observatory is suitable for this study because the Galactic centre (constellation of Sagittarius), lies well in the field of view of the experiment. The angular resolution of the SD of Auger depends on the number of tanks activated by the shower and it is better than one degree at high-energy.

Some of the Auger results [6] on the observed and the expected number of events in the direction of the Galactic centre are shown in Table 1 and 2.

Table 1. The number of observed and of expected events and the corresponding ratios are listed for different angular windows in the direction of the Galactic centre (energy less than 1 EeV)

Angular window (degrees)	$N_{observed}$ / $N_{expected}$	Ratio (errors: stat, syst)
5	1667/1606	1.04 ± 0.04 ± 0.01
10	6427/6437	1.00 ± 0.02 ± 0.01
20	25350/25454	0.99 ± 0.01 ± 0.01

Table 2. The number of observed and of expected events and the corresponding ratios are listed for different angular windows in the direction of the Galactic centre (energy between 1 and 10 EeV)

Angular window (degrees)	$N_{observed}$ / $N_{expected}$	Ratio (errors: stat, syst)
5	425/393	1.08 ± 0.07 ± 0.01
10	1662/1578	1.05 ± 0.04 ± 0.01
20	6365/6252	1.02 ± 0.02 ± 0.01

Clearly in the Auger data there is no indication of a statistically significant excess in the direction of the Galactic centre.

THE MASS COMPOSITION

The most direct method for studying the mass composition is based on the measurement of the longitudinal profile of the showers. It is well known that for a given energy protons are more penetrating than light/medium nuclei which interact essentially as a collection of nucleons. The depth of the maximum of the observed shower profile X_{max}, as measured by the fluorescence telescopes, is well correlated with the particle mass. The principle of the method is indicated in Fig.3.

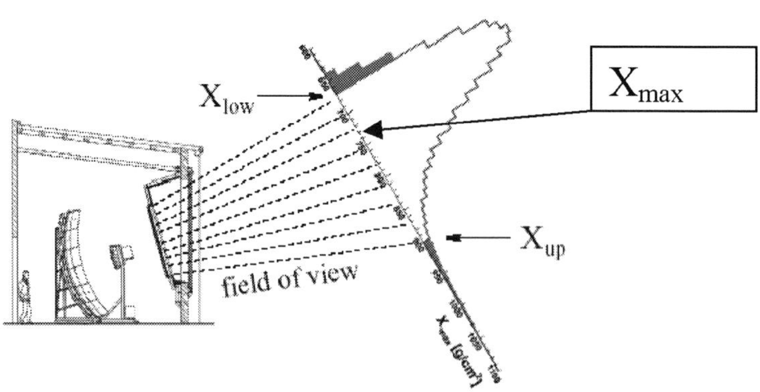

FIGURE 3. Sketch of the measurement of the longitudinal development of the shower and of the quantity X_{max} as measured by a fluorescence telescope of the Auger Observatory .

A compilation of earlier data on X_{max} for energies above 10^{14} eV is shown in Fig.4 where expectations from simulation programs are also given for Fe nuclei, protons and photons. The value of X_{max} for protons is about 100 g cm^{-2} larger than for iron.

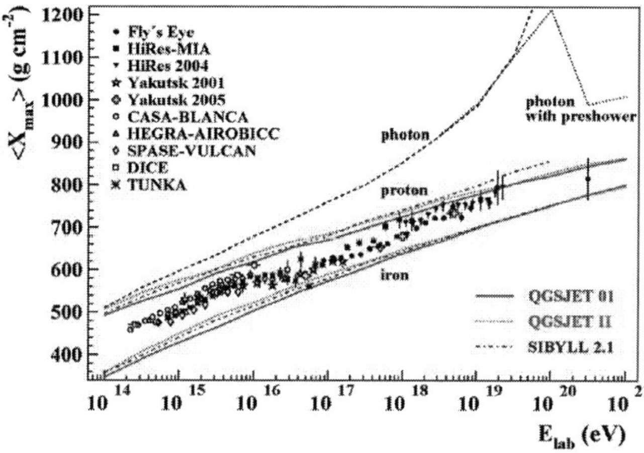

FIGURE 4. Compilation of earlier data on the quantity X_{max} as a function of energy. Prediction of various simulation programs for incident photons, protons and iron nuclei are also shown.

These earlier data indicate that above the knee the average mass becomes heavier approaching iron. However, above 10^{18} eV the trend of these data seems to suggest a change of regime towards proton dominance.

Recent data by Auger [7] are presented in Fig. 5 and compared to previous results from Fly's Eye and HiRes. The fluorescence detector of Auger measures X_{max} with systematic uncertainty less than 15 g cm^{-2}. A comparison of the Auger data to the predictions of various simulation programs is shown in Fig. 6. In spite of the still low statistics, the data indicate some change of regime at 2-3 EeV where the slope (elongation rate) changes. At the highest energies the trend is intermediate between protons and Fe nuclei with a mean mass number of about 5.

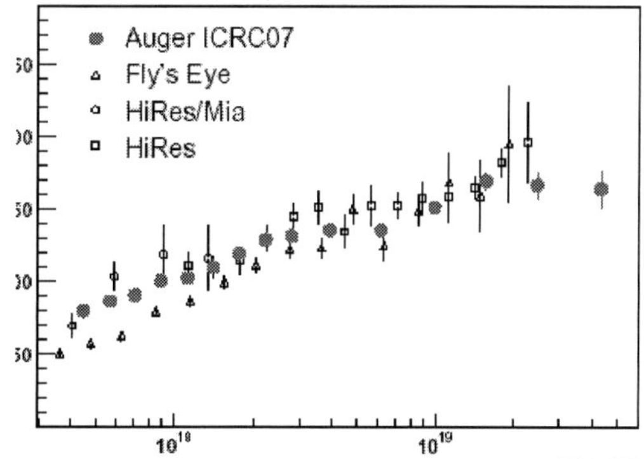

FIGURE 5. Recent Auger data on the quantity X_{max} are plotted as a function of energy and compared to the results of other experiments.

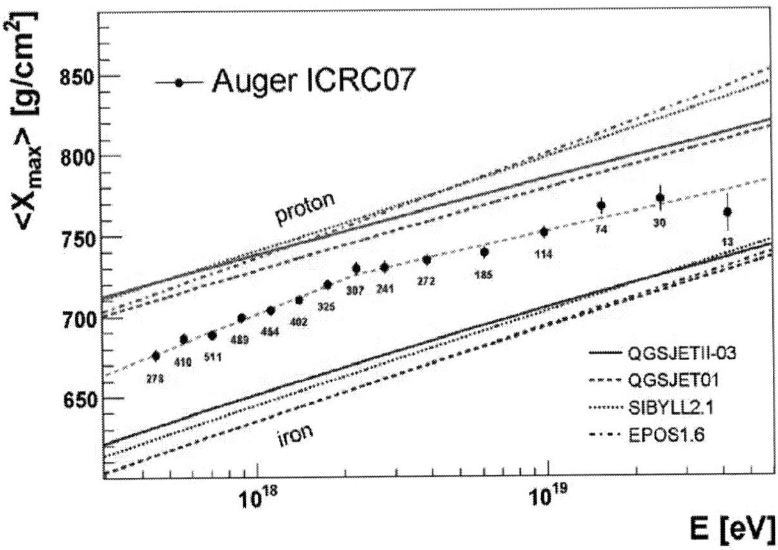

FIGURE 6. The Auger data on the quantity X_{max} are plotted as a function of energy and compared to predictions of simulation programs for protons and iron nuclei. The number of events for each data point is also shown. Quoted errors are statistical.

THE ENERGY SPECTRUM

An important feature of the spectrum in the energy region above 10^{19} eV is the mechanism suggested by Greisen, Zatsepin and Kuzmin, known as GZK cutoff and due to the interactions of the cosmic rays with the low energy photons of the Cosmic Microwave Background. Protons with energy above the threshold for photoproduction of pions ($\sim 4\text{x}10^{19}$ eV) will lose energy as they travel in space. The value of the energy where an integral power-law spectrum would be reduced to one half is $5.3\text{x}10^{19}$ eV [8]. The energy loss per interaction is about 15 – 20 %.
At $\sim 5\text{x}10^{19}$ eV most of the observed particles must have come from sources within 100 Mpc. Production of electron-positron pairs is also present but it is less effective than photo-pion production. However, this process is expected [8] to be responsible for a feature related to the so-called "ankle", a shallow minimum (or "dip") in the plot of the flux times E^3 which is centered at energies of a few 10^{18} eV.

In the past there was a controversy on the actual presence of the suppression due to the GZK cutoff. The AGASA data did not show a suppression, contrary to the preliminary data of HiRes. The experimental situation is now clarified by the final data of HiRes [9], shown in Fig. 7 and by the data of Auger. The HiRes data clearly show a steepening of the spectrum above $10^{19.6}$ eV with a fitted value of the spectral index $\gamma = 5.1 \pm 0.7$. The steepening agrees with expectations from GZK cutoff.

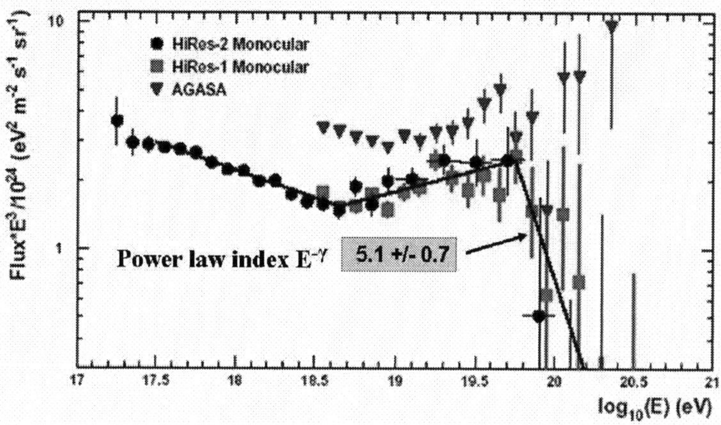

FIGURE 7. The final HiRes results on the energy spectrum are presented as Flux x E^3 and compared to the earlier AGASA data. The steepening due to the GZK cutoff is clearly seen. In addition the shallow minimum centered around $10^{18.6}$ eV is also evident.

The method used by Auger to measure the energy spectrum exploits the hybrid nature of the experiment with the aim of using the data itself rather than simulations. For each event the tanks of the SD measure the particle density expressed in units of VEM (Vertical Equivalent Muons) and the times of arrival which are used to determine the axis of the shower. The dependence of the particle density on the distance from the shower axis is fitted by a lateral distribution function (LDF). An example is shown in Fig.8. The LDF fit allows determining the particle density S(1000), expressed in units of VEM at the distance of 1000 m from the axis. This quantity is a good energy estimator [10] in the sense that it is strongly correlated with the energy of the cosmic ray and almost independent of the mass.

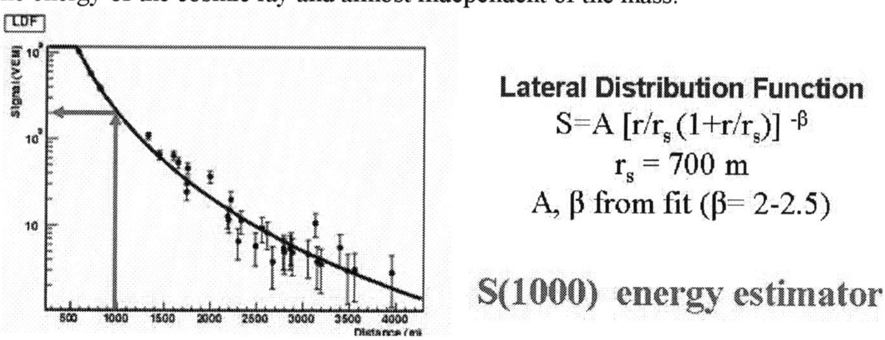

FIGURE 8. A high-energy event from Auger that has activated 34 tanks. The observed particle densities, expressed in units of VEM are shown in the left panel as a function of the distance from the axis of the shower together with the fit using a standard form for the Lateral Distribution Function.

The energy estimator S(1000) depends on the zenith angle because the effective atmosphere thickness seen by showers before reaching ground changes with the zenithangle. The value of S(1000) corresponding to the median zenith angle of 38^0 is

used as reference and the zenith angle dependence of the energy estimator is determined assuming that the arrival directions are isotropically distributed. This procedure is traditionally called "Constant intensity cut".
The absolute calibration of S(1000) is derived from the hybrid events using the calorimetric energy measured by the FD which is then corrected for the missing energy (neutrinos and muons) using the mean value between proton and iron (uncertainty about 4% at 10^{19} eV). This absolute calibration, which defines the energy scale, is at present affected by a systematic error of about $\pm$ 20%, mainly due to uncertainties on the fluorescence yield and on the calibration of the FD telescopes.

The energy calibration, obtained from the subset of hybrid events (see Fig.9) is then used for the full set of events with higher statistics as measured by the SD.

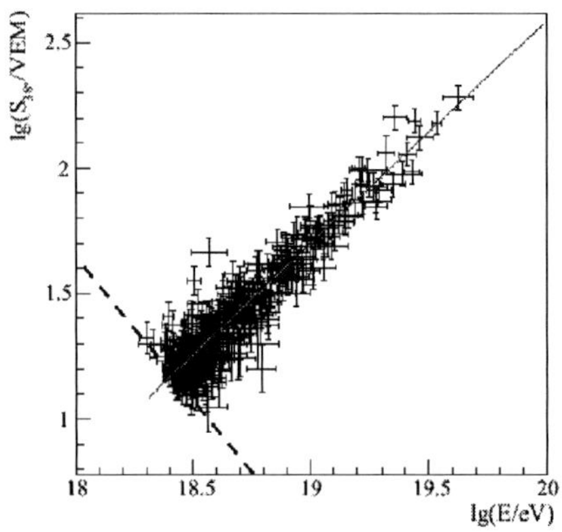

387 hybrid events are used to establish the absolute calibration of the energy estimator S as shown in the left panel.

The fitted line corresponds to $E_{FD} = K \times S^B$ with B= 1.10 $\pm$ 0.02

FIGURE 9. Calibration of the energy estimator S(1000) using the energy from the FD.

Three different measurements of the energy spectra were obtained by the Auger Collaboration as explained in detail in ref.11. The SD "vertical" spectrum is for zenith angle $\theta < 60^0$, the SD "inclined" is for $60^0 < \theta < 80^0$. The SD spectra starts at the energy of $3\text{x}10^{18}$ eV where the efficiency goes to a plateau. The hybrid spectrum refers to FD events with at least one SD tank and starts at a lower energy (10^{18} eV). The three spectra are consistent within statistics, as shown in Fig.10 and therefore were combined in a single spectrum which is presented in Fig.11.

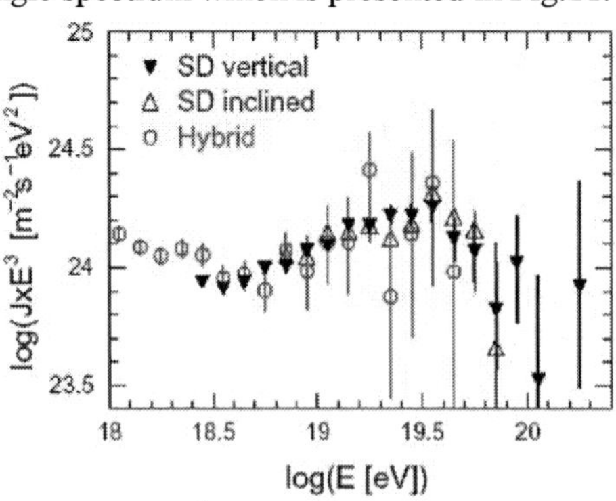

FIGURE 10. The Auger spectra (flux x E^3) from the SD and from hybrid events.

The "ankle", a barely noticeable undulation in Fig.11, appears as a clear shallow minimum in Fig.10 with shape and position similar to the predictions of ref.8. Above $\sim 10^{19.6}$ eV, the steepening expected from the GZK mechanism is also evident.

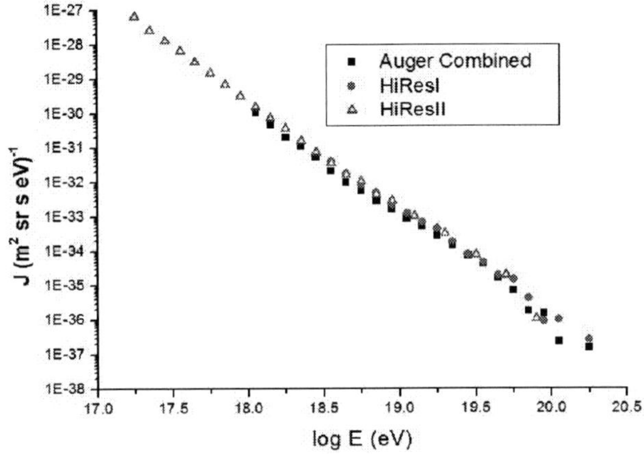

FIGURE 11. The Auger energy spectrum is shown together with the HiRes results.

The structures present in the Auger spectrum are better analysed by taking the relative difference of the data with respect to the reference form $J_s = A\,E^{-2.6}$. The result is presented in Fig.12. Numerical values of the spectral index γ are given in Table 3.

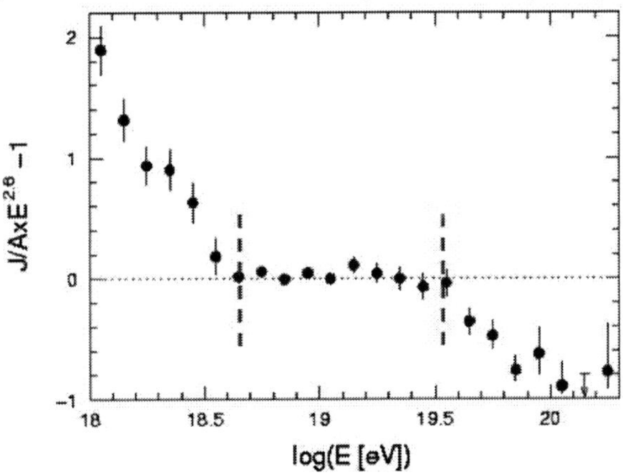

FIGURE 12. The relative difference of the Auger energy spectrum with respect to the form $E^{-2.6}$. The vertical dashed lines separate the energy intervals where power law fits are made.

Table 3. Numerical values of the spectral index γ of the power law fits in the different energy intervals. The energy values E_{ankle} and E_{GZK} correspond to the position of the breaks.

	Auger	**HiRes**
E_{ankle} **(eV)**	4.5×10^{18}	4.5×10^{18}
E_{GZK} **(eV)**	3.5×10^{19}	5.6×10^{19}
γ $(E < E_{ankle})$	3.30 ± 0.06	
γ **for** $(E_{ankle} < E < E_{GZK})$	2.62 ± 0.03	2.81 ± 0.03
γ **for** $(E > E_{GZK})$	4.1 ± 0.4	5.1 ± 0.7

CONCLUSIONS

- The non observation of an excess from the Galactic centre might not be surprising because the Galactic magnetic fields may destroy the directional correlation.
- The direct measurement of the mass composition with X_{max} require substantial statistical improvement to become really meaningful.
- The steepening of the energy spectrum observed by Auger and HiRes is evidence for the presence of the GZK mechanism and therefore for extra-galactic origin of the particles with energy above a few 10^{19} eV.
- Accurate studies of the interplay between mass composition, shape of the "ankle" and shape of the GZK steepening are of great interest and should provide relevant information on the production of the cosmic rays of very high energy, in particular on the local densities of the extragalactic sources.

REFERENCES

1. M. Nagano and A.A.Watson, *Rev. Mod. Phys.* **72,** 689 (2000)
2. H. Ohoka et al., *Nucl. Instr. and Methods*, **A385**, 268 (1997)
3. J.H.Boyer et al., *Nucl. Instr. And Methods*, **A482**, 457 (2002)
4. J.Abraham et al., *Nucl. Instr. and Methods* **A523,** 50 (2004)
5. M.Teshima et al., *Proc. ICRC2001* **1**, 337 (2001)
6 J.Abraham et al, *Astropart. Phys.* **27**, 244 (2007) and E.M.Santos, *Proc. ICRC2007*
7. M.Unger (Auger Collaboration), *Proc ICRC2007*
8. V.Berezinsky et al., *Phys. Rev.* **D74**, 043005 (2006) and refs therein
9. R.U.Abbasi et al., *astro-ph*/0703099v1 (2007) and refs therein
10. D.Newton et al., *Astropart. Phys.* **26**, 414 (2007).
11. Contributions of P.Facal San Luis, L.Perrone, M.Roth and T.Yamamoto (Auger Collaboration) to *Proc. ICRC2007*

Heavy Ion Collisions and New Forms of Matter

Larry McLerran

Riken Brookhaven Center and Physics Department, Brookhaven National Laboratory, Upton, NY 11973 USA

Abstract. I discuss forms of high energy density matter in QCD. These include the Color Glass Condensate, the Glasma and the Quark Gluon Plasma. These all might be studied in ultra-relativistic heavy ion collisions, and the Color Glass Condensate might also be probed in electron-hadron collisions. I present the properties of such matter, and some aspects of what is known of their properties.

Keywords: High Energy Density Matter, Heavy Ion Collisions
PACS: 21.65.Qr, 12.38.-t

INTRODUCTION

There is a generic similarity between the physical conditions in the early stages of heavy ion collisions, and those of the early universe. This is illustrated in Fig. 1 . The Color Glass Condensate describes the initial quantum mechanical wavefunctions which initiate the collision. This is similar to the initial wavefunction in cosmology. At the moment of the collision there is the initial singularity, analogous to that of the quantum gravity and inflationary stages of cosmology, where quantum fluctuations are important. These fluctuations eventually evolve into the large scale fluctuations in the matter distribution produced in such collisions, analogous to the fluctuations produced in inflationary cosmology which form the seeds for matter fluctuations which eventually develop galaxies and clusters of galaxies. After the singularity there is a Glasma phase which has topological excitations analogous to those associated with baryon number violation in electroweak theory. The Quark Gluon Plasma is analogous to both the electroweak and QCD thermal phases of expansion in cosmology, and the deconfinement transition is analogous to that which generates masses for electroweak bosons. In cosmology there are a variety of phase transitions which occur at various times, corresponding to the confinement-deconfinement transition of QCD.

There is a new paradigm: Various forms of matter control the high energy limit of QCD. These forms of matter have intrinsically interesting properties, as well as properties which once understood have extensions to other areas of physics such as cosmology.

It is the purpose of this talk to describe for you the properties of these high energy density forms of matter, how they appear in high energy physics phenomenology, and what is known experimentally about these forms of matter. For reasons of space, these topics are covered only in the most generic terms.

CP957, *Particles, Strings, and Cosmology, 13th International Symposium*
edited by A. Rajantie, C. Contaldi, P. Dauncey, and H. Stoica

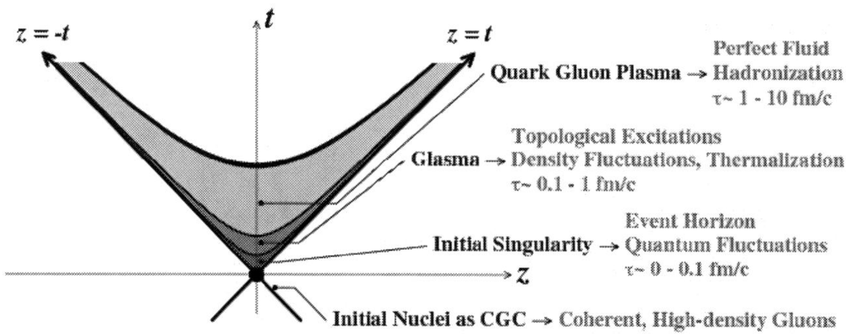

FIGURE 1. A schematic picture of the evolution of matter produced in the heavy ion collisions.

THE INITIAL WAVEFUNCTION

In the leftmost part of Fig. 2, I show the various Fock space components for a baryon wavefunction. At low energies, the dominant states for physical processes have three quarks and a few gluons. In high energy collisions, many particles are produced. These ultimately arise from components of the hadron wavefunction which have many gluons in them. These components make a gluon wall of longitudinal extent of $1/\Lambda_{QCD}$. The density of gluons in the transverse plane grows as the energy increases, as is also shown Fig. 2

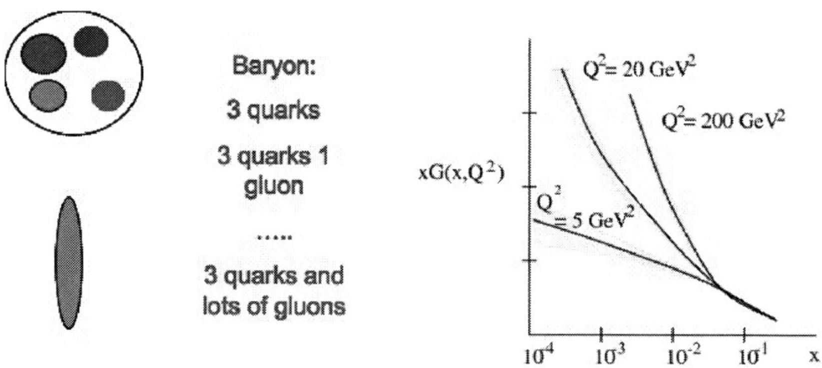

FIGURE 2. The leftmost figure on this slide illustrates the wavefunction for a gluon as a function of energy. The rightmost figure illustrates the experimentally measured gluon distribution functions as a function of x.

These gluons must become very tightly packed together. They form a high density, highly coherent condensate of gluons, the Color Glass Condensate (CGC).[1]-[3] Because the typical separation of gluons is small, at high enough energy $\alpha_S << 1$, and the system is weakly coupled. Due to the coherence, it is also strongly interacting. An example of coherence greatly amplifying a very weak interaction is given by gravity: The intrinsic strength of the gravitational force is very small, but due to the coherent superposition of forces arising from individual particles, it becomes a large force.

There is a typical momentum scale Q_{sat}, and gluons with momentum less than this scale have maximal phase space density,

$$\frac{dN}{d^2p_T d^2r_T dy} \sim \frac{1}{\alpha_S} \tag{1}$$

As one goes to higher energies, the saturation momentum increases, as more gluons are added to the CGC.

The theory of the CGC has provided both a rich phenomenology as well as first principles understanding within QCD for the high energy limit. [3] Some of the successes of this theory include

- Scaling properties in electron-hadron scattering.
- Elastic and almost elastic electron-hadron scattering.
- Nuclear size dependence of quark and gluon distributions.
- Distributions of produced particles in hadron and nuclear collisions.
- Scaling properties of hadron-hadron collisions.
- Long range momentum correlations for particles produced in hadron-hadron collisions.
- Intuitive understanding of high energy limit for cross sections.
- The origin of the multiparticle excitations which control high energy scattering (Pomeron, Reggeon, Odderon).

This is a very active area of research. It has as one of its intellectual goals, the unification of the description of all strong interaction processes at high energy and as such involves different phenomena: electron-hadron, hadron-hadron, hadron-nucleus and nucleus-nucleus collisions.

THE INITIAL SINGULARITY AND THE GLASMA

Before the collision of two hadrons, two sheets of Colored Glass approach one another. Because the phase space density of gluons is large, the gluons can be treated as classical fields. The fields are Lorentz boosted Coulomb fields, that is Lienard-Wiechart potentials, which are static in the transverse plane of the hadrons and have $E \perp B \perp \hat{z}$, where z is the direction of motion. When the classical equations for the evolution of these fields

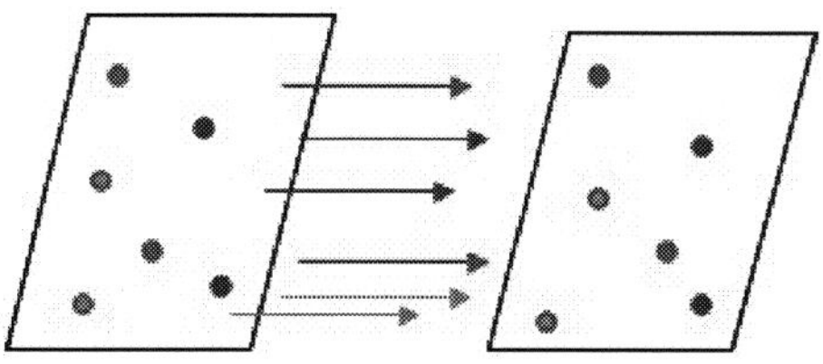

FIGURE 3. After the collision of two sheets of colored glass, longitudinal electric and magnetic fields associated with the Glasma form.

is solved, in a very short time scale of order $\Delta t \sim \frac{1}{Q_s} e^{-\kappa/\alpha_S}$, the fields change from purely transverse to purely longitudinal. This is because the collisions generate color electric and magnetic monopole charge densities of opposite sign on the two sheets. This description is reminiscent of the flux tube models of color electric fields which are used in phenomenological descriptions of low energy strong interactions. At high energies, there are both color electric and magnetic field because the electric and magnetic fields were of equal magnitude in the CGC, and because of the electric-magnetic duality of QCD.

Fields with a non-zero $\vec{E} \cdot \vec{B}$ carry a topological charge. In QCD, they are associated with anomalous mass generation and chiral symmetry violation. In electroweak theory, such fields may be responsible for generating the baryon asymmetry of the universe. In QCD, they may generate the masses of those particles which constitute the visible matter of the universe. Each field configuration violates CP. An experimental discovery of the effects of such fields would be of great importance. Theoretical ideas for experimental signatures are sketchy.[4] These initial longitudinal fields at the time of production evolve into transverse fields and eventually form a Quark Gluon Plasma. The matter from production until the formation of the Quark Gluon Plasma is called the Glasma.[5]-[7]

The initial Glasma fields are unstable, and after a time scale of order $1/Q_s$, instabilities begin to become of the order of the original classical fields. [8]-[11]The origin of these Weibel instabilities was originally found for plasmas close to thermal equilibrium by Mrowczinski. Quantum fluctuations in the original wavefunction can grow by these instabilities, and eventually overwhelm the longitudinal electric and magnetic Glasma fields. Perhaps these fields form a chaotic or turbulent liquid which might thermalize and isotropize the system.[12]

The amplification of quantum fluctuations to macroscopic magnitude is reminiscent of inflation in the early universe. These quantum fluctuations expand to size scale larger than the event horizon during inflation, and are imprinted into the fabric of space-time. At much later times, the event horizon size scale becomes of the order of galactic size scales. Ultimately, these fluctuations drive gravitationally unstable modes which form galaxies and clusters of galaxies.

In heavy ion collisions, analogous fluctuations in the hadronic wavefunction might appear as momentum dependent fluctuations. Such fluctuations might be frozen into the final state distribution of particles.

THE STRONGLY INTERACTING QUARK GLUON PLASMA

Some of the first data from RHIC showed that there was strong radial and elliptic flow, and strong jet quenching. [13] The flow data could be described using perfect fluid hydrodynamic equations. Although there is a good deal of uncertainty due to lack of precise knowledge of the initial state, final decoupling, equations of state and the magnitude of viscous corrections, it is nevertheless remarkable that perfect fluid hydrodynamic computations can describe the data. Such a description was not possible for data at lower energies. [14] Such agreement involves comparison of almost limitless numbers of transverse and longitudinal momentum distributions of produced particles as a function of the impact parameter of the nuclear collisions. The impact parameter of such collisions is determined by measurements of the particle multiplicity both in the forward and central regions of the collision (at low longitudinal and high longitudinal momentum in the center of mass frame). The agreement between data and experiment seems to be improved by the more careful treatments of late time evolution in the form of hadrons, and using Color Glass initial conditions.[14]

Perfect fluid hydrodynamics is the limit where interactions are very small and viscosities are zero. This suggests that the Quark Gluon Plasma is well thermalized and fairly strongly interacting. The degree to which the viscosity may be treated as zero is subject to a better understanding of the initial conditions. Color Glass Condensate initial condition allow for larger viscosities than the standard phenomenological choices.

Jet quenching is the disappearance of high transverse momentum particles because of interactions with the produced media. The production of high transverse momentum particles is described well within perturbative QCD. One needs only know the distribution functions of quarks and gluons inside hadrons and nuclei. This can be done by deep inelastic scattering combined with additional data on hadron-nucleus collisions..

There is suppression of jet production in gold-gold collisions by a factor of four out to transverse momenta of about 20 GeV, corresponding to several GeV. Such a large energy loss is stunning and was not anticipated.[13], [15]

These observations have led to a consensus that one has produced a strongly interacting Quark Gluon Plasma, and that to a fair or good or perhaps excellent approximation, the system is thermalized. A well thermalized system is a perfect fluid. The outstanding issue is how perfect is the perfect fluid.

This may be parameterized by the dimensionless ratio of viscosity to entropy density, η/s. As argued by Son and collaborators,[16] $N = 4$ supersymmetric Yang-Mills theory satisfies a bound

$$\frac{\eta}{s} \geq \frac{1}{4\pi} \tag{2}$$

with the lower bound approached for infinite coupling. This is a beautiful result which leads to new insight. Although $N = 4$ supersymmetric theories bear no direct relation to our physical world, this bound was also more generally argued as a consequence that in the strong coupling limit of QCD, the limiting small viscosity is reached when the mean free path of a particle become equal to its deBroglie wavelength. Below this limit, scattering is difficult due to quantum mechanical coherence.

Is η/s close to the minimum value for temperatures accessible at RHIC in the QGP? There are reasons to be skeptical: (1) Some computations argue that one does not need extremely large cross sections within transport computations to describe the elliptic flow data.[17] (2) Flow results depend on initial conditions, about which there is not yet consensus, the way that the equation of state is treated, and how one coalesces quarks and gluons into hadrons. There is much work now on initial conditions, so this has potential for resolution. If the initial conditions are more like those predicted from the Color Glass Condensate, one generates more elliptic flow, which could be reduced by increasing the viscosity. [18](3) New realistic computations with relativistic hydro and viscosity are being done. The preliminary results from these hydro computations indicate that varying viscosity can be compensated by various other uncertainties in the computations.[19]-[20]

STRINGS AND ADSCFT

There has been some discussion at the literature about string theory and its relationship to heavy ion physics.[16],[21] AdSCFT is a mathematical trick which allows one to compute the properties of $N = 4$ supersymmetric Yang-Mills theory in terms of gravitational theory in curved space. The strong coupling limit of N=4 SUSY corresponds to weak coupling gravity, allowing strong coupling computations for the Yang-Mills theory.

N= 4 supersymmetric Yang-Mills is not QCD. There are many caveats:

- It has no mass scale and is conformally invariant.
- It has no confinement and no running coupling constant.
- It is supersymmetric.
- It has no chiral symmetry breaking or mass generation.
- It has six scalar and fermions in the adjoint representation.

The interesting applications of this correspondence for QCD are in the strong coupling limit. Even in lowest order strong coupling computations it is very speculative to make relationships between this theory and QCD, because of the above. It is much more difficult to relate non-leading computations to QCD.

It may be possible to correct some or all of the above problems, or, for various physical problems, some of the objections may not be relevant. As yet there is neither consensus nor compelling argument for the conjectured fixes or phenomena which would insure that the $N = 4$ supersymmetric Yang Mills results would reliably reflect QCD.

Further, these computations are applied to the QGP at temperatures above the deconfinement transition temperature, where the validity of strong coupling limit is arguable. Lattice computations give reliable computations of the properties of the QGP, and indicate the coupling is of intermediate to weak strength in this region. There are improved weak coupling computations which agree with lattice data at temperatures more than a few times T_c.[22]

The AdSCFT correspondence, is probably best thought of as a discovery tool with limited resolving power. An example is the η/s computation. The discovery of the bound on η/s could be argued be verified by an independent argument, as a consequence of the deBroglie wavelength of particles becoming of the order of mean free paths. It is a theoretical discovery but its direct applicability to heavy ion collisions remains to be shown.

Certainly, computations in lattice gauge theory or in weak coupling provide reliable QCD computations of a limited set of physical phenomena. One has had hope for analytic methods in the strong coupling dynamics of QCD, and we should continue to have such hope.

ACKNOWLEDGMENTS

I gratefully acknowledge my colleagues Edmond Iancu, Dima Kharzeev, Genya Levin, Al Mueller, and Raju Venugopalan for their insights in formulating the material presented here.

This manuscript has been authorized under Contract No. DE-AC02-98CH0886 with the US Department of Energy.

REFERENCES

1. L. D. McLerran and R. Venugopalan, *Phys. Rev.* **D49**, 2233(1994); 3352 (1994); **D50**, 2225 (1994).
2. E. Iancu, A. Leonidov and L. D. McLerran, *Nucl. Phys.* **A692**, 583 (2001); E. Ferreiro E. Iancu, A. Leonidov and L. D. McLerran, *Nucl. Phys.* **A710**,373 (2002);J. Jalilian-Marian, *Phys. Rev,.* **C70** 027902 (2004) E. Iancu and L. McLerran, *Phys. Lett.* **B510**, 145 (2001); J. Jalilian-Marian, A. Kovner, L. McLerran and H. Weigert, *Phys. Rev.* **D55** (1997), 5414; J. Jalilian-Marian, A. Kovner, A. Leonidov and H. Weigert, *Nucl. Phys.* **B504** (1997), 415; *Phys. Rev.* **D59** (1999), 014014.
3. E. Iancu and R. Venugopalan, Publisehd in QGP3, Eds. R. C. Hwa and X. N. Wang, World Scientific. hep-ph/0303204
4. D. Kharzeev, R. D. Pisarski and M. H. G. Tytgat, *Phys. Rev. Lett.* **81**, 512 (1998); hep-ph/9804221
5. A. Kovner, L. McLerran and H. Weigert, *Phys. Rev.* **D52** 3809 (1995); 6231 (1995).
6. A. Krasnitz and R. Venugopalan, *Nucl. Phys.* **B557** 237 (1999); *Phys. Rev. Lett.* **84** 4309 (2000); Phys. Rev. Lett. **86** 1717 (2001); A. Krasnitz, Y. Nara and R. Venugopalan, *Phys. Rev. Lett.* **87** 192302 (2001); *Nucl. Phys.* **A717** 268 (2003).
7. T. Lappi, *Phys. Rev.* **C67** 054903 (2003); T. Lappi and L. Mclerran, *Nucl. Phys.* **A772** 200 (2006).
8. S. Mroczynski, *Phys. Lett.* **B214** 587 (1988); **B314** 118 (1993); **B363**, 26 (1997).
9. P. Romatschke and M. Strikland, *Phys. Rev.* **D68** 036004 (2003); **D70** 116006 (2004).
10. P. Arnold, J. Lenaghan, and G. Moore, *JHEP* **0308** 002 (2003); P. Arnold, J. Lenaghan, G. Moore and L. Yaffe, *Phys. Rev. Lett.* **94** 072302 (2005).
11. P. Romatshke and R. Venugopalan, *Phys. Rev. Lett.* **96** 062302 (2006) . A. Dumitru and Y. Nara, *JHEP* **0509** 041 (2005); A. Dumitru, Y. Nara and M. Strikland, *Phys. Lett.* **B621** 89 (2005).
12. M. Asakawa, S. A. Bass and B. Muller, *Porg. Theor. Phys.* **116**, 725 (2007); *Phys. Rev. Lett.* **96**, 252301 (2006).
13. Reports of the Star, Phenix, Phobos and Brahms experimental collaborations in *Nucl. Phys* **A757** (2005) Brahms Collaboration p 1; Phobos Collaboration p 28; Star Collaboration p 102; Phenix Collaboration 184.
14. For a theoretical review of the properties of the QGP and it effects in heavy ion collisions see Workshop on New Discoveries at RHIC: The Current Case for the Strongly Interactive Quark Gluon Plasma, Brookhaven, May 14-15 , 2004, *Nucl. Phys.* **A750** (2005) .
15. J. D. Bjorken, Fermilab-Pub-82/59-THY (1982) and erratum. (unpublished); D. Appel, *Phys. Rev.* **D33** 717 (1986).
16. G. Policastro, D. T. Son and A. O. Starinets, *Phys. Rev. Lett.* **87** 081601 (2001); P. Kovtun, D. T. Son and A. O. Starinets, *Phys. Rev. lett.* **94** 11601 (2005).
17. V. Greco, C. M. Ko and P. Levai, *Phys. Rev.* **C68** 034904 (2003).
18. H. Drescher, A. Dumitru, C. Gombeaud and J. Ollitrault, arXiv:0704.3353[nucl-th].
19. U. Heinz, private communication.
20. P. Romatschke and U. Romatschke, arXiv:0706.1522[nucl-th].
21. I. R. Klebanov, *Nucl. Phys.* **B496** 231 (1997).
22. J. P. Blaizot and E. Iancu, *Phys. Rev. Lett.* **70**, 3376 (1993); *Nucl. Phys.* **B417**, 608 (1994); J. P. Blaizot, E. Iancu and A. Rebhan, *Phys. Rev. lett.* **83**, 2906 (1999); *Phys. Lett.* **B470**, 181 (1999); *Phys. Rev.* **D63**, 065003 (2001).

Neutrinos and the origin of fermion mass structure.

Graham G. Ross

Rudolf Peierls Centre for Theoretical Physics,
Department of Physics,
University of Oxford.

Abstract. The pattern of neutrino masses and mixings is characteristically different from those observed in the quark sector. I discuss why this should be the case and what implications this has for the origin of quark and lepton masses, mixings and CP violation.

INTRODUCTION

As discussed in Dave Wark's lecture our present data is consistent with just 3 flavours of neutrinos. However one should note that, although the recent MiniBoone measurement of $\nu_\mu \to \nu_e$ oscillation provides no evidence for a sterile neutrino component, it is not in clear disagreement with the LSND evidence for sterile neutrino(s). The reason is that LSND measured $\overline{\nu}_\mu \to \overline{\nu}_e$ oscillation and, due to the new CP violating phases that are expected to be present if there are additional sterile neutrinos, the LSND and MiniBoone results can be made consistent in (3+2) and (3+3) schemes[1]. However in these fits the tension between appearance/disappearance experiments that was there when fitting the LSND result remains and I will assume just 3 neutrino flavours in the rest of this talk[1].

The pattern of quark and lepton masses and mixing angles is shown in Figure 1. The structure shown assumes that, like the quarks and charged leptons, the neutrinos have an hierarchical structure, although the data is consistent too with a near degenerate structure with mean mass slightly less than 1eV. Clearly the lepton structure is quite different from the quarks. Apart from the fact the neutrinos are extremely light the mixing angles in the lepton sector are large[3], in contrast to the small mixing angles observed in the quark sector. This is illustrated more clearly in Figure 2 where one may see that for the normal mass hierarchy the heaviest state, the atmospheric neutrino, is approximately an equal admixture of ν_μ and ν_τ while the solar neutrino is a nearly equal mixture of ν_e, ν_μ and ν_τ. The form of the MNS mixing matrix shown for this precise pattern is also shown and corresponds to what is known as tri-bi-maximal mixing[4]. As we shall discuss such a form strongly suggests a non-Abelian family symmetry which can relate the proportion of different flavours of neutrino in the mass eigenstates. The fact the mixing seems to be quantized (tri-bi-maximal) further suggests a discrete non-Abelian structure.

---

[1] It is possible that observation of supernovae neutrinos will provide much more sensitive measures of sterile neutrinos in the future[2].

CP957, *Particles, Strings, and Cosmology, 13th International Symposium*
edited by A. Rajantie, C. Contaldi, P. Dauncey, and H. Stoica

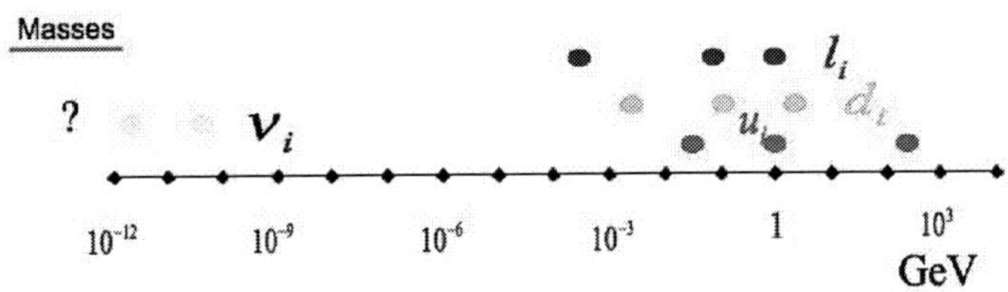

Mixing

$$V_{CKM} \approx \begin{pmatrix} 1 & 0.218-0.224 & 0.002-0.005 \\ 0.218-0.224 & 1 & 0.032-0.048 \\ 0.004-0.015 & 0.03-0.048 & 1 \end{pmatrix} \quad V_{MNS} = \begin{pmatrix} 0.72-0.89 & 0.45-0.69 & <0.2 \\ 0.24-0.58 & 0.39-0.76 & 0.52-0.84 \\ 0.24-0.58 & 0.39-0.76 & 0.53-0.84 \end{pmatrix}$$

FIGURE 1. Quark and lepton masses and mixing angles.

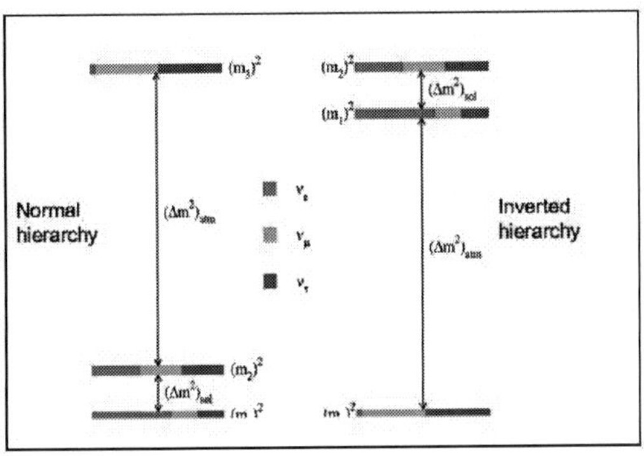

$$V_{MNS} \approx \begin{pmatrix} \sqrt{\frac{2}{3}} & \frac{1}{\sqrt{3}} & \sim 0 \\ -\frac{1}{\sqrt{6}} & \frac{1}{\sqrt{3}} & -\frac{1}{\sqrt{2}} \\ -\frac{1}{\sqrt{6}} & \frac{1}{\sqrt{3}} & \frac{1}{\sqrt{2}} \end{pmatrix}$$

FIGURE 2. The composition of the neutrino mass eigenstates for the case of a normal and an inverted hierarchy showing near tri-bi-maximal mixing and the form of the mixing matrix for this case. The shading indicates the neutrino flavour with ν_μ, ν_e, and ν_τ being represented by progressively darker shades and ordered left to right.

THE ORIGIN OF NEUTRINO MASSES

Majorana masses

The first questions to answer is what is the origin of neutrino mass and why are they so much lighter than the quarks and charged leptons? A very plausible reason is because neutrinos are the only neutral states and hence are the only states that can have a Majorana mass. However, as Weinberg pointed out as long ago as 1979[5], the only Majorana mass term consistent with the Standard Model gauge symmetry requires the neutral components of two Higgs doublet fields, H, and hence appears only through

	$SU(2)_L$ singlets		$SU(2)_L$ triplets
(A)	$\nu_i H^0 - l_i H^+$	(D)	$[\nu_i H^+, (\nu_i H^0 + l_i H^+)/\sqrt{2}, l_i H^0]$
(B)	$\nu_i l_j - l_i \nu_j$	(E)	$[\nu_i \nu_j, (\nu_i l_j + l_i \nu_j)/\sqrt{2}, l_i l_j]$
(C)	$H_1^+ H_2^0 - H_1^0 H_2^+$	(F)	$[H^+ H^+, \sqrt{2} H^+ H^0, H^0 H^0]$

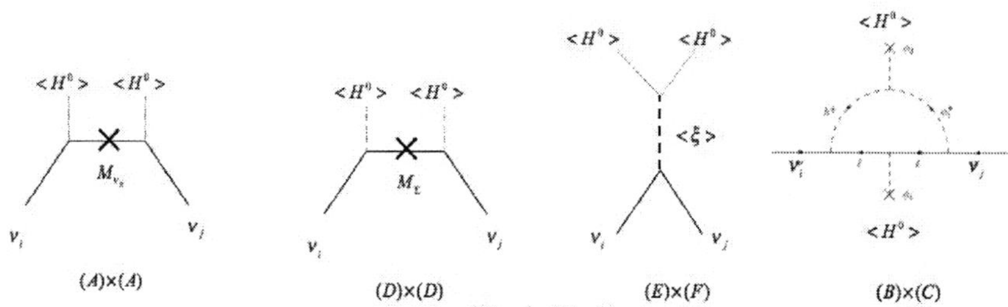

FIGURE 3. Possible origins of Majorana neutrino mass.

a higher dimensional term in the effective Lagrangian, $\frac{f_{ij}}{\Lambda}\nu_i\nu_j H^0 H^0$, giving $M_{ij}^{\nu} = f_{ij}\frac{<H^0>^2}{\Lambda}$. If the mediator scale Λ associated with the operator is large compared to the electroweak scale the mass will be small. Of course one wants to know what generates the higher dimensional term and here there are several possibilities as indicated in Figure 3.

The first three "see-saw" graphs[6] correspond to different $SU(2)$ properties of the heavy messenger state responsible for generating the higher operator. However, as may be seen from the fourth graph, it is also possible that the smallness of neutrino masses may be due to a radiative origin of the Majorana mass. The classic, "Type 1" see-saw corresponds to the first graph with a heavy fermion exchange with quantum numbers corresponding to a right-handed neutrino. The introduction of such right handed neutrinos restores the symmetry between quarks and leptons. Moreover the right-handed neutrino being a singlet under the Standard Model group can acquire a Majorana mass which is not controlled by the electroweak breaking scale and hence can be very large, explaining why it has not been observed to date and also explaining why the left-handed neutrino masses are anomalously small. The scale of mass is given by $m_\nu = O(\frac{<H^0>^2}{M_{\nu_R}})$. To get a neutrino mass of $0.1eV$ one needs $m_{\nu_R} \sim 10^{13} GeV \left(\frac{<H^0>}{10^2 GeV}\right)^2$.

Although this provides a very plausible origin for the smallness of neutrino mass it is nor the only possibility. For example it could be that the couplings or vacuum expectation values associated with the see-saw diagram are very small and the intermediate state is correspondingly lighter. For example in supersymmetry with $R-$parity violation the see-saw diagram can occur due to electroweak doublet sneutrino vevs[7] and the intermediate state can be a neutralino with mass $m_N \sim 10^3 GeV \left(\frac{<\tilde{\nu}>}{1MeV}\right)^2$.

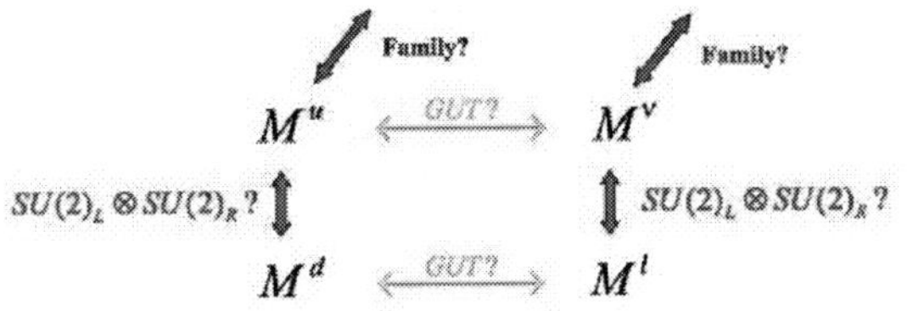

FIGURE 4. Possible symmetries capable of relating fermion masses.

Dirac masses

Yet another possibility capable of naturally explaining the small neutrino masses follows in models with large extra dimensions[8]. In this case a very small Dirac neutrino mass is generated if the right handed neutrino can propagate in the bulk while the left-handed neutrinos are confined to a four dimensional brane. The Dirac coupling between the left and right handed states is very small due to the right-handed neutrino flux spreading into the extra dimensions in the same way that the gravitational coupling is small in these models due to the graviton flux spreading into the extra dimensions. In fact the same suppression applies in both cases leading to the Dirac neutrino mass of the form $m_\nu = \frac{\lambda <H^0> M_*}{M_{Planck}} \simeq \lambda .10^{-4} eV \left(\frac{M_*}{TeV}\right)$ where M_* is the Planck scale in the extra dimension for the case we have one additional space dimension. Actually it is not necessary to have additional dimensions to motivate small Dirac masses. For example this can happen if the Dirac mass term arises through a higher dimension operator and the suppression follows if the messenger mass is large in a manner similar to the Majorana mass see-saw mechanism[9]. Interestingly baryogenesis through leptogenesis is still possible in this case even though lepton number is conserved by the Dirac neutrino mass[10]. The existence of these possibilities emphasizes how important it is to determine whether neutrino masses are Majorana or Dirac. The future generation of double beta decay experiments will ultimately be able to probe down to the scale of the largest mass squared difference needed to explain atmospheric neutrino oscillation.

SYMMETRIES

Although there are several mechanisms capable of explaining in a natural way why neutrinos have a very small mass there remains the questions what explains the various hierarchies of quark and lepton masses and why are the mixing angles so different in the quark and lepton sectors? The most promising explanation of these structures is that there are symmetries relating the Yukawa couplings responsible, after spontaneous symmetry breaking, for the fermion masses. For the case there are right handed neutrinos the possible symmetries are illustrated in Figure 4. They consist of Grand Unified symmetries which may relate the Yukawa couplings within a single family and family symmetries which can relate couplings involving different families.

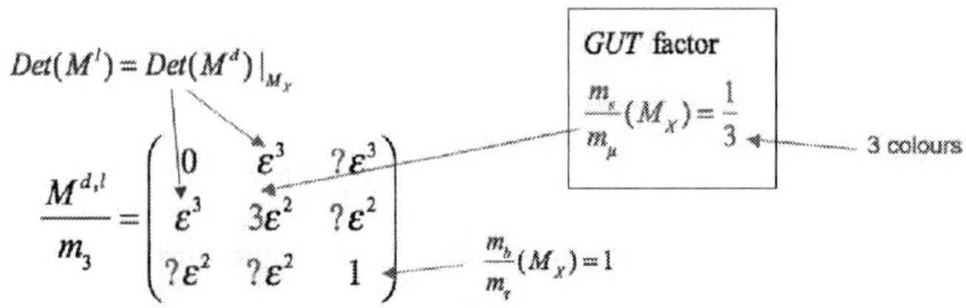

FIGURE 5. GUT mass relations. The hierarchical structure of the mass matrix is determined in terms of powers of the small parameter $\varepsilon = 0.15$. The entries with question marks do not contribute in leading order to the mass eigenvalues. The "texture zero" in the $(1,1)$ element leads to a successful prediction for the Cabibbo angle and may result from an underlying family symmetry.

Grand Unified symmetries

There is considerable circumstantial evidence for an underlying GUT. A complete family of fermions fits neatly into a 16 dimensional representation of $SO(10)$ and the strong weak and electromagnetic gauge couplings unify with remarkable precision provided there is low scale supersymmetry. In fact there are several good mass relations that may be derived from an underlying GUT. These are illustrated in Figure 5 for the case the down quark and charged lepton (Dirac) mass matrices are symmetric. The natural expectation is that the mass matrix elements of the down quark and charged lepton mass matrices should be equal and this works well for the elements in the $(3,3)$, $(1,2)$ and $(2,1)$ positions leading to the approximate relations $m_b(M_X) = m_\tau(M_X)$[11] and $Det\left(M^l\right)|_{M_X} = Det\left(M^d\right)|_{M_X}$ where M_X is the GUT scale. After including radiative corrections these relations can be in good agreement with the measured masses[12]. In the supersymmetric case the corrections are sensitive to the SUSY spectrum and $\tan\beta$. Continuing the measured masses up to the GUT scale a recent analysis [12] gives $m_b(M_X)/m_\tau(M_X) = \{1.0, 0.6, 1.0\}$, $Det\left(M^l\right)|_{M_X}/Det\left(M^d\right)|_{M_X} = \{0.6, 0.3, 1.04\}$ for a SUGRA spectrum with $\tan\beta = 2$, $\tan\beta = 30$ and for an anomaly mediated spectrum respectively. However equality for the down quark and charged lepton $(2,2)$ matrix elements is not consistent with the measured masses. What is required is for the lepton element to be a factor of 3 greater than the down quark element. As first shown by Georgi and Jarlskog[13] this is possible in specific GUT theories if a different group invariant dominates the $(2,2)$ element with the factor of 3 being just the number of colours. It leads to the relation $3m_s/m_\mu(M_X) = 1$ to be compared to the measured result $3m_s/m_\mu(M_X) = \{0.9, 0.82, 1.05\}$.

Family symmetries

Going beyond GUTs, fermion masses may also be related via family symmetries. A simple example applies if the family symmetry forces the $(1,1)$ matrix element to be zero as in Figure 5. This, with the symmetric structure assumed here, leads to a precise prediction[14] for the Cabibbo angle in terms of the first two generation down quark masses which is in remarkable agreement with experiment. Can family symmetries explain the difference in mixing angles in the quark and lepton sectors? As emphasized in Figure 2 the most significant feature of the neutrino mixing is the near bi-tri-maximal mixing. This strongly suggests that a family symmetry is non-Abelian in order to relate the proportion of different family neutrino components in a mass eigenstate. Moreover the quantization of the mixing angles suggests that the non-Abelian symmetry be discrete!

Triggered by these observations there has been intense activity studying mass relations which can follow from (discrete) non-Abelian symmetries[15]. The big question is how can such a symmetry lead to near tri-bi-maximal mixing? To discuss this it is useful to write down the effective Lagrangian for the light neutrinos that describes this mixing pattern. It is given by

$$L^{\nu}_{Majorana,eff} = a\nu_i\theta^i_{23}\nu_j\theta^j_{23} + b\nu_i\theta^i_{123}\nu_j\theta^j_{123} \tag{1}$$

where $\theta_{23} = (0,1,-1)$ and $\theta_{123} = (1,1,1)$ and a and b are constants. This is to be compared to the leading term for quarks and charged leptons responsible for generating the heavy third generation masses

$$L^{q,l}_{Dirac} = c\psi_i\theta^i_3\psi^c_j\theta^j_3 + ... \tag{2}$$

where $\theta_3 = (0,0,1)$. One sees that the problem is first to explain why the dominant breaking in the third direction does not apply to the neutrinos and then to explain what forces the elements of the θs to have their special form. Surprisingly the first problem has an easy answer if the see-saw mechanism is responsible for the neutrino effective Lagrangian. To see how this works consider the following Lagrangian which generates Dirac neutrino masses

$$L^{\nu}_{Dirac} = \alpha\nu_i\theta^i_3\nu^c_j\theta^j_3 + \beta\nu_i\theta^i_{23}\nu^c_j\theta^j_{23} + \gamma\nu_i\theta^i_{123}\nu^c_j\theta^j_{123} \tag{3}$$

where, with $\alpha > \beta > \gamma$, the largest coupling is to θ_3 as is the case for the quarks and charged leptons. We similarly allow the Majorana mass matrix, M_M, for the right handed neutrinos to have an hierarchical form. To illustrate the point we take M_M diagonal, $M_M = Diag(M_1, M_2, M_3)$ with $M_3 \gg M_2, M_1$. In a realistic model the hierarchical form of M_M will be controlled by the same family symmetries that generate the hierarchical structure for the Dirac masses and may be non-diagonal[16]. With this the effective Lagrangian has the form

$$L^{\nu}_{Majorana,eff} \simeq \frac{\beta^2}{M_2}\nu_i\theta^i_{23}\nu_j\theta^j_{23} + \gamma^2\left(\frac{1}{M_1} + \frac{1}{M_2}\right)\nu_i\theta^i_{123}\nu_j\theta^j_{123} + \frac{\alpha^2}{M_3}\nu_i\theta^i_3\nu_j\theta^j_3 \tag{4}$$

The important point is that the see-saw mechanism has the Majorana mass in the denominator so that if the third generation dominance should be much greater for the Majorana masses than the Dirac masses than the third term in eq(4) will be small[17]. In this case one can get the desired mixing pattern of eq(1).

There remains the question what forces the elements of the θs to have their special form? This can be answered if the θs are "familon" fields transforming under a family symmetry. Then vacuum alignment can arrange for the special form of their vacuum expectation values (vevs) breaking the family symmetry. For the case of a discrete non-Abelian family symmetry this vacuum alignment can arise naturally with the choice of $(0,0,1)$ or $(1,1,1)$ vev depending on the sign of various terms in the effective potential for the familon fields[18]. Typically this requires the discrete symmetry should contain a permutation symmetry in order to obtain the $(1,1,1)$ alignment. Realistic models in which symmetries guarantee the sequential domination and vacuum alignment have been constructed[16].

EXPERIMENTAL SIGNALS OF FERMION MASS MODELS

If we are ever to know if any of these ideas for generating the observed masses and mixing angles is true it will be necessary to have experimental tests.

Mixing angles

In the first instance these models give predictions for various mixing angles. For example, as mentioned above, the $(1,1)$ texture zero gives a prediction[14] for the Cabibbo angle

$$|V_{us}| = |\sqrt{\frac{m_d}{m_s}} - e^{i\delta}\sqrt{\frac{m_u}{m_c}}|$$

which continues to be in excellent agreement with experiment. For the case of the tri-bi-maximal mixing schemes just discussed there is a prediction for the third mixing angle, θ_{13}, because it comes entirely from the charged lepton diagonalization. For it the $(1,1)$ texture leads to the related prediction

$$\theta_{13} \simeq \frac{\theta_{12}}{\sqrt{2}} \simeq \sqrt{\frac{m_e}{2m_\mu}} \simeq 3^0.$$

For the case that θ^e_{13} and θ^ν_{13} are small, which applies to the tri-bi-maximal models as well as a wider class of models, one can derive another sum rule[19] : $\theta_{12} - \theta_{13}\cos\delta = \theta^\nu_{12}(= 35^0$ for tri-bi-maximal mixing).

SUSY spectrum and FCNC

In supersymmetric models the SUSY spectrum will be constrained by the family symmetries too. Measurement of this spectrum will give important tests of the existence of such a family symmetry. For example for the symmetries discussed above the leading soft mass term is constrained to have the family independent form $m^2_\phi\,(\phi_i)^\dagger\,\phi_i$ where ϕ_i is a scalar partner of a quark or a lepton and m^2_ϕ is a SUSY breaking mass. This is the structure needed to suppress the new SUSY contributions to flavour changing neutral currents (FCNC) and provides an elegant origin for the degeneracy. We know the family symmetry must be strongly split to get the heavy third family. The leading effect of such breaking is suppressed relative to the leading term by powers of the familon field vevs, $m_\phi^2 \left(\phi_i \theta_3^i\right)^\dagger \phi_j \theta_3^j$[20]. If the family symmetry is continuous there will also be $D-$term contributions to the soft masses. These have the potential to generate unacceptably large FCNC but are under control for a subset of SUGRA and gauge mediated models[21].

In addition to soft masses there are other soft terms, in particular the trilinear A-terms. In a theory with family symmetries these are also potentially dangerous as in general they are not diagonalized by the same rotations that diagonalize the fermion masses and so can lead to FCNC processes. Detailed studies [22] show that the constraints on FCNC strongly limit the (poorly constrained) quark and lepton mass matrix elements giving rise to mixing in the right handed quark sector (such mixing is unobservable in non-SUSY models but generates observable effects in the squark and slepton sector). The mass matrix symmetric case considered above has these FCNC under control, but only just and the expectation is that there will be FCNC within a factor of 10 of the present bounds.

Family gauge bosons, familons, CP.

There are further possible indications for a family symmetry. If the symmetry is continuous there will be new gauge interactions coupling to the family charge. Unfortunately we do not know at what scale these symmetries are broken as we only know the ratio of the familon vevs to the underlying messenger scale and the latter is unknown. The scale could be close to the Planck scale making the new gauge interactions unobservable or it could be at the TeV scale leading to a wealth of new phenomena[2].

If the family symmetry is discrete there will be no new gauge bosons associated with the family symmetry but there may be light Pseudo Goldstone familon states resulting from the spontaneous breaking of approximate family global symmetries. Just how heavy such states are is very model dependent and the phenomenology of such familon states remains to be explored.

Perhaps most interesting are the implications for CP violation in family symmetry models. In SUSY models there is a potential SUSY CP problem because new CP violating phases can be present which must be very small ($< O(10^{-2})$) if they are to be consistent with the bounds on dipole electric moments[23]. The family symmetry models offer an attractive way to solve this problem because in them CP violation may arise through spontaneous breaking once the flavon fields acquire their vevs. In this case CP violation occurs in the flavour changing sector where it is observed. CP violating effects in the flavour conserving sector arise only via two flavour changing vertices suppressed by the square of a small mixing angle and hence are suppressed to a level

---

[2] In the latter case the success of gauge coupling unification is problematic as the family symmetry models require new vectorlike quark and lepton states which affect gauge coupling running.

consistent with present bounds[22] but should be present within a factor of 10 of present limits.

SUMMARY AND CONCLUSIONS

In my opinion the data on neutrino masses and mixing angles is "maximally attractive" in the sense that the disparity between the structures found in the quark and lepton sectors is very suggestive of a new paradigm for the underlying theory capable of describing the data. If the explanation is to be provided by an extension of the symmetries of the Standard Model then the data strongly suggests an underlying spontaneously broken non-Abelian family group structure. Furthermore the near tri-bi-maximal mixing structure observed can elegantly be explained if the group is discrete rather than continuous. Surprisingly the difference between the quark and lepton sectors can readily be explained if neutrino masses are generated by the see-saw mechanism with sequential dominance. In this case the Yukawa couplings responsible for the Dirac masses are consistent with a quark-lepton symmetry with a similar form in the down quark, the up quark, the charged lepton and the neutrino sectors, as is expected if there is an underlying Grand Unified group (perhaps only at the string scale) commuting with the family symmetry group. If such a scenario is correct there will be many associated measurable signals in flavour changing and CP violating processes and in the SUSY scalar sector if low energy supersymmetry is realised in nature.

REFERENCES

1. M. Maltoni and T. Schwetz, arXiv:0705.0107 [hep-ph]; S. Goswami and W. Rodejohann, arXiv:0706.1462 [hep-ph]; J. T. Goldman, G. J. . Stephenson and B. H. J. McKellar, Phys. Rev. D **75** (2007) 091301 [arXiv:nucl-th/0703023].
2. S. Choubey, N. P. Harries and G. G. Ross, arXiv:hep-ph/0703092; Phys. Rev. D **74** (2006) 053010 [arXiv:hep-ph/0605255].
3. See M. Maltoni, T. Schwetz, M. A. Tortola and J. W. F. Valle, New J. Phys. **6** (2004) 122 [arXiv:hep-ph/0405172] and references therein.
4. P. F. Harrison, D. H. Perkins and W. G. Scott, Phys. Lett. B **530** (2002) 167 [arXiv:hep-ph/0202074].
5. S. Weinberg, Phys. Rev. Lett. **43** (1979) 1566.
6. P. Minkowski, Phys Lett 67B 421 (1977);
M. Gell-Mann, P. Ramond and R. Slansky in Sanibel Talk, CALT-68-709, Feb 1979, and in *Supergravity* (North Holland, Amsterdam 1979); T. Yanagida in *Proc. of the Workshop on Unified Theory and Baryon Number of the Universe*, KEK, Japan, 1979; S.L.Glashow, Cargese Lectures (1979);
R. N. Mohapatra and G. Senjanovic, Phys. Rev. Lett. **44** (1980) 912;
J. Schechter and J. W. Valle, Phys. Rev. D **25** (1982) 774.
7. A. Masiero and J. W. F. Valle, Phys. Lett. B **251** (1990) 273; J. C. Romao, C. A. Santos and J. W. F. Valle, Phys. Lett. B **288** (1992) 311; B. de Carlos and P. L. White, Phys. Rev. D **54** (1996) 3427 [arXiv:hep-ph/9602381].
8. K. R. Dienes, E. Dudas and T. Gherghetta, Nucl. Phys. B **557** (1999) 25 [arXiv:hep-ph/9811428]; A. E. Faraggi and M. Pospelov, Phys. Lett. B **458** (1999) 237 [arXiv:hep-ph/9901299]; N. Arkani-Hamed, S. Dimopoulos, G. R. Dvali and J. March-Russell, Phys. Rev. D **65** (2002) 024032 [arXiv:hep-ph/9811448].
9. See S. Abel, A. Dedes and K. Tamvakis, Phys. Rev. D **71** (2005) 033003 [arXiv:hep-ph/0402287]; D. G. Cerdeno, A. Dedes and T. E. J. Underwood, JHEP **0609** (2006) 067 [arXiv:hep-ph/0607157], and references therein.

10. K. Dick, M. Lindner, M. Ratz and D. Wright, Phys. Rev. Lett. **84** (2000) 4039
11. A. J. Buras, J. R. Ellis, M. K. Gaillard and D. V. Nanopoulos, Nucl. Phys. B **135** (1978) 66.
12. G. Ross and M. Serna, arXiv:0704.1248 [hep-ph].
13. H.Georgi and C.Jarlskog, Phys.Lett.B86 (1979) 297; H.Georgi and D.V.Nanopoulos, Nucl.Phys. B159 (1979) 16; J.Harvey, P.Ramond and D.B.Reiss, Phys.Lett. B92 (1980) 309; Nucl.Phys. B199 (1982) 223.
14. R. Gatto, G. Sartori and M. Tonin, Phys. Lett. B **28**, 128 (1968).
15. S. Pakvasa and H. Sugawara, Phys. Lett. B **73**, 61 (1978). J. Kubo, A. Mondragon, M. Mondragon and E. Rodriguez-Jauregui, Prog. Theor. Phys. **109**, 795 (2003) [Erratum-ibid. **114**, 287 (2005)] [arXiv:hep-ph/0302196]. Prog. Theor. Phys. **114**, 287 (19yy); F. Caravaglios and S. Morisi, arXiv:hep-ph/0503234; W. Grimus and L. Lavoura, JHEP **0508**, 013 (2005) [arXiv:hep-ph/0504153], JHEP **0601**, 018 (2006) [arXiv:hep-ph/0509239]; R. N. Mohapatra, S. Nasri and H. B. Yu, Phys. Lett. B **639**, 318 (2006) [arXiv:hep-ph/0605020]; R. Jora, S. Nasri and J. Schechter, Int. J. Mod. Phys. A **21**, 5875 (2006) [arXiv:hep-ph/0605069]; O. Felix, A. Mondragon, M. Mondragon and E. Peinado, AIP Conf. Proc. **917**, 383 (2007) [Rev. Mex. Fis. **S52N4**, 67 (2006)] [arXiv:hep-ph/0610061]; K. S. Babu, A. G. Bachri and Z. Tavartkiladze, arXiv:0705.4419 [hep-ph]; E. Ma and G. Rajasekaran, Phys. Rev. D **64**, 113012 (2001) [arXiv:hep-ph/0106291]; K. S. Babu, E. Ma and J. W. F. Valle, Phys. Lett. B **552**, 207 (2003) [arXiv:hep-ph/0206292]; K. S. Babu and X. G. He, arXiv:hep-ph/0507217. A. Zee, Phys. Lett. B **630**, 58 (2005) [arXiv:hep-ph/0508278]; S. Morisi, M. Picariello and E. Torrente-Lujan, Phys. Rev. D **75**, 075015 (2007) [arXiv:hep-ph/0702034]; M. Hirsch, A. S. Joshipura, S. Kaneko and J. W. F. Valle, arXiv:hep-ph/0703046; M. C. F. Chen and K. T. Mahanthappa, Phys. Lett. B **652**, 34 (2007) [arXiv:0705.0714 [hep-ph]]; A. Aranda, arXiv:0707.3661 [hep-ph]; C. Hagedorn, M. Lindner and R. N. Mohapatra, JHEP **0606**, 042 (2006) [arXiv:hep-ph/0602244]; Y. Koide, Eur. Phys. J. C **50**, 809 (2007) [arXiv:hep-ph/0612058], arXiv:0705.2275 [hep-ph]; I. de Medeiros Varzielas and G. G. Ross, Nucl. Phys. B **733**, 31 (2006) [arXiv:hep-ph/0507176]; E. Ma, Mod. Phys. Lett. A **21**, 2931 (2006) [arXiv:hep-ph/0607190], Mod. Phys. Lett. A **21**, 1917 (2006) [arXiv:hep-ph/0607056], arXiv:0709.0507 [hep-ph], arXiv:hep-ph/0701016, arXiv:0705.0327 [hep-ph]; D. B. Kaplan and M. Schmaltz, Phys. Rev. D **49**, 3741 (1994) [arXiv:hep-ph/9311281]; M. Schmaltz, Phys. Rev. D **52**, 1643 (1995) [arXiv:hep-ph/9411383]; C. Hagedorn, M. Lindner and F. Plentinger, Phys. Rev. D **74**, 025007 (2006) [arXiv:hep-ph/0604265]; M. Frigerio and E. Ma, arXiv:0708.0166 [hep-ph]; G. Altarelli, arXiv:0705.0860 [hep-ph]; C. S. Lam, arXiv:0708.3665 [hep-ph]; C. Luhn, S. Nasri and P. Ramond, J. Math. Phys. **48** (2007) 073501 [arXiv:hep-th/0701188], Phys. Lett. B **652** (2007) 27 [arXiv:0706.2341 [hep-ph]], arXiv:0709.1447 [hep-th].
16. I. de Medeiros Varzielas, S. F. King and G. G. Ross, Phys. Lett. B **648** (2007) 201 [arXiv:hep-ph/0607045]; S. F. King and M. Malinsky, Phys. Lett. B **645**, 351 (2007) [arXiv:hep-ph/0610250]; F. Feruglio, C. Hagedorn, Y. Lin and L. Merlo, Nucl. Phys. B **775**, 120 (2007) [arXiv:hep-ph/0702194].
17. S. F. King, Phys. Lett. B **439** (1998) 350 [arXiv:hep-ph/9806440], Nucl. Phys. B **562** (1999) 57 [arXiv:hep-ph/9904210], Nucl. Phys. B **576** (2000) 85 [arXiv:hep-ph/9912492]; G. Altarelli, F. Feruglio and I. Masina,Phys. Lett. B **472** (2000) 382 [arXiv:hep-ph/9907532]; A. Y. Smirnov,Phys. Rev. D **48** (1993) 3264 [arXiv:hep-ph/9304205].
18. G. Altarelli and F. Feruglio, Nucl. Phys. B **720** (2005) 64 [arXiv:hep-ph/0504165], Nucl. Phys. B **741** (2006) 215 [arXiv:hep-ph/0512103]; I. de Medeiros Varzielas, S. F. King and G. G. Ross, Phys. Lett. B **644** (2007) 153 [arXiv:hep-ph/0512313].
19. S. Antusch and S. F. King, Phys. Lett. B **591** (2004) 104 [arXiv:hep-ph/0403053]; S. Antusch, P. Huber, S. F. King and T. Schwetz, JHEP **0704** (2007) 060 [arXiv:hep-ph/0702286].
20. M. R. Ramage and G. G. Ross, JHEP **0508** (2005) 031 [arXiv:hep-ph/0307389].
21. I. de Medeiros Varzielas and G. G. Ross, arXiv:hep-ph/0612220.
22. G. G. Ross and O. Vives, Phys. Rev. D **67** (2003) 095013 [arXiv:hep-ph/0211279]; S. Antusch, S. F. King and M. Malinsky, arXiv:0708.1282 [hep-ph].
23. S. Abel, S. Khalil and O. Lebedev, Nucl. Phys. B **606** (2001) 151 [arXiv:hep-ph/0103320].

Gauge-gravity duality and heavy-ion collisions

D. T. Son

Institute for Nuclear Theory, University of Washington, Seattle, WA 98915-1550, USA

Abstract. We review recent progress in applying the AdS/CFT correspondence in problems relevant for the physics of relativistic heavy-ion collisions.

Keywords: gauge gravity duality, heavy ion collisions
PACS: 1.10.Wx, 04.70.Dy, 11.25.Tq, 47.75.+f

INTRODUCTION

At very high densities or high temperatures ($T \gg \Lambda_{\text{QCD}}$) quarks and gluons become "deconfined" ad we have states of matter very different from ordinary matter. High density matter can be found in the core of neutron stars; there is a large literature on the subject that will not be reviewed here. Matter at very high temperature is called the quark-gluon plasma. It did exist briefly in the early Universe and is recreated in collisions of heavy-ions. The heavy-ion experiments are being carried out at the Relativistic Heavy Ion Collider (RHIC) at Brookhaven National Lab with gold nuclei at the energy of 200 GeV per nucleus.

The findings of RHIC can be summarized as follows. The initial fireball created in heavy-ion collisions have extremely high-energy density, much larger than the energy density of QCD at the crossover from the confined phase to the deconfined phase. The "stuff" that is created shows remarkable collective behavior. One observes its flow under pressure gradient; this phenomenon is called the "elliptic flow," which is the azimuthal asymmetry of the final-particle momentum spectrum in collisions with nonzero impact parameters. One also finds that high-energy jets lose considerable energy if they have to move through the medium. As the result, one sometimes find single-jet events, in which only one of the two jets that were produced back-to-back manages to get out of the medium.

On the other hand, attempts for a theoretical description of the medium in the language of a weakly-interacting system of quarks and gluons have not been too successful. The medium is tentatively given the name "strongly-interacting quark-gluon plasma" (sQGP), with the expectation that its characteristics will be measured experimentally with greater precision and its properties understood theoretically with greater clarity.

Theoretically, we have the following tentative picture of what happens in a heavy ion collion. Right after the collision one has an initial state that can be described at the partonic level (see McLerran's talk in this volume). Then the partons reach thermal equilibrium, after which the system undergoes a stage of expansion which can be described by hydrodynamics. At some moment the particles in the medium stop interacting with each other. This moment is called "freezout," and after this moment

CP957, *Particles, Strings, and Cosmology, 13th International Symposium*
edited by A. Rajantie, C. Contaldi, P. Dauncey, and H. Stoica

particles fly to the detectors.

It should be said that the picture above is only tentative, and many details is still to be worked out. One does not know how the initial partonic state reaches thermal equilibrium, and what is the time required for this thermalization process. Embarassingly, this problem has not been solved even for collisions at asymptotically high energy, where perturbative QCD should apply during the thermalization process.

Why perturbative QCD cannot be more successful in describing heavy-ion collisions? The main reason is that the typical energy scale is not much larger than $\Lambda_{\mathrm{QCD}} \sim 200$ GeV. The problem is made worse by the fact that perturbative expansion for QCD thermodynamics is not well-behaved for any realistic temperature. The expansion parameter at finite temperature is g, not α_s, and is never numerically small. Therefore one has to use the numerical simulations on a lattice to find the thermodynamics of QCD (analytical methods based on partial resummation of the perturbative series have also been suggested). However there are quantities that are fundamentally difficult to extract from lattice data. These are the kinetic coefficients like the shear and bulk viscosities, or in general any quantities that are given by some Green's function in the Minkowski domain (for example the dilepton production rate, which is related to the current-current correlator at time-like momenta).

STRING THEORY AS A TOOL

The paucity of theoretical tools to study strongly coupled gauge plasmas has lead many researchers to explore the AdS/CFT correspondence [1, 2, 3] as an alternative method to study such theory.

The AdS/CFT correspondence has its roots in string theory. It came out of the study of nonperturbative objects called D-branes. It is a conjectural equivalence between the $\mathcal{N}=4$ supersymmetric Yang-Mills theory and type IIB string theory on the $\mathrm{AdS}_5\times\mathrm{S}^5$ background metric

$$ds^2 = \frac{r^2}{R^2}(-dt^2+d\vec{x}^2)+\frac{R^2}{r^2}dr^2+R^2d\Omega_5^2 \tag{1}$$

The metric is a direct product of a five-dimensional space of negative curvature (AdS_5) that is spanned by the coordinates $t,\vec{x},r$ and a five dimensional sphere. R is the radius of curvature. The AdS/CFT dictionary establishes an exact mapping between quantities in the field-theory ($\mathcal{N}=4$ SYM theory) and the string-theory side of the duality. In particular, the 't Hooft coupling in field theory is related to a dimensionless ratio of the curvature radius and the string length,

$$g^2N_c = \left(\frac{R}{l_{\mathrm{st}}}\right)^4 \tag{2}$$

The string length l_{st} is related to the string tension α' by $l_{\mathrm{st}}=\alpha'^{-1/2}$; it can be understood as the typical quantum fluctuations of the string

There is an important consequence of Eq. (2). When the 't Hooft coupling of field theory is large, in the dual gravity picture the curvature scale R is much larger than the string length l_{st}. In this case it is possible to limit oneself to the low-energy limit of string theory, which is supergravity. Large N_c means that it is actually classical supergravity, which is known very well. Therefore, AdS/CFT correspondence allows us to treat the large 't Hooft coupling limit of string theory, where no other method exists.

Finite-temperature AdS/CFT correspondence

One of the first applications of the AdS/CFT correspondence (which was actually found before the precise AdS/CFT conjecture is formulated) is for the thermodynamics of the $\mathcal{N}=4$ SYM theory.

At zero coupling, the pressure of a $\mathcal{N}=4$ SYM plasma can be found by just enumerating all particles in the plasma: 6 real scalars, 4 Weyl fermions, and 1 massless gauge bosons, all in the adjoint representation of the SU(N_c) group,

$$P_0=(N_c^2-1)\frac{\pi^2}{90}T^4\left(6+4\times 2\times\frac{7}{8}+1\times 2\right)=\frac{\pi^2}{6}(N_c^2-1)T^4 \tag{3}$$

At infinite 't Hooft coupling, on the other hand, the entropy density can be found by identifying the entropy density of a plasma with that of the dual black hole,

$$ds^2=\frac{r^2}{R^2}(-fdt^2+dx^2+dy^2+dz^2)+\frac{R^2}{r^2 f}dr^2,\qquad f=1-\frac{r_0^4}{r^4}, \tag{4}$$

The entropy turns out to be equal to $3/4$ of the ideal gas value. Therefore the effect of interactions amounts to a 25% reduction of the pressure.

THE VISCOSITY

The standard textbook definition of the shear viscosity is as follows. Take two large parallel plates separated by a distance d. The space between the two plates is filled with a fluid. Let one plate moves relative to the other with a velocity v. Then the drag force acting on a unit area of the plate is

$$\frac{F}{A}=\eta\frac{v}{d} \tag{5}$$

which defines the viscosity η. It is measured in kg m^{-1} s^{-1}. In d spatial dimensions, shear viscosity η is measured in kg m^{2-d} s^{-1}. The volume density of entropy s is measured in m^{-d} (in units where the Boltzmann constant k_B is set to one). The ratio η/s thus has the dimension of kg m^2 s^{-1}, i.e., the same as the Planck constant $\hbar$. This observation is more than merely a curiosity, as we shall see shortly.

VISCOSITY FROM DUAL GRAVITY DESCRIPTION

We now apply the AdS/CFT correspondence to find the viscosity of a strongly-coupled plasma. Consider a field theory dual to a black-brane metric, for example, the $\mathcal{N}=4$

supersymmetric Yang-Mills theory dual to the metric (4). (Our discussion is not tied to any specific form of the metric.) All black branes have an event horizon ($r = r_0$ for the metric (4)), which is extended along several spatial dimensions (x, y, z in the case of (4)). The dual field theory is at a finite temperature, equal to the Hawking temperature of the black brane.

The shear viscosity of the dual theory can be computed from gravity in a number of approaches[4, 5, 6]. Here we use Kubo's formula, which relates viscosity to equilibrium correlation functions. In a rotationally invariant field theory,

$$\eta = \lim_{\omega \to 0} \frac{1}{2\omega} \int dt\, d\vec{x}\, e^{i\omega t} \langle [T_{xy}(t,\vec{x}), T_{xy}(0,\mathbf{0})] \rangle \,. \tag{6}$$

Here T_{xy} is the xy component of the stress-energy tensor (one can replacc T_{xy} by any component of the traceless part of the stress tensor). We shall now relate the right hand side of (6) to the absorption cross section of low-energy gravitons.

According to the AdS/CFT correspondence[7], the stress-energy tensor $T_{\mu\nu}$ couples to metric perturbations at the boundary. Following Klebanov[8], let us consider a graviton of frequency ω, polarized in the xy direction, and propagating perpendicularly to the brane. In the field theory picture, the absorption cross section of the graviton by the brane measures the imaginary part of the retarded Greens function of the operator coupled to h_{xy}, i.e., T_{xy},

$$\sigma_{\rm abs}(\omega) = -\frac{2\kappa^2}{\omega} \operatorname{Im} G^{\rm R}(\omega) = \frac{\kappa^2}{\omega} \int dt\, d\vec{x}\, e^{i\omega t} \langle [T_{xy}(t,\vec{x}), T_{xy}(0,\mathbf{0})] \rangle \,, \tag{7}$$

where $\kappa = \sqrt{8\pi G}$ appears due the normalization of the graviton's action. Comparing (6) and (7), one finds

$$\eta = \frac{\sigma_{\rm abs}(0)}{2\kappa^2} = \frac{\sigma_{\rm abs}(0)}{16\pi G} \,. \tag{8}$$

The absorption cross section $\sigma_{\rm abs}$, on the other hand, is calculable classically by solving the linearized wave equation for h^x_y. It can be shown (see Appendix) that under rather general assumptions the equation for h^x_y is the same as that of a minimally coupled scalar. The absorption cross section for the scalar is constrained by a theorem[9, 10], which states that in the low-frequency limit $\omega \to 0$ this cross section is equal to the area of the horizon, $\sigma_{\rm abs} = a$. Since $s = a/4G$, one immediately finds that

$$\frac{\eta}{s} = \frac{\hbar}{4\pi} \,, \tag{9}$$

where $\hbar$ is now restored. This shows that the ratio η/s does not depend on the concrete form of the metric within the assumptions of[9, 10].

Indeed, explicit calculations of the viscosity using the AdS/CFT correspondence or the "membrane paradigm" technique show that the ratio η/s is the $1/(4\pi)$ for Dp[4, 6], M2 and M5[11] branes and $\mathcal{N} = 2^*$ deformations of the D3 metric[6, 12].

A VISCOSITY BOUND

Dual gravity description of gauge theories is valid in the regime of infinitely strong coupling. As Eq. (9) shows, in this regime the ratio η/s appears to be universal (independent of the coupling constant and other microscopic details of the theory). Let us now argue that the ratio η/s approaches infinity in the limit of vanishing coupling.

The entropy density s of a weakly coupled system is proportional to the number density of quasiparticles n,

$$s \sim n. \tag{10}$$

The shear viscosity is proportional to the product of the energy density and the mean free time (time between collisions) τ

$$\eta \sim n\varepsilon\tau, \tag{11}$$

where ε is the average energy per particle (which is of the order of the temperature T). Therefore

$$\frac{\eta}{s} \sim \varepsilon\tau. \tag{12}$$

Now, in order for the quasiparticle picture to be valid, the width of the quasiparticles must be small compared to their energies, i.e., one should have

$$\frac{\hbar}{\tau} \ll \varepsilon \tag{13}$$

which means that

$$\frac{\eta}{s} \gg \hbar. \tag{14}$$

The observation that η/s is a constant in strongly coupled theories with gravity dual and is large in weakly coupled theories prompts us to formulate the "viscosity bound" conjecture: in any finite-temperature field theory, the ratio of shear viscosity to entropy density cannot be smaller than the value of this ratio in theories with gravity duals:

$$\frac{\eta}{s} \geq \frac{\hbar}{4\pi}. \tag{15}$$

As we have seen, the bound (15) can be understood as a consequence of the uncertainty principle: the product of the energy and the mean free time of a quasiparticle cannot be smaller than $\hbar$. The precise numerical coefficient $1/(4\pi)$ cannot, however, be obtained from the uncertainty principle alone.

CHECKS OF THE VISCOSITY BOUND

The viscosity bound has a verifiable consequence for thermal $\mathcal{N}=4$ supersymmetric Yang-Mills theory. It implies any value of the 't Hooft coupling $g^2_{\mathrm{YM}}N$ the ratio η/s is larger than $1/(4\pi)$. In particular, it implies that the first correction in the strong-coupling expansion of η/s is *positive*.

Corrections in inverse power of 't Hooft coupling correspond to string-theory corrections to supergravity. For the entropy density, it has been computed by Gubser, Klebanov and Tseytlin:[13]

$$s = \frac{\pi^2}{2} N_c^2 T^3 \left[1 + \frac{15}{8} \zeta(3) (2 g_{\rm YM}^2 N_c)^{-3/2} \right]. \tag{16}$$

The correction to η, and hence η/s, was found by Buchel, Liu and Starinets:[14]

$$\eta = \frac{\pi}{8} N_c^2 T^3 \left[1 + \frac{75}{4} \zeta(3) (2 g_{\rm YM}^2 N_c)^{-3/2} \right], \tag{17}$$

$$\frac{\eta}{s} = \frac{1}{4\pi} \left[1 + \frac{135}{8} \zeta(3) (2 g_{\rm YM}^2 N_c)^{-3/2} \right]. \tag{18}$$

The first correction to η/s is positive, in agreement with the conjectured bound.

Moreover, since the viscosity bound does not contain the speed of light c, one can try to check it on nonrelativistic laboratory liquids. First let us check the conjecture for the most ubiquitous fluid—water. Under normal conditions ($P = 0.1$ MPa, $T = 298.15$ K) the viscosity of water is $\eta \approx 0.89 \times 10^{-3}$ Pa s and the entropy density is $s \approx 2.8 \times 10^{29}$ m^{-3}. The ratio η/s is 380 times larger than $\hbar/(4\pi)$. Using standard tables[15] one can find η/s for many liquids and gases at different temperatures and pressures. The viscosity bound is well satisfied all substances. Liquid helium reaches the smallest value of η/s, but this value still exceeds the bound by a factor of about 10.

Recently there has been a suggestion of a possible counter example to the viscosity bound above [16]. One consider a large N_c gauge theory with $N_f = O(N_c)$ light flavors and one heavy flavor. Then there will be N_f types of heavy-light mesons. Taking an appropriate limit it is shown in Ref. [16] that the ratio η/s can be decreased to zero by using the Gibbs mixing entropy to increase s while keeping η fixed. However in this example it can be shown that η and s exists in different ranges of length scales. The system is metastable with a lifetime of only during times of order a power of $\ln(N_f)$, so η can be measured only for length scales smaller than this power of $\ln(N_f)$. On the other hand the thermodynamic entropy exists only for volumes large enough to contains at least one meson of each species, i.e., larger than N_f. Therefore there is no region of length scales where both η and s are meaningful concepts.

OTHER APPLICATIONS

The gauge/gravity duality has also been recently applied to the problem of finding the drag force acting on a quark moving through a finite-temperture plasma. First one inserts quarks in the fundamental representation into the theory. How to do that has been known for some time: one inserts N_f space-filling $D7$-branes into the $D3$ metric. A moving quark in a plasma is mapped into a string with one end on the $D7$ brane, with its tail being dragged along the horizon. The problem is then mapped onto a extremizing the string worldsheet action, and can be solved. For details see Refs. [17, 18, 19]. Another related work is Ref. [20].

CONCLUSION

It is useful to compare the viscosity bound with two other bounds widely discussed in the literature: the entropy bound (e.g., in its covariant formulation[21]) which states that the entropy of a region of space is limited by the area A of the region's boundary,

$$S \leq \frac{c^3}{\hbar G}\frac{A}{4}, \tag{19}$$

and the Bekenstein bound[22], which states that the entropy of a system is limited by the product of its linear size R and mass M,

$$S \leq \frac{c}{\hbar} 2\pi RM. \tag{20}$$

The viscosity bound can also be written as an upper bound on the entropy,

$$S \leq \frac{4\pi}{\hbar}\eta V, \tag{21}$$

where V is the total volume. It is similar to the Bekenstein bound in the sense that it covers non-gravitating systems despite having the origin in the theory of gravitation. The viscosity bound stands apart as the only bound that does not involve the speed of light. Because of this feature, it is relevant for non-relativistic systems, in contrast to the other bounds. One may hope that it is possible to relate the viscosity bound to the two other bounds, and eventually to the generalized second law of thermodynamics. It will be important to learn whether the viscosity bound suffers from the same problems as the other bounds such as the species problem.

The bound (15) is most useful for strongly interacting systems where no theoretical calculation of viscosity is possible. One of such systems is the quark-gluon plasma (QGP) created in heavy ion collisions which behaves in many respects as a droplet of a liquid. As the viscosity of the QGP is not computable reliably except at inaccessibly high temperatures, the information provided by the conjectured bound is extremely valuable. There are experimental hints that the viscosity of the QGP at temperatures achieved by the Relativistic Heavy Ion Collider is surprisingly small, possibly close to saturating the viscosity bound[23]. Further investigations may reveal whether the QGP conforms to the viscosity bound.

Another possible application of the viscosity bound is trapped atomic gases. By using the Feshbach resonance, strongly interacting Fermi gases of atoms have been created recently. These gases have been observed to behave hydrodynamically[24]. Currently the viscosity bound is the only source of information about the viscosity of these gases. It would be very interesting to test the bound experimentally using these gases.

ACKNOWLEDGMENTS

This work is supported, in part, by DOE grant No. DE-FG02-00ER41132.

REFERENCES

1. J. M. Maldacena, Adv. Theor. Math. Phys. **2**, 231 (1998).
2. S. S. Gubser, I. R. Klebanov and A. M. Polyakov, Phys. Lett. B **428**, 105 (1998).
3. E. Witten, Adv. Theor. Math. Phys. **2**, 253 (1998).
4. G. Policastro, D. T. Son and A. O. Starinets, *Phys. Rev. Lett.* **87**, 081601 (2001).
5. G. Policastro, D. T. Son and A. O. Starinets, *J. High Energy Phys.* **0209**, 043 (2002).
6. P. Kovtun, D. T. Son and A. O. Starinets, *J. High Energy Phys.* **0310**, 064 (2003).
7. O. Aharony, S. S. Gubser, J. M. Maldacena, H. Ooguri and Y. Oz, *Phys. Rep.* **323**, 183 (2000).
8. I. R. Klebanov, *Nucl. Phys.* **B496**, 231 (1997).
9. S. R. Das, G. W. Gibbons and S. D. Mathur, *Phys. Rev. Lett.* **78**, 417 (1997).
10. R. Emparan, *Nucl. Phys.* **B516**, 297 (1998).
11. C. P. Herzog, *J. High Energy Phys.* **0212**, 026 (2002).
12. A. Buchel and J. T. Liu, hep-th/0311175.
13. S. S. Gubser, I. R. Klebanov and A. A. Tseytlin, *Nucl. Phys.* **B534**, 202 (1998).
14. A. Buchel, J. T. Liu and A. O. Starinets, hep-th/0406264.
15. E. W. Lemmon, M. O. McLinden and D. G. Friend, *NIST Chemistry WebBook* (http://webbook.nist.gov); J. D. Cox, D. D. Wagman and V. A. Medvedev, *CODATA Key Values for Thermodynamics* (http://www.codata.org).
16. T. D. Cohen, Phys. Rev. Lett. **99**, 021602 (2007).
17. C. P. Herzog, A. Karch, P. Kovtun, C. Kozcaz and L. G. Yaffe, JHEP **0607**, 013 (2006).
18. J. Casalderrey-Solana and D. Teaney, Phys. Rev. D **74**, 085012 (2006).
19. S. S. Gubser, Phys. Rev. D **74**, 126005 (2006).
20. H. Liu, K. Rajagopal and U. A. Wiedemann, Phys. Rev. Lett. **97**, 182301 (2006).
21. R. Bousso, *J. High Energy Phys.*, **9907**, 004 (1999).
22. J. D. Bekenstein, *Phys. Rev.* **D23**, 287 (1981).
23. E. Shuryak, *Prog. Part. Nucl. Phys.* **53**, 273 (2004).
24. K. M. O'Hara et al., *Science* **298**, 2179 (2002).

Is the cosmological "constant" a nonlocal quantum residue of discreteness of the causal set type?

Rafael D. Sorkin

Perimeter Institute, 31 Caroline Street North, Waterloo ON, N2L 2Y5 Canada
Department of Physics, Syracuse University, Syracuse, NY 13244-1130, U.S.A.

Abstract. The evidence for an accelerating Hubble expansion appears to have confirmed the heuristic prediction, from causal set theory, of a fluctuating and "ever-present" cosmological term in the Einstein equations. A more concrete phenomenological model incorporating this prediction has been devised and tested, but it remains incomplete. I will review these developments and also mention a possible consequence for the dimensionality of spacetime.

INTRODUCTION

The inference from causet theory of a fluctuating cosmological constant Λ is possibly the earliest theoretical prediction of a non-zero Λ; and yet relatively little work has been devoted to developing it. In its original form [1], the prediction yielded only an order of magnitude estimate for Λ, namely that its current value should be about $\pm 10^{-120}$ in Planck units. When evidence started accumulating for a Λ of just this size, it seemed time to try to embed the original prediction in a more complete model. The resulting model, as elaborated by Scott Dodelson, Patrick Greene, Maqbool Ahmed and me, had some ad hoc elements, but it realized concretely the original implication of an "ever-present Λ", that is one whose (fluctuating) magnitude is always comparable to the Hubble scale. [2] [3]

My talk today will review these developments in the hope that people will be encouraged to carry the underlying idea further. I will explain where the original heuristic prediction came from, and I will describe a concrete model that arose from it. Before doing so, however I want to mention the two greatest weaknesses that the model had.

The first weakness was a certain arbitrariness in how one interprets the idea of a varying Λ. The second was the imposition of spatial homogeneity and isotropy on the model, ie the assumption that the metric takes the FRW form. The first weakness has been largely overcome in subsequent work by Maqbool Ahmed but the second still needs to be tackled.

CP957, *Particles, Strings, and Cosmology, 13th International Symposium*
edited by A. Rajantie, C. Contaldi, P. Dauncey, and H. Stoica

LAMBDA FORGOTTEN

The cosmological constant originated as a non-solution to a non-problem, *viz* the fact that the Einstein equation

$$\frac{1}{\kappa}G^{ab} + \Lambda g^{ab} = T^{ab} \tag{1}$$

does not admit a static cosmos as a solution if the Λ-term is omitted. This, of course, was not a real problem because the cosmos is not static. Nor does the inclusion of Λ solve this non-problem, because the resulting static cosmos is unstable to collapse or unbounded expansion, as is well known. In view of this inauspicious beginning, the introduction of Λ must have seemed unmotivated, and the prejudice grew up among cosmologists that Λ was 0. With few exceptions, they either set $\Lambda = 0$, or they never even bothered to mention it at all, despite many indications from quantum field theory and quantum gravity that things could not possibly be that simple.

LAMBDA REMEMBERED

Within the last decade or so, all that has changed, and the origin of the so called "dark energy" is recognized as a central question of astronomy. Although many lines of thought and observation seem to have contributed to the change (including, for example, the problem that without Λ (or with $\Lambda < 0$), the cosmos seemed to be younger than some of its contents), the two most persuasive arguments concerned the CMB relic radiation[1] and the observations of distant supernovas. Let us briefly recall both of these arguments.

SPATIAL FLATNESS AND MISSING MASS

As I understand it, this argument proceeds from the angular size of the so called "acoustic peaks" in the CMB to the spatial flatness of the universe (in our vicinity) to the need for a new term in the Einstein equation. The density and temperature of the medium that emitted the CMB photons resulted from the initial gravitational collapse and subsequent fluid oscillation of that medium. This process produced bright regions of approximately known diameter and distance from us, which correspond to the peaks one sees in plots of CMB brightness vs. angular size. But since the *apparent* angular size of a distant object is greater in spherical space than in flat space (and correspondingly less in hyperbolic space), we can infer the radius of curvature of our cosmos, assuming it to be spatially homogeneous and isotropic (Friedmann cosmos). The best fit is to zero curvature, i.e. to spatial flatness. But for such a cosmos, the $a = b = 0$ component of eq. (1) reads, if we take $\Lambda = 0$,

$$3\left(\frac{\dot{a}}{a}\right)2 = \rho \tag{2}$$

---

[1] CMB = cosmic microwave background radiation. It was already detected as early as 1940 [4], but its significance was not appreciated, and it was forgotten.

where $\rho = T^{00}$ is the total energy density in matter (including the so-called dark matter inferred from galactic rotation curves and gravitational lensing). Also, a is the scale-factor or "radius of the universe", $\dot{a} = da/d\tau$ with τ = proper time, and I've set $\kappa \equiv 8\pi G = 1$. By comparing this equality with various, more or less direct measurements of ρ, one finds that the latter would have to be tripled in order to satisfy (2). With Λ restored on the other hand, one obtains instead of (2) the equation

$$3\left(\frac{\dot{a}}{a}\right)2 - \Lambda = \rho \tag{3}$$

from which Λ can be determined once ρ and $H = \dot{a}/a$ are known. The conclusion seems to be that either Λ is nonzero or something rather like it is out there, carrying twice the effective mass-density of non-gravitational matter.

DIM DISTANT SUPERNOVAS

Observations of supernovas (of type IA) have yielded the most direct evidence for a positive Λ, because they let us deduce the value of the "acceleration" $\ddot{a}$ from a plot of luminosity vs. redshift, and the analysis depends only on the well understood behavior of electromagnetic fields in curved spacetime. This yields for the "luminosity distance" or normalized "dimness" d_L of a known source that is not too far away, the equation

$$Hd_L = z + \frac{1}{2}\left(1 + \frac{a\ddot{a}}{\dot{a}\dot{a}}\right) z2 + O(z3) \tag{4}$$

where $H = \dot{a}/a$ is the Hubble constant and z the redshift of the light from the supernova. Clearly, one can deduce both $\dot{a}$ and $\ddot{a}$ if one knows how d_L varies with z. Moreover one sees that, other things being equal, a larger $\ddot{a}$ will produce an image that is *dimmer* at equal redshift. The graph of d_L as a function of z thus shifts upward, and this is what has been seen.

Actually, the sign of this effect can be deduced from the equivalence principle, using only what we know of the Doppler shift in flat spacetime. According to the equivalence principle, spacetime *is* flat in our neighborhood, whence what is usually described as the expansion of space we can reinterpret locally as the flow of galaxies away from us. In such a coordinate system, two supernovas of equal brightness are at equal spatial distance from us. Hence their light was emitted at the same time t_0 in the past. Now let these two stars be in two different spacetimes with the same expansion rate H but different accelerations $\ddot{a}$. In the cosmos with *greater* acceleration, the expansion rate at t_0 was smaller, whence the recession rate of the supernova was smaller, whence its redshift z was also less. Equivalently, the supernova appears dimmer at the same redshift.

To complete the argument, observe that a positive Λ serves to increase $\ddot{a}$. This follows from the $a = b = 1$ component of (1), and can be remembered from the fact that, for pure $\Lambda > 0$, one obtains a de Sitter cosmos, in which a grows exponentially with proper time τ. By contrast, in a cosmos with $\Lambda = 0$, the expansion necessarily slows down, responding to the gravitational attraction of the galaxies for each other.

THE Λ PUZZLE

The evidence we have just reviewed points to a cosmological constant of magnitude, $\Lambda \approx 10^{-120}\kappa^{-2}$, and this raises two puzzles:[2] Why is Λ so small without vanishing entirely, and Why is it so near to the critical density $\rho_{critical} = 3H2$ that appears in eqs. (2) and (3)? Is the latter just a momentary occurrence in the history of the universe (which we are lucky enough to witness), or has it a deeper meaning?

Clearly both puzzles would be resolved if we had reason to believe that $\Lambda \approx H2$ always. In that case, the smallness of Λ today would merely reflect the large age of the cosmos. But such a Λ would conflict with our present understanding of nucleosynthesis in the early universe and of "structure formation" more recently. (In the first case, the problem is that the cosmic expansion rate influences the speed with which the temperature falls through the "window" for synthesizing the light nuclei, and thereby affects their abundances. According to (3) a positive Λ at that time would have increased the expansion rate, which however is already somewhat too big to match the observed abundances. In the second case, the problem is that a more rapid expansion during the time of structure formation would tend to oppose the enhancement of density perturbations due to gravitational attraction, making it difficult for galaxies to form.) But neither of these reasons excludes a *fluctuating* Λ with typical magnitude $|\Lambda| \sim H2$ but mean value $\langle\Lambda\rangle = 0$. The point now is that *such fluctuations can arise as a residual, nonlocal quantum effect of discreteness*, and specifically of the type of discreteness embodied in the causal set.

FEATURES OF CAUSET THEORY NEEDED IN THE FOLLOWING

In order to explain this claim, I will need to review some basic aspects of causet theory. [5] According to the causal set hypothesis, the smooth manifold of general relativity dissolves, near the Planck scale, into a discrete structure whose *elements* can be thought of as the "atoms of spacetime". These atoms can in turn be thought of as representing "births", and as such, they carry a relation of ancestry that mathematically defines a *partial order*, $x \prec y$. Moreover, in our best dynamical models [6], the births happen sequentially in such a way that the number n of elements plays the role of an auxiliary time-parameter. (In symbols, $n \sim t$.)[3]

Two basic assumptions complete the kinematic part of the story by letting us connect up a causet with a continuum spacetime. One posits first, that the underlying microscopic order $\prec$ corresponds to the macroscopic relation of before and after, and second, that the number of elements N comprising a region of spacetime equals the *volume* of that region

[2] I prefer the word puzzle or riddle to the word problem, which suggests an inconsistency, rather than merely an unexplained feature of our theoretical picture.

[3] It is an important constraint on the theory that this auxiliary time-label n should be "pure gauge" to the extent that it fails to be determined by the physical order-relation $\prec$. That is, it must not influence the dynamics, this being the discrete analog of general covariance.

in fundamental (i.e. Planckian) units. (In slogan form: geometry = order + number.) The equality between number N and volume V is not precise however, but subject to Poisson fluctuations, whence instead of $N = V$, we can write only

$$N \sim V \pm \sqrt{V} \,. \tag{5}$$

(These fluctuations express a "kinematical randomness" that seems to be forced on the theory by the noncompact character of the Lorentz group.)

To complete the causet story, one must provide a "dynamical law" governing the birth process by which the causet "grows" (the discrete counterpart of equation (1)). This we still lack in its quantum form, but for heuristic purposes we can be guided by the classical sequential growth (CSG) models referred to above; and this is what I have done in identifying n as a kind of time-parameter.

EVER-PRESENT Λ

We can now appreciate why one might expect a theory of quantum gravity based on causal sets to lead to a fluctuating cosmological constant. Let us assume that at sufficiently large scales the effective theory of spacetime structure is governed by a gravitational path-integral, which at a deeper level will of course be a sum over causets. That n plays the role of time in this sum suggests that it must be held fixed, which according to (5) corresponds to holding V fixed in the integral over 4-geometries. If we were to fix V exactly, we'd be doing "unimodular gravity", in which setting it is easy to see that V and Λ are conjugate to each other in the same sense as energy and time are conjugate in nonrelativistic quantum mechanics. [This conjugacy shows up most obviously in the Λ-term in the gravitational action-integral, which is simply

$$-\Lambda \int \sqrt{-g}\, d^4x = -\Lambda V \,. \tag{6}$$

It can also be recognized in canonical formulations of unimodular gravity [7], and in the fact that (owing to (6)) the "wave function" $\Psi(^3g;\Lambda)$ produced by the unrestricted path-integral with parameter Λ is just the Fourier transform of the wave function $\Psi(^3g;V)$ produced at fixed V.] In analogy to the $\Delta E \Delta t$ uncertainty relation, we thus expect in quantum gravity to obtain

$$\Delta\Lambda\, \Delta V \sim \hbar \tag{7}$$

Remember now, that even with N held exactly constant, V still fluctuates, following (5), between $N + \sqrt{N}$ and $N - \sqrt{N}$; that is, we have $N \sim V \pm \sqrt{N} \Rightarrow V \sim N \pm \sqrt{V}$, or $\Delta V \sim \sqrt{V}$. In combination with (7), this yields for the fluctuations in Λ the central result

$$\Delta\Lambda \sim V^{-1/2} \tag{8}$$

Finally, let us *assume* that, for reasons still to be discovered, the value about which Λ fluctuates is strictly zero: $\langle\Lambda\rangle = 0$. (This is the part of the Λ puzzle we are *not* trying to solve.[4]) A rough and ready estimate identifying spacetime volume with the Hubble

[4] But see the ansatz (10) below, which yields $\langle\Lambda\rangle = 0$ automatically.

scale H^{-1} then yields

$$V \sim (H^{-1})4 \sim H^{-4} \Rightarrow \Lambda \sim V^{-1/2} \sim H2 \sim \rho_{critical}$$

(where I've used that $\Lambda = \Lambda - \langle\Lambda\rangle$ since $\langle\Lambda\rangle = 0$). In other words, Λ would be "ever-present" (at least in 3+1 dimensions).

A CONCRETE MODEL INCORPORATING EQUATION (8)

In trying to develop (8) into a more comprehensive model, we not only have to decide exactly which spacetime volume 'V' refers to, we also need to interpret the idea of a varying Λ itself. Ultimately the phenomenological significance of V and Λ would have to be deduced from a fully developed theory of quantum causets, but until such a theory is available, the best we can hope for is a reasonbly plausible scheme which realizes (8) in some recognizable form.

As far as V is concerned, it pretty clearly wants to be the volume to the past of some hypersurface, but which one? If the local notion of "effective Λ at x" makes sense, and if we can identify it with the Λ that occurs in (8), then it seems natural to interpret V as the volume of the past of x, or equivalently (up to Poisson fluctuations) as the number of causet elements which are ancestors of x:

$$V = \text{volume}(\text{past}(x)) .$$

One could imagine other interpretations,[5] but this seems as simple and direct as any.

As far as Λ is concerned, the problems begin with eq. (1) itself, whose divergence implies (at least naively[6]) that Λ = constant. The model of [2] and [3] addresses this difficulty in two stages. First it limits itself to spacetimes of the FRW form, i.e. it assumes that the cosmos is spatially homogeneous and isotropic (which of course requires for consistency that Λ *also* be spatially homogeneous). Having assumed this, we might as well assume in addition that space is flat ($k = 0$), since that simplifies the equations and matches current data. The Einstein equations (1) then reduce (with $\kappa = 1$) to a pair of ordinary differential equations known as the Friedmann equations:

$$3(\dot{a}/a)2 = \rho + \rho_\Lambda \tag{9a}$$

$$2\ddot{a}/a + (\dot{a}/a)2 = -(p + p_\Lambda) , \tag{9b}$$

where $\rho_\Lambda \equiv \Lambda$ and $p_\Lambda \equiv -\Lambda$ (corresponding to the familiar "equation of state" of Λ, $p = -\rho$.)

Now in the usual case where Λ is time-independent, equation (9b) is a consequence of (9a); and conversely the two equations are incompatible when $\dot{\Lambda} \neq 0$, this being

[5] For example, interpretations in which Λ is not a spacetime field at all, but must be understood more nonlocally.

[6] Naively because it neglects the circumstance that a fluctuating Λ would be something like a stochastic Brownian function that need not even have a derivative.

precisely the difficulty with which we began. To bypass this incompatibility we are forced to modify the Friedmann equations.[7] The most straightforward way of doing so is to retain only one of them, or possibly some other linear combination of (9a) and (9b). In reference [2] we followed this approach by adopting (9a) as our "dynamical guide" and discarding (9b). This choice is appealing because the resulting dynamics is easy to simulate, and because it admits an alternative description in which *neither* (9a) nor (9b) is compromised, but instead the "equation of state of Λ" is modified in a simple, local manner. Fortunately, changing one's "guide" by adopting a different linear combination of (9a) and (9b) appears to alter nothing qualitatively [3], so let me ignore such possibilities for now. Then our dynamical scheme is just (9) with

$$\begin{aligned} \rho_\Lambda &= \Lambda \\ p_\Lambda &= -\Lambda - \dot{\Lambda}/3H. \end{aligned}$$

Finally, to complete our model and obtain a closed system of equations, we need to specify Λ as a (stochastic) function of V, and we need to choose it so that $\Delta\Lambda \sim V^{-\frac{1}{2}}$. But this is actually easy to accomplish, if we begin by observing that (with $\kappa = \hbar = 1$) $\Lambda = S/V \approx S/N$ can be interpreted as the action per causet element that is present even when the spacetime curvature vanishes. (As one might say, it is the action that an element contributes just by virtue of its existence.[8]) Now imagine that each element contributes (say) $\pm\hbar$ to S, with a random sign. Then S is just the sum of N independent random variables, and we have

$$S/\hbar \sim \pm\sqrt{N} \sim \pm\sqrt{V/\ell 4}\,,$$

where $\ell \sim \sqrt{\hbar\kappa}$ is the fundamental time/length of the underlying theory, which thereby enters our model as a free phenomenological parameter. This in turn implies, as desired, that

$$\Lambda = S/V \sim \pm\frac{\hbar/\ell 2}{\sqrt{V}} \tag{10}$$

We have thus arrived at an ansatz that, while it might not be unique, succeeds in producing the kind of fluctuations we were seeking. Moreover, it lends itself nicely to simulation by computer.

NUMERICAL SIMULATION

Mathematically, our model is defined in the first place by eq. (9a), and secondly by the ansatz for Λ described above, according to which $S = V\Lambda$ is the sum of $N = V/\ell 4$ independent random contributions, where V is the volume of the spacetime region within the past light cone of any point in the hypersurface on which Λ is being evaluated.

---

[7] Notice in this connection that (9b) will not even be well defined if (9a) holds and Λ is a function of Brownian motion type.

[8] More properly one should probably think of each element as contributing a multiplicative *phase* $\exp(iS)$. However, our analysis here is only being used to suggest a simple ansatz for Λ, and this ansatz can stand on its own in the present, phenomenological context.

Strictly speaking, this scheme is not consistently defined since it mixes discrete variables with derivatives; however with $N \sim 10^{240}$ elements currently to our past, we are so close to the continuum that we can safely treat our model as defined by a pair of stochastic differential equations,

$$\frac{da}{a} = \sqrt{\frac{\rho+\Lambda}{3}} d\tau \tag{11}$$

$$V d\Lambda = V d(S/V) = dS - \Lambda \dot{V} d\tau \tag{12}$$

where (11) is just a rewriting of (9a). (Perhaps, though, we should call these integro-differential equations, inasmuch as V, and therefore also the "stochastic driving term" dS, depends on the whole past-history of $a(\tau)$.)

At this point we could try to give our equations a more precise mathematical meaning, but it is just as easy to pass directly to a finite-difference form of them suitable for a computer. I suspect that the scheme described below corresponds to the so-called Itô form of the stochastic system (11)-(12). If so, one might consider also the Stratonovich alternative, but we have not done so.

In references [2] and [3] our model was simulated as follows. Let a_i be the cosmic scale-factor at the ith step and similarly for N_i, V_i, S_i and Λ_i. Let $\rho = \rho_{matter} + \rho_{radition}$, with ρ_{matter} taken to be "dust" scaling like $1/a3$. Begin at the "Planck time" with the appropriate ratio of $\rho_{matter}/\rho_{radition}$ to end up matching our current universe. Then evolve by iterating the following steps.

- $a_{i+1} = a_i + a_i\sqrt{\frac{\rho_i+\Lambda_i}{3}}(\tau_{i+1} - \tau_i)$
- Given a_{i+1}, compute V_{i+1} using[9]

$$V(\tau) = \frac{4\pi}{3}\int_0^\tau d\tau' a(\tau')3 \left(\int_{\tau'}^{\tau} \frac{d\tau''}{a(\tau'')}\right)3$$

- $N_{i+1} = V_{i+1}/\ell 4$
- $S_{i+1} = S_i + \xi\sqrt{N_{i+1} - N_i}$ (ξ gaussian with unit variance)
- $\Lambda_{i+1} = S_{i+1}/V_{i+1}$

(The random variable ξ is gaussian thanks to the central limit theorem. It has unit variance because our ansatz arbitrarily took each causet element to contribute $\pm\hbar = \pm 1$ to S. More justified would be $\pm\sigma$ with σ of order unity, but we may omit this new parameter since it coalesces with ℓ in its effect on the model.)

An extensive discussion of the simulations can be found in [3] and [2]. The most important finding was that "tracking behavior" was indeed observed: the absolute value of Λ follows $\rho_{radiation}$ very closely during the era of radiation dominance, and then follows ρ_{matter} when the latter dominates. Secondly, the simulations confirmed that Λ fluctuates with a "coherence time" which is $O(1)$ relative to the Hubble scale. Thirdly, a range of present-day values of Ω_Λ is produced, and these are $O(1)$ when $\ell 2 = O(\kappa)$. (Notice in this connection that the variable Λ of our model cannot simply be equated to the observational parameter Λ^{obs} that gets reported on the basis of supernova observations, for

[9] By rearranging the formulas, one can avoid recomputing this whole integral at each iteration.

example, because Λ^{obs} results from a fit to the data that presupposes a constant Λ, or if not constant then a deterministically evolving Λ with a simple "equation of state". It turns out that correcting for this tends to make large values of Ω_Λ more likely [3].) Fourthly, the Λ-fluctuations affect the age of the cosmos (and the horizon size), but not too dramatically. In fact they tend to increase it more often than not. Finally, the choice of (9a) for our specific model seems to be "structurally stable" in the sense that the results remain qualitatively unchanged if one replaces (9a) by some linear combination thereof with (9b), as discussed above.

I should emphasize though, that all these results come from simulations with the free parameter ℓ closer to 3 than to 1 (ie to $\sqrt{\kappa}$). If one lowers ℓ much beyond this, the negative fluctuations in Λ typically grow so large that the simulation cannot continue long enough for the cosmos to reach its present size. (The square root in (11) becomes imaginary. In the "linear combination" version of the model, the evolution can continue, but the cosmos recollapses to a singularity.) This creates a tension in the model between the need for the cosmos to reach its present size and the need for the present value of $\Omega_\Lambda = \Lambda/3H2$ to be as big as it seems to be.

SUMMARY AND OUTLOOK

Heuristic reasoning rooted in the basic hypotheses of causal set theory predicted $\Lambda \sim \pm 1/\sqrt{V}$, in agreement with current data. But a fuller understanding of this prediction awaits the "new QCD" ("quantum causet dynamics"). Meanwhile, a reasonably coherent phenomenological model exists, based on simple general arguments. It is broadly consistent with observations but a fuller comparison is needed. It solves the "why now" problem: Λ is "ever-present". It predicts further that $p_\Lambda \neq -\rho_\Lambda$ ($w \neq -1$) and that Λ has probably changed its sign many times in the past.[10] The model contains a single free parameter of order unity that must be neither too big nor too small.[11] In principle the value of this parameter is calculable, but for now it can only be set by hand.

In this connection, it's intriguing that there exists an analog condensed matter system the "fluid membrane", whose analogous parameter is not only calculable in principle from known physics, but might also be measurable in the laboratory! [9]

That our model so far presupposes spatial homogeneity and isotropy is no doubt its weakest feature. Indeed, the ansatz on which it is based strongly suggests a generalization such that Λ-fluctuations in "causally disconnected" regions would be independent of each other; and in such a generalization, spatial inhomogeneities would inevitably arise. Such inhomogeneities were a source of worry in [2] and their potential to disagree badly with the isotropy of the CMB brightness has recently been emphasized in [10] and

[10] It also tends to favor the existence of something, say a "sterile neutrino", to supplement the energy density at nucleosynthesis time. Otherwise, we might have to assume that Ω_Λ had fluctuated to an unusually small value at that time. It also carries the implication that "large extra dimensions" will not be observed at the LHC [8].

[11] unless we want to try to make sense of imaginary time (= quantum tunneling?) or to introduce new effects to keep the right hand side of (9) positive (production of gravitational waves? onset of large-scale spatial curvature or "buckling"?).

[11]. On the other hand they could also act as a new type of source for density fluctuations. Without a generalized model allowing for spatial inhomogeneities, one cannot do better than guessing. Let me just note that such a model would evidently have to replace the Einstein equation by some sort of "stochastic PDE", just as our homogeneous model led to a stochastic form of the Friedmann equations.

CLOSING REMARKS

In itself the smallness of Λ is a riddle and not a problem. But in a fundamentally discrete theory, recovery of the continuum *is* a problem, and I think that the solution of this problem will also explain the smallness of Λ. (The reason is that if Λ were to take its "natural", Planckian value, the radius of curvature of spacetime would also be Planckian, but in a discrete theory such a spacetime could no more make sense than a sound wave with a wavelength smaller than the size of an atom. Therefore the only kind of spacetime that can emerge from a causet or other discrete structure is one with $\Lambda \ll 1$.) One can also give good reasons why the emergence of a manifold from a causet must rely on some form of nonlocality. The size of Λ should also be determined nonlocally then, and this is precisely the kind of idea realized in the above model.

One pretty consequence of this kind of nonlocality is a certain restoration of symmetry between the very small and the very big. Normally, we think of G (gravity) as important on large scales, with $\hbar$ (quantum) important on small ones. But we also expect that on still smaller scales G regains its importance once again and shares it with $\hbar$ (quantum gravity). If the concept of an ever-present Λ is correct then symmetry is restored, because $\hbar$ rejoins G on the largest scales in connection with the cosmological constant.

Finally, let me mention a "fine tuning" that our model has not done away with, namely the tuning of the spacetime dimension to $d = 4$. In any other dimension but 4, Λ could not be "ever-present", or rather it could not remain in balance with matter. Instead, the same crude estimates that above led us to expect $\Lambda \sim H2$, lead us in other dimensions to expect either matter dominance ($d > 4$) or Λ-dominance ($d < 4$). Could this be a dynamical reason favoring $3+1$ as the number of noncompact dimensions? [12] [10]

A LAST WORD

The cosmological constant is just as constant as Hubble's constant.

ACKNOWLEDGMENTS

It's a pleasure to thank Barbara and Syd Bulman-Fleming for acquainting me with reference [4]. Research at Perimeter Institute for Theoretical Physics is supported in part by the Government of Canada through NSERC and by the Province of Ontario through MRI. This research was partly supported by NSF grant PHY-0404646.

REFERENCES

1. Rafael D. Sorkin, "A Modified Sum-Over-Histories for Gravity", reported in *Highlights in gravitation and cosmology: Proceedings of the International Conference on Gravitation and Cosmology, Goa, India, 14-19 December, 1987*, edited by B. R. Iyer, Ajit Kembhavi, Jayant V. Narlikar, and C. V. Vishveshwara, see pages 184-186 in the article by D. Brill and L. Smolin: "Workshop on quantum gravity and new directions", pp 183-191 (Cambridge University Press, Cambridge, 1988).

   ———, "On the Role of Time in the Sum-over-histories Framework for Gravity", paper presented to the conference on The History of Modern Gauge Theories, held Logan, Utah, July 1987, published in *I*nt. J. Theor. Phys. **33**, 523-534 (1994).

   ———, "First Steps with Causal Sets", in R. Cianci, R. de Ritis, M. Francaviglia, G. Marmo, C. Rubano, P. Scudellaro (eds.), *General Relativity and Gravitational Physics* (Proceedings of the Ninth Italian Conference of the same name, held Capri, Italy, September, 1990), pp. 68-90 (World Scientific, Singapore, 1991).

   ———, "Forks in the Road, on the Way to Quantum Gravity", talk given at the conference entitled "Directions in General Relativity", held at College Park, Maryland, May, 1993, published in *I*nt. J. Th. Phys. **36**, 2759–2781 (1997), `gr-qc/9706002`,
   `http://www.physics.syr.edu/~sorkin/some.papers/`
   (Syracuse University Preprint: SU-GP-93-12-2).

   ———, "Discrete Gravity", a series of lectures to the *First Workshop on Mathematical Physics and Gravitation*, held Oaxtepec, Mexico, Dec. 1995 (unpublished).

   Y. Jack Ng and H. van Dam, "A small but nonzero cosmological constant", *I*nt. J. Mod. Phys D. **10**, 49 (2001) `hep-th/9911102`.

2. Maqbool Ahmed, Scott Dodelson, Patrick Greene and Rafael D. Sorkin, "Everpresent Λ", *P*hys. Rev. D **69**, 103523 (2004), `astro-ph/0209274`,
   `http://www.physics.syr.edu/~sorkin/some.papers/`.

3. Maqbool Ahmed, Doctoral dissertation (Syracuse University, 2006)

   Maqbool Ahmed and Rafael D. Sorkin, "Everpresent Lambda II: Structural Stability and Better Comparison with Observations" (in preparation)

4. Andrew McKellar, "The problem of possible molecular identification for interstellar lines", *P*ublications of the Astronomical Society of the Pacific **53**, 233-235 (1941) `http://adsbit.harvard.edu/cgi-bin/nph-iarticle_query?1941PASP...53..233M`

5. Luca Bombelli, Joohan Lee, David Meyer and Rafael D. Sorkin, "Spacetime as a Causal Set", *P*hys. Rev. Lett. **59**, 521-524 (1987).

   Rafael D. Sorkin, "Causal Sets: Discrete Gravity (Notes for the Valdivia Summer School)", in *Lectures on Quantum Gravity* (Series of the Centro De Estudios Científicos), proceedings of the Valdivia Summer School, held January 2002 in Valdivia, Chile, edited by Andrés Gomberoff and Don Marolf (Plenum, 2005) `gr-qc/0309009`

   Fay Dowker, "Causal sets and the deep structure of Spacetime", in *100 Years of Relativity - Space-time Structure: Einstein and Beyond*" ed Abhay Ashtekar (World Scientific, to appear) `gr-qc/0508109`.

   Joe Henson, "The causal set approach to quantum gravity" `gr-qc/0601121`.

6. David P. Rideout and Rafael D. Sorkin, "A Classical Sequential Growth Dynamics for Causal Sets", *P*hys. Rev. D **61**, 024002 (2000) `gr-qc/9904062`.

   Madhavan Varadarajan and David Rideout, "A general solution for classical sequential growth dynamics of Causal Sets" *P*hys. Rev. D **73**, 104021 (2006) `gr-qc/0504066`.

7. M. Henneaux and C. Teitelboim, "The Cosmological Constant and General Covariance", *P*hys. Lett. B **222**, 195 (1989)

8. Rafael D. Sorkin, "Big extra dimensions make Λ too small", in the proceedings of the Second International Workshop "DICE 2004" held September, 2004, Piombino, Italy, edited by Hans-Thomas Elze `gr-qc/0503057` *B*razilian Journal of Physics **35**, 280-283 (2005)

9. Joseph Samuel, Supurna Sinha, “Surface Tension and the Cosmological Constant” *Phys. Rev. Lett.* **97**, 161302 (2006), arXiv:cond-mat/0603804
10. John D. Barrow, “A Strong Constraint on Ever-Present Lambda”, *Phys.Rev. D* **75**, 067301 (2007), arXiv:gr-qc/0612128
11. Joe Zuntz, "The CMB in a Causal Set Universe" (in preparation)
12. Rafael D. Sorkin, “Why the Cosmological Constant can’t quite vanish”, Talk delivered at the Third Eastern Gravity Meeting, held March 1999 at Cornell University, Ithaca, NY.

Emergent Quantum Mechanics and Emergent Symmetries

Gerard 't Hooft

Institute for Theoretical Physics
Utrecht University
and

Spinoza Institute
Postbox 80.195
3508 TD Utrecht, the Netherlands

e-mail: g.thooft@phys.uu.nl
internet: http://www.phys.uu.nl/~thooft/

Abstract. Quantum mechanics is 'emergent' if a statistical treatment of large scale phenomena in a locally deterministic theory requires the use of quantum operators. These quantum operators may allow for symmetry transformations that are not present in the underlying deterministic system. Such theories allow for a natural explanation of the existence of gauge equivalence classes (gauge orbits), including the equivalence classes generated by general coordinate transformations. Thus, local gauge symmetries and general coordinate invariance could be emergent symmetries, and this might lead to new alleys towards understanding the flatness problem of the Universe.

STARTING POINTS

The 19th century view of the physical world was that all dynamical laws of Nature were *deterministic*, which means that every single event in the Universe was assumed to be an irrevocable consequence of configurations in its past; according to Laplace the fate of every single atom at all times in the future has been sealed right from the beginning, by the boundary conditions of the Universe at $t = 0$. A theory of the world would be most satisfactory if it not only uniquely fixes the local dynamical laws, but also provides for one particular unavoidable choice of the boundary conditions, not only in the spacelike directions, but also at time $t = 0$, the moment that would now be called the "Big Bang". This would have been the ultimate deterministic scenario. Today it is usually dismissed as being totally at odds with what has been learned from the theory of quantum mechanics. Today, it certainly seems as if the only theories we have are the ones where the outcomes of most experiments can only be predicted in terms of probabilities. The root of this situation is the occurrence of a large number of non-commuting operators, describing the values of measurable quantities.

The observation we wish to investigate further is that non-commuting operators do not necessarily imply a breakdown of determinism. One may perfectly well introduce non-commuting operators if the evolution law itself is still deterministic. For instance,

CP957, *Particles, Strings, and Cosmology, 13th International Symposium*
edited by A. Rajantie, C. Contaldi, P. Dauncey, and H. Stoica

in studying the planetary system, it would be totally legal to introduce an operator that interchanges the positions of Mars and Venus, or an operator that changes the tilt of the rotation axis of the Earth. Such operators would 'violate Bell's inequalities' just like the ones we use in ordinary quantum mechanics[1][2]. Of course, one must ask what the circumstances may be that would necessitate such a step, in a theory that nevertheless is deterministic. If we study the evolution of planetary systems or bouncing billiard balls, non-commuting operators never appear to play any substantial role[1]. A completely satisfactory answer to this question has not yet been found, but there do exist systems where one could strongly suspect that exactly this happens; these theories are *complemented* by the addition of non-commuting operators.

We expect the non-commuting operators to begin playing more essential roles in systems that have a *dynamical vacuum solution*. This is one dynamical, chaotic solution that may appear to be completely static at large distance scales, but microscopically it is not static at all. For simplicity, we refer to this microscopical scale as the 'Planck scale'. According to our scenario, it will be fundamentally impossible to recover deterministic rules of behavior at large scales, just because things are surrounded by a chaotic vacuum already at the Planck scale (or thereabouts). So we should not be surprised at having an *effective* theory at large scales that produces statistical fluctuations in its answers to what might be expected to be observed at large scales. We suspect that the introduction of non-commuting operators may be essential to understand the statistical features at large scales. This is suspected to be what we call quantum mechanics today.

It should be emphasized that, in this view, we are not opting for a quantum mechanics that would be only "approximate". Non-commuting operators evolve quantum mechanically even in deterministic systems (think of how the Earth-Mars exchange operator would evolve in time). Quantum mechanics as a description of the evolution laws of operators is probably exactly valid in the same sense that the laws of thermodynamics follow exactly from classical (or quantum) mechanics at atomic scales. In this work, we furthermore claim that the mathematical features of quantum mechanics indeed can arise naturally, in particular the need to introduce wave functions, whose amplitudes squared are to be interpreted as probabilities. There are problems, however, and it will be important to confront these directly, rather than try to hide them. Thus, quantum mechanics as it is known today has become the theory enabling us to produce the *best possible predictions* for the future, given as much information as we can give about the system's past, in any conceivable experimental setup. Quantum mechanics is *not* a description of the actual course of events between past and future.

Quantum mechanics will exactly reproduce the statistical features of Nature at a local scale, in our laboratories. The only effect our present considerations will have on the pursuit of an improved, accurate theory of quantum gravity and cosmology is that the universe in itself is required to be controlled by equations beyond quantum mechanics. The argument for this is simple. Quantum mechanics has been tailored by us to describe the statistical outcomes of experiments when repeated many times, locally in some laboratory. We may well assume this theory to be exact in describing local

[1] There is one important exception: the infinitesimal displacement operators, which are needed to define, for instance, Lyapunov exponents; see also Section

statistics. The entire Universe, however (in particular when we are talking about a closed universe), is itself an 'experiment' carried out only once, and all events in it are unique. The question whether a single event took place or not can only be answered by 'yes' or 'no', but there is no probabilistic answer. A theory that yields 'maybe' as an answer should be recognized as an inaccurate theory. If this is what we should believe, then only deterministic theories describing the entre cosmos should be accepted. There can be no 'quantum cosmology'.

New light might be shed precisely on those problems that are connected to our attempts to describe Nature at the Planck scale, as well as at the cosmological scale. A notorious difficulty is the mystery of the cosmological constant, which appears to be fantastically small yet not zero.[3] We think that indeed we have something to say about that. This problem cannot go away as long as we hold on to quantum mechanics as a primary foundation of our theories rather than an emergent effect.

If the vacuum state indeed is a complicated dynamical solution of local deterministic equations, this implies among others that two systems can never be completely isolated from one another. They are connected by the dynamical degrees of freedom of the surrounding vacuum. This should explain why isolated systems appear to feature stochastic behavior; they simply aren't isolated at all.

We shall furthermore argue that *symmetries* such as rotation symmetry, translation, Lorentz invariance, but also local gauge symmetries and coordinate reparametrization invariance, might all be *emergent*, together with quantum mechanics itself. They are exact locally, but they are not properties of the underlying ToE; they certainly are not a property of the boundary conditions of the Universe. We furthermore conjecture that *information* is not conserved in the deterministic description[4]. This assumption appears to be necessary in order to obtain a chaotic yet relatively ordered vacuum state, and also was found to be desired in order to understand the positivity of the hamiltonian.

The notion of 'free will' is given a meaning that deviates from most standard views on the subject;[5][6][7] ours[8] appears to go back as far as Benedict de Spinoza: an individual's actions are completely dictated by laws of Nature, yet this does not exempt us from our responsibilities for our actions. The 'free will postulate' is presently seen as an axiom in the reconstruction of the interpretation of quantum mechanics. This postulate will not be needed in the form usually given. We replace it by the condition that a theory or model of Nature should give appropriate responses for every conceivable initial state. So, in investigating our model, we have the 'freedom' to choose our initial state at will; whatever that state is, the model should tell me what will happen next. This is referred to as the 'unconstrained initial state' requirement.[9]

MATHEMATICS: EMERGENT SYMMETRY ON A LATTICE

Consider a system with deterministic laws[10][11]:

$$\frac{\mathrm{d}}{\mathrm{d}t}\vec{x}(t)=\vec{f}(\vec{x})\ , \tag{1}$$

where $\vec{x}$ represents a set of variables such as the positions and the velocities of the planets at a given time, and $\vec{f}$ can be any ordinary vector function. It allows us to introduce quantum operators such as

$$\hat{\vec{p}} = -i\frac{\partial}{\partial\vec{x}}\,; \tag{2}$$

$$\hat{H} = \hat{\vec{p}}\cdot\vec{f}(\vec{x})+\vec{g}(\vec{x})\,. \tag{3}$$

The latter operator is not the standard energy observable for things such as planets, but the generator of their time evolution:

$$\frac{\mathrm{d}}{\mathrm{d}t}\vec{x}(t) = -i[\vec{x}(t),\hat{H}] = \vec{f}(\vec{x})\,. \tag{4}$$

Thus, we have a description of a deterministic system in a quantum mechanical language. This language allows us to identify emergent symmetries, but not yet many rigorous examples are known or understood. One interesting example of an emergent symmetry, or rather the emergent enhancement of a given symmetry, is the following.[12] Consider a model of spins on a discrete lattice in one space, one time dimension (extension of this model to more dimensions is straightforward). The sites on the lattice are indicated by the integers (x,t), where t is a discretized time variable, but only the *even* sites are occupied by a single boolean variable $\sigma_{x,t}$. So,

$$\sigma_{x,t} = \pm 1\,, \qquad x+t = \text{even}\,. \tag{5}$$

The equation of motion is chosen to be

$$\sigma_{x+1,t+1} = \sigma_{x,t}\,\sigma_{x+2,t}\,\sigma_{x+1,t-1}\,. \tag{6}$$

This system has an obvious classical translation symmetry:

$$\begin{pmatrix} x \\ t \end{pmatrix} \to \begin{pmatrix} x+\delta x \\ t+\delta t \end{pmatrix}\,; \qquad \delta x+\delta t = \text{even}\,. \tag{7}$$

We now claim that the quantum symmetry is larger. To see this, we call $\sigma_{x,t} = \sigma^3_{x,t}$, and we introduce the quantum operators $\sigma^1_{x,t}$ on the odd sites: $x+t = \text{odd}$. On a Cauchy surface, here defined to be the lines $t=t_0$ and $t=t_0-1$, the operator σ^1_{x,t_0} is defined to switch the variables $\sigma^3_{x,t-1}$, $\sigma^3_{x-1,t}$, and $\sigma^3_{x+1,t}$, leaving the others fixed. One quickly establishes how these few switches propagate in time, both to the future and to the past: the operator σ^1_{x,t_0} switches all variables $\sigma^3_{y,t}$ for which $|y-x|<|t-t_0|$.

By inspection then, one finds that the switches produced by four σ^1 operations that are arranged in an elementary square cancel each other out completely, hence

$$\sigma^1_{x,t}\,\sigma^1_{x-1,t+1}\,\sigma^1_{x+1,t-1}\,\sigma^1_{x,t+2} = 1\,. \tag{8}$$

This is exactly the same equation as the equation of motion for the σ^3 *operators*, eq. (6)! Note, that the σ^1 operators are only defined on the odd lattice sites. The transformation

$$\sigma^3_{x,t} \leftrightarrow \sigma^1_{x+1,t}\,, \tag{9}$$

is our first example of an emergent symmetry due to the introduction of a non-commutative operator such as σ^1. It adds the translations over odd distances in space-time to the symmetry translations (7) that we already had.

HARMONIC OSCILLATORS

One way to continue our discussion is to imagine elementary building blocks for a realistic model. A realistic model of the universe could consist of a large number of tiny segments that are mutually interacting with their neighbors. These tiny segments could be discrete, like the lattice elements of the previous section, but it is better to re-introduce a time continuum. Even if time were to be discrete in some model, one can still imagine filling up the time segments between the discrete events, even if no events take place in these continua. One now first formulates how the segments would interact if they were not coupled to their neighbors, and then adds the interaction, assuming all of these to be deterministic.

If every segment has only a finite number of different states to choose from, then inevitably a non-interacting segment would evolve in a cyclic motion: it would be periodic. So, prior to the inclusion of the interaction, our non-interacting universe will consist of many segments moving periodically. Let us first describe that motion using our new "quantum" language. Let a segment have a period T, and call the periodic variable φ, taking values in the segment $[0, 2\pi)$. Our quantum hamiltonian is taken to be[2]

$$H = \omega p ; \qquad \omega = \frac{2\pi}{T} ; \qquad p = -i\frac{\partial}{\partial\varphi} . \tag{10}$$

Because of the periodicity, we must have

$$e^{-iHT} = 1 \quad \rightarrow \quad H = \frac{2\pi n}{T} = \omega n ; \qquad n = 0, \pm 1, \pm 2, \dots . \tag{11}$$

This is exactly the spectrum of a quantum harmonic oscillator, except for the emergence of negative energy states. The states with $n < 0$ are forbidden, for some reason. Of course, we also do not have the 'vacuum energy' $\frac{1}{2}\omega$, which would have emerged if we would have required $e^{-iHT} = -1$. This minus sign is as harmless as an overall addition of $\frac{1}{2}\omega$ to the hamiltonian, but perhaps it will mean something in a more complete theory; we will ignore it for the time being.

The disappearance of the negative energy modes is very troublesome, however. In a single oscillator, one might still say that energy is conserved, and once it is chosen to be positive, it will stay positive. However, when two or more of these systems interact, they might exchange energy, and we will have to explain why the real world appears to have interactions only in such a way that only positive energy states are occupied. This is a very difficult problem, and, disguised one way or another, it keeps popping up throughout our investigations. It still has not been solved in a completely satisfactory manner, but we can try to handle this difficulty, and one then reaches a number of quite interesting conclusions. In short, our problem is this: in a deterministic theory, one can reproduce quantum-like mathematics in a multitude of ways, but in many cases one encounters hamiltonian functions that are either periodic (in case time is taken to be

---

[2] the function $g(\varphi)$ in Eq. (3) is taken to be zero here.

discrete), or not bounded from below (when time is continuous). Can the real world nevertheless be approximated by, or rather exactly reproduced in, some deterministic model? What then causes the hamiltonian of the real world to be bounded below, with a very special lowest energy state, the 'vacuum', as a result? Without this positivity of H, we would not have thermodynamics. The hamiltonian is conjugated to time. Is there something about time that we are not handling correctly?

To get some ideas, consider two periodic systems, with periods $T_1 = 2\pi/\omega_1$ and $T_2 = 2\pi/\omega_2$. It would be described by the hamiltonian

$$H = \omega_1 n_1 + \omega_2 n_2 \,. \tag{12}$$

As soon as interactions are added (which can easily be imagined to be deterministic as well), transition amplitudes will be proportional to $\langle n_1, n_2 | H^{\text{int}} | m_1, m_2 \rangle$, where $n_{1,2}$ and $m_{1,2}$ take the entire range of integral values between $-\infty$ and ∞. This would make the state $|0,0\rangle$ unstable. States with *both* n_1 and n_2 negative would not cause a problem; we could interpret these as the bra states. It is the states $n_1 < 0,\ n_2 > 0$, or vice versa, which must somehow be excluded.

A very crude, but nevertheless suggestive observation is the following. Let us write the time evolution operator as $e^{-i(H_1 t_1 + H_2 t_2)}$. Since φ is proportional to time, we could actually replace t_i here by φ_i/ω_i, but the present notation is more suggestive. Then, if E_i are the eigenstates of H_i,

$$E_1 t_1 + E_2 t_2 = \tfrac{1}{2}\Big((E_1 + E_2)(t_1 + t_2) + (E_1 - E_2)(t_1 - t_2)\Big) \,, \tag{13}$$

so that we see an 'uncertainty relation' not only saying that $E_i\,\delta t_i \approx \frac{1}{2}$, but also

$$2\delta(t_1 + t_2) \approx 1/(E_1 + E_2) \,, \qquad 2\delta(t_1 - t_2) \approx 1/|E_1 - E_2| \,. \tag{14}$$

Since we demand only to have states with $E_1 E_2 \geq 0$, one easily derives

$$E_1 + E_2 \geq |E_1 - E_2| \quad \to \quad \delta(t_1 + t_2) \leq \delta(t_1 - t_2) \,, \tag{15}$$

The spread in relative time differences must be larger than the spread in average time! This might indicate that there is information loss: only the information about average time must play a role in the interactions, whereas information about relative time, or relative positions, is gradually lost.

INFORMATION LOSS

The prototype model with information loss is a simple automaton having four states $\{i\}$, $i = 1, \cdots, 4$, with the following evolution rule (here discrete in time):

$$\{1\} \to \{2\} \to \{3\} \to \{1\} \,; \quad \{4\} \to \{2\} \,. \tag{16}$$

In such a simple model, one could decide once and for all that state $\{4\}$ is superfluous; after one time step it never is reached anymore. In a complicate universe, however, it is impossibly difficult to distinguish the states, like state $\{4\}$, that can never be reached, from the states that continue to reappear in a periodic motion; what is more, the universe is too large anyway to ever reach a perfectly periodic mode. Rather than removing the superfluous states, it is technically more convenient to combine states into equivalence classes. Two states are equivalent iff within a definite amount of time these two states evolve into exactly the same state. Thus, states $\{1\}$ and state $\{4\}$ form one equivalence class, which we denote as the Dirac ket state $|1\rangle$. Thus, our model is unitary only in terms of these equivalence classes: $|1\rangle \rightarrow |2\rangle \rightarrow |3\rangle \rightarrow |1\rangle$.

In our model of two interacting periodic automata, as described in the previous section, we now assume that, *before* the interaction is switched on, information loss is to be added to the system, such that information concerning $\varphi_1 - \varphi_2$ is erased, while information concerning $\varphi_1 + \varphi_2$ is preserved[3]. Only then interactions can be admitted. After the interaction, one then again obtains a periodic system, so that the procedure can be repeated to describe all interactions.

Some of the details of this process have been further described in Ref[11]. One not only sees the need for information loss to occur in the deterministic system, one also sees that the information loss will be huge. There are two places in the real world where we expect this information loss to play decisive roles. One is the physics of black holes. According to the holographic principle, the amount on information in this universe should grow no faster than one bit of information per Planckian surface element, as in black holes. The surface area of the universe is huge, so there is a gigantic amount of information that is not lost, but in Planckian units, the volume of the universe is far bigger, and all bits and bytes in the 3-dimensional bulk apparently do disappear.

The other place where one might suspect information loss to take place is in non-Abelian gauge theories. indeed, the information equivalence classes do resemble the gauge equivalence classes. In practice, they seem to play the same role: information concerning the gauge parameters of a theory does not seem to propagate in time. Rather than saying that the gauge degrees of freedom are “unphysical", one might conclude that the gauge parameters contain information that, somewhere during the evolution of our fields, got lost: two states in a different gauge may evolve both into the same state in the same gauge. Needless to say that this is only a conjecture, but it seems to make sense.

EMERGENT SYMMETRIES

Thus, unitarity of the quantum evolution only holds in terms of the gauge equivalence classes, as is routinely taken for granted in gauge theories. Local gauge invariance thus could be interpreted as an emergent symmetry. What about other symmetries such as translation, rotation, and Lorentz invariance? There is one easy way to elevate discrete symmetries into continuous ones using quantum operators. Consider for example a discrete displacement operator U, defined on wave functions $|\{\psi(x)\}\rangle$ by

$$U|\{\psi(x)\}\rangle = |\{\psi(x-1)\}\rangle \,, \tag{17}$$

[3] the coefficients of proportionality indeed have to be adapted, see ref[11].

or by the commutation rule

$$xU = U(x+1)\ . \tag{18}$$

Its eigenstates are defined by

$$U|p,r\rangle = e^{ip}|p,r\rangle\ , \tag{19}$$

where $-\pi < p \leq \pi$, and r describes other quantum numbers. A fractional displacement operator can then be defined by

$$U(a) = e^{iap}\ , \tag{20}$$

where now a can be any (real or complex) number. Symmetry under discrete displacements then also implies symmetry under the fractional displacements (20), which is unitary for real a. The only price paid for this extension is that, for fractional a, the transformation transforms "ontological" states into quantum superpositions, even if the underlying dynamics is deterministic.

Thus, our world could be deterministic and discrete at the Planck scale, but since we only know about continuous symmetries, we are always dealing with quantum superpositions.

If the non-Abelian gauge classes are indeed information equivalence classes, an even more daring conjecture could be that the same could be true for the classes of the coordinate transformations in general relativity. Their being continuous could be an emergent phenomenon as described above, and their gauge orbits could be the information equivalence classes. This could mean that the "ontological" theory could be defined on an 'ordinary' flat space-time. The nice feature of this conjecture would be that it would explain the strange preference of our Universe for flat spacial dimensions: dark matter and dark energy cancel out (in the spacelike direction).

A SIMPLE MODEL

In this section, we describe a simple model from which we can derive an existence theorem:

> *For any quantum system there exists at least one deterministic model that reproduces all its dynamics after prequantization.*

To keep the argument transparent, we consider only a finite dimensional subspace of Hilbert space. Let Schrödinger's equation be

$$\frac{\mathrm{d}\psi}{\mathrm{d}t} = -iH\psi\ ; \qquad H = \begin{pmatrix} H_{11} & \cdots & H_{1N} \\ \vdots & \ddots & \vdots \\ H_{N1} & \cdots & H_{NN} \end{pmatrix}\ . \tag{21}$$

We now claim that this system, with the same Hamiltonian, is reproduced by the following deterministic system. There are two degrees of freedom, ω and φ, of which the

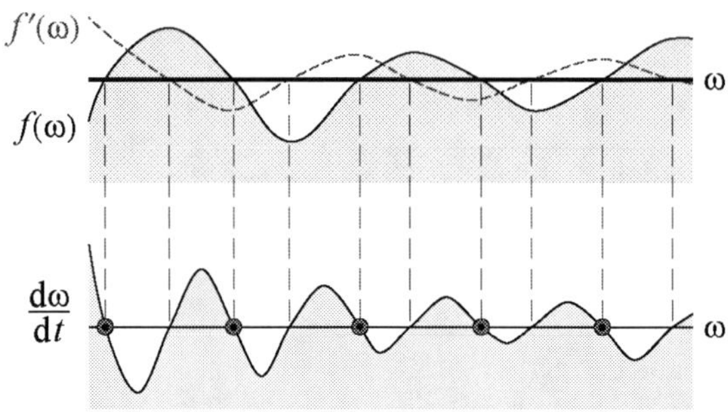

FIGURE 1. Above: sketches of $f(\omega)$ and of $f'(\omega)$ (dotted). Below: a sketch of the product $-\kappa f(\omega)\, f'(\omega)$. The red dots indicate its attractive zeros, which now coincide with the zeros of f.

latter is periodic: $\varphi \in [0, 2\pi)$, or $\psi(\omega, \varphi) = \psi(\omega, \varphi + 2\pi)$. Now, take as classical equations of motion:

$$\frac{\mathrm{d}\varphi(t)}{\mathrm{d}t} = \omega(t) \quad ; \tag{22}$$

$$\frac{\mathrm{d}\omega(t)}{\mathrm{d}t} = -\kappa\, f(\omega) f'(\omega)\,, \qquad f(\omega) = \det(H - \omega)\,. \tag{23}$$

The function $f(\omega)$ has zeros exactly at the eigenvalues of H. Its derivative, the function $f'(\omega)$, has zeros between the zeros of f, see Fig. 1.

This way, we achieve the situation that all eigenvalues ω_i, $i = 1, \cdots, N$ are attractive zeros, as can be read off from the Figure. Depending on the initial configuration, one of the eigenvalues of the matrix H is rapidly approached. If $\omega_{i+\frac{1}{2}}$ is defined to be the zero of $f'(\omega)$ between ω_i and ω_{i+1}, then the regions $\omega_{i-\frac{1}{2}} < \omega < \omega_{i+\frac{1}{2}}$ are the equivalence classes associated to the final state $\omega = \omega_i$. Since the convergence towards ω_i is exponential in time, one can say that for all practical purposes the limit value ω_i is soon reached, at which point the system indeed enters into a limit cycle. At this cycle, the variable φ is periodic in time with period $T = 2\pi/\omega$. The pre-quantum states can be Fourier transformed in φ:

$$\psi(\omega, \varphi) = \sum_n e^{in\varphi} \psi_n(\omega)\,. \tag{24}$$

At $t \to \infty$, we have

$$\psi(\omega, \varphi, t) \to \sum_n \psi_n(\omega_i, 0)\, e^{in(\varphi - \omega_i t)}\,. \tag{25}$$

At this point, we can argue that ω_i is the ontological energy. In this particular model, we can be even more explicit. The quantum number n is absolutely conserved, so, we can use the superselection rule that the universe settles for one particular value for n, say, $n = 1$. Then, the states $\psi_1(\omega_i)$ can be exactly identified with the energy eigenstates of the original quantum system. This completes our mathematical mapping. In this model, interference is a formality: we are always free to consider probabilistic superpositions, described by superimposing wave functions ψ_i with different i.

REFERENCES

1. J. S. Bell, "On the Einstein-Rosen-Podolsky paradox," *Physica* **1** (1964) 195, Peprinted as chapter 2 of Ref. [2]; "Free variables and local causality," *Epistemological Letters* Feb. 1977. Reprinted as chapter 12 of Ref. [2].
2. J. S. Bell, *Speakable and unspeakable in quantum mechanics* (Cambridge Univ. Press, Cambridge, 1987).
3. T. Padmanabhan, *Dark Energy and Gravity*, Invited Review, arXiv:0705.2533, *Gen. Rel. Grav.* to be publ., Issue on Dark Energy, edited by G.F.R.Ellis et al.
4. G. 't Hooft, *Quantum Gravity as a Dissipative Deterministic System, Class. Quant. Grav.* **16** (1999) 3263, gr-qc/9903084.
5. J. Conway and S. Kochen, quant-ph/0604079, *Found. Phys.* **36** (2006) 1441;
6. R. Tumulka, "Comment on the 'Free Will Theorem'", *Found. Phys.* **37** (2007) 186.
7. A. Bassi and G.C. Ghirardi, "The Conway-Kochen argument and relativistic GRW models", quant-ph/0610209, *Found. Phys.* **37** (2007) 169..
8. J. Conway, in *New Scientist*, 6 May 2006, p. 8: "Free will - you only think you have it".
9. G. 't Hooft, *The Free-Will Postulate in Quantum Mechanics*, quant-ph/0701097.
10. *Quantummechanical behaviour in a deterministic model, Found. Phys. Lett.* **10** (1997) 105; quant-ph/9612018.
11. *The mathematical basis for deterministic quantum mechanics*, quant-ph/0604008; to be publ. in Proceedings "Dice 2006", Sptember 11-17, 2006, Piombino, Italy, H.-Th. Elze et al, eds.
12. G. 't Hooft, *Quantum discrete theory by Hilbert Space Extension, Nucl. Phys.* **B342** (1990) 471.

Quantum Resolution of Cosmological Singularities using AdS/CFT

Neil Turok

DAMTP, Centre for Mathematical Sciences, Wilberforce Rd, Cambridge CB3 0WA

Abstract. I review recent work with Ben Craps and Thomas Hertog[1] on the quantum resolution of cosmological singularities using the AdS/CFT correspondence. In particular, we study a five dimensional cosmology, showing that one can define a consistent quantum evolution of the dual boundary theory across the classical singularity. An intriguing consequence of the instability, approximate scale-invariance, and asymptotic freedom of the boundary theory is that an approximately scale-invariant spectrum of stress-energy perturbations is generated. I briefly comment on four dimensional generalizations and the implications for cosmological perturbations in the bulk.

Keywords: Cosmological Singularity, AdS/CFT Correspondence, Cosmological Perturbations
PACS: 98.80.-k,11.10.Kk,11.25.Yb

INTRODUCTION

Over the past decade, a series of powerful new observations has transformed our understanding of the structure of the universe on lagre scales, revealing a remarkably simple picture. The universe is well-described by a spatially flat background geometry with the simplest possible perturbations: Gaussian, Linear, Adiabatic, Scalar and Scale-Invariant, or GLASSy perturbations for short. The astonishing simplicity of this result cries out for a simple theoretical explanation.

For the last 25 years, inflationary theory has dominated cosmology, and some slow-roll inflationary models are consistent with a flat, GLASSy universe. However, these models are not without their problems. First, they are based on the assumption that the universe started out inflating. There is no accepted explanation for this initial state. Its specialness may be quantified by the gravitational entropy of de Sitter space, roughly the inverse of the the effective cosmological constant in Planck units. The gravitational entropy during inflation is roughly a googol times smaller than that of a vacuous state with today's value of the dark energy. The corresponding probability for inflation is around $e^{-10^{100}}$! An alternate estimate is based on assuming a uniform measure on phase space, with the assumption that observationally indistinguishable universes (universes which are very flat today) should be identified. This measure gives a probability of roughly e^{-3N} for N inflationary e-foldings[2], also distressingly small for phenomenologically acceptable numbers of e-foldings.

Second, as is well-known, inflationary models require a significant degree of fine tuning, with dimensionless parameters in the Lagrangian typically tuned to 10^{-10} or less in order to suppress the density perturbations to an acceptable level. The fine tuning is greatly exacerbated by having to account simultaneously for super-dense inflationary energy and for today's dark energy, whose density is a googol times smaller. Even in

CP957, *Particles, Strings, and Cosmology, 13th International Symposium*
edited by A. Rajantie, C. Contaldi, P. Dauncey, and H. Stoica

a landscape-type model, with a huge number of possible vacua, it is hard to see why the initial conditions for the universe were "aimed" so precisely to achieve such a tiny, positive value of the dark energy density. This conundrum has driven many to accept that the universe we see might have been anthropically selected from a "multiverse" in which all the landscape states are realized. However, no satisfactory measure on the multiverse has yet been proposed and in its absence no reliable predictions are possible.

The most fundamental failure of conventional approaches to inflation, however, is that they fail to deal with the cosmological singularity. It hardly needs saying that any cosmological model in which the initial singularity remains unresolved rests on shaky foundations. Therefore, it seems to me that *none* of the proposed landscape vacua make sense as cosmologies unless the initial singularity, which they all possess, is resolved.

I believe that these considerations provide a more than adequate motivation for a radical re-think, and over the past few years, Paul Steinhardt and myself, and others, have been attempting to construct radically different cosmological scenarios. In the cyclic model[3], for example, the standard cosmological puzzles are solved by physical processes occurring before the (last) big bang, and high energy inflation is not required. In such a scenario, there is far more time available and technically natural mechanisms exist whereby the dark energy can relax to a tiny value over many cycles, with no need for anthropic selection.[4]

Finally, if string theory really predicts a landscape of vacua, then the requirement of a smooth passage through the singularity may well provide a crucial selection mechanism influencing or even determining the final configuration of the extra dimensions. Like everything else in the universe, they too came out of the initial singularity!

The key challenge for a cyclic model is whether passage through the singularity is possible. Some progress on this has been made within heterotic M-theory, where it was shown[5] that the lowest energy states, represented by M2-branes wrapped around the M-theory dimension, dimensionally reduce to fundamental strings whose equation of motion is generically regular across the end-of-the-world brane collision. However, a fully quantum formulation of the problem including the higher KK modes of the winding M2-branes, is currently beyond reach.

It is natural to turn for guidance to the AdS/CFT correspondence[6], which has emerged as an extremely powerful tool for the study of quantum gravity. The correspondence is now believed to provide a complete, non-perturbative description of string theory in certain situations. Hertog and Horowitz[7], and others[8] have discussed how the correspondence might be extended to deal with cosmological singularities.

In essence, there are only two possibilities: either time begins (or ends) at cosmological singularities, or it does not. In either case, a departure from the laws of classical general relativity is needed. If cosmological singularities mark the beginning or end of time, an even more radical departure is required: there can be no notion of time or even causality there. A simpler and more conservative possibility, which we have pursued, is that in curing the perturbative UV infinities of quantized general relativity, string/M-theory might provide a consistent rule for dynamical evolution through cosmological singularities. In this situation, as I will explain, there is the possibility of an extremely natural explanation for the GLASSy nature of cosmological perturbations, as the inevitable result of (i) the scaling symmetry of the dual theory residing on the conformal boundary of AdS in the AdS/CFT correspondence, (ii) the instability of the dual theory when cosmological evolution in the bulk is allowed and (iii) the asymptotic freedom of the coupling governing the instability in the dual theory. The latter implies that the perturbations have a slightly red spectrum, since the coupling which drives them grows logarithmically with scale. The flatness of the universe may also be explained in such a model, even without the inflationary effect of dark energy, as a consequence of the extra expansion created by the release of radiation at the bang.[9] Dark energy would still, however, be required to ensure the stability of the cycling solution.

AN ADS^5 COSMOLOGY

Ten-dimensional type IIB supergravity compactified on a five-sphere can be consistently truncated, in several inequivalent ways, to five dimensional gravity coupled to a single scalar field ϕ, describing one of the possible quadrupole distortions of the five-sphere, whose mass saturates the Breitenlohner-Freedman (BF) bound $m^2 = -4/R^2_{AdS}$. In all solutions asymptotic to the AdS$_5$ metric

$$ds^2 = R^2_{AdS}\left(-(1+r^2)dt^2 + \frac{dr^2}{1+r^2} + r^2 d\Omega_3\right), \tag{1}$$

at large r, the scalar field decays as

$$\phi(r) \sim \frac{\alpha \log r}{r^2} + \frac{\beta}{r^2}, \qquad r \to \infty. \tag{2}$$

In order to define a unique time evolution of the bulk, one has to specify the boundary conditions. One class of choices is

$$\alpha = f\beta. \tag{3}$$

The standard supersymmetric boundary condition is $f = 0$, which respects the full AdS symmetry group. For this choice, the pure AdS solution is a stable ground state and the bulk theory is conjectured to be dual to $N = 4$ supersymmetric Yang-Mills theory, a conformal field theory (CFT) defined on the conformal boundary $\mathbb{R} \times S^3$ of AdS^5.

More interesting for cosmology is the case where $f > 0$, elaborated by Witten.[10] The set of bulk scalars which saturate the BF bound couple to boundary operators $\mathscr{O}$ consisting of linear combinations of the gauge-invariant quantities $\mathrm{Tr}[\Phi^i\Phi^j - (1/6)\delta^{ij}\Phi_k^2]$ constructed from the six adjoint scalar fields Φ_i appearing in the $N = 4$ theory. The function α appearing (3) plays the role of a source for $\mathscr{O}$ in the boundary theory, while β corresponds to the expectation value of $\mathscr{O}$. Boundary conditions with $f > 0$ correspond to the following deformation of the CFT,

$$S \to S + \frac{f}{2}\int \mathscr{O}^2, \tag{4}$$

which breaks supersymmetry and yields a classical potential that is unbounded from below. The cosmological singularity which develops in the bulk corresponds to a singularity on the boundary, at which $\langle \mathscr{O} \rangle$ runs to infinity at the same, finite time.

Because of the "wrong" sign of the potential in (4), the coupling f turns out to be asymptotically free. Therefore, the quantum corrections it induces are under excellent perturbative control for large $\mathcal{O}$, *i.e.*, near the singularity. The renormalized one-loop (Coleman-Weinberg) effective potential reads, schematically,

$$V = -\frac{\mathcal{O}^2}{\ln(\mathcal{O}/M^2)}, \tag{5}$$

where the coupling f has been replaced by an arbitrary mass scale M through "dimensional transmutation".[11] Perturbation theory is accurate for $\mathcal{O} > M^2$, and (5) becomes increasingly accurate as the singularity is approached. We shall ignore the effects of gauge interactions. This is obviously valid at weak 't Hooft coupling. However, Witten has shown, using the AdS/CFT correspondence, that the effective potential is still given by (5) at large 't Hooft coupling. So it is plausible that our description will hold in that case too. According to the AdS/CFT correspondence, $Ng_{YM}^2 = (R_{AdS}/\ell_s)^4$, so if one wants to begin with a bulk which is well-described by supergravity, with modest α' corrections, one must work at large 't Hooft coupling.

For some choices of $\mathcal{O}$, the boundary scalar fields run to infinity along isolated valleys in field space. In the following discussion, we focus on the most unstable direction, ignoring the orthogonal, massive directions. The effective potential is then schematically of the form

$$V(\phi) \sim R_{AdS}^{-2}\phi^2 - \frac{\phi^4}{\ln(\phi/M)} \equiv R_{AdS}^{-2}\phi^2 - \lambda\phi^4, \tag{6}$$

where the first term arises from the usual coupling to the Ricci scalar on the S^3, required by conformal invariance. If M is taken to be much smaller than R_{AdS}^{-1}, and we study field values ϕ beyond the barrier created by the first term in (6), then the logarithm is large and we are always in the regime of weak coupling λ.

A crucial point is that the dual theory lives on a finite space, the S^3. This means that the spatially homogeneous mode of ϕ is really quantum mechanical and the quantum spread of its wavefunction must be taken into account. Even if one starts the system off in a Gaussian wavepacket, *i.e.*, localized around some initial ϕ, in an arbitrarily short time $t > 0$ the wavefunction will spread out to $\phi = \infty$ and, if suitable boundary conditions are not imposed, there will be a violation of unitarity. Therefore, if one wants to study this theory at all, there is no option but to impose unitarity-preserving boundary conditions at $\phi = \infty$ and this is what we shall do. The choice of boundary condition in effect halves the Hilbert space of wavefunctions and keeps only the subspace on which the Hamiltonian is self-adjoint. The restricted, unitary theory is known as the "self-adjoint" extension.[12]

The quantum mechanical nature of the homogeneous mode of ϕ will also turn out to be crucial to obtaining a finite answer for the total energy generated by quantum mechanical "particle creation" in the inhomogeneous modes. The quantum spread in the homogeneous mode means that the probability to find the homogeneous mode at infinity is actually zero and as we shall see, this means that for generic final values of the homogeneous mode, at a given final time, the relevant classical solution is complex and misses the singularity by a finite amount. The equation governing the inhomogeneous modes is therefore nonsingular and this leads to a natural UV cutoff in the quantum particle creation.

ULTRALOCALITY

Is it possible to construct a "self-adjoint extension" for quantum field theory? The first argument in favor is that spatial gradients become dynamically irrelevant near the singularity. One can show that generic solutions solutions of the full nonlinear classical field equation are described by two arbitrary functions of the spatial coordinates on the S^3, representing the local value of the time at which ϕ hits infinity, $t_s(\mathbf{x})$, and the energy density on the singular hypersurface, $\rho_s(\mathbf{x})$. Such a simple description, with no "BKL chaos" as there would be in the bulk supergravity theory, suggests the possibility of a smooth local matching condition for the field ϕ across the singularity.

The fact that spatial gradients become unimportant means that different spatial points decouple and the quantum field theory becomes described by an independent set of quantum mechanical systems each of which can be made unitary by the same "self-adjoint extension."

THE SELF-ADJOINT EXTENSION

The construction of the self-adjoint extension for the quantum theory rests on the fact that the semiclassical approximation becomes increasingly accurate as the singularity is approached. From the Schrödinger equation, one reads off the momentum,

$$p_\phi \sim \sqrt{2V_3(E-U(\phi))} \sim \sqrt{\lambda}\phi^2 V_3, \qquad \phi \to \infty, \tag{7}$$

where V_3 is the volume of the S^3, $U(\phi) = V_3 V(\phi)$ is the effective potential and E the energy. The condition for the semiclassical approximation to be valid is that $p_\phi^{-2} dp_\phi/d\phi$ is small, which is clearly true at large ϕ.

In the method developed by Reed and Simon in the 1970's[12], and elaborated in Ref.[13], the self-adjoint extension is implemented by restricting the energy eigenstates to those with the behavior

$$\Psi_E \sim \lambda^{-\frac{1}{4}}\phi^{-1}\cos\left(\int p_\phi d\phi + \alpha\right), \qquad \phi \to \infty, \tag{8}$$

where α is an arbitrary, energy-independent phase. These wavefunctions are clearly normalizable and it is straightforward to check that the Hamiltonian is self-adjoint between any two of them. The complete Hilbert space for the system consists of arbitary linear combinations of these allowed stationary states.

For our problem, we are interested in dynamical states, where ϕ starts out at nearly zero potential energy, $\phi \sim \lambda^{-\frac{1}{2}} R_{AdS}^{-1}$, and runs off to infinity with the wavepacket centered on the zero-energy classical scaling solution $\phi \sim \lambda^{-\frac{1}{2}}(t_s - t)^{-1}$. These states are better described in a basis of initially Gaussian wavepackets, with their width chosen to minimize the quantum uncertainty over the duration of the "bounce," a time found from (6) to be of order R_{AdS}. The corresponding minimal spread is $\Delta\phi \sim R_{AdS}^{-1}$: this corresponds to a time-delay of $\Delta t \sim \sqrt{\lambda} R_{AdS}$. The quantum spread of the wavepacket is only important for times close to the singularity, $|t-t_s| \sim \Delta t$ but as we shall see, it plays a crucial role there.

Since the semiclassical approximation becomes exact at large ϕ, we can describe the bounce of ϕ off infinity using the semiclassical expansion to solve the Schrödinger equation, using complex classical solutions.

COMPLEX CLASSICAL SOLUTIONS: ELEMENTARY EXAMPLE

Consider, for example, a free particle constrained to the half-line $x>0$ by a brick wall located at $x=0$. The semiclassical approximation is exact in this case since the Hamiltonian is quadratic and the boundary condition may be implemented by the method of images. If the initial wavefunction is a Gaussian wavepacket centred on $x=x_c>0$ and with momentum $p_c<0$, the initial condition for the corresponding complex classical solution reads

$$\frac{x}{2L}+i\frac{pL}{\hbar}=\frac{x_c}{2L}+i\frac{p_cL}{\hbar}, \qquad t=t_i, \tag{9}$$

where $L=\Delta x$ is the uncertainty in x in the initial state. The final condition is just $x=x_f$, positive and real, at the final time t_f. One constructs the semiclassical wavefunction as the sum of two terms:

$$\Psi \sim A\left(e^{\frac{iS_1}{\hbar}}-e^{\frac{iS_2}{\hbar}}\right), \tag{10}$$

where S_1 is the classical action for the complex trajectory satisfying (9) and S_2 is that for the "image" trajectory, with x_c and p_c reversed so that the wavepacket starts out behind the brickwall. The classical actions have to include a boundary term required to ensure that $\delta S=0$ is equivalent to the equations of motion for the chosen complex boundary condition (9). The classical solutions are straightforwardly found since they are both just straight lines in the complex x-plane.

The resulting probability density (dropping the subscripts on x_f and t_f) is just

$$|\Psi|^2(x,t)=C\left(e^{-\frac{(x-x_c-p_c(t-t_i))^2}{2\sigma^2}}+e^{-\frac{(x+x_c+p_c(t-t_i))^2}{2\sigma^2}}-2e^{-\frac{x^2+(x_c+p_c(t-t_i))^2}{2\sigma^2}}\cos\left(\frac{2x\hat{p}_cL^2}{\hbar\sigma^2}\right)\right), \tag{11}$$

with $C=1/\sqrt{2\pi\sigma^2}$, $\sigma^2=L^2+\frac{\hbar^2t^2}{4L^2}$ the growing dispersion of the Gaussian and $\hat{p}_0(t)=p_c-x_c\hbar^2(t-t_i)/(4L^4)$. To minimize the spread of the wavefunction over some time $t-t_i$ of interest, we should choose $L=\sqrt{\hbar(t-t_i)/2}$.

At early times, the probability density (11) is well-approximated by the first term alone, namely a slowly spreading Gaussian centered on the first classical trajectory. When the wavepacket hits the wall, at $t\sim -x_c/p_c$, interference between the incident and reflected waves leads to strongly oscillatory behavior in both space and time. As time runs forward the second classical solution, obeying the "image" initial conditions, now determines the dominant piece of the wavefunction. The probability once again assumes a smooth Gaussian form but moving away from the wall. Similar behavior is encountered in the "bounce from infinity" in our problem.

EVOLUTION OF A GAUSSIAN WAVEPACKET IN $-\lambda\phi^4$

We solve for the evolution of an initial Gaussian wavepacket, centred on a classical trajectory of nearly zero energy, in a similar manner. The initial condition (9) is adopted, with L representing the initial quantum spread of the homogeneous mode. The initial coordinate is $\phi_c \sim \lambda^{-\frac{1}{2}} R_{AdS}^{-1}$ and for minimal quantum mechanical spreading, we adopt a spread of $\Delta\phi \sim R_{AdS}^{-1}$.

Since we wish to resolve the behavior at $\phi = \infty$, we change variables to $\chi \sim \lambda^{-\frac{1}{2}}\phi$. Ignoring the running of the coupling, the real classical zero-energy scaling solution is then just $\chi = t$. The semiclassical wavefunction is then constructed just as in (10), as the sum of two terms each determined by a complex classical solution. Again ignoring the running of the coupling, the theory has an exact symmetry under $\phi \to -\phi$ and one can easily show that if we define the second, "image" classical solution to have initial conditions in which χ and $\dot{\chi}$ are reversed so that ϕ starts from "behind" the singularity and emerges from it, the semiclassical wavefunction

$$\Psi(t,\phi) \sim A_1 e^{\frac{iS_1}{\hbar}} + A_2 e^{-2i\alpha} e^{\frac{iS_2}{\hbar}}, \tag{12}$$

with the amplitude factors A_1 and A_2 determined by the semiclassical expansion, enjoys the same asymptotic behaviour in ϕ as (8), and hence belongs within the domain of the self-adjoint extension.

A NATURAL UV CUTOFF FOR MOST ϕ_f

We have seen in the brick wall example how the wavefunction "bounces." In that case, it is obvious that the probability to find the particle at the wall is always zero. Likewise, in our example, the probability to find the homogeneous component of the field ϕ at infinity is always zero since the wavefunction dies as ϕ^{-1}. This fact is reflected in the behavior of the complex classical solutions describing the wavefunction once it has bounced back from infinity. The solutions which determine the wavefunction at values of ϕ ahead or behind the real classical solution (with a bounce from infinity assumed) miss the singularity at $\chi = 0$ (or $\phi = \infty$) by passing above or below it in the complex χ-plane, by an amount proportional to their difference from the classical solution. For typical final values, this is of order $\Delta\phi$, the spread of the Gaussian wavepacket.

For each value of ϕ_f, the spatially inhomogeneous modes evolve in the appropriate complex classical background. As explained, ϕ generically misses the singularity in that solution so that the term $V_{,\phi\phi}$ which appears in the linearized equation of motion for the inhomogeneous modes is smooth and bounded. This means that for generic ϕ_f there is automatically a UV cutoff in the production of modes at high k. Generically, the complex classical trajectory misses the singularity at $\chi = 0$ by an amount $\mathrm{Im}(\chi) \sim \lambda^{\frac{1}{2}} R_{AdS}$, and this results in an exponential suppression of particle production for $k > \lambda^{-\frac{1}{2}} R_{AdS}^{-1}$.

PARTICLE PRODUCTION AND THE RUNNING COUPLING

When the running of the coupling λ is ignored, we find that our construction of the self-adjoint extension yields precisely zero particle production. Including the running coupling is subtle, because the presence of the logarithm introduces a branch point at $\phi = \infty$, spoiling the construction of the self-adjoint extension via the method of images. This problem may be circumvented by carefully defining the effective potential with a finite UV cutoff in Euclidean momentum, for which the method of images still works, and only taking the limit of infinite cutoff at the end of the calculation. Details are explained in Ref.[1]. Our final expression for the energy density in quantum-created particles is:

$$\rho_c = \int \frac{d^3\mathbf{k}}{(2\pi)^3} k \frac{\pi^2}{\ln^2(k/M)} e^{-4k\varepsilon}, \tag{13}$$

where $\varepsilon \sim \lambda^{\frac{1}{2}} R_{AdS}$ is the imaginary part of χ when the "image" classical solution passes near to the origin. It is this latter solution which determines the dominant part of the wavefunction in the final state, *i.e.*, when the wavepacket has returned near to its starting value ϕ_c. The energy in created particles is (13) is finite, and of order R_{AdS}^{-4}. In comparison, the potential energy at the time of the bounce is of order $-\lambda\phi^4 \sim -\lambda^{-3} R_{AdS}^{-4}$. It follows that for small λ the energy density in particles is far less than the kinetic energy in ϕ so the backreaction is small and the field ϕ returns to nearly its starting value. After N bounces, the energy in particles is roughly NR_{AdS}^{-4}. One can determine the smallest value of ϕ attained after the N'th bounce, ϕ_{min}, by setting the energy in particles equal to $-V(\phi_{min})$, since the total energy is always zero. Because the notion of the real bouncing background fails once ϕ_{min} grows larger than $\lambda^{-1} R_{AdS}^{-1}$, we estimate the number of bounces to be $N \sim \lambda^{-3}$.

SCALE INVARIANCE FROM SCALE INVARIANCE

One of the most intriguing aspects of these calculations is that they provide an automatic mechanism for generating scale-invariant stress-energy perturbations in the boundary field theory. As discussed, the boundary theory corresponding to an interesting cosmology is unstable because the quartic coupling is negative, and hence generates a spectrum of fluctuations by the ekpyrotic mechanism. As the scalar field ϕ rolls downhill, it gives the inhomogeneous modes an increasing tachyonic mass, causing modes of successively higher k to cease oscillating and enter a "growing mode" solution instead. Once the background has passed through the bounce, one has a spectrum of spatially dependent modes which are excited states about the final adiabatic vacuum. The boundary field theory is classically scale-invariant. The quantum theory breaks that symmetry and induces anomalous dimensions but these are small because the coupling λ is asymptotically free. A lengthy but straightforward computation of the two-point correlation of the "improved" stress-energy tensor in the final state, yields

$$\langle \delta T_{00}(r,t) \delta T_{00}(0,t) \rangle = \frac{1}{(\ln Mr)^2} \frac{a_0}{r^6 t^2}$$

$$\langle \partial_i \delta T_{0i}(r,t) \partial_i \delta T_{0i}(0,t) \rangle = (\ln Mr)^2 \left(\frac{a_1}{r^8 t^2} + \frac{a_2}{r^6 t^4} \right)$$
$$\langle \partial_i \partial_j \tilde{\delta T}_{ij}(r,t) \partial_k \partial_l \tilde{\delta T}_{kl}(0,t) \rangle = \frac{1}{(\ln Mr)^2} \left(\frac{a_3}{r^{10} t^2} + \frac{a_4}{r^8 t^4} + \frac{a_5}{r^6 t^4} \right), \qquad (14)$$

with a_i constant coefficients, for $r < t$, and similar expressions for $r > t$. The only violations of naive ("engineering dimensions") scale-invariance is due to the logarithms and these violations are small at weak coupling. Similarly, the dimensionless fluctuations in the energy density can be written as

$$\langle \frac{\delta\rho(r,t)}{P+\rho} \frac{\delta\rho(0,t)}{P+\rho} \rangle \sim \frac{1}{(\ln Mr)^2 \ln Mt} f(r/t), \qquad (15)$$

where f is a dimensionless function. The asymptotic freedom of the boundary theory, coupled to the fact that the particle production is proportional to a positive power of the running coupling λ, means that the spectrum of stress-energy fluctuations weakens in the UV, *i.e.*, it naturally possesses a small red tilt.

To be sure, the correlators we have computed so far are boundary correlators and hence not the quantities of direct interest in the bulk cosmology. We need to translate the boundary correlators into the bulk, which is not straightforward, and this work is still in progress. Similar calculations were performed in Ref. [14]. Since the only scale which enters the translation is R_{AdS}, and since scale-invariance is "holographic" in the sense that the fluctuations on a planar subspace of a space with scale-invariant fluctuations are also scale-invariant, it seems plausible that the bulk curvature perturbations will also turn out to be scale-invariant. If so, this mechanism will provide a rather compelling alternative to inflationary theory, since the scale-invariance arises here from profound properties, namely the exact conformalsymmetry of the static dual CFT, and the asymptotic freedom of the deformed, dynamical dual, of string theory.

ACKNOWLEDGMENTS

I thank Ben Craps and Thomas Hertog for a wonderful collaboration and the organisers and participants of PASCOS-07 for such a thoroughly enjoyable meeting.

REFERENCES

1. B. Craps, T. Hertog and N. Turok, in preparation.
2. G. W. Gibbons and N. Turok, "The measure problem in cosmology," arXiv:hep-th/0609095.
3. P. J. Steinhardt and N. Turok, "Cosmic evolution in a cyclic universe," Phys. Rev. D **65**, 126003 (2002) [arXiv:hep-th/0111098].
4. P. J. Steinhardt and N. Turok, "Why the cosmological constant is small and positive," Science **312**, 1180 (2006) [arXiv:astro-ph/0605173].
5. N. Turok, M. Perry and P. J. Steinhardt, "M theory model of a big crunch / big bang transition," Phys. Rev. D **70**, 106004 (2004) [Erratum-ibid. D **71**, 029901 (2005)] [arXiv:hep-th/0408083].
6. J. M. Maldacena, "The large N limit of superconformal field theories and supergravity," Adv. Theor. Math. Phys. **2**, 231 (1998) [Int. J. Theor. Phys. **38**, 1113 (1999)] [arXiv:hep-th/9711200].
7. T. Hertog and G. T. Horowitz, "Holographic description of AdS cosmologies," JHEP **0504**, 005 (2005) [arXiv:hep-th/0503071].

8. S. Elitzur, A. Giveon, M. Porrati and E. Rabinovici, "Multitrace deformations of vector and adjoint theories and their holographic duals," JHEP **0602**, 006 (2006) [arXiv:hep-th/0511061].
9. J. K. Erickson, S. Gratton, P. J. Steinhardt and N. Turok, "Cosmic perturbations through the cyclic ages," Phys. Rev. D **75**, 123507 (2007) [arXiv:hep-th/0607164].
10. E. Witten, "Multi-trace operators, boundary conditions, and AdS/CFT correspondence," arXiv:hep-th/0112258.
11. S. R. Coleman and E. Weinberg, "Radiative Corrections As The Origin Of Spontaneous Symmetry Breaking," Phys. Rev. D **7**, 1888 (1973).
12. M. Reed and B. Simon, "Methods Of Modern Mathematical Physics. 2. Fourier Analysis, Selfadjointness," *New York 1975, 361p*
13. M. Carreau, E. Fahri, S. Gutmann, P. F. Mende, "The Functional Integral for Quantum Systems with Hamiltonians Unbounded from Below," Ann. Phys. **204** (1990) 186.
14. S. Gratton and N. Turok, "Cosmological perturbations from the no boundary Euclidean path integral," Phys. Rev. D **60**, 123507 (1999) [arXiv:astro-ph/9902265].

PARTICLES

CP Violation Results from the NA48 experiments

Giuseppina Anzivino

Department of Physics, University of Perugia, via A. Pascoli, Perugia (I) [1]

Abstract.
The main goal of the NA48 experiments at the CERN SPS has been the search for direct CP violation (CPV) in kaon decays. In this article two selected results will be presented. In the neutral kaon sector, NA48 determined η_{+-}, related to the parameters of indirect and direct CPV ($\eta_{+-} = \varepsilon + \varepsilon'$), from data collected during a dedicated run in 1999 with a pure K_L beam. Using a sample of 47000 $K_L \to \pi^+\pi^-$ and 5 million $K_L \to \pi^\pm e^\mp \nu$, we extract $|\eta_{+-}| = (2.223 \pm 0.012) \times 10^{-3}$. In the charged kaon decays, NA48/2 has measured the asymmetry $A_g = (g^+ - g^-)/(g^+ + g^-)$ of the linear slope parameter g in the Dalitz plot of $K^\pm \to 3\pi$ decays from data collected in 2003 and 2004. Any non-zero value of A_g would be evidence for direct CPV. A new technique of asymmetry measurement, involving simultaneous K^+ and K^- beam and a large data sample allowed a result of unprecedented precision. The charge asymmetry parameter was measured to be $A_g^c = (-1.5 \pm 2.1) \times 10^{-4}$ with 3.11×10^9 $K^\pm \to \pi^\pm\pi^+\pi^-$ decays, and $A_g^n = (1.8 \pm 1.8) \times 10^{-4}$ with 9.13×10^7 $K^\pm \to \pi^\pm\pi^0\pi^0$ decays. The results are compatible with the SM predictions, i.e. no evidence for direct CP violation at the order of 10^{-4} has been found.

Keywords: CP Violation, Kaon Decays
PACS: 13.20.Eb,13.25.Es

EXPERIMENTAL SET-UP

The data for η_{+-} measurement were taken during a dedicated run in 1999 with a pure K_L beam produced by a 450 GeV primary proton beam hitting a beryllium target. The neutral beam, after passing the final collimator, entered the decay volume which was followed by the NA48 detector. Two simultaneous focused kaon beams of opposite charge, with a momentum of (60 ± 3) GeV/c were used for the measurement of the charge asymmetry. The main components of the detector, described in detail in [1], are a Liquid Krypton Calorimeter (LKr) with tower read-out and a magnetic spectrometer. The calorimeter is an almost homogeneous ionization chamber with an active volume of $\sim 10\ m^3$ and a 27 X_0 thickness; it fully contains electromagnetic showers up to energies of 100 GeV and has an energy resolution of $\sim 1\%$ for photons of 20 GeV, and a spatial resolution of $\sim 1\ mm$. The magnetic spectrometer, consisting of four drift chambers and a dipole magnet located between the second and the third chamber, has a spatial resolution of 100 μm.

---

[1] on behalf of the NA48 and NA48/2 Collaborations:Cagliari, Cambridge, CERN, Chicago, Dubna, Edinburgh, Ferrara, Firenze, Mainz, Northwestern, Perugia, Pisa, Saclay, Siegen, Torino, Warsaw, Wien.

CP957, *Particles, Strings, and Cosmology, 13th International Symposium*
edited by A. Rajantie, C. Contaldi, P. Dauncey, and H. Stoica

CP VIOLATION PARAMETER η_{+-}

The observable η_{+-} is related to the parameters of indirect and direct CPV ($\eta_{+-} = \varepsilon + \varepsilon'$) and defined as the CP violating amplitude ratio of the neutral kaon decaying into two charged pions: $\eta_{+-} = A(K_L \to \pi^+\pi^-)/A(K_S \to \pi^+\pi^-)$. NA48 has determined $|\eta_{+-}|$ via the measurement of the ratio of the decay rates

$$R = \Gamma(K_L \to \pi^+\pi^-)/\Gamma(K_L \to \pi^\pm e^\mp \nu)$$

denoted as $\Gamma_{K2\pi}/\Gamma_{Ke3}$. The analysis is based on 47000 $K_L \to \pi^+\pi^-$ and five million $K_L \to \pi^\pm e^\mp \nu$ decays. The selection is based on the requirement of good 2-track events for the two decays; then additional cuts are applied to suppress the dominant backgrounds to $K_L \to \pi^+\pi^-$, namely semileptonic K_{e3} and $K_{\mu3}$ decays (Fig.1, left). For the selection of K_{e3} events, a cut on the E/p is applied, where E is the energy deposited in the LKr and p is the track momentum measured in the magnetic spectrometer (Fig.1, right).

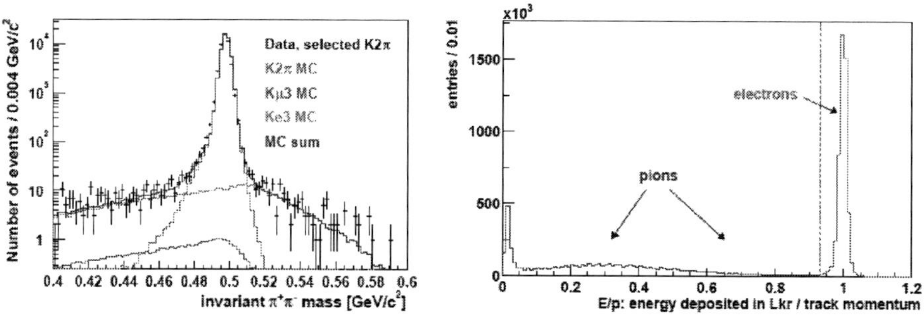

FIGURE 1. Left: Distribution of the invariant $\pi^+\pi^-$ mass. Right: The E/p distribution for K_{e3} events.

The final result, after applying all the corrections is

$$\frac{\Gamma(K_L \to \pi^+\pi^-)}{\Gamma(K_L \to \pi^\pm e^\mp \nu)} = (4.835 \pm 0.022_{stat} \pm 0.016_{syst}) \times 10^{-3} = (4.835 \pm 0.027) \times 10^{-3}.$$

Using the obtained ratio $\Gamma_{K2\pi}/\Gamma_{Ke3}$, we determined the $K_L \to \pi^+\pi^-$ branching ratio, after subtraction of the contribution of the CP conserving direct emission process $K_L \to \pi^+\pi^-\gamma(DE)$ (~ 0.2) and including the CP violating inner bremsstrahlung component $K_L \to \pi^+\pi^-\gamma(IB)$:

$$BR(K_L \to \pi^+\pi^-) = (1.941 \pm 0.019) \times 10^{-3}.$$

Finally we determine η_{+-}, using our result for the branching ratio of $K_L \to \pi^+\pi^-$ decay

$$|\eta_{+-}| = \sqrt{\frac{BR(K_L \to \pi^+\pi^-)}{BR(K_S \to \pi^+\pi^-)} \cdot \frac{\tau_{KS}}{\tau_{KL}}} = (2.223 \pm 0.012) \times 10^{-3}.$$

The results are in good agreement with recent measurements by KTeV [2] and KLOE [3] and all the three experiments contradict the 2004 PDG value [4].

CP VIOLATION IN $K^{\pm} \rightarrow 3\pi$ DECAYS

The $K^{\pm} \rightarrow 3\pi$ decays are usually described in terms of two Dalitz variables u and v: $u = s_3 - s_0/m_\pi^2$ and $v = s_2 - s_1/m_\pi^2$, where $s_i = (p_K - p_{\pi i})^2$, p_K being the four-momentum of the decaying kaon, and $s_0 = \sum s_i/3$ $(i = 1,2,3)$, index $i = 3$ for the odd pion (the one with different charge from the other two). The matrix element can be expressed as a polynomial expansion of u and v:

$$|\mathcal{M}(u,v)|^2 \propto 1 + gu + hu^2 + kv^2$$

where g, h, k are the linear and quadratic Dalitz plot parameters ($|h|, |k| \ll |g|$). Complementary with ε'/ε, the CP observable in the charged kaon sector is the asymmetry

$$A_g = (g^+ - g^-)/(g^+ + g^-) \approx \Delta g/2g$$

of the linear slope parameter g (g^+ refers to K^+, g^- to K^-). Since there is no mixing in decays of charged kaons, any non-zero value of A_g would reflect evidence for direct CPV. SM predictions for the charge asymmetry lay in the range 10^{-6} – 10^{-5} [5], while theoretical calculations involving processes beyond the SM do not exclude substantial enhancements of A_g [6]. The goal of NA48/2 was to measure A_g with an accuracy of few 10^{-4}.

The measurement of is based on the comparison of the u distributions of K^+ and K^- decays. The density of $K^{\pm}$ events in the Dalitz plot is projected onto the u axis to obtain one dimensional distributions $N^{\pm}(u)$. Then the ratio

$$R(u) = N^+(u)/N^-(u) \sim 1 + \frac{\Delta g \cdot u}{1 + gu + hu^2}$$

is computed. Detector acceptance effects as well as variation in the response are canceled by design in NA48/2, being the two charged beams superimposed in space and present at the same time. This results in a large cancelation of systematic effects in the asymmetry measurement. However, the magnetic fields of the beam line (achromats, focusing quadrupoles, etc.) and of the spectrometer (dipole) are intrinsic source of charge asymmetry in the detector. In order to compensate for left-right asymmetries, the polarities of the relevant magnets along the beam line and the detector have been periodically reversed during the data taking. In two weeks an almost complete cycle, called super-sample, was completed. Each super-sample, treated as an independent unit in the analysis, contains four $K^{\pm} \rightarrow 3\pi$ samples with different combinations of beam line and spectrometer magnet polarities. The product of the four possible combinations forms a quadruple ratio

$$R_4(u) = R(u)_{US} \cdot R(u)_{UJ} \cdot R(u)_{DS} \cdot R(u)_{DJ}$$

where the subscript $U(D)$ refers to a configuration in which K^+ beam runs through the upper (lower) beam path in the achromatic magnet sets, while the subscript $S(J)$ refers to the spectrometer magnet polarity in which decay products, having the same charge as the corresponding beam, are deflected to the right (left) in the spectrometer (towards the Saleve (Jura) mountain). Finally, Δg was extracted by fitting the quadruple ratio with a

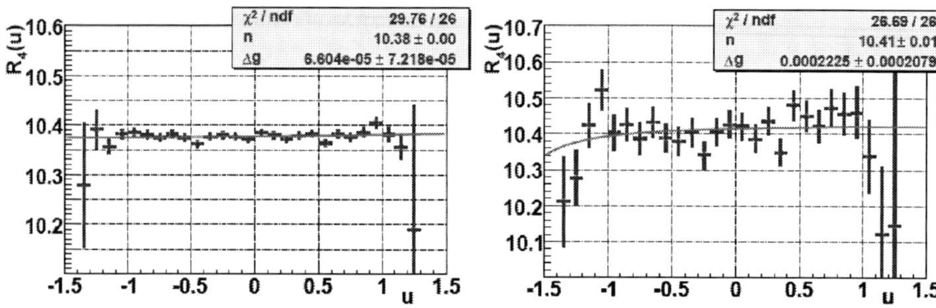

FIGURE 2. The quadruple ratio in bins of u for charged (left) and neutral (right) mode, with the result of the fit to extract Δg.

function $f(u) = n \cdot (1 + \frac{\Delta g u}{1+gu+hu^2})^4$ and converted to the CP violating charge asymmetry, using the value of the Dalitz plot slope g^c, recently measured by NA48/2 [7], for the charged mode and the PDG value of g^n for the neutral mode. In Fig.2 the final result for the fit is shown.

From the data samples taken in 2003 and 2004, 3.11×10^9 $K^{\pm} \rightarrow \pi^{\pm}\pi^{+}\pi^{-}$ and 9.13×10^7 $K^{\pm} \rightarrow \pi^{\pm}\pi^{0}\pi^{0}$ were selected. The results for the asymmetries are

$$A_g^c = (-1.5 \pm 1.5_{stat} \pm 0.9_{trig} \pm 1.1_{syst}) \times 10^{-4} = (-1.5 \pm 2.1) \times 10^{-4}$$

$$A_g^n = (1.8 \pm 1.7_{stat} \pm 0.9_{syst}) \times 10^{-4} = (1.8 \pm 1.8) \times 10^{-4}.$$

The charge asymmetry parameter A_g was determined, as proposed, with a total uncertainty of $\sim 2 \times 10^{-4}$ for each mode, which is an improvement in accuracy over the previous measurements [4] by more than an order of magnitude. NA48/2 precision is limited mainly by the available statistics. The results are compatible with the SM predictions, i.e. no evidence for direct CP violation at the order of 10^{-4} has been found. Due to the high precision achieved, the result can be used to constrain extensions of the SM predicting enhancements of the CP violating effects.

REFERENCES

1. V. Fanti et al., NA48 Collaboration, *Nucl. Instrum. and Meth.* ***A574***, Elsevier, Amsterdam, 2007, pp. 433–471.
2. T. Alexopoulos et al., KTeV Collaboration, *Phys. Rev.* ***D70***, APS, Maryland, 2004, pp. 092006-1–092006–20.
3. F. Ambrosino et al., KLOE Collaboration, *Phys. Lett.* ***B638***, Elsevier, Amsterdam, 2006, pp. 140–145.
4. S.Eidelman et al., Particle Data Group, *Phys. Lett.* ***B592***, Elsevier, Amsterdam, 2004, 1.
5. E. Gámiz, J. Prades, I. Scimemi, JHEP 0310, 2003, pp 042-088;
   G. Fäldt, E. P. Shabalin, *Phys. Lett.* ***B635***, Elsevier, Amsterdam, 2006, pp. 295-298.
6. E. P. Shabalin ITEP preprint 8-98, 1998;
   G. D'Ambrosio, G. Isidori, G. Martinelli *Phys. Lett.* ***B480***, Elsevier, Amsterdam, 2000, pp. 64-170.
7. R. J. Batley et al., NA48/2 Collaboration, *Phys. Lett.* ***B649***, Elsevier, Amsterdam, 2007, pp. 349-358.

Search for New Physics Beyond the Standard Model at BaBar

Matthew Barrett
Representing the BaBar Collaboration.

Brunel University, Uxbridge, UB8 3PH, United Kingdom of Great Britain and Northern Ireland

Abstract.
A review of selected recent BaBar results are presented that illustrate the ability of the experiment to search for physics beyond the standard model. The decays $B \to \tau\nu$ and $B \to s\gamma$ provide constraints on the mass of a charged Higgs. Searches for Lepton Flavour Violation could provide a clear signal for beyond the standard model physics. Babar does not observe any signal for New Physics with the current dataset.

Keywords: Beyond Standard Model BaBar
PACS: 13.20.He

INTRODUCTION

The BaBar detector (which is described fully elsewhere [1]) is a particle detector recording events produced from electron - positron collisions at a centre of mass energy of 10.58 GeV, provided by PEP-II. This energy was selected to produce large numbers of B mesons. In addition to being a "B-factory", PEP-II also produces large numbers of charmed mesons and tau lepton pairs.

BaBar follows the pattern of most particle physics detectors, being built from a number of subsystems comprising layers around the interaction point. The innermost Silicon Vertex Tracker, is surrounded by the Drift Chamber. These systems provide primary tracking information for the experiment. The next subsystem is the Detector of Internally Reflected Cherenkov Radiation, which is used for particle identification. An Electromagnetic calorimeter provides all the calorimetry for the detector. A 1.5T magnet provides the prerequisite magnetic field, and the Instrumented Flux Return of this magnet also provides the muon identification.

The Babar detector had recorded an integrated luminosity of 391fb^{-1} as of the end of June 2007. The actual luminosity used in each of the analyses covered in this document are detailed in the discussion of the analysis.

THE DECAY $B \to \tau\nu$

In the standard model the Branching fraction for the process $B \to \ell\nu$ is given by:

$$\mathscr{B}(B \to \ell\nu) = \frac{G_F^2 m_B}{8\pi} m_\ell^2 \left(1 - \frac{m_\ell^2}{m_B^2}\right)^2 f_B^2 |V_{ub}|^2 \tau_B \quad (1)$$

CP957, *Particles, Strings, and Cosmology, 13th International Symposium*
edited by A. Rajantie, C. Contaldi, P. Dauncey, and H. Stoica

where G_F is the Fermi constant, m_B is the B meson mass, m_ℓ is the lepton mass, f_B is the B meson decay constant, V_{ub} is a CKM matrix element, and τ_B is the B meson mean lifetime. Helicity suppression leads to a ratio between the $\tau : \mu :$ e modes of $1 : 5\times10^{-3} : 10^{-7}$, thus $B \to \tau\nu$ is the most accessible mode to search for purely leptonic decays of the B meson. The Standard Model prediction is $\mathscr{B}(B \to \tau\nu) \sim 1.6 \times 10^{-4}$.

Beyond the standard model, there can be an additional contribution from a charged Higgs boson, replacing the W Boson. In the Two Higgs Doublet Model (2HDM) [2] the branching fraction is modified in the following way:

$$\mathscr{B}^{2HDM} = \mathscr{B}^{SM}\left(1 - \frac{m_B^2 \tan^2\beta}{m_{H^\pm}^2}\right)^2 \tag{2}$$

where m_H is the mass of the charged Higgs boson, and $\tan\beta$ is the ratio of the vacuum expectation values of the Higgs doublets.

The analysis of this decay is experimentally challenging due to the presence of two or three neutrinos in the final state. This leads to a lack of kinematic constraints on the signal B meson. The other B meson in the decay is fully reconstructed, and known as B_{reco}. The properties of the signal B meson can thus be inferred from the properties of B_{reco}. The type of decays used to reconstruct this particle are referred to as the "tag". BaBar carries out analyses of this mode using two statistically independent tags — hadronic tags consist the decay $B \to D^{(*)}X$, where X is up to 6 light hadrons ($\pi^\pm$, π^0, $K^\pm$, or K_S^0). Semileptonic tags reconstruct the decay $B \to D\ell\nu X$ where only the neutrino is missing, and X is a π^0, γ or nothing.

The reconstruction of the τ lepton is done for four modes: $\tau \to e\nu\nu$, $\tau \to \mu\nu\nu$, $\tau \to \pi\nu$, $\tau \to \rho(\pi\pi)\nu$, $\tau \to a_1(\pi\pi\pi)\nu$. These modes make up about 70% of the total τ lepton branching fraction.

The most important variable for distinguishing signal from background is E_{Extra}, which is the sum of all the energy in the calorimeter that is not attributed to one of the reconstructed particles in the event. A cut on this variable of less than approximately 0.3GeV (mode dependent), forms a signal region in which the expected number of background events is calculated using studies of Monte Carlo (MC) and off-resonance data (data taken at a centre of mass energy 40MeV lower than the nominal value, which is below the threshold for the production of B mesons).

Each BaBar analysis performs a maximum likelihood fit to extract the yield of events in the signal region for each mode, incorporating probability density functions (PDFs) for signal events, and the expected background. This likelihood fit can be used to extract the branching fraction measurement.

For the semileptonic tags analysis[3] the preliminary branching fraction is B($B \to \tau\nu$)= $(0.9 \pm 0.6(\text{stat}) \pm 0.1(\text{syst})) \times 10^{-4}$, where the first error is statistical and the second is systematic. Since this measurement does not have a high statistical significance, an upper limit is calculated. The upper limit is 1.7×10^{-4} at a 90% confidence level.

The hadronic analysis [4] reports a preliminary branching fraction of B = $(1.8^{1.0}_{0.9}$ (stat.+bkg.) ± 0.3(syst.)$)\times10^{-4}$, where the first error incorporates the statistical and background systematics, and the second error is the remaining systematics. The 90% confidence level upper limit on this measurement is $B < 3.4 \times 10^{-4}$. Both of these analyses used a luminosity corresponding to 383 million $B\bar{B}$ meson pairs.

Since the two measurements for this decay are statistically independent, they can be combined. This is achieved by extending the maximum likelihood fit to encompass all the yields. This results in an overall branching fraction measurement of $\mathscr{B}(B \to \tau\nu) = (1.20^{+0.40}_{-0.38}(stat.)^{+0.29}_{-0.30}(bkgsyst.) \pm 0.22(syst.)) \times 10^{-4}$. The Belle collaboration has published a result for this mode of $\mathscr{B}(B \to \tau\nu) = (1.79^{+0.56}_{-0.49}(stat.)^{+0.39}_{-0.46}(syst.)) \times 10^{-4}$ [5].

The branching fraction result can also be used to constrain new physics involving charged a Higgs, and combined with constraints from the analysis of $B \to s\gamma$.

THE DECAY $B \to s\gamma$

The decay $B \to s\gamma$ is a flavour changing neutral current, which is forbidden at tree level in the SM. It proceeds via a penguin loop, involving a W boson, and u-type quark, which is dominated by the contribution from the t quark. Beyond the SM the W boson could be replaced with a charged Higgs boson, or the W boson and quark could be replaced by supersymmetric chargino and squark. In rare decays proceeding through penguin loops, contributions from new physics could prove significant.

There are two methods for studying the decay $B \to s\gamma$ at BaBar — Fully Inclusive, and Sum of Exclusive. The fully inclusive method involves only reconstructing the photon, this is theoretically clean, but there are high backgrounds. The sum of Exclusive method involves reconstructing a subset of final states arising from the s quark — X_s. This method has a lower background, but there is an additional X_s fragmentation systematic, and missing X_s modes. These proceedings discuss a fully inclusive analysis.

The signature for this decay is an isolated high energy photon. Any photons compatible with decays from π^0, η, or ρ decays are vetoed. This analysis is carried out using the recoil method described for the aforementioned $B \to \tau\nu$ analysis, and uses hadronic tags. It was carried out using 210 fb^{-1} of data.

The SM theoretical prediction for $\mathscr{B}(B \to s\gamma$ is $(3.16 \pm 0.23) \times 10^{-4}$ [6] for photon energy $E_\gamma > 1.6$ GeV. The BaBar preliminary measurement is $(3.66 \pm 0.85 \pm 0.59) \times 10^{-4}$ for $E_\gamma > 1.9$ GeV — using an extrapolation technique [7], this gives a preliminary extrapolation of $(3.91 \pm 0.91 \pm 0.63) \times 10^{-4}$ for $E_\gamma > 1.6$ GeV. The experimental errors are statistical and systematic respectively. There is no significant discrepancy between the theoretical and experimental measurements, given the large errors on the measurement.

The branching fraction measurements from $B \to \tau\nu$ and $B \to s\gamma$ give complementary exclusions on the mass of a charged Higgs in the m_H–$\tan\beta$ plane.

LEPTON FLAVOUR VIOLATION

Lepton Flavour Violation (LFV) only enters the standard model through neutrino oscillations. However the Branching fraction of LFV τ decays due to neutrino oscillation is predicted to be $\sim \mathscr{O}(10^{-54})$, therefore any observation of LFV decays would be a signal

TABLE 1. Branching fraction upper limits of LFV τ decays at 90% confidence level.

	BaBar			Belle		
Mode	$\mathscr{B}_{UL}$ (10^{-7})	$\mathscr{L}(\text{fb}^{-1})$	Reference	$\mathscr{B}_{UL}$ (10^{-7})	$\mathscr{L}(\text{fb}^{-1})$	Reference
$\tau \to e\gamma$	1.1	232.2	[8]	1.2	535.0	[9]
$\tau \to \mu\gamma$	0.7	232.2	[10]	0.5	535.0	[9]
$\tau \to \ell\pi^0\eta\eta'$	1.1–1.6	339.0	[11]	0.8–1.2	401.0	[12]
$\tau \to \ell\ell\ell$	1–3	91.5	[13]	2–4	87.1	[14]
$\tau \to \ell hh$	1–5	221.4	[15]	2–8	158.0	[16]

of new physics. New physics contributions, involving Higgs bosons or sparticles could lead to measurable LFV branching fractions in τ decays.

There are active searches for LFV in the decay modes: $\tau \to \ell\gamma$, and $\tau \to \ell\ell\ell$. SM backgrounds in the search for $\tau \to \ell\gamma$ include $\tau \to \ell\nu_\ell\nu_\tau\gamma$ and $e^+e^- \to \mu^+\mu^-\gamma$. Bhabha and di-muon events are the major SM backgrounds for $\tau \to \ell\ell\ell$. Unlike most τ decays these LFV modes have no neutrinos in the £nal state, and thus have no missing momentum. Current limits on LFV in τ decays are $\sim \mathscr{O}(10^{-7})$, and BaBar measurements, together with Belle measurements are shown in Table 1.

SUMMARY

BaBar is able to search for new physics beyond the standard model through analysis of rare decay modes. BaBar is continuing this search, and will continue taking data until September 2008. So far no clear signal for beyond the SM physics has been observed, but the full BaBar dataset will have increased sensitivity to rare decays. Indirect searches for new physics are complementary to direct searches, such as those in the proposed LHC physics programme, and will still offer an exciting opportunity to search for new physics, even if direct searches find no new beyond the SM particles.

REFERENCES

1. B. Aubert *et. al. (BaBar Collaboration) Nucl. Instrum. Meth., A479:1-116, 2002.*
2. *W.S.Hou PRD 48 2342 (1993)*
3. *arXiv:0705.1820v1 [hep-ex]*
4. *arXiv:0706.2030v1 [hep-ex]*
5. *K. Ikado et. al. (Belle Collaboration) PRL 97, 251802 (2006)*
6. *M. Misiak et. al. PRL98.022002(2007)*
7. *arXiv:hep-ph/0507253v3*
8. *B. Aubert et. al. (BaBar Collaboration) PRL96(2006)41801*
9. *K. Abe et. al. (Belle Collaboration) hep-ex/0609049*
10. *B. Aubert et. al. (BaBar Collaboration) PRL95(2005)41802*
11. *B. Aubert et. al. (BaBar Collaboration) PRL(2007)061803*
12. *K. Abe et. al. (Belle Collaboration) hep-ex/0609013*
13. *B. Aubert et. al. (BaBar Collaboration) PRL92(2004)121801*
14. *K. Abe et. al. (Belle Collaboration) PLB589(2004)103*
15. *B. Aubert et. al. (BaBar Collaboration) PRL95(2005)191801*
16. *K. Abe et. al. (Belle Collaboration) PLB640(2006)138*

Cosmological effects of neutrino mixing

M. Blasone*, A. Capolupo†, S. Capozziello** and G. Vitiello*

*Istituto Nazionale di Fisica Nucleare (INFN), Gruppo Collegato di Salerno and DMI, Università di Salerno, Fisciano (SA) - 84084 Italy
†Department of Physics and Astronomy, University of Leeds, Leeds LS2 9JT UK
** Dipartimento di Scienze Fisiche, Università di Napoli "Federico II" and INFN Sez. di Napoli, Compl. Univ. Monte S. Angelo, Ed.N, Via Cinthia, I-80126 Napoli, Italy.

Abstract.
We report on the recent result that a contribution to the dark energy can be achieved by the vacuum condensate induced by neutrino mixing phenomenon.

Keywords: Neutrino mixing, quantum field theory, dark energy
PACS: 98.80.Cq, 98.80. Hw, 04.20.Jb, 04.50+h

The energy content of the neutrino-antineutrino pair condensate in the vacuum can have effects on cosmic scale [1] and, in particular, can be interpreted as dynamically evolving dark energy [2] that, at present epoch, assumes the behavior and the value of the observed cosmological constant. This result is obtained by studying the neutrino mixing phenomenon [3] - [11] in the framework of the quantum field theory (QFT) formalism, see [12] - [19] and refs. therein cited. Indeed, the unitary inequivalence between the massive neutrino vacuum and the flavor vacuum leads to new oscillation formulas and to a non-zero contribution to the vacuum energy.

We summarize the main features of the QFT formalism for the neutrino mixing (for a detailed review see [17]). The Pontecorvo mixing transformations for two Dirac neutrino fields are

$$\nu_e(x) = \nu_1(x)\cos\theta + \nu_2(x)\sin\theta \tag{1}$$

$$\nu_\mu(x) = -\nu_1(x)\sin\theta + \nu_2(x)\cos\theta\,, \tag{2}$$

where $\nu_e(x)$ and $\nu_\mu(x)$ are the fields with definite flavors, θ is the mixing angle and ν_1 and ν_2 are the fields with definite masses $m_1 \neq m_2$:

$$\nu_i(x) = \frac{1}{\sqrt{V}}\sum_{\mathbf{k},r}\left[u^r_{\mathbf{k},i}\,\alpha^r_{\mathbf{k},i}(t) + v^r_{-\mathbf{k},i}\,\beta^{r\dagger}_{-\mathbf{k},i}(t)\right]e^{i\mathbf{k}\cdot\mathbf{x}}, \qquad i = 1,2, \tag{3}$$

with $\alpha^r_{\mathbf{k},i}(t) = \alpha^r_{\mathbf{k},i}e^{-i\omega_{k,i}t}$, $\beta^{r\dagger}_{\mathbf{k},i}(t) = \beta^{r\dagger}_{\mathbf{k},i}e^{i\omega_{k,i}t}$, and $\omega_{k,i} = \sqrt{\mathbf{k}^2 + m_i^2}$. The operators $\alpha^r_{\mathbf{k},i}$ and $\beta^r_{\mathbf{k},i}$, $i = 1,2$, $r = 1,2$ annihilate the vacuum state $|0\rangle_{1,2} \equiv |0\rangle_1 \otimes |0\rangle_2$.

The mixing transformation Eqs.(1), (2) can be written as [12]:

$$\nu^\alpha_e(x) = G^{-1}_\theta(t)\;\nu^\alpha_1(x)\;G_\theta(t) \tag{4}$$

$$\nu^\alpha_\mu(x) = G^{-1}_\theta(t)\;\nu^\alpha_2(x)\;G_\theta(t) \tag{5}$$

CP957, *Particles, Strings, and Cosmology, 13th International Symposium*
edited by A. Rajantie, C. Contaldi, P. Dauncey, and H. Stoica

where $G_\theta(t)$ is the mixing generator that, at finite volume, is an unitary operator, $G^{-1}_\theta(t) = G_{-\theta}(t) = G^\dagger_\theta(t)$, preserving the canonical anticommutation relations. $G^{-1}_\theta(t)$ maps the Hilbert spaces for ν_1 and ν_2 fields $\mathscr{H}_{1,2}$ to the Hilbert spaces for the flavor fields $\mathscr{H}_{e,\mu}$. In particular, we have $|0(t)\rangle_{e,\mu} = G^{-1}_\theta(t)\; |0\rangle_{1,2}$, where $|0\rangle_{e,\mu}$ is the vacuum for $\mathscr{H}_{e,\mu}$, which is referred to as the flavor vacuum. In the infinite volume limit the flavor vacuum $|0(t)\rangle_{e,\mu}$ turns out to be unitarily inequivalent to the vacuum for the massive neutrinos $|0\rangle_{1,2}$ [12]. For a discussion on QFT favor states see [12] - [19] and [20].

The flavor annihilators, relative to the fields $\nu_e(x)$ and $\nu_\mu(x)$, are

$$\alpha^r_{\mathbf{k},\sigma}(t) \equiv G^{-1}_\theta(t)\; \alpha^r_{\mathbf{k},i}(t)\; G_\theta(t), \tag{6}$$

$$\beta^r_{\mathbf{k},\sigma}(t) \equiv G^{-1}_\theta(t)\; \beta^r_{\mathbf{k},i}(t)\; G_\theta(t), \tag{7}$$

with $(\sigma, i) = (e,1), (\mu,2)$. In the reference frame such that $\mathbf{k} = (0,0,|\mathbf{k}|)$ they are explicitly:

$$\alpha^r_{\mathbf{k},e}(t) = \cos\theta\; \alpha^r_{\mathbf{k},1}(t) + \sin\theta\; \left(|U_\mathbf{k}|\; \alpha^r_{\mathbf{k},2}(t) + \varepsilon^r\; |V_\mathbf{k}|\; \beta^{r\dagger}_{-\mathbf{k},2}(t)\right) \tag{8}$$

and similar for $\alpha^r_{\mathbf{k},\mu}$, $\beta^r_{\mathbf{k},e}$, $\beta^r_{\mathbf{k},\mu}$, where $|U_\mathbf{k}| \equiv u^{r\dagger}_{\mathbf{k},i} u^r_{\mathbf{k},j} = v^{r\dagger}_{-\mathbf{k},i} v^r_{-\mathbf{k},j}$, $|V_\mathbf{k}| \equiv \varepsilon^r\; u^{r\dagger}_{\mathbf{k},1} v^r_{-\mathbf{k},2} = -\varepsilon^r\; u^{r\dagger}_{\mathbf{k},2} v^r_{-\mathbf{k},1}$ with $i,j = 1,2$ and $i \neq j$. We have $|U_\mathbf{k}|^2 + |V_\mathbf{k}|^2 = 1$. Extension of the QFT formalism to three neutrinos has been also done [12] - [19]. In the present report we consider only the two neutrino case.

The number of condensed neutrinos is given by

$${}_{e,\mu}\langle 0|\alpha^{r\dagger}_{\mathbf{k},i}\alpha^r_{\mathbf{k},i}|0\rangle_{e,\mu} = {}_{e,\mu}\langle 0|\beta^{r\dagger}_{\mathbf{k},i}\beta^r_{\mathbf{k},i}|0\rangle_{e,\mu} = \sin^2\theta\; |V_\mathbf{k}|^2 , \quad i = 1,2. \tag{9}$$

The Bogoliubov coefficient $|V_\mathbf{k}|^2$ is zero for $m_1 = m_2$, it has a maximum at $|\mathbf{k}|^2 = m_1 m_2$ and goes to zero for large momenta (i.e. for $|\mathbf{k}|^2 \gg m_1 m_2$) as $|V_\mathbf{k}|^2 \approx \frac{(\Delta m)^2}{4k^2}$. It acts as a "form factor" (in the $\mathbf{k}$ space) controlling the neutrino vacuum condensate. The mixing of neutrinos contributes to the dark energy exactly because of the non-zero value of $|V_\mathbf{k}|^2$ [1]: its behavior at very high momenta, together with the Lorentz invariance of the vacuum condensate at the present time, leads to a contribution of the dark energy that has the very tiny value of the cosmological constant [2].

Let us calculate the contribution ρ^{mix}_{vac} of the neutrino mixing to the vacuum energy density. The calculation is performed in a Minkowski space-time but it can be easily extended to curved space-times. The treatment here presented is a good approximation in the present epoch of that in FRW space-time since, at present epoch, the characteristic oscillation length of the neutrino is much smaller than the universe curvature radius. When computations are carried on in a curved background, the mixing mechanism gives a time dependent dark energy, leading, however, to the same final result we obtain in the flat space-time.

The Lorentz invariance of the vacuum implies that the vacuum energy-momentum tensor density is equal to zero: $\mathscr{T}^{vac}_{\mu\nu} = \langle 0| : \mathscr{T}_{\mu\nu} : |0\rangle = 0$, (: ... : denotes the normal ordering). The (0,0) component of $\mathscr{T}_{\mu\nu}(x)$ for the fields ν_1 and ν_2 is $: \mathscr{T}_{00}(x) := \frac{i}{2} : \left(\bar{\Psi}_m(x)\gamma_0 \overleftrightarrow{\partial_0}\, \Psi_m(x)\right) :$, where $\Psi_m = (\nu_1, \nu_2)^T$. In terms of the annihilation and creation operators of fields ν_1 and ν_2 the (0,0) component of the energy-momentum tensor $T_{00} = \int d^3x \mathcal{T}_{00}(x)$ is given by $: T^{00}_{(i)} := \sum_r \int d^3\mathbf{k}\, \omega_{k,i} \left(\alpha^{r\dagger}_{\mathbf{k},i} \alpha^r_{\mathbf{k},i} + \beta^{r\dagger}_{-\mathbf{k},i} \beta^r_{-\mathbf{k},i} \right)$, with $i = 1,2$. Note that $T^{00}_{(i)}$ is time independent.

In the early universe epochs, when the Lorentz invariance of the vacuum condensate is broken, ρ^{mix}_{vac} presents also space-time dependent condensate contributions [2]. Then ρ^{mix}_{vac} and the contribution of the neutrino mixing to the vacuum pressure p^{mix}_{vac} are given respectively by:

$$\rho^{mix}_{vac} = -\frac{1}{V}\, \eta_{00}\ {}_{e,\mu}\langle 0 | \sum_i : T^{00}_{(i)} : |0\rangle_{e,\mu}\,, \tag{10}$$

$$p^{mix}_{vac} = -\frac{1}{V}\, \eta_{jj}\ {}_{e,\mu}\langle 0 | \sum_i : T^{jj}_{(i)} : |0\rangle_{e,\mu}\,, \tag{11}$$

where $: T^{jj}_{(i)} := \sum_r \int d^3\mathbf{k} \frac{k^j k^j}{\omega_{k,i}} \left(\alpha^{r\dagger}_{\mathbf{k},i} \alpha^r_{\mathbf{k},i} + \beta^{r\dagger}_{-\mathbf{k},i} \beta^r_{-\mathbf{k},i} \right)$ (no summation on the index j is intended). This implies that the adiabatic index is $w = p^{mix}_{vac} / \rho^{mix}_{vac} \simeq 1/3$ when the cut-off is chosen to be $K \gg m_1, m_2$ [2].

At the present epoch, the breaking of the Lorentz invariance is negligible and then ρ^{mix}_{vac} comes from space-time independent condensate contributions (i.e. the contributions carrying a non-vanishing $\partial_\mu \sim k_\mu = (\omega_k, k_j)$ are missing). Thus, the energy-momentum density tensor of the vacuum condensate is given by

$${}_{e,\mu}\langle 0 | : T_{\mu\nu} : |0\rangle_{e,\mu} = \eta_{\mu\nu} \sum_i m_i \int \frac{d^3x}{(2\pi)^3}\ {}_{e,\mu}\langle 0 | : \bar{\nu}_i(x) \nu_i(x) : |0\rangle_{e,\mu} = \eta_{\mu\nu}\, \rho^{mix}_{\Lambda}. \tag{12}$$

Since $\eta_{\mu\nu} = diag(1,-1,-1,-1)$ and, in a homogeneous and isotropic universe, the energy-momentum tensor is $T_{\mu\nu} = diag(\rho, p, p, p)$, then the state equation is $\rho^{mix}_{\Lambda} = -p^{mix}_{\Lambda}$. This means that ρ^{mix}_{Λ} contributes today to the dynamics of the universe by a cosmological constant behavior [2]. We obtain

$$\rho^{mix}_{\Lambda} = \frac{2}{\pi} \sin^2\theta \int_0^K dk k^2 \left[\frac{m_1^2}{\omega_{k,1}} + \frac{m_2^2}{\omega_{k,2}} \right] |V_{\mathbf{k}}|^2, \tag{13}$$

where K is the cut-off on the momenta. Notice the presence of the $|V_{\mathbf{k}}|^2$ factor in the integrand (13): ρ^{mix}_{Λ} would be zero for $|V_{\mathbf{k}}|^2 = 0$ for any $|\mathbf{k}|$, as it is in the usual quantum mechanical Pontecorvo formalism.

For neutrino masses of order of $10^{-3} eV$ and $\Delta m^2 \approx 10^{-5} eV^2$ we have $\rho^{mix}_{\Lambda} = 2.9 \times 10^{-47} GeV^4$, which is the observed dark energy value, even for a value of the cut-off of order of the Planck scale $K = 10^{19} GeV$.

In conclusion, the vacuum condensate generated by neutrino mixing can be interpreted as an evolving dark energy that, at present epoch, behaves as the cosmological constant, giving rise to its observed value. The result is achieved even when a cut-off K of the order of Planck scale is considered. Such a result links together dark energy with the neutrino masses.

So far we have considered the case of mixing of lighter neutrino generations (ν_1 and ν_2). When the three generation mixing is considered we obtain a value for ρ_Λ^{mix} which differs of about 5 order of magnitude from the observed dark energy value. We observe that even in such a case our result represents a dramatic improvement with respect to the 123 order of magnitude of disagreement obtained in other theoretical approaches, which makes our treatment worth to be studied and discussed. Moreover, our approach to the dark energy problem does not need the *ad hoc* introduction of exotic fields and particles. Finally, we mention that one plausibility argument which is put forward in the literature [21] to justify the exclusion of heavier neutrinos from the dark energy contribution is based on the abundance of electrons in the galaxy, implying the dominance of mixing and oscillations of lighter neutrinos. However, a deeper theoretical understanding of such an exclusion is needed and we are presently working on this.

We thank Arttu Rajantie and the organizers of PASCOS 07 for giving us the opportunity to present this work at the PASCOS Conference. We also acknowledge INFN for financial support.

REFERENCES

1. M. Blasone, A. Capolupo, S. Capozziello, S. Carloni and G. Vitiello, *Phys. Lett. A* **323**, 182 (2004).
2. A. Capolupo, S. Capozziello and G. Vitiello, *Phys. Lett. A* **363**, 53 (2007);
   A. Capolupo, S. Capozziello and G. Vitiello, hep-th/0705.0319 ;
   A. Capolupo, S. Capozziello and G. Vitiello, *J. Phys. Conf. Ser.* 67:012032, (2007).
3. B. Pontecorvo, *Sov. Phys. JETP* **6**, 429 (1957) [*Zh. Eksp. Teor. Fiz.* **33**, 549 (1957)].
4. Z. Maki, M. Nakagawa and S. Sakata, *Prog. Theor. Phys.* **28**, 870 (1962).
5. K.Fujii, *Nuovo Cim.* **34**, 722 (1964).
6. V. N. Gribov and B. Pontecorvo, *Phys. Lett. B* **28**, 493 (1969).
7. S. M. Bilenky and B. Pontecorvo, *Phys. Rept.* **41**, 225 (1978).
8. S.M. Bilenky and S.T. Petcov, *Rev. Mod. Phys.* **59**, 671 (1987).
9. R. N. Mohapatra and P. B. Pal, *World Sci. Lect. Notes Phys.* **41**, 1 (1991).
10. L. Wolfenstein, *Phys. Rev. D* **17**, 2369 (1978).
    S.P. Mikheev and A. Yu. Smirnov, *Sov. J. Nucl. Phys.* **42**, 913 (1985).
11. C. Giunti, C. W. Kim and U. W. Lee, *Phys. Rev.* D **44**, 3635 (1991).
12. M. Blasone and G. Vitiello, *Annals Phys.* **244**, 283 (1995).
13. M. Blasone, P.A. Henning and G. Vitiello, Phys. Lett. B **451**, 140 (1999); in *Particles, Strings and Cosmology*, Proceedings of PASCOS 98, P. Nath Editor, World Scientific, Singapore (1999) p.306.
14. K. Fujii, C. Habe and T. Yabuki, Phys. Rev. D **59**, 113003 (1999); *Phys. Rev. D* **64**, 013011 (2001).
15. C. R. Ji and Y. Mishchenko, *Annals Phys.* **315**, 488 (2005).
16. M. Blasone, A. Capolupo and G. Vitiello, *Phys. Rev. D* **66**, 025033 (2002).
17. A. Capolupo, Ph.D. Thesis [hep-th/0408228].
18. M. Blasone, A. Capolupo, F. Terranova and G. Vitiello, *Phys. Rev. D* **72**, 013003 (2005).
19. M. Blasone, A. Capolupo, C. R. Ji and G. Vitiello, hep-ph/0611106.
20. M. Blasone, A. Capolupo, C. R. Ji and G. Vitiello, this volume.
21. N. Mavromatos and M. Sakellariadou, hep-th/0703156.

Flavor states of mixed neutrinos

M. Blasone*, A. Capolupo†, C. R. Ji** and G. Vitiello*

*Istituto Nazionale di Fisica Nucleare (INFN), Gruppo Collegato di Salerno and DMI, Università di Salerno, Fisciano (SA) - 84084 Italy
†Department of Physics and Astronomy, University of Leeds, Leeds LS2 9JT UK
** Department of Physics, North Carolina State University, Raleigh, NC 27695-8202, USA

Abstract.
By resorting to previous results on flavor mixing in Quantum Field Theory, we show how to consistently define flavor states of mixed neutrinos as eigenstates of the flavor charge operators.

Keywords: Neutrino mixing, weak interactions
PACS: 14.60.Pq; 11.15.Tk, 12.15.Ff

The issue of a proper definition of flavor states for mixed neutrinos has been object of discussion in recent years [1]-[13]: although the usual Pontecorvo states [14, 15] represent a valid tool for describing the main physical features of neutrino oscillations, it has been clear since some time that conceptual problems arise in connection with a proper definition of flavor states. In fact, it was even stated [15] that it is impossible to construct such states and a formalism has been developed with the aim to avoid their use in the calculation of oscillation probabilities [9].

The root of such difficulties has been found by studying flavor mixing at level of quantum fields. It has then emerged [1] that the vacuum for neutrino fields with definite masses turns out to be unitarily inequivalent to the vacuum for the flavor neutrino fields. The condensate structure of the flavor vacuum leads to a modification of flavor oscillation formulas [2, 5, 11, 8], exhibiting new features with respect to the usual quantum mechanical ones [9, 16]. Further developments include Lorentz invariance violation [17] and neutrino mixing contribution to the dark energy of the Universe [18].

In the Standard Model, flavor neutrinos are produced in charged current weak interaction processes, like $W^+ \to e^+ + \nu_e$. At tree level, flavor charge is strictly conserved in the vertex; the presence of mixing allows for the possibility of violation of lepton number via loop corrections. Such corrections are however extremely small and practically unobservable. Therefore one can assume the tree level as a good approximation of the real processes in which neutrinos are created. This is what is done in practice: neutrinos are identified by the observation of the corresponding charged leptons, *assuming* flavor conservation in production/detection vertices. It is therefore essential to have a clear understanding of the neutrino flavor charges and states in the presence of mixing.

One could nevertheless insist in *declaring* that flavor states are not to be used because everything can be written in terms of neutrino fields with definite masses and consequently solely neutrino mass eigenstates are being used. This is, however, like "to sweep the dust under the carpet", since in this way one operates a selection in favor of the Hilbert space of the mass eigenstates. Such a choice has no mathematical basis, because of the existence of a separate Hilbert space (the one for the flavor eigenstates). It is

CP957, *Particles, Strings, and Cosmology, 13th International Symposium*
edited by A. Rajantie, C. Contaldi, P. Dauncey, and H. Stoica

also lacking of physical motivations: indeed, in practice, it adopts the Pontecorvo states which are not eigenstates of the flavor charges, in contrast with the observed conservation of lepton number in the neutrino production/detection vertices.

We therefore devote the present report to: a) properly define flavor charges and states and b) to estimate how much lepton charge is violated by Pontecorvo states.

Let us start by considering the charges for flavor (mixed) neutrinos [6, 13]. We discuss here the case of mixing between two generations. Extension to three generations and beyond can also be done [7, 8].

For our purposes, the relevant Lagrangian density terms are

$$\mathscr{L} = \mathscr{L}_0 + \mathscr{L}_{int}, \tag{1}$$

where $\mathscr{L}_0$ is the free lepton Lagrangian:

$$\mathscr{L}_0 = \left(\bar{\nu}_e, \bar{\nu}_\mu\right)\left(i\gamma_\mu\partial^\mu - M_\nu\right)\begin{pmatrix}\nu_e\\ \nu_\mu\end{pmatrix} + \left(\bar{e}, \bar{\mu}\right)\left(i\gamma_\mu\partial^\mu - M_l\right)\begin{pmatrix}e\\ \mu\end{pmatrix}, \tag{2}$$

including the neutrino non-diagonal mass matrix M_ν and the mass matrix of charged leptons M_l:

$$M_\nu = \begin{pmatrix} m_{\nu_e} & m_{\nu_{e\mu}} \\ m_{\nu_{e\mu}} & m_{\nu_\mu}\end{pmatrix} \quad ; \quad M_l = \begin{pmatrix} m_e & 0 \\ 0 & m_\mu \end{pmatrix}. \tag{3}$$

$\mathscr{L}_{int}$ is the charged current weak interaction Lagrangian:

$$\mathscr{L}_{int} = \frac{g}{2\sqrt{2}}\left[W_\mu^+(x)\,\bar{\nu}_e(x)\,\gamma^\mu\,(1-\gamma^5)\,e(x) + W_\mu^+(x)\,\bar{\nu}_\mu(x)\,\gamma^\mu\,(1-\gamma^5)\,\mu(x) + h.c.\right]. \tag{4}$$

Of course, $\mathscr{L}_0$ can be diagonalized in terms of neutrino fields ν_1, ν_2, with definite masses m_1, m_2. Here, however, we are interested in the construction of flavor charges and therefore we consider the above Lagrangian in terms of flavor fields. To this end, we observe that $\mathscr{L}$ is invariant under the global phase transformations

$$e(x) \to e^{i\alpha} e(x), \qquad \nu_e(x) \to e^{i\alpha}\nu_e(x), \tag{5}$$

together with

$$\mu(x) \to e^{i\alpha}\mu(x), \qquad \nu_\mu(x) \to e^{i\alpha}\nu_\mu(x). \tag{6}$$

These are generated by

$$Q_e(t) = \int d^3\mathbf{x}\, e^\dagger(x) e(x), \qquad Q_{\nu_e}(t) = \int d^3\mathbf{x}\, \nu_e^\dagger(x)\nu_e(x), \tag{7}$$

$$Q_\mu(t) = \int d^3\mathbf{x}\, \mu^\dagger(x)\mu(x), \qquad Q_{\nu_\mu}(t) = \int d^3\mathbf{x}\, \nu_\mu^\dagger(x)\nu_\mu(x), \tag{8}$$

respectively. The invariance of the Lagrangian is then expressed by

$$[Q_l^{tot}, \mathscr{L}] = 0, \tag{9}$$

which guarantees the conservation of the total lepton number. Here, Q_l^{tot} is the total Noether (flavor) charge:

$$Q_l^{tot} = Q_{\nu_e}(t) + Q_{\nu_\mu}(t) + Q_e(t) + Q_\mu(t) \;=\; Q_e^{tot}(t) \,+\, Q_\mu^{tot}(t)\,, \tag{10}$$

$$Q_e^{tot}(t) = Q_{\nu_e}(t) + Q_e(t)\,, \qquad Q_\mu^{tot}(t) = Q_{\nu_\mu}(t) + Q_\mu(t)\,. \tag{11}$$

Note that the presence of the mixed neutrino mass term, i.e. the non-diagonal mass matrix M_ν, prevents the invariance of $\mathscr{L}$ under the separate phase transformations (5) and (6). Consequently, the flavor charges of Eq.(11) are time dependent.

However, family lepton numbers are still good quantum numbers if the neutrino production/detection vertex can be localized within a region much smaller than the region where flavor oscillations take place. This is what happens in practice, since tipically the spatial extension of the neutrino source/detector is much smaller than the neutrino oscillation length.

We thus look for the flavor neutrino states as eigenstates of the neutrino flavor charges Q_{ν_e} and Q_{ν_μ}. These operators may be expressed in terms of the (conserved) charges for the neutrinos with definite masses Q_{ν_1} and Q_{ν_2} in the following way [13]:

$$Q_{\nu_e}(t) = \cos^2\theta\; Q_{\nu_1} + \sin^2\theta\; Q_{\nu_2} + \sin\theta\cos\theta \int d^3\mathbf{x} \left[\nu_1^\dagger(x)\nu_2(x) + \nu_2^\dagger(x)\nu_1(x)\right] , \tag{12}$$

$$Q_{\nu_\mu}(t) = \sin^2\theta\; Q_{\nu_1} + \cos^2\theta\; Q_{\nu_2} - \sin\theta\cos\theta \int d^3\mathbf{x} \left[\nu_1^\dagger(x)\nu_2(x) + \nu_2^\dagger(x)\nu_1(x)\right] . \tag{13}$$

Notice that the last term in the above expressions forbids the construction of eigenstates of the $Q_{\nu_\sigma}(t)$, $\sigma = e,\mu$, in the Hilbert space $\mathscr{H}_{1,2}$ for the fields with definite masses. Indeed the vacuum $|0\rangle_{e,\mu}$ for the mixed field operators ν_σ turns out to be orthogonal to the vacuum state $|0\rangle_{1,2}$ for the fields with definite masses ν_j, with $j = 1,2$ [1]. One can also show that the above neutrino flavor charge operators are diagonal in the ladder operators α_σ, β_σ, for the neutrino flavor fields:

$$:: Q_{\nu_\sigma}(t) :: \;=\; \sum_r \int d^3\mathbf{k} \left(\alpha^{r\dagger}_{\mathbf{k},\sigma}(t)\alpha^r_{\mathbf{k},\sigma}(t) \,-\, \beta^{r\dagger}_{-\mathbf{k},\sigma}(t)\beta^r_{-\mathbf{k},\sigma}(t)\right), \qquad \sigma = e,\mu \tag{14}$$

where :: ... :: denotes normal ordering with respect to $|0\rangle_{e,\mu}$. This makes straightforward the definition (at reference time $t = 0$) of flavor neutrino and antineutrino states as:

$$|\nu^r_{\mathbf{k},\sigma}\rangle \equiv \alpha^{r\dagger}_{\mathbf{k},\sigma}|0\rangle_{e,\mu} \quad ; \qquad |\bar{\nu}^r_{\mathbf{k},\sigma}\rangle \equiv \beta^{r\dagger}_{\mathbf{k},\sigma}|0\rangle_{e,\mu}\,, \qquad \sigma = e,\mu. \tag{15}$$

which are thus by construction eigenstates of the operators in Eq.(14) at $t = 0$.

Let us now turn to the quantum mechanical (Pontecorvo) flavor neutrino states [14]:

$$|\nu^r_{\mathbf{k},e}\rangle_P \;=\; \cos\theta\, |\nu^r_{\mathbf{k},1}\rangle \,+\, \sin\theta\, |\nu^r_{\mathbf{k},2}\rangle\,, \tag{16}$$

$$|\nu^r_{\mathbf{k},\mu}\rangle_P \;=\; -\sin\theta\, |\nu^r_{\mathbf{k},1}\rangle \,+\, \cos\theta\, |\nu^r_{\mathbf{k},2}\rangle\,. \tag{17}$$

They are clearly *not* eigenstates of the flavor charges [10] as can be explicitly seen by using Eqs.(12) and (13). In order to estimate how much the flavor charge is violated in the usual quantum mechanical states, one can take the expectation values of the flavor charges on the above Pontecorvo states. Considering for example an electron neutrino state, we obtain:

$$ {}_P\langle \nu^r_{\mathbf{k},e}| : Q_{\nu_e} : |\nu^r_{\mathbf{k},e}\rangle_P = \cos^4\theta + \sin^4\theta + 2|U_{\mathbf{k}}|\sin^2\theta\cos^2\theta < 1, \qquad (18) $$

$$ {}_P\langle \nu^r_{\mathbf{k},e}| : Q_{\nu_\mu} : |\nu^r_{\mathbf{k},e}\rangle_P = 2\,(1-|U_{\mathbf{k}}|)\,\sin^2\theta\cos^2\theta > 0, \qquad (19) $$

for any $\theta \neq 0$, $m_1 \neq m_2$, $\mathbf{k} \neq 0$ and where : ... : denotes normal ordering with respect to the vacuum state $|0\rangle_{1,2}$. In the relativistic limit, $|U_{\mathbf{k}}| \to 1$ [1, 18] and the Pontecorvo states are a good approximation for the exact charge eigenstates Eq.(15).

We thank Arttu Rajantie and the organizers of PASCOS-07 for giving us the opportunity to present this work at the PASCOS conference. We also acknowledge INFN for financial support.

REFERENCES

1. M. Blasone and G. Vitiello, Annals Phys. **244**, 283 (1995).
2. M. Blasone, P. A. Henning and G. Vitiello, Phys. Lett. B **451**, 140 (1999); in *Particles, Strings and Cosmology*, Proceedings of PASCOS 98, P. Nath Editor, World Scientific, Singapore (1999) p.306.
3. M. Zralek, Acta Phys. Polon. B **29**, 3925 (1998).
4. K. Fujii, C. Habe and T. Yabuki, Phys. Rev. D **59**, 113003 (1999); Phys. Rev. D **64**, 013011 (2001).
5. M. Binger and C.R. Ji, Phys. Rev. D **60**, 056005 (1999).
6. M. Blasone, P. Jizba and G. Vitiello, Phys. Lett. B **517**, 471 (2001).
7. C.R. Ji, Y. Mishchenko, Phys. Rev. D **64**, 076004 (2001); Phys. Rev. D **65**, 096015 (2002).
8. M. Blasone, A. Capolupo and G. Vitiello, Phys. Rev. D **66**, 025033 (2002).
9. M. Beuthe, Phys. Rept. **375**, 105 (2003).
10. M. Blasone, A. Capolupo, F. Terranova and G. Vitiello, Phys. Rev. D **72**, 013003 (2005).
11. C. R. Ji and Y. Mishchenko, Annals Phys. **315**, 488 (2005)
12. C. Giunti, J. Phys. G **34**, R93 (2007)
13. M. Blasone, A. Capolupo, C. R. Ji and G. Vitiello, [hep-ph/0611106].
14. S. M. Bilenky and B. Pontecorvo, Phys. Rept. **41**, 225 (1978).
15. C. W. Kim and A. Pevsner, *Neutrinos in physics and astrophysics*, Harwood Academic Press, USA, (1993).
16. C. Giunti and C. W. Kim, *Fundamentals of Neutrino Physics and Astrophysics,* Oxford University Press, Oxford, UK, (2007).
17. M. Blasone, J. Magueijo and P. Pires-Pacheco, Europhys. Lett. **70**, 600 (2005).
18. A. Capolupo, S. Capozziello and G. Vitiello, *Phys. Lett. A* **363**, 53 (2007); M. Blasone, A. Capolupo, S. Capozziello and G. Vitiello, this volume.

Status and future of the CDMS experiment: CDMS-II to SuperCDMS

T. Bruch for the CDMS Collaboration

Physik Institut, University of Zurich, Winterthurerstrasse 190, 8057 Zurich, Switzerland

Abstract. The Cryogenic Dark Matter Search experiment (CDMS-II) employs low-temperature Ge and Si detectors to detect WIMPs via their elastic scattering interactions with the target nuclei. No Dark Matter signal has been observed so far, resulting in a limit on the spin-independent WIMP-nucleon elastic scattering cross-section with a minimum of 1.6 x 10^{-43} cm^2 at a WIMP mass of 60 GeV c^{-2}. To increase the sensitivity, new one inch thick detectors have been developed which will be used in the SuperCDMS phase. SuperCDMS will be operated at SNOLAB with an expected sensitivity on the spin-independent WIMP-nucleon elastic scattering cross-section of 1 x 10^{-45} cm^2 at the 25 kg stage.

Keywords: CDMS, Dark Matter, Direct Detection, astroparticlephysics
PACS: 95.35.+d, 95.30.Cq, 82.25.Oj, 29.40.Wk, 14.80.Ly

I. The CDMS-II experiment

The Cryogenic Dark Matter Search experiment searches for non-luminous, non-baryonic WIMPs [1], that could form the majority of the matter in the universe [1, 2]. The kinetic energy imparted to a nucleus in an elastic WIMP-nucleon scattering event would range from a few keV to tens of keV. The small recoil energy, coupled with an expected low event rate, means that it is vital to suppress backgrounds effectively. Active and passive shielding are used to reduce backgrounds produced outside the experimental apparatus, leaving decays of radioactive contamination inside of the shielding as the dominant natural background. Gammas and electrons from these decays interact electromagnetically and can be discriminated against nuclear recoils. The remaining background comes from muon induced and spallation neutrons, which produce nuclear-recoil events identical to WIMPs. The 780 m (2090 mwe[2]) of rock overburden at the Soudan Underground Laboratory reduces the surface muon flux by a factor of 5 x 10^4. Monte Carlo simulations predict a total unvetoed rate of 0.05 $\pm$ 0.01 muon induced neutrons kg^{-1} y^{-1}. The muon induced neutrons, as well as neutrons produced by (α ,n) reactions and spallation from the contamination will ultimately limit the sensitivity of the experiment. The estimates for the neutron rates induced by the above processes are small enough to provide a neutron background free operation of the CDMS-II experiment for current and future runs at the Soudan site.

---

[1] **W**eakly **I**nteracting **M**assive **P**articles
[2] meters water equivalent

CP957, *Particles, Strings, and Cosmology, 13th International Symposium*
edited by A. Rajantie, C. Contaldi, P. Dauncey, and H. Stoica

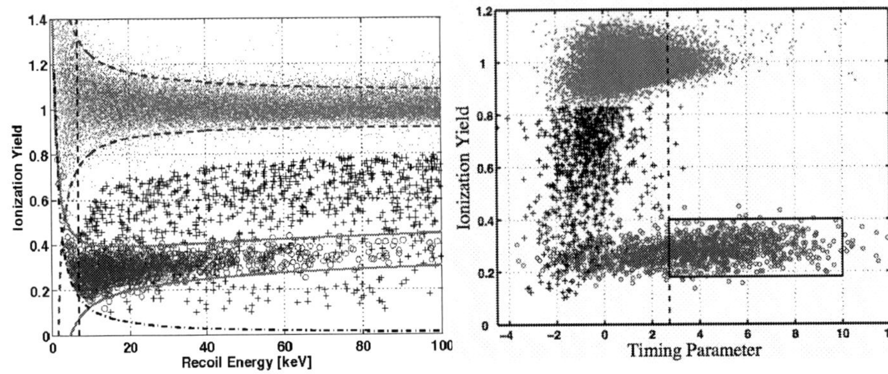

FIGURE 1. Ionization yield versus recoil energy (left) and phonon-timing parameter (right) for calibration data in a germanium detector. Three different classes of events are shown: ^{133}Ba gamma-calibration bulk (red dots) and surface (black crosses) events as well as nuclear recoils (blue circles) from ^{252}Cf neutron-calibration. In the left plot the nuclear- (red line) and electron-recoil band (blue/dashed) are shown with the analysis threshold (dashed/vertical) of 7 keV and the ionization threshold (black/dot dash). The box in the right plot indicates the background free, approximate signal region.

CDMS ZIP Detectors. Each CDMS ZIP (Z-sensitive Ionization- and Phonon-mediated) [3] detector is a cylindrical high-purity Ge or Si crystal, 1cm thick and 7.6 cm in diameter. A single Ge (Si) ZIP has a mass of 250 g (100 g). Two concentric ionization electrodes and four independent athermal phonon sensors are photolithographically patterned onto each crystal. A particle interacting in a ZIP detector can interact with an electron in the crystal, or with a nucleus. The interaction deposits energy into the crystal through charge excitations (electron-hole pairs) and lattice vibrations (phonons). The simultaneous ionization and phonon measurement with a ZIP detector not only allows an accurate measurement of the recoil energy independent of recoil type, but also distinguishes between the two types of recoils (electron-recoil and nuclear-recoil). The dimensionless ionization yield parameter, y, is the ratio of ionization energy to recoil energy, $y = E_Q/E_R$. Nuclear recoils produce fewer charge pairs, and hence less ionization energy, E_Q, than do electron recoils of the same recoil energy. The ionization yield for electron- and nuclear-recoils is determined by calibrations with a ^{133}Ba and a ^{252}Cf source providing the bands shown in figure 1. The simultaneous measurement of ionization and recoil energy therefore provides the possibility to identify and reject electron-recoil background events with $> 99.99\%$ efficiency. Since WIMPs are expected to interact with the target nuclei, the nuclear-recoil band defines the signal region.

Low-yield surface events. Events interacting in the first few microns of the crystal have an incomplete charge collection. Hence the ionization yield parameter is suppressed for such events, and they may leak into the signal region, mimicking nuclear recoils. These events, referred to as surface events, can be identified as a third population between the electron and nuclear band in figure 1 (left plot). Surface events are the most dangerous background and can limit the sensitivity of the experiment. To discriminate between surface and nuclear recoil events the timing properties of the phonon signals are used. The two parameters used in the analysis are the delay of the slower phonon signal with respect to the ionization signal and the risetime of the leading phonon sensor (which is the one with the largest signal amplitude). Since surface events have faster delays and risetimes than bulk nuclear-recoils, these two parameters can be used to cut out surface events and select nuclear recoils. This cut is defined on low-yield ^{133}Ba and ^{252}Cf calibration data as shown in figure 1 (right plot). In this figure the three populations can easily be identified and a background-free signal window is defined in the ionization yield - timing parameter space. Only single scatter nuclear-recoils in this signal window are considered as WIMP candidates.

Results from the CDMS-II setup

The CDMS-II setup was run with one tower containing four Ge and two Si ZIPs (Run118) and in a two tower setup with six Ge and six Si ZIPs (Run119). All cuts designed to select nuclear recoils were determined in a blind manner from *in situ* calibrations with external radioactive sources, without any prior knowledge of the event distribution in the signal region. A 64 keV and and 10.5 keV event were found in the signal region in the first and second run respectively, consistent with the estimated systematic background [4]. These results can be combined and interpreted as an upper limit on the spin-independent WIMP-nucleon elastic scattering cross-section (σ_{WN}) shown in figure 2. The first two runs with five towers (Run123/124) with 19 Ge and 11 Si ZIPs in the CDMS-II setup lasted from October 2006 to July 2007 accumulating 650 kg days of Ge raw exposure. After a short maintenance period Run125 was started, which is expected to accumulate an additional 1000 kg days of raw exposure until the end of 2008.

II. CDMS-II to SuperCDMS

To further increase the sensitivity of the CDMS experiment the total accumulated exposure has to be increased. This can be achieved by increasing the detector mass and the runtime of the experiment. For the SuperCDMS setup new 1 inch thick detectors have been developed and tested, providing an increase of a factor of 2.54 in mass with respect to the 1cm thick detectors used in the CDMS-II setup. The redesign of the phonon readout, which maximizes the active phonon collection area, and the hydrogenation of the amorphous silicon layer between the crystal and the electrodes, show improvements in the discrimination between surface events and nuclear-recoils. In addition, an improved analysis for a better surface event discrimination is in hand and being applied in a reanalysis of Run118/119 data. The first two super towers consisting each of six 1 inch thick detectors will be installed at the Soudan site at the end of 2008 to demonstrate the improved discrimination capabilities, and test if SCDMS can be operated with a background free signal region.

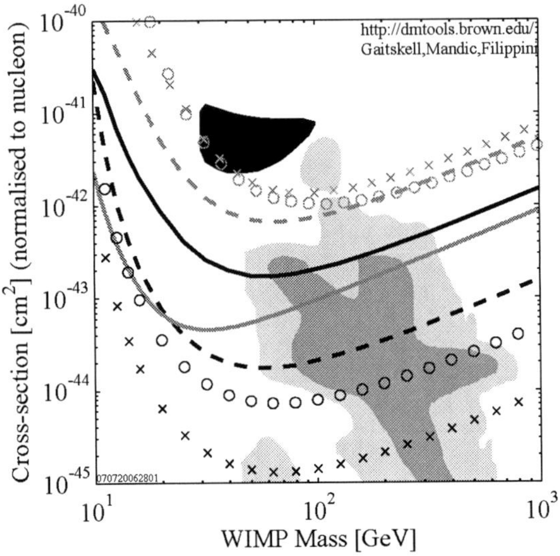

FIGURE 2: Upper limits on the WIMP-nucleon spin-independent cross-section; CDMS two previous runs combined (black/line) [4]; XENON (gray/line)[5]; ZEPLIN II [6] (gray/dashed); WARP [7] (gray/circle) and EDELWEISS (gray/cross) [8]. Also shown are the projected sensitivities for CDMS-II final (black/dashed); the two super tower run at Soudan (black/circle) and SCDMS 25 kg (black/cross). Filled region indicate CMSSM models (68% CL dark gray and 90% CL light gray)[10] and an interpretation of the DAMA/NaI annual modulation signal (black) [9].

At the SCDMS 25 kg stage seven super towers will be installed and operated at SNOLAB. SCDMS 25 kg aims to reach a sensitivity of 1 x 10^{-45} cm^2 at a WIMP mass of 60 GeV, as shown in figure 2.

CONCLUSIONS

The CDMS-II setup is currently running with 30 ZIPs and aims to accumulate roughly 1600 kg days of WIMP search exposure until the end of 2008. If no WIMP candidate events are found, it will reach a sensitivity of $\sigma_{WN} \sim 1$ x 10^{-44} cm^2. With the improvements on the hardware and the analysis the future phases of the CDMS experiment should stay background free. The next upgrade of the CDMS experiment to SuperCDMS 25 kg will be installed at SNOLAB increasing the sensitivity by one order of magnitude.

REFERENCES

1. D.N. Spergel et al., Astrophys. J. Suppl. 148, 175 (2003).
2. G.Jungman et al., Phys. Rep. 267, 195 (1996).
3. T. Saab et al., Nucl.Instr. and Meth. A444 (2000) 300.
4. D. S. Akerib et al., Phys.Rev. Lett. 96, 011302 (2006).
5. J. Angle et al., (2007), arXiv:0706.0039 .
6. G. J. Alner et al., (2007), arXiv:astro-ph/0701858 .
7. P. Benetti et al., (2007), arxiv:astro-ph/0701286 .
8. V. Sanglard et al., Phys. Rev. D71, 122002 (2005).
9. R. Bernabei et al., Riv. Nuovo Cim. 26N1, 1 (2003).
10. L. Roszkowski, R. Ruiz de Austri and R. Trotta, (2007), arXiv:0705.2012

Prospects for (non-SUSY) new physics with first LHC data

Jonathan Butterworth, on behalf of the ATLAS and CMS collaborations

Department of Physics and Astronomy, University College London, Gower St. WC1E 6BT, UK

Abstract. The ATLAS and CMS experiments will take first data soon. I consider here the prospects for new physics (excluding SUSY) with a few fb^{-1} of data. This means processes with signal cross sections of a few 100 fb or less, with clear and fairly simple signatures - precision comparison of data to Standard Model tails will take longer, needing more luminosity and very good understanding of detector calibrations, resolutions and trigger efficiencies. The approach I take here is signature rather than model based, but examples of models will be given.

Keywords: Beyond the standard model, LHC
PACS: 12.60.-i; 13.85.Qk

INTRODUCTION

The Large Hadron Collider (LHC) will provide proton-proton collisions at 14 TeV - a factor of seven above previous colliders. This is not just "any old" new energy regime either. It will give us the first direct look at physics far above the electroweak-symmetry-breaking scale. Thus if the Standard Model (SM) is treated as a Higgsless low-energy effective theory, calculations of SM cross sections violate unitarity within the experimental reach of the experiments. Thus the experiments must see either the SM Higgs, or other new physics, or both. This underlines the need for the experiments to be alert to the widest range of possible exotic signatures. In this brief review I discuss prospects for observing some of these signatures with early data, and to which models they are sensitive.

VARIOUS RESONANCES

One of the simplest places to start is looking for anomalous, in particular resonant, production of dileptons. The CMS collaboration have recently [1] presented a sensitivity study of $\mu^+\mu^-$ resonances above the Drell-Yan background. Such signals can arise for example from Z' bosons present in grand unification models [2]. With even 0.1 fb^{-1} there is sensitivity to a 1 TeV Z' in all the models considered, even when a realistic estimate of the detector alignment achievable with this luminosity is included. With 1 fb^{-1} there is still sensitivity up to around 2.5 TeV, dependent upon the specific model - see Fig. 1 (left). Such studies also give an indication of sensitivity to graviton resonances in Randall-Sundrum models, for example. One might begin to distinguish spin-1 from spin-2 resonances with 50 fb^{-1}.

CP957, *Particles, Strings, and Cosmology, 13th International Symposium*
edited by A. Rajantie, C. Contaldi, P. Dauncey, and H. Stoica

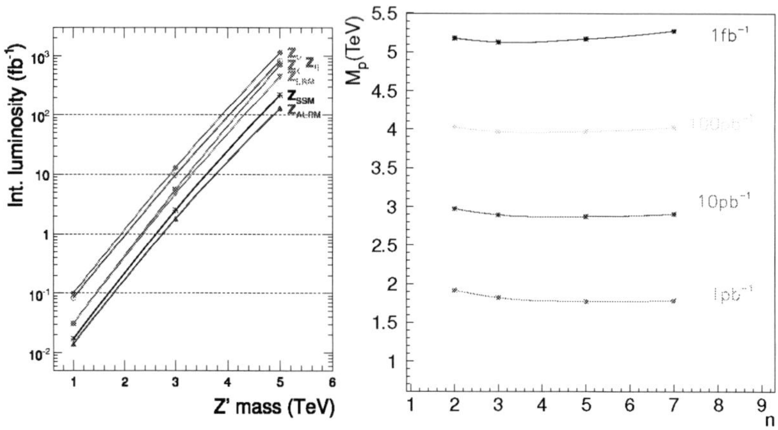

FIGURE 1. (l) CMS 5σ discovery reach as a function of integrated luminosity as a function of Z' mass for various models with $Z' \rightarrow \mu^+\mu$ (from [1]) (r) ATLAS discovery contours in the plane of Planck mass (M_P) and number of extra dimensions (n) for black hole masses more than 1 TeV above M_P (from [5])

Graviton resonances are an example where one also expects to see resonances in the diphoton channel. Here the CMS experiment expects [1, 3] to have sensitivity up to around 3.5 TeV with 10 fb^{-1}.

Kaluza-Klein excitations of the gluon in TeV^{-1} extra dimension models are an example of a model producing resonances decaying to quarks. ATLAS has carried out a study [4] of 1 TeV resonances decaying to $t\bar{t}$, where the W from one of the tops decays to hadrons, the other to (e or μ) + ν. There is sensitivity here with around fb^{-1}, and the mass reach eventually is expected to be up to around 3.3 TeV.

MISSING MOMENTUM

It is natural to benchmark signatures with missing transverse momentum plus leptons against models with high mass W' bosons, just as the dilepton case is benchmarked against Z' models. Discovery limits are typically based on a measured transverse mass distribution, after a selection requiring the presence of a high transverse momentum (p_T) lepton. An example study from CMS [1, 6] using a reference model [7] shows sensitivity up to around 3.5 TeV in the W' mass with 1 fb^{-1} of data.

Multiple lepton signatures in association with missing p_T arise naturally in models where a complex decay chain ensues from a massive particle decay. For example in universal extra dimension models [8], Kaluza-Klein excitations of the gluon occur, carrying a conserved quantum number which plays a similar role to R-parity in the minimal supersymmetric SM. That is, decays follow through a cascade (which may produce leptons) ending in the lightest Kaluza-Klein state, which cannot decay and carries off "missing" momentum. A study from CMS [9] shows sensitivity after a few fb^{-1} of up to around 700 GeV in the mass scale associated with the size of the extra dimensions in the minimal universal extra dimension model.

JETS AND VECTOR BOSONS

Jets themselves will give the highest statistics probe of the highest energy scales at the LHC (unless they are suppressed by other exotic physics such as TeV black holes). The major limitation on the sensitivity is likely to be from systematic errors at first associated with detector, where understanding the energy measurement for jets of several TeV will require much detailed study. In the end, as the detectors become better understood, theoretical uncertainties on the QCD cross section will also play a role. That said, both these effects can be reduced by considering the ratio of jets in different rapidity regions. In this case, a CMS study [1] indicates that the dijet mass ratios could be measured up to 6.5 TeV with an accuracy of better than 50%, giving sensitivity to new physics above 15 TeV if the effects are manifested as effective contact interactions.

A qualitatively new feature expected at the LHC is the abundance of particles with mass O(100) GeV and p_T of several hundred GeV decaying to hadrons - SM candidates being the W, Z, top and Higgs, but with other exotica also possible. The boost means that the decay products will often appear in a single jet. Thus measuring the single-jet mass, and studying substructure of the jet for signs of the heavy particle decay, provides information which can be exploited in the search for and diagnosis of new physics. This was first studied in the case of vector boson scattering at high masses [10] and has since been proposed and/or studied in several other cases [11, 12, 13, 14, 15, 16], including new physics searches ranging from superheavy quarks to Kaluza-Klein graviton excitations, as well as supersymmetric decay chains and SM measurements. Preliminary studies from ATLAS show that single-jet mass resolutions of a few GeV should be obtainable, which is good enough to make the method useful. For example, t' candidates may be observable using this technique with only 2.5 fb^{-1} of data [11].

Vector boson scattering, where a W or Z is radiated from a quark in each proton and they then interact, is a process of great interest over the whole range of centre-of-mass energy accessible at the LHC. It is characterised by the W/Z decay products and two high rapidity "tag" jets travelling close to the beam directions. At low masses, WW and ZZ scattering are search channels for the SM Higgs. At high masses, unitarity is violated in this process without a Higgs, so it represents the "no lose" channel at the LHC. If a Higgs is seen, measuring this process is vital to show that the Higgs mechanism actually operates. If no Higgs is seen, some other new physics must appear in this channel to regulate the cross section at the highest masses.

Of course, in its extreme this is a very high luminosity topic, where the kinematic limit is probed only with the very high luminosity (a luminosity upgrade is in fact required to exploit it fully). However, diboson resonances in this channel are a generic feature of models involving new strong interactions around 1 TeV [17]. For example, studies by CMS and at ATLAS of WZ and WW resonances respectively [1, 18, 19] indicate that first sensitivity will be obtained after a few fb^{-1}.

MULTIPLE JETS AND LEPTONS

In extra dimension models, the Planck scale may be lowered far enough to be accessible at the LHC. This would lead to copious production of mini black-holes, which would decay very rapidly via Hawking radiation to "flavour democratic" high multiplicity states. Cuts on multiplicity and event shapes (such as circularity) can be used to extract as signal. There are large uncertainties in the phenomenology, but the cross sections may be enormous, and there are cases where this physics would be visible with very low luminosities [5], down even to 1 pb^{-1}, once the detector and backgrounds are sufficiently well understood! - see Fig.1 (right).

SUMMARY

Many surprises are possible in the early days of LHC running. A "bonfire of models" is certain, but some (perhaps only one?) will survive. I have not discussed at all the enormous complexity of the detectors and the huge challenges involved in understanding them, and in understanding the SM at 14 TeV. The first things we see will be features of new detectors - so buyer beware rumours and blogs! Nevertheless, there will be some really exciting physics.

ACKNOWLEDGMENTS

Thanks to especially to G. Brooijmans, S. Eno and F. Ledroit, as well as to the organisers for a stimulating and fun conference.

REFERENCES

1. CMS TDR Vol II: Physics Performance, Tech. Rep. CERN/LHCC 2006-021, CERN (2006).
2. A. Leike, *Phys. Rept.* **317**, 143–250 (1999), hep-ph/9805494.
3. M. C. Lemaire, V. A. Litvin, and H. Newman, Tech. Rep. CERN-CMS-NOTE-2006-051 (2006).
4. B. Salvachúa, L. March, and E. Ros, Tech. Rep. ATL-PHYS-PUB-2006-002, CERN (2006).
5. J. Tanaka, T. Yamamura, S. Asai, and J. Kanzaki, *Eur. Phys. J.* **C41**, 19–33 (2005), hep-ph/0411095.
6. C. Hof, T. Hebbeker, and K. Hoepfner, Tech. rep. (2006), cERN-CMS-NOTE-2006-117.
7. G. Altarelli, B. Mele, and M. Ruiz-Altaba, *Z. Phys. C* **45**, 109 (1989).
8. T. Appelquist, H.-C. Cheng, and B. A. Dobrescu, *Phys. Rev.* **D64**, 035002 (2001), hep-ph/0012100.
9. M. Kazana, *Acta Phys. Polon.* **B38**, 449–458 (2007).
10. J. M. Butterworth, B. E. Cox, and J. R. Forshaw, *Phys. Rev.* **D65**, 096014 (2002).
11. B. Holdom, *JHEP* **03**, 063 (2007), hep-ph/0702037.
12. W. Skiba, and D. Tucker-Smith, *Phys. Rev.* **D75**, 115010 (2007), hep-ph/0701247.
13. B. Lillie, L. Randall, and L.-T. Wang (2007), hep-ph/0701166.
14. S. Fleming, A. H. Hoang, S. Mantry, and I. W. Stewart (2007), hep-ph/0703207.
15. U. Baur, and L. H. Orr (2007), arXiv:0707.2066[hep-ph].
16. J. M. Butterworth, J. R. Ellis, and A. R. Raklev, *JHEP* **05**, 033 (2007), hep-ph/0702150.
17. C. T. Hill, and E. H. Simmons, *Phys. Rept.* **381**, 235–402 (2003), hep-ph/0203079.
18. P. Kreuzer, Tech. Rep. CERN-CMS-NOTE-2006-135 (2006).
19. E. Stefanidis, *Study of WW scattering in the absence of a light Higgs boson using the ATLAS Detector at the LHC*, Ph.D. thesis, UCL, London (2007).

Sparticle spectra from Large-Volume String Compactifications

Joseph P. Conlon

DAMTP, Centre for Mathematical Sciences, Wilberforce Road, Cambridge, CB3 0WA, UK

Abstract. Large-volume models are a promising approach to stabilising moduli and generating the weak hierarchy through TeV-supersymmetry. I describe the pattern of sparticle mass spectra that arises in these models.

Keywords: Flux Compactifications, Supersymmetry
PACS: 11.25.Mj, 11.25.Wx, 12.60.Jv, 14.80.Ly

INTRODUCTION

TeV supersymmetry is one of the most promising ideas for physics beyond the standard model. It stabilises the electroweak hierarchy by cancelling the quadratic divergences to the Higgs mass and can provide a dark matter candidate. It also gives a dynamical explanation of electroweak symmetry breaking, through the radiative running of the up-type Higgs mass. It is furthermore consistent with the absence of strong new physics corrections to electroweak precision observables.

It is desirable to realise TeV supersymmetry in string compactifications. For many years string compactifications suffered from the moduli problem. The moduli, that parametrise the size and shape of the extra dimensions, are massless at tree-level. The moduli vevs determine the properties of the vacuum and so it is necessary to fix the moduli before one can do phenomenology. The understanding of supersymmetry breaking in string theory is also related to the problem of how to fix moduli.

The problem of how to fix moduli is now solved. An important ingredient in this is the use of fluxes. Fluxes which have non-trivial profiles in the extra dimensions source potentials for the moduli [1, 2]. This is because fluxes carry a potential energy which depends on the size and shape of the extra dimensions. There exist various proposals to fix moduli. One of the most promising is that of the large-volume models [3]. This is because in addition to fixing moduli these also generate supersymmetry broken at a hierarchically small scale.

These models are defined in supergravity by the following Kähler potential and superpotential,

$$K = -2\ln((T_b+\bar{T}_b)^{3/2} - (T_s+\bar{T}_s)^{3/2} + \xi), \quad (1)$$

$$W = \int G_3 \wedge \Omega + Ae^{-a_s T_s}, \quad (2)$$

CP957, *Particles, Strings, and Cosmology, 13th International Symposium*
edited by A. Rajantie, C. Contaldi, P. Dauncey, and H. Stoica

where $\mathscr{V} = (T_b + \bar{T}_b)^{3/2} - (T_s + \bar{T}_s)^{3/2}$ is the compactification volume. The resulting scalar potential is

$$V = \frac{8(a_sA)^2\sqrt{\tau_s}e^{-2a_s\tau_s}}{3\mathscr{V}} - \frac{4(a_sA)W_0\tau_s e^{-a_s\tau_s}}{\mathscr{V}^2} + \frac{\xi W_0^2}{\mathscr{V}^3} \tag{3}$$

This potential admits a minimum at exponentially large values of the volume, $\mathscr{V} \sim e^{a_s\tau_s}$, $\tau_s \sim \frac{\xi}{g_s}$. As the volume is exponentially large, the gravitino mass is exponentially small:

$$m_{3/2} = e^{K/2}W = \frac{W_0}{\mathscr{V}} \ll M_P.$$

It therefore makes sense to study the phenomenology of these models. This requires computing the soft terms at the high scale, evolving these soft terms down to the low scale and then studying the resulting supersymmetric particle spectrum.

SOFT TERMS

The expressions for the soft scalar masses and the trilinear A-terms are given by

$$\begin{aligned}
\tilde{m}^2_{\alpha\bar{\beta}} &= (m^2_{3/2}+V_0)\tilde{K}_{\alpha\bar{\beta}} - \bar{F}^{\bar{m}}F^n\left(\partial_{\bar{m}}\partial_n\tilde{K}_{\alpha\bar{\beta}} - (\partial_{\bar{m}}\tilde{K}_{\alpha\bar{\gamma}})\tilde{K}^{\bar{\gamma}\delta}(\partial_n\tilde{K}_{\delta\bar{\beta}})\right) \\
A'_{\alpha\beta\gamma} &= e^{\hat{K}/2}F^m\left[\hat{K}_m Y_{\alpha\beta\gamma} + \partial_m Y_{\alpha\beta\gamma} - \left((\partial_m\tilde{K}_{\alpha\bar{\rho}})\tilde{K}^{\bar{\rho}\delta}Y_{\delta\beta\gamma} + (\alpha\leftrightarrow\beta) + (\alpha\leftrightarrow\gamma)\right)\right]
\end{aligned} \tag{4}$$

Here $\tilde{K}_{\alpha\beta}$ is the Kähler metric for the chiral matter fields. In the dilute flux limit, where the magnetic flux on the D7 branes is dilute compared to the size of the cycle, this is given for the large volume model by [4]

$$\tilde{K}_{\alpha\bar{\beta}} = \frac{\tau_s^{1/3}}{\mathscr{V}^{2/3}}k_{\alpha\bar{\beta}}(\phi), \tag{5}$$

where $k_{\alpha\bar{\beta}}(\phi)$ is an unknown function of the complex structure moduli. In this case the soft terms take a specified form in terms of F^s, the F-term for the small modulus.

More realistically, however, there will be corrections due to magnetic fluxes. Magnetic fluxes are necessary in order to render the matter spectrum chiral. Magnetic flux on a 2-cycle can be thought of as being a D5-brane embedded in a D7 brane. Two D5 branes intersect in a pointlike fashion, which is known to generate chiral matter. The magnetic flux will alter the form of the matter metrics and thus of the soft terms. It will also alter the gauge couplings, as the full expression for the gauge couplings is

$$f_a = \frac{T}{4\pi} + h_a(F)S. \tag{6}$$

We assume the Standard Model is realised by a stack of D7 branes on the small cycle. In the dilute flux limit the gauge couplings are given by

$$f_2 = f_3 = T_s, f_1 = k_Y T_s. \tag{7}$$

The factor k_Y accounts for the fact that there is no canonical normalisation for U(1) gauge factors. The magnetic fluxes correct the gauge couplings away from the dilute flux limit. We interpret this as giving the correct gauge couplings at 10^{11}GeV (as the weak and strong couplings cannot unify with a lower string scale).

In the dilute flux limit the soft terms are

$$M_i = \frac{F^s}{2\tau_s}, \tag{8}$$

$$m_\alpha^2 = \frac{1}{3}\left(\frac{F^s}{2\tau_s}\right)^2, \tag{9}$$

$$A_{\alpha\beta\gamma} = -\left(\frac{F^s}{2\tau_s}\right) Y_{\alpha\beta\gamma}. \tag{10}$$

To go beyond the dilute flux limit we assume that the matter metrics are in fact modified to

$$\tilde{K}_{\alpha\bar{\beta}} = \frac{(\tau_s + \varepsilon_i)^{1/3}}{\mathcal{V}^{2/3}} k_{\alpha\bar{\beta}}(\phi), \tag{11}$$

ε parametrises the effects of magnetic fluxes. It is regarded as universal within a particular species (e.g for all U_R) and non-universal for fields of different gauge charges. In string theory fields of the same species come from intersections of the same branes, whereas fields of different gauge charges come from intersections of different brane stacks.

To study the resulting phenomenology, we perform the following procedure [5]. The gaugino masses including the effects of magnetic fluxes are given by

$$M_3 = \frac{F^s}{2(\tau_s + h_3 S)}, \tag{12}$$

$$M_2 = \frac{F^s}{2(\tau_s + h_2 S)}, \tag{13}$$

$$M_1 = \frac{F^s}{2(\tau_s + h_1 S)}. \tag{14}$$

We view these as perturbations about a common value, $M = F^s/2\tau_s$. The ratio of M_3 and M_2 is equivalent to the ratio of gauge couplings, and so is fixed at the high scale. To estimate M we use $(M_2 + M_3)/2$. M_1 is a perturbation about M, and is allowed to vary by twenty per cent at the high scale. To normalise the overall spectrum, we take a fixed value for M_3 at the high scale of 500 GeV. This fixes M_2 and also M. The soft masses are also allowed to vary by up to twenty per cent from their dilute-flux values.

The resulting range of particle spectra is shown in figure 1, corresponding to the generation of 100 particle spectra satisfying collider and phenomenological constraints and with the gluino mass normalised to 900 GeV. This gives a measure of the expected supersymmetric mass spectrum that will arise in the large volume scenario.

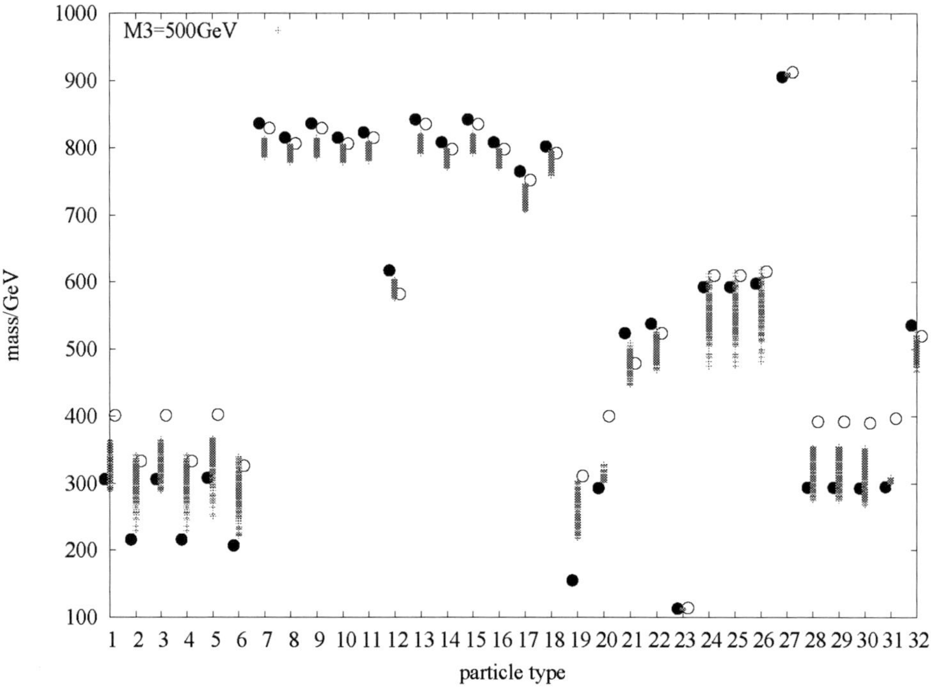

FIGURE 1. Particle spectra with the gluino mass normalised to $\sim$ 900 GeV and 20% high scale uncertainties in other parameters. For comparative purposes, an SPS1a spectrum with the same value of the gluino mass is also shown using black dots. The hollow circles show a spectrum coming from soft terms in the dilute flux limit. Particles 1-6 are the left and right handed sleptons, particles 7-18 are the squarks, 19 is the lightest supersymmetric particle (LSP) $\tilde{\chi}_1^0$, 20-22 are the remaining $\tilde{\chi}_i^0$, 23-26 the Higgs particles, 27 the gluino, 28-30 the sneutrinos, 31 and 32 are the charginos $\tilde{\chi}_{1,2}^{\pm}$.

ACKNOWLEDGMENTS

I thank my collaborators Chun-Hay Kom, Kerim Suruliz, Ben Allanach and Fernando Quevedo. I thank Trinity College, Cambridge for financial support.

REFERENCES

1. S. B. Giddings, S. Kachru and J. Polchinski, Phys. Rev. D **66** (2002) 106006 [arXiv:hep-th/0105097].
2. S. Kachru, R. Kallosh, A. Linde and S. P. Trivedi, Phys. Rev. D **68**, 046005 (2003) [arXiv:hep-th/0301240].
3. V. Balasubramanian, P. Berglund, J. P. Conlon and F. Quevedo, JHEP **0503**, 007 (2005) [arXiv:hep-th/0502058].
4. J. P. Conlon, D. Cremades and F. Quevedo, JHEP **0701**, 022 (2007) [arXiv:hep-th/0609180].
5. J. P. Conlon, C. H. Kom, K. Suruliz, B. C. Allanach and F. Quevedo, arXiv:0704.3403 [hep-ph].

Recent Results From HERA

François Corriveau

IPP / McGill University, Department of Physics, 3600 University, Montréal, Canada, H3A2T8
on behalf of the H1 and ZEUS collaborations

Abstract. A brief overview of the HERA physics program is given, covering deep inelastic scattering observations from the H1 and ZEUS experiments and results from recent polarisation data.

Keywords: proton, structure function, electroweak, QCD, quarks, polarisation
PACS: 12.15.Dm, 12.38.Qk, 14.20.Dh

INTRODUCTION

The only existing electron-hadron collider, HERA (Hadron Electron Ring Accelerator), is located at the DESY research center in Hamburg, Germany. As of the time of this presentation, HERA had stopped its operations just a few days before, on June 30th, 2007, thus concluding a rich physics program which had started taking data in 1992.

With 27.5 GeV electrons or positrons colliding on 920 GeV protons, HERA did not only provide the tools to investigate the structure of the proton down to the attometer scale (10^{-18} m) through deep inelastic scattering (DIS), it also opened new avenues to study Quantumchromodynamics (QCD) and numerous hadronic final states in great details, to search for special indications of matter like the leptoquarks or excited fermions, as well as e.g. to attempt understanding diffraction processes or rare phenomena.

After the 2001 HERA upgrade, the luminosity collected was quadrupled and spin rotator devices before and after the H1 [1] and ZEUS [2] experiments aligned the electron/positron spins longitudinally with the beam direction, thus providing polarisation and allowing yet more innovative tests of the Standard Model.

DEEP INELASTIC SCATTERING

Lepton-hadron collisions proceed via the exchange of a virtual photon γ or one of the three heavy intermediate vector bosons Z^0, W^+ and W^-. Events mediated by a γ or a Z^0 are labelled neutral current (NC) events, while the others become charged current (CC) events. At the lowest order, the outgoing lepton is typically balanced in p_T by an ejected quark hadronizing into a jet. Three parameters can characterize a DIS event: the square of the exchanged momentum, Q^2, the x Bjorken scaling variable (associated to the fraction of the proton momentum participating to the the reaction) and the y scaling "elasticity" variable. Each parameter can be calculated from the scattered lepton data in the NC events, or the jet observables in all events.

One of the foremost measurement results obtained in the HERA program is the DIS cross section as function of Q^2 for both NC and CC events and for either e^-p or e^+p

CP957, *Particles, Strings, and Cosmology, 13th International Symposium*
edited by A. Rajantie, C. Contaldi, P. Dauncey, and H. Stoica

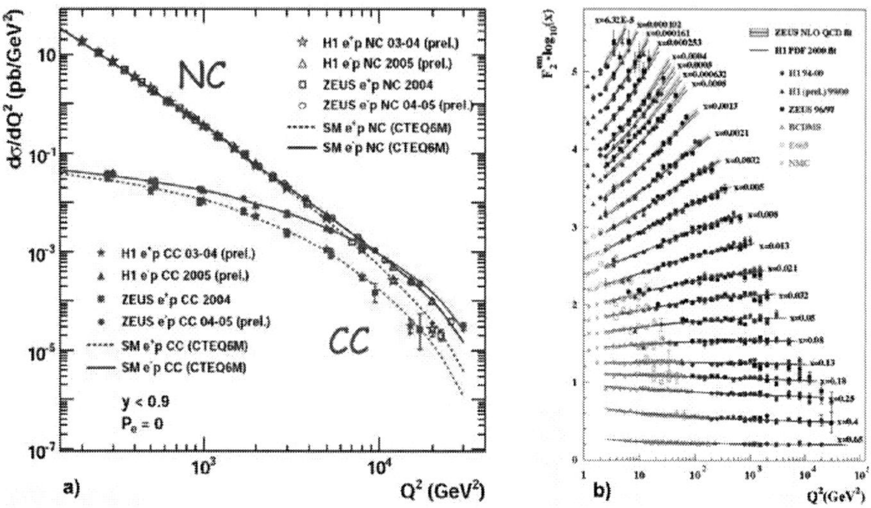

FIGURE 1. a) Cross section as function of Q^2. b) F_2 structure function results as function of Q^2 and x.

collisions. Fig. 1a shows those distributions and their comparison to the Standard Model predictions. One main feature is that the NC distributions are dominated by photon exchange, especially at low Q^2 values. At high Q^2 values, the weak M_Z-associated contributions appear and the CC and NC cross sections actually become similar, thus confirming electroweak unification.

Yet more information can be extracted when the double differential cross section is computed as a function on the two independent variables x and Q^2:

$$\frac{d^2\sigma(e^\pm)}{dxdQ^2} = \frac{2\pi\alpha^2}{xQ^4}[Y_+F_2(x,Q^2) - y^2F_L(x,Q^2) \mp Y_-xF_3(x,Q^2)] \tag{1}$$

where $Y_\pm = 1 \pm (1-y)^2$. The structure functions F_2, xF_3 and F_L are respectively the main term, the parity-violating term and a longitudinal term, which should be zero except for contributions from gluon radiation. All terms are themselves combinations of the individual parton density functions, i.e. of quarks and gluons.

Fig. 1b summarizes the HERA results for F_2, extending the low x and low Q^2 results from previous fixed targets experiments and vindicating the validity of perturbative QCD methods down to $x \sim 10^{-4}$ or $Q \sim 1\ GeV^2$. Scaling is clearly seen for high x values, but violated in the low x range, thus enabling extraction of the gluon density distributions. QCD next-to-leading order analyses of all experimental data lead to the determination of the parton momentum distributions for the proton valence quarks, the sea quarks and the gluons, thus mapping the proton structure. More detailed results can be found e.g. in [3]. The strong coupling constant α_s can also be deduced from this analysis.

The DIS NC data has been used to extract the xF_3 structure function. Its main feature, as seen from Fig. 1a, is the increasing cross section divergence between e^-p and e^+p results at large Q^2 values. Using Eq. 1 to calculate the difference leads to xF_3. Coupling

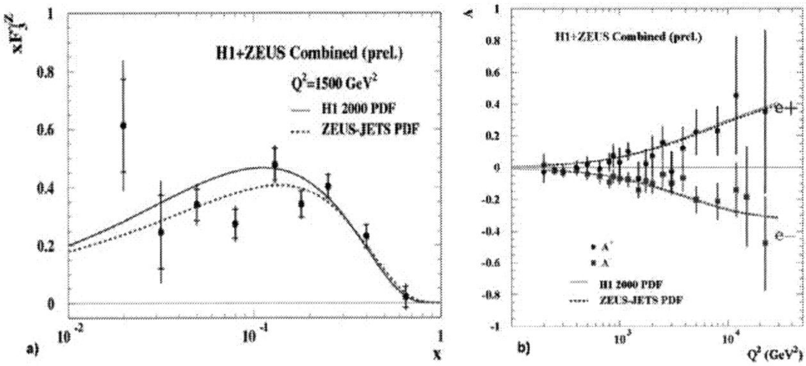

FIGURE 2. a) $xF_3^{\gamma Z}$ structure function as function of x. b) CC cross sections as function of polarisation.

strength considerations make the interference term $xF_3^{\gamma Z}$ dominant over the xF_3^{Z} term. The combined H1 and ZEUS results for $xF_3^{\gamma Z}$ are plotted in Fig. 2a.

POLARISATION RESULTS

With the advent of higher luminosities and e^-/e^+ polarisation in HERA, overall statistics were increased and new polarisation dependent measurements were achieved. Parity violation due to γZ interference in NC events was unambiguously demonstrated by calculating the asymmetry between left- and right-handed cross sections:

$$A^{\pm} = \frac{2}{P_e^+ - P_e^-} \cdot \frac{\sigma^{\pm}(P_e^+) - \sigma^{\pm}(P_e^-)}{\sigma^{\pm}(P_e^+) + \sigma^{\pm}(P_e^-)} \tag{2}$$

Fig. 2b shows the global results where the e^-p and e^+p data have been combined. The difference between reversed polarisations for either type of lepton can in turn be used to extract the structure function $F_2^{\gamma Z}$. Once $F_2^{\gamma Z}$ is accessible and expressed in terms of the parton densities, it also yields detailed information about the electroweak couplings, e.g. those of the light quarks u and d to the Z [4].

Turning to CC events, the very strong dependence of the total cross sections on the polarisation is shown in Fig. 3. Extrapolating the e^-p data to full positive polarisation, where the cross section should vanish, is a critical test of the Standard Model and unique to HERA. The results are compatible with zero within approximately 6 *pb* (68% CL), inclusive of all statistical and systematic errors. Searches are also ongoing for evidence of right-handed currents and will lead to mass limits for possible W_R bosons.

SPECIAL TOPICS

F_L explores the small-x physics and the gluon density distributions in a domain where the theoretical uncertainties are large. The only practical way (see Eq. 1) to get at it is to

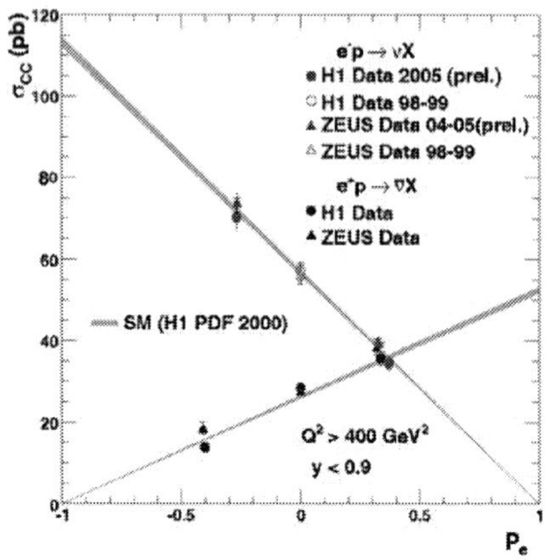

FIGURE 3. CC cross sections as function of polarisation.

make measurements at different values of y, hence at different center of mass energies for the same sets of values of x and Q^2. Runs at lower proton energies were recently successfully done at 575 and 460 GeV. Stay tuned for the analysis results to come!

Several other means exist to test and probe the Standard Model: by looking for contact interactions, search for large extra dimensions, for isolated high p_T leptons, etc.. One of those methods is to assume that a quark's charge is not punctual but distributed in space. Introducing a form factor allows the HERA experiments to set a limit of $0.67 \cdot 10^{-3}\ fm$ to the average radius of the quark charge [5].

CONCLUSIONS AND OUTLOOK

With polarisation and high luminosity, the HERA II program has enabled us to study electroweak physics at large Q^2 and to deepen our understanding of the proton structure. 23 years after conception, HERA stopped its 15 years of operation with 1 fb^{-1} cumulated by both H1 and ZEUS experiments. The results on parton distribution functions and fragmentation processes will have a considerable impact on the LHC physics, in particular at low x values. A wealth of topics are still being investigated on QCD, diffraction, exotic processes and other searches.

REFERENCES

1. The H1 Collaboration, I. Abt *etal.*, Nucl. Instrum. Meth. A 386 (1997) 310, ibid., 348.
2. The ZEUS Collaboration, U. Holm (ed.), *The ZEUS Detector Status Report* (unpublished), DESY (1993), available on http://www-zeus.desy.de/bluebook/bluebook.html.
3. The ZEUS Collaboration, S. Chekanov *etal.*, Eur. Phys. J. C 42 (2005) 1-16.
4. The H1 Collaboration, C. Diaconu, *H1 Results*, presented at the 15th International Workshop on Deep Inelastic Scattering (DIS2007), in Munich, Germany, April 2007.
5. The ZEUS Collaboration, M. Corradi, *Preview of ZEUS results*, Idem DIS2007.

Light-Nuclei identification with PAMELA space telescope: preliminary in-flight results

G. De Rosa on behalf of PAMELA Collaboration

Universita di Napoli Federico II, Dipartimento di Scienze Fisiche Complesso Universitario di Monte S. Angelo, via Cintia I-80126 Napoli, ITALY

Abstract. The observational objectives of the PAMELA space telescope, launched the 15th of June 2006, are to measure the spectra of antiprotons, positrons and nuclei in a wide range of energies, to search for primordial antimatter as well as the determination of light nuclei fluxes from hydrogen to oxygen in a wide energy range and with very high statistics. Data gathered with the PAMELA instrument will deal with a wide range of fundamental issues like the understanding of the acceleration, propagation and trapping mechanism of cosmic rays in our Galaxy. In this paper preliminary results on the charge discrimination capabilities of PAMELA for light nuclei will be presented.

Keywords: Cosmic Rays, Antimatter, Space Experiment
PACS: 98.70.Sa, 96.50.sb

INTRODUCTION

The PAMELA experiment [1] is a space-borne apparatus mounted on board of the Russian Resurs-DK1 satellite, launched on the 15th of June 2006 from the Baykonur cosmodrome. The satellite flies at an inclination of 70°), with an altitude of 350-600 km. The expected mission length is of 3 years. After a short commissioning phase, PAMELA has been in a nearly continuous data taking mode since July 11th 2006.

The main scientific goal of the Pamela investigation is the search for evidence of dark matter and of heavy antinuclei covering wide energy range (between 50 MeV and 270 GeV for positrons and between 80 MeV and 190 GeV for antiprotons, up 200 GeV/n for light nuclei) and with a sensitivity unreachable by previous experiments ([2], [3], [4]). No atmospheric overburden allows to greatly reduce the systematic errors of the balloon measurements. Additionally, PAMELA will measure light nuclear component of Galactic cosmic rays in a wide energy range (100 MeV/n-200 GeV/n) with very high statistics. Since the ratios of the spallogenic nuclei (B, Be, and Li) to mostly primary nuclei such as C are sensitive to the encountered amount of matter and to the nuclei lifetime before escaping from the Galaxy, PAMELA measurement will be particularly important in constraining propagation models.

THE PAMELA INSTRUMENT

PAMELA is a multi-purpose device consisting of a number of redundant detectors capable of identifying particles, providing charge, mass, rigidity and beta informations over a very wide energy range.

CP957, *Particles, Strings, and Cosmology, 13th International Symposium*
edited by A. Rajantie, C. Contaldi, P. Dauncey, and H. Stoica

The core of the instrument is a permanent magnet equipped with a microstrip silicon tracker providing particle charge, rigidity and incoming angle information. The tracking system consists of six 300 μm thick silicon sensors segmented into micro-strips on both sides. The mean magnetic field inside the magnet cavity is 0.43 T. Particle momentum is determined on the basis of the deflection in the magnetic field measured by the silicon detectors. A Time-of-Flight (ToF) system, made of three double-layers of plastic scintillator strips, provides the fast trigger to the experiment, the rejection of albedo particles, the velocity (β=v/c) and energy loss (dE/dx) measurements and allows particle identification at low energies up to Oxygen. Lepton/hadron identification is performed by an imaging silicon-tungsten electromagnetic calorimeter mounted below the spectrometer. A scintillation shower tail catcher and a neutron detector, consisting of ^{3}He counters and placed at the bottom of the device, improve the positron and antiprotons separation. An anticounter system is used offline to reject false triggers coming from interactions with the satellite. More details about the instrument and the entire mission can be found in reference [1].

NUCLEI IDENTIFICATION

Light nuclei measurements are preformed with three different sub-detectors (ToF, tracker and calorimeter), with different efficiencies, resolutions and Z ranges. Charge measurement, obtained independently by the three detectors, will allow the determination of the relative abundances of nuclei in different Z ranges and, for the subsample of events for which the particle momentum is measured by the spectrometer, will be possible to reconstruct the energy spectrum.

Nuclei identification with Tracker

Ionization energy loss measurements in the six silicon planes of the magnetic spectrometer allow the measurement of the absolute charge of traversing particles in independent way. In Figure 1 the ionization loss obtained as the mean of the six tracker planes measurement are shown as function of rigidity. Accurate studies are in progress to improve the charge separation. Notice that, since the tracker system design was optimized for the detection of relativistic Z=1 particles, that is to search for antiprotons and positrons, its performance are slightly compromised for high Z values and a loss of linearity is expected.

Nuclei identification with Calorimeter

The reconstruction of the energy released in the calorimeter can be used to identify incident nuclei. In particular, for nuclei not interacting in the first calorimeter layer it is possible use multiple energy losses in the planes before the interaction point to derive the charge of incident particle. Figure 2 shows the charge separation obtained by using

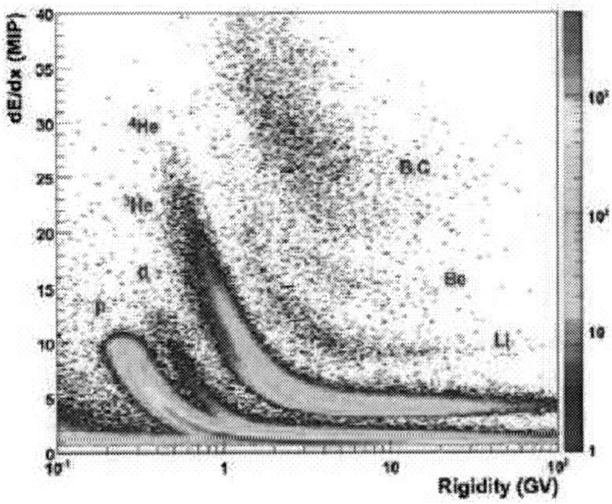

FIGURE 1. Ionizzation loss as measured by the Tracker as a function of rigidity.

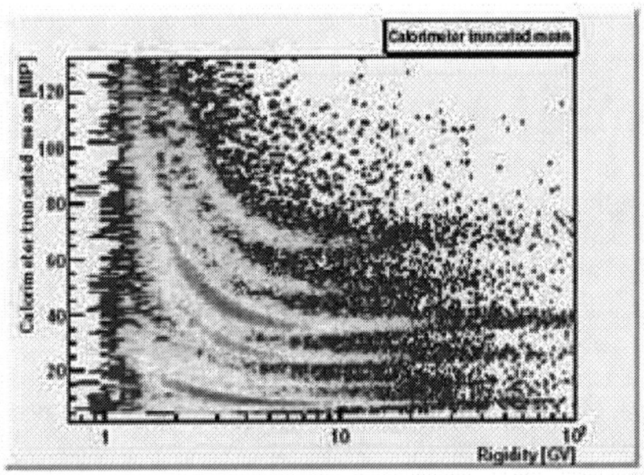

FIGURE 2. Ionization loss in the calorimeter with the truncated mean method, using three points and at least four dE/dx measurements, as a function of rigidity. Charge bands are visible up to Oxygen. A very lose cut was used to reject Helium nuclei which appear as a diffused region at about 2 GV and 5/6 MIP energy release.

the method of the truncated mean to determine the ionization loss in the calorimeter. More sophisticated methods are under investigation.

Nuclei identification with ToF

The charge of an incident particle is derived from six independent ionization energy loss measurements using the six TOF layers. After rejecting events with signals recorded in the anticoincidence system and events with multiple hit on scintillator layers, for

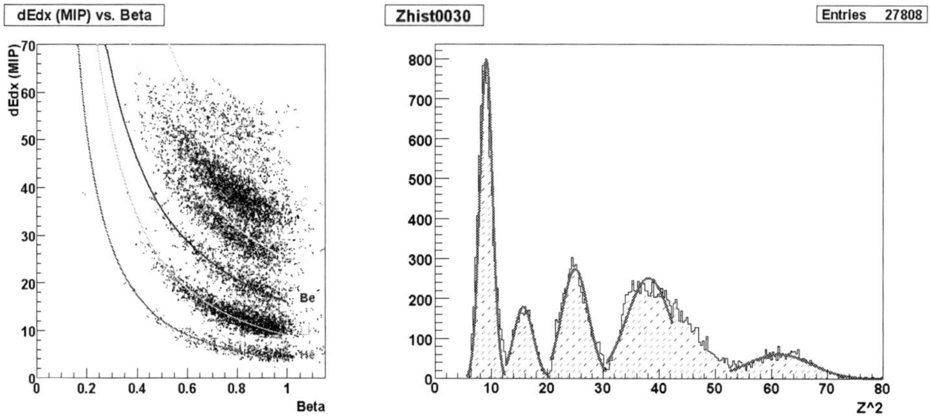

FIGURE 3. Left Panel: Ionization energy loss as function of particle velocity for one ToF paddle. Solid curves show the theoretical behavior; Right Panel: Preliminary distribution of Z^2 for light nuclei ($3 \leq Z \leq 8$) as reconstructed by one PMT of the ToF system. Where possible, gaussian fi t is superimposed to the charge peaks.

which the reconstruction of released energy and velocity is more difficult, to evaluate the energy loss in each layer, charge calibration has to be performed. Using samples of relativistic helium selected by the tracker with known rigidity and trajectory, the correction factors to normalize the signals of all PMTs, to compensate for variations in gain of the PMTs and to calculate the attenuation lengths of the paddles can be evaluated. After the correction of the ADC also for the incident angle of the trajectory of the particle was possible to define the energy release in terms of minimum ionizing particle (mip). By plotting for each PMT this preliminary (no Time-walk corrections) charge versus β, different families of nuclei separates into different bands and selection of Z>2 particle candidates is possible by applying a loose dE/dx cut. For each scintillator counter we combine the preliminary charge obtained from the two PMTs at the two sides of a paddle to determine the average charge. In Figure 1 (Left Panel) the energy loss as a function of β is shown for one of the 24 scintillation counters. The charge bands obtained can be fitted for different families and Z distribution can be extrapolated from the curves after normalizing for the velocity. As an example to show the validity of the method, in Figure 3 (Right Panel) a very preliminary Z^2 distribution as obtained for just one PMT is shown. These preliminary results are very promising about the capability of the instrument to detect nuclei at least up to Oxigen.

REFERENCES

1. P. Picozza et al., "PAMELA: a Payload for Antimatter Matter Exploration and Light-nuclei Astrophysics", Astroparticle Physics 27 (2007), 296-315.
2. J. Mitchell et al., Phys. Rev. Lett. 76 (1996) 3057.
3. M. Boezio et al., Astrophys. J. 487 (1997) 415.
4. S. Orito et al., Phys. Rev. Lett. 84 (2000) 1078.

Precision Standard Model Measurements at the Tevatron

Tommaso Dorigo
for the CDF and DØ collaborations

Dipartimento di Fisica "G.Galilei", Via Marzolo 8, 35131 Padova, Italy

Abstract.
The CDF and DØ collaborations at the Tevatron have produced exquisite precision measurements on high-P_T physics with their large datasets of $p\bar{p}$ collisions. The top quark is being studied in great detail, and a precision of 1.1% in the measurement of its mass has been achieved. The large datasets of W and Z boson decays have allowed the most precise measurement of the W mass to date;; moreover, associated production of pairs of vector bosons have been observed and measured. The precise knowledge of top and W masses is providing decisive new input for the allowed mass range of a standard model Higgs boson, as well as for the parameter space of benchmark scenarios in supersymmetric theories.

Keywords: Standard Model, top quark, Higgs boson, electroweak production, hadron collisions
PACS: 12.38.Qk, 13.85.Qk, 13.85.Rm, 14.65.Ha, 14.80.Bn, 14.80.Cp

THE FACILITIES

The Tevatron is the highest energy collider in the world, producing proton-antiproton interactions at a center-of-mass energy of 1.96 TeV. A peak luminosity of $2.9 \times 10^{32} cm^{-2} s^{-1}$ has been recently surpassed, and the integrated luminosity of 3.2 fb^{-1} has been delivered so far to the CDF and DØ experiments. That number is expected to double by the end of 2009.

An overview of the CDF and DØ detectors for Run II (2001-present) can be found in[1]. In what follows their most relevant features for standard model tests are briefly mentioned.

Both detectors are all-purpose, near-hermetic devices consisting of a tracker immersed in a solenoidal field and an outer shell of calorimeters and muon chambers. In DØ an excellent system of silicon microstrips has been installed in Run II. Six barrels of silicon sensors organized in four concentrical layers provide coverage for central tracks, while sixteen silicon disks cover the forward region. A similar device consisting of seven barrels of silicon strips is located in the core of CDF.

Outside the silicon barrels CDF features a large gas tracking chamber, while DØ has a compact scintillating fiber tracker. Central tracks are reconstructed with very high efficiency in both detectors, and the combination with silicon detector information allows the determination of track impact parameter with sufficient precision to reconstruct B-hadron decay and enable $\sim 45\%$-efficient tagging of b-quark-originated jets, with fake rates well below 1%.

Calorimeters are finely segmented in projective towers and are divided in a inner

CP957, *Particles, Strings, and Cosmology, 13th International Symposium*
edited by A. Rajantie, C. Contaldi, P. Dauncey, and H. Stoica

electromagnetic and an outer hadronic section. Electrons within the pseudorapidity interval $|\eta| < 2.0$ are identified with high purity and efficiency, with a resolution of about $15\%/\sqrt{E}$ in both detectors, and hadronic jets are reconstructed with resolutions better than $100\%/\sqrt{E}$. Muon chambers cover the rapidity region $|\eta| < 1.5$ in CDF and $|\eta| < 2.0$ in DØ .

Data acquisition proceeds through a three-level trigger system which is capable of reducing the 2.5 MHz collision rate by selecting online the most interesting events, writing full detector information to tape at 100 Hz rate. Most of the high-P_T measurements discussed in this paper use datasets collected by triggers identifying isolated electrons and muons.

TOP QUARK MEASUREMENTS

The heaviest quark is a particularly interesting laboratory for perturbative QCD, given its large mass and short lifetime. The large datasets collected by CDF and DØ enable many precision measurements of top quark properties. Among them, the top mass M_t is the most urgent to determine precisely, since it represents a crucial input to the global fits to electroweak observables which verify the internal consistency of the theory and provide a prediction for the unknown mass of the Higgs boson.

There are by now tens of determinations of M_t, using different techniques and all the significant final states of top decay. The most recent precise result from CDF is based on 1.7 fb^{-1} of data in the single-lepton final state of top pair production. 307 events are selected to contain a high-P_T lepton, missing transverse energy, and four jets, at least one of which is tagged to originate from b-quark hadronization. The dijet mass from the hadronic W boson decay is used to reduce the uncertainty arising from the jet energy scale, using two-dimensional probability density functions to compare the reconstructed top quark mass and hadronic W mass distributions in data and simulations. The top mass and the jet energy scale are thus determined simultaneously. CDF finds $M_t = 171.6 \pm 2.1 \pm 1.1$ GeV, where the first error is the statistical plus energy scale uncertainty, and the second accounts for other systematics.

Recent results from DØ include a measurement performed in the very clean sample of dilepton decays of top quark pairs, which consists of 57 candidates in 1 fb^{-1}. Two different weighting methods are used to extract a top mass from the unconstrained final state, which includes two undetectable neutrinos from the leptonic W decays. The combination of the two results yields $M_t = 173.7 \pm 5.4$(stat.) ± 3.4 (syst.) GeV.

The latest average of CDF and DØ dterminations of M_t is based on datasets including less than 1 fb^{-1} of data. The top mass is measured at $M_t = 170.9 \pm 1.1 \pm 1.5$ GeV. Before the end of Run II the two experiments are likely to reach a 1 GeV precision on the top quark mass.

The search for electroweak production of single top quarks is still ongoing. The combined cross section of the contributing processes is only two times smaller than that of QCD pair production, but the signal has a much less characteristic signature, and is buried in the large background of QCD $W+$ jets events. Both DØ and CDF have seen evidence for single top events in their data, with statistical significances exceeding the level of 3σ. A definitive observation will most likely be achieved next year.

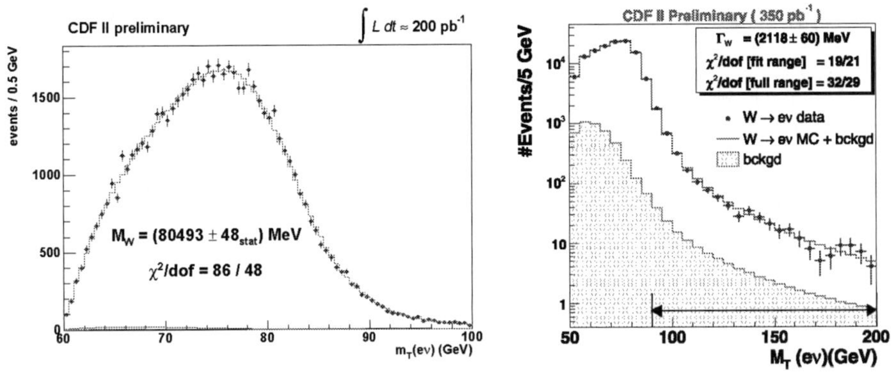

FIGURE 1. Left: the W transverse mass distribution computed by CDF from electron-neutrino pairs (blue points) and the result of a likelihood fit (red histogram). Right: a fit to the tail of the W transverse mass distribution for muon-neutrino pairs allows the extraction of the natural width of the W boson.

VECTOR BOSON PHYSICS

The samples of leptonic W and Z boson decays collected by CDF and DØ now count millions of events. These datasets have allowed very precise measurements of electroweak processes and observables, including the world's most precise value for the W boson mass ($M_W = 80413 \pm 48$ MeV), which is the result of a long and detailed analysis of only 200 pb^{-1} of electron and muon CDF data (see figure 1, left). By the end of Run II the Tevatron experiments expect to reduce the uncertainty on M_W below 20 MeV.

Another quantity measured with high precision is the natural width of the W boson, which is determined with a fit to the tail of the transverse mass distribution of lepton and missing energy. Again CDF has the most precise measurement of this quantity: $\Gamma_W = 2032 \pm 71$ MeV, in good agreement with standard model predictions (see figure 1, right).

Very rare diboson signals have also started to emerge: all associated production processes involving pairs of W and Z bosons have now been detected, with cross sections in excellent agreement with theoretical predictions. These signatures are studied in detail given the importance of multi-lepton signals for searches of the Higgs boson and new physics. Table 1 lists the Run II results obtained by CDF and DØ on diboson cross sections.

IMPLICATIONS

Global fits to electroweak observables in the context of the standard model have recently started to point to a very small value of the Higgs boson mass. This trend is largely due to the precise measurements of the top quark and W boson masses obtained at the Tevatron. The most recent fits predict a Higgs mass $M_H = 76^{+33}_{-24}$ GeV, well below the

TABLE 1. Measured inclusive cross sections for diboson production processes at the Tevatron. The column labeled "NLO" shows the standard model prediction computed at next-to leading order; when multiple uncertainties are quoted in the CDF and DØ measurements, they refer in order to statistical, systematic, and luminosity errors.

	NLO pred.*	CDF	DØ
$\sigma_{p\bar{p}\to WW+X}$	12.4 ± 0.8 pb	$13.6\pm2.3\pm1.6\pm1.2$ pb	$13.8^{+4.3+1.2}_{-3.8-0.9}\pm0.9$ pb
$\sigma_{p\bar{p}\to WZ+X}$	3.7 ± 0.3 pb	$4.3^{+1.3}_{-1.0}\pm0.2\pm0.3$ pb	$2.7^{+1.7}_{-1.3}$ pb
$\sigma_{p\bar{p}\to ZZ+X}$	1.4 ± 0.1 pb	$0.75^{+0.71}_{-0.54}$ pb	-

* J.M.Campbell and R.K.Ellis, Phys. Rev. D60 (1999) 113006.

lower experimental bound set by the LEP II experiments; the upper limit on M_H has been set at 144 GeV, at 95% confidence level[2].

If one examines the contribution of each input measurement to the above result one notices that the strongest downward pulls on M_H are due to the lepton asymmetry measurements by SLD[3] and the world average value of M_W ($M_W = 80398\pm25$ MeV). There is some tension in the mixture of results, but the fit is reasonably good: the standard model still holds, although in the next few years the increased precision of the Tevatron measurements –especially on M_W– could strain it to the breaking point. On the other hand, in the context of minimal supersymmetric models the overall agreement of all observables appears very good for a wide chunk of the parameter space of a few benchmark scenarios. In particular, within the NUHM (a non-universal Higgs model), it has been shown that one can fit well also B-physics and dark matter constraints to the lightest supersymmetric particle, obtaining a coherent picture[4]. Many options are still on the table, but the ongoing searches for the Higgs boson at the Tevatron, and those due to start soon at the LHC, are likely to clarify matters.

CONCLUSIONS

The Tevatron experiments are producing exquisite new results on high-P_T physics and are bringing standard model tests to a level of precision which meets or exceeds that of electron-positron colliders. The top quark mass is now known with 1.1% precision, and the W boson mass with 0.04% precision. These results are contributing to narrower margins of consistency of global fits to electroweak observables. CDF and DØ will continue to produce excellent physics through 2009, and possibly after that. They expect to reach a 0.7% precision on the top mass and a 0.02% precision on the W mass.

REFERENCES

1. T. Lecompte and H.T. Diehl, *The CDF and DØ upgrades for Run II*, Ann. Rev. Nucl. Part. Sci 50 (2000) 71.
2. The LEP Electroweak Workin Group, http://lepewwg.web.cern.ch/LEPEWWG/ .
3. R.S.Panvini, Int. J. Mod. Phys. A16 S1A (2001) 292.
4. J. Ellis, S. Heinemeyer, K.A. Olive, A.M. Weber, and G. Weiglein, J. High En. Phys. 08 (2007) 083.

Results from the XENON10 Dark Matter search experiment at Gran Sasso Laboratories

Alfredo Davide Ferella

RWTH Aachen University, Physics Department, Sommerfeldstr. 14, 52074 Aachen, Germany
on behalf of the XENON collaboration

Abstract.
One of the most appealing questions in astroparticle physics concerns the nature of the Dark Matter (DM) particles and their possible interaction with ordinary matter. In order to find a solution to such problem, a double phase liquid-gas Xenon Time Projection Chamber has been proposed and built as a suitable detector for the direct detection of DM in the form of Weakly Interacting Massive Particles (WIMPs). The possibility of simultaneously measuring the scintillation and the ionization signals, and a 3D event localization feature, represent a powerful background discrimination method. During the Autumn-Winter 2006-2007 a prototype of 15 kg sensitive mass of Liquid Xenon (XENON10) has been successfully run at the Gran Sasso Underground Laboratory. The technique and the results of the XENON10 detector are presented.

Keywords: Dark Matter; WIMP; WIMP search
PACS: 07.20.Mc; 95.35.+d; 29.40.Mc; 95.55.Vj

Understanding the nature of dark matter might both help to solve the cosmological questions of the matter abundance in the Universe and open new frontiers in particle physics, since it may involve new particles with masses and cross sections characteristic of the electroweak scale. Such Weakly Interacting Massive Particles (WIMPs) represent a generic class of dark matter candidates [3, 4]. If WIMPs are the dark matter, their density in the Milky Way halo may allow them to be detected in laboratory experiments by looking for the nuclear recoils produced in elastic WIMP-nuclei collisions [5, 4]. A WIMP with a typical mass between 10 GeV - 10 TeV will deposit a nuclear recoil energy below 100 keV in a terrestrial detector. The XENON experiment aims to search for interactions of WIMPs off xenon nuclei.

XENON10, the first liquid xenon (LXe) 3D position sensitive time projection chamber (TPC) of the XENON project, is a new dark matter direct search experiment with the aim of observing the small energy released after a WIMP scatters off a xenon nucleus [1]. The simultaneous detection of ionization and scintillation allows to distinguish nuclear recoils (as produced by WIMPs and neutrons) from the dominant electron recoil background (as originated from photon and electron interactions) [6, 7].

The TPC is enclosed in a SS vessel, insulated by a vacuum cryostat also in SS. The total mass of SS is $\sim$ 180 kg. The TPC works in dual phase (liquid-gas) and is filled with 22 kg of ultra-pure liquid xenon, the active volume being defined by a teflon cylinder with an inner diameter of 20 cm and a height of 15 cm (the total active mass is 15 kg of LXe). The TPC is equipped with four wire meshes, two in the liquid and two in the gas phase. The bottom mesh serves as a cathode and the next one, positioned just below the liquid level, forms, together with a series of field shaping rings, the 15 cm drift region.

CP957, *Particles, Strings, and Cosmology, 13th International Symposium*
edited by A. Rajantie, C. Contaldi, P. Dauncey, and H. Stoica

The two last meshes, along with the one below the liquid level, serve to define the gas proportional scintillation region. The active xenon volume is viewed by 89 Hamamatsu R8520-06AL 1 inch square PMTs, 35 mm high. The bottom array of 41 PMTs is located below the cathode and mainly detects the prompt light signal (in the following labeled S1). The 48 PMTs of the top array are located above the liquid, detecting the proportional light signal (labeled S2) which is created by the collision of extracted electrons with Xe atoms in the gas phase. The hit pattern of S2 in the top array is used to reconstruct the x-y position of an event with few mm resolution. The z-coordinate is calculated from the time difference between the pulses of direct and proportional light (with a maximum of 80 μs and a resulting z-resolution of <1 mm).

The results from XENON10 running at Gran Sasso during the period between October 6, 2006 and February 14, 2007 are reported in reference [8].

In order to determine the response of the detector the following calibrations were performed. (1) A total of $\sim 10^4$ electron recoil events, in the energy range of interest (4.5 to 26.9 keV), were collected with ^{137}Cs source. (2) The detector's response to nuclear recoils was obtained from 12 hours irradiation in-situ, using a 200 n/s AmBe source. (3) Energy calibration was obtained with an external ^{57}Co gamma ray source and with gamma rays from metastable Xe isotopes produced by neutron activation of a 450 g Xe sample, introduced into the detector after the WIMP search data taking.

$S1$ for 122 keV gamma rays has an average light yield of L_y of 3.0 ± 0.1(syst)± 0.1(stat) pe/keVee in the fiducial volume at the drift field of 0.73 kV/cm. The nuclear recoil energy (in unit of keV) can be calculated as, $E_{nr} = S1/L_y/\mathscr{L}_{eff} \cdot S_e/S_r$. $\mathscr{L}_{eff}$ is the nuclear-recoil scintillation efficiency relative to that of 122 keV gamma rays at zero drift field ([9, 10]), S_e and S_r are the scintillation quenching factors, due to the electric field, for electron and nuclear recoils ([11]).

Background rejection is based on the ionization/scintillation ($S2/S1$) ratio, which is different for nuclear and electron recoils in LXe. Figure 1 shows the energy dependence of the logarithm of this ratio for electron recoils from ^{137}Cs gamma ray and for nuclear recoils from AmBe fast neutron calibrations. For each energy bin, the nuclear recoil acceptance window is defined between $\Delta \text{Log}_{10}(\text{S2}/\text{S1}) = \mu$ and $\Delta \text{Log}_{10}(\text{S2}/\text{S1}) = \mu - 3\sigma$ (see Fig. 2), where $\Delta \text{Log}_{10}(\text{S2}/\text{S1})$ is the subtraction of the mean $\text{Log}_{10}(\text{S2}/\text{S1})$ of nuclear recoil band from the electron recoil band.

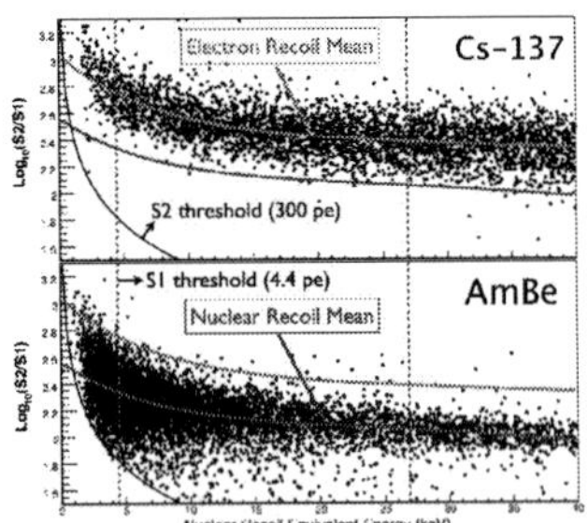

FIGURE 1. $\text{Log}_{10}(\text{S2}/\text{S1})$ as a function of energy for electron recoils (top) and nuclear recoils (bottom) from calibration data. The two curves are the mean $\text{Log}_{10}(\text{S2}/\text{S1})$ values of the electron recoil (upper, red) and nuclear recoil (lower, blue) bands.

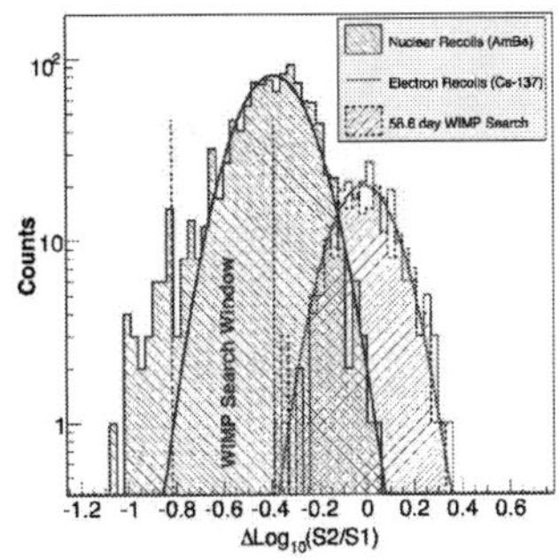

FIGURE 2. Distributions of $\Delta\text{Log}_{10}(\text{S2}/\text{S1})$ for nuclear recoils, electron recoils and WIMP search data for one energy bin. The nuclear recoil acceptance (WIMP Search Window) is defined by the two vertical lines, which are the -3σ and mean from a Gaussian fit to the nuclear recoil $\Delta\text{Log}_{10}(\text{S2}/\text{S1})$ distribution.

Together with the statistical events leaking from the electron recoil band into the nuclear recoil acceptance window, we observed anomalous leakage events in the ^{137}Cs calibration data and un-blinded WIMP search data. These events were identified to be multiple-scatter events with one scatter in the LXe below the cathode and a second scatter in the active LXe volume above the cathode. The $S2$ signal from this type of event is from the interaction in the active volume only, while the $S1$ signal is the sum of the two $S1$'s in both the active and non-active volume. Two types of cuts in the primary blind analysis, one using the $S1$ signal asymmetry between the top and bottom PMT arrays and the other using the $S1$ hit pattern, defined as $S1_{RMS} = \sqrt{\frac{1}{n}\Sigma(S1_i - \overline{S1})^2}$ $(i = 1,n)$, on either the bottom or the top PMT array, are defined to remove these anomalous events. A large fraction of events that leaked into the WIMP-signal window are of this type of background and could be removed by the cuts discussed above.

The 3D position sensitivity of the XENON10 detector gives additional background suppression with fiducial volume cuts [12]. Due to the high stopping power of LXe, the background rate in the central part of the detector is lower (0.6 events/keVee/kg/day) than that near the edges (3 events/keVee/kg/day). The fiducial volume is defined to be within 15 to 65 μs drift time window and with radius less than 8 cm (out of 10 cm) in XY (total mass of 5.4 kg).

Figure 4 shows the 90% C.L. upper limits on WIMP-nucleon cross sections as a function of WIMP mass by using the "maximum gap" method in [13] and using the standard assumptions for the galactic halo [14].

This work was funded by NSF Grants No. PHY-03-02646 and No. PHY-04-00596, the CAREER Grant No. PHY-0542066, the DOE Grant No. DE-FG02-91ER40688, by the Volkswagen Foundation (Germany) and by the FCT Grant No POCI/FIS/60534/2004 (Portugal). We would like to thank the LNGS/INFN staff and engineers for their continuous help and support.

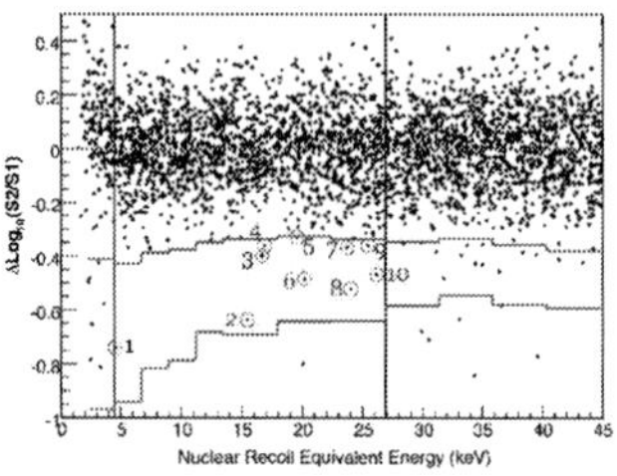

FIGURE 3. Results from 58.6 live-days of WIMP-search in 5.4 kg of LXe. WIMP search window defined between the two vertical lines (4.5 to 26.9 keV) and blue lines (50% nuclear recoil acceptance).

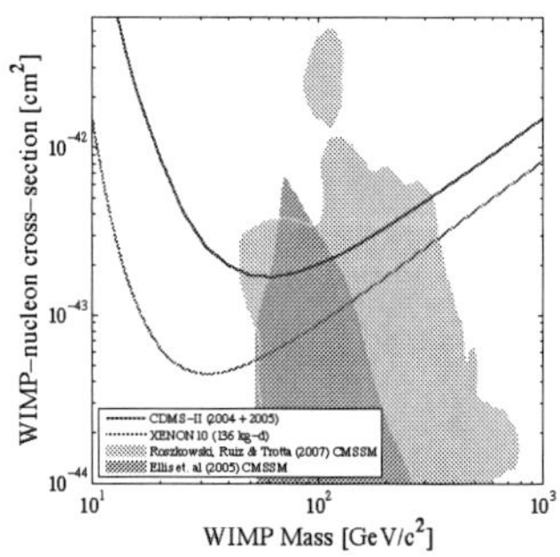

FIGURE 4. Spin-independent WIMP-nucleon cross-section upper limits (90% C.L.) versus WIMP mass. Shown curves are for the previously best published limit (upper, blue) [15] and the XENON10 result (lower, red). The shaded area is from [16] (yellow), [17] (maroon) and [18] (brown).

REFERENCES

1. E. Aprile et al. (XENON) New Astr. Rev. **49** (2005) 289
   J. Angle et al., astro-ph/0609714, Proceedings of the 7th UCLA Symposium on ÓSources and Detection of Dark Matter and Dark Energy in the UniverseÓ (2006).
2. D.N. Spergel et al, Astrophys. J. Suppl. **148** (2003) 175 and astro-ph/0603449.
3. B.W. Lee and S. Weinberg, Phys. Rev. Lett. **39** (1977) 165.
4. G. Jungman, M. Kamionkowski, and K. Griest, Phys. Rep. **267** (1996) 195.
5. M. W. Goodman and E. Witten, Phys. Rev. D **31** (1985) 3059.
6. E. Aprile et al., Phys. Rev. D **72** (2005) 072006.
7. E. Aprile et al., Phys. Rev. Lett. **97** (2006) 081303.
8. J. Angle et al. arXiv:0706.0039 [astro-ph]. Submitted to Phys. Rev. Lett.
9. E. Aprile et al., Phys. Rev. **D72**, 072006 (2005).
10. V. Chepel et al., Astropart. Phys. **26**, 58 (2006).
11. E. Aprile et al., Phys. Rev. Lett. **97**, 081302 (2006).
12. P. Sorensen et al., Proc. 6th Intl. Workshop for Identification of Dark Matter, World Scientific (2006).
13. S. Yellin, Phys. Rev. D, **66**, 032005 (2002).
14. J.D. Lewin and P.F. Smith, Astropart. Phys. **6**, 87 (1996).
15. D. S. Akerib *et al.* (CDMS Collaboration), Phys. Rev. Lett. **96**, 011302 (2006).
16. A. Bottino *et al.*, Phys. Rev. D **69**, 037302 (2004).
17. J. Ellis *et al.*, Phys. Rev. D **71**, 095007 (2005).
18. L. Roszkowski, R. Ruiz de Austri and R. Trotta, arXiv:0705.2012 [hep-ph].

O'Raifeartaigh models with spontaneous R-symmetry breaking

Luca Ferretti

SISSA/ISAS and INFN, Trieste, Italy

Abstract. O'Raifeartaigh models with general R-charge assignments can have vacua where both supersymmetry and R-symmetry are spontaneously broken. Most of these vacua are metastable because the potential shows a runaway behaviour. We explain the relation between runaway directions and R-symmetry.

INTRODUCTION

In supersymmetric model building, the supersymmetry-breaking sector is the most elusive one: finding generic and natural models of supersymmetry breaking has been a theoretical challenge for many years. The recent discovery of metastable supersymmetry-breaking vacua in simple theories like $\mathcal{N} = 1$ SQCD [1] opens new and promising directions towards complete models including both the supersymmetric and the supersymmetry-breaking sector.

The simplest example of metastable SUSY-breaking vacuum is the Intriligator-Seiberg-Shih model, which is the low-energy theory of $\mathcal{N} = 1$ SQCD with $N_c < N_f < \frac{3}{2}N_c$ in the Seiberg-dual picture. This model corresponds to a weakly-gauged O'Raifeartaigh models: SUSY is broken by the Nelson-Seiberg argument [2] near the origin of field space, while the gauge dynamics restores the usual supersymmetric vacua far away in field space, so the non-supersymmetric vacuum is metastable but parametrically long-lived.

O'Raifeartaigh models are interesting because they can be effective theories of strong-coupling gauge theories, as in the ISS model. R-symmetry plays a crucial role in these models: in fact the Nelson-Seiberg argument states that the existence of an R-symmetry is a necessary condition for supersymmetry breaking. However, R-symmetry must be broken in order to give an high mass (greater than about 100 GeV) to gauginos. It is possible to break R-symmetry explicitly, but this implies that the SUSY-breaking vacua are metastable and there can be some tension between the requirement of long-lived vacua and the bounds on gaugino masses, depending on the model. The spontaneous breaking of R-symmetry seems to be an interesting possibility, neglecting the problem of the existence of a massless Goldstone boson (R-axion).

Most of the O'Raifeartaigh models featured in the literature have fields whose R-charges can be chosen to be either 2 or 0. Shih noted that in these models R-symmetry cannot be spontaneously broken, while spontaneous breaking often occurs in models with fields with $R \neq 2,0$ [3]. The simplest model which breaks R-symmetry sponta-

CP957, *Particles, Strings, and Cosmology, 13th International Symposium*
edited by A. Rajantie, C. Contaldi, P. Dauncey, and H. Stoica

neously for some values of its parameters is:

$$W = fX + \lambda X\phi_{(1)}\phi_{(-1)} + m_1\phi_{(3)}\phi_{(-1)} + \frac{1}{2}m_2\phi_{(1)}^2 \tag{1}$$

where $R(X) = 2$ and $R(\phi_{(k)}) = k$. The flat direction parametrized by X is lifted by quantum corrections and R-symmetry is broken if the resulting vacuum has $\langle X\rangle \neq 0$.

In this contribution we consider a generalization of the model (1) with generic R-charge assignment. The superpotential has the form

$$W = fX + \frac{1}{2}(M^{ij} + N^{ij}X + Q_a^{ij}Y_a)\phi_i\phi_j \tag{2}$$

where M, N, Q_a are generic symmetric complex matrices. The vacua of these models break R-symmetry for a wide range of parameters of the superpotential. Then we show that most of the Wess-Zumino models with field s with $R \neq 2, 0$ have runaway directions, both supersymmetric and non-supersymmetric, and explain the relation between the runaway behaviour and the R-symmetry of the theory.

SPONTANEOUS R-SYMMETRY BREAKING

We consider the model (2) with the condition $\det M \neq 0$. This is a sufficient condition for supersymmetry breaking. In fact the equations for a SUSY vacuum $\partial_a W = 0$ are

$$f + \frac{1}{2}N^{ij}\phi_i\phi_j = 0 \tag{3}$$

$$\frac{1}{2}Q_a^{ij}\phi_i\phi_j = 0 \tag{4}$$

$$(M^{ij} + N^{ij}X)\phi_j = 0 \tag{5}$$

and cannot be solved simultaneously. To prove this it is sufficient to note that if $\det(M + NX + Q_aY_a) \neq 0$ the only solution for (5) is $\phi_i = 0$ which cannot satisfy (3). R-symmetry implies that $\det(M + NX + Q_aY_a) = \det M$, therefore SUSY is spontaneously broken [3].

The classical flat directions parametrized by X, Y_a correspond to $\phi_i = 0$ and are local minima of the potential for small f. These flat directions are lifted by the 1-loop Coleman-Weinberg potential. The quadratic part of the 1-loop potential at the lowest nonzero order in $|\hat{M}^{-2}f\hat{N}|$ reduces to [4]

$$V_{quad} = \frac{f^2}{32\pi^2}\mathrm{Tr}\int_0^\infty dv\, v^3\left[\mathcal{M}_1(v)\mathcal{M}_1^\dagger(v) - \mathcal{M}_2(v)\mathcal{M}_2^\dagger(v)\right] \tag{6}$$

with

$$\mathcal{M}_1(v) = \frac{1}{\sqrt{v^2+\hat{M}^2}}\left(\hat{N}\frac{\sqrt{2}v}{v^2+\hat{M}^2}\hat{Y}\right)\frac{1}{\sqrt{v^2+\hat{M}^2}} \tag{7}$$

$$\mathcal{M}_2(v) = \frac{1}{\sqrt{v^2+\hat{M}^2}}\left(\hat{N}\frac{\hat{M}}{v^2+\hat{M}^2}\hat{Y} + \hat{Y}\frac{\hat{M}}{v^2+\hat{M}^2}\hat{N}\right)\frac{1}{\sqrt{v^2+\hat{M}^2}} \tag{8}$$

where

$$\hat{M}=\begin{pmatrix}0 & M^{\dagger}\\ M & 0\end{pmatrix},\ \hat{N}=\begin{pmatrix}0 & N^{\dagger}\\ N & 0\end{pmatrix},\ \hat{Y}=\begin{pmatrix}0 & (NX+Q^{a}Y_{a})^{\dagger}\\ NX+Q^{a}Y_{a} & 0\end{pmatrix} \tag{9}$$

The two terms are generally of the same order, but the contribution of the first term is always positive, while the second term always gives a negative contribution. For models containing only fields with $R=2,0$ the trace of the second term is always zero because of the R-symmetry constraints on the form of the matrices M,N,Q_{a}: this means that R-symmetry is unbroken. However, for models with general R-charge assignment the contribution of $\mathscr{M}_{2}$ is generally non-zero and V_{quad} can be positive or negative.

If the whole expression is negative for some choice of (X,Y_{a}) then the classical vacuum $X=0,Y_{a}=0$ is unstable and there can be an R-symmetry breaking vacuum along one of these tachyonic directions. This is what happens for a wide range of parameters M,N,Q_{a}.

RUNAWAY DIRECTIONS

The SUSY-breaking vacuum of the model (1) is metastable because of the existence of a runaway direction [3]:

$$\phi_{(1)}=-\frac{f}{\lambda\phi_{(-1)}},\ X=\frac{m_{2}f}{\lambda^{2}\phi_{(-1)}^{2}},\ \phi_{(3)}=\frac{m_{2}f^{2}}{m_{1}\lambda^{2}\phi_{(-1)}^{3}},\ \phi_{(-1)}\to 0 \tag{10}$$

It is interesting to note that the runaway direction can be seen as a rescaling of fields

$$\varphi(\varepsilon)=\varepsilon^{-R(\varphi)}\varphi(0)\quad ,\quad \varepsilon\to 0 \tag{11}$$

which is a complexified R-symmetry transformation.

In fact most of the Wess-Zumino models with an R-symmetry and generic R-charge assignment have runaway directions [4]. To show this, it is useful to note that the vacuum equations (and the F-terms) can be classified by their R-charge:

$$\partial_{i}W=0,\ R(\varphi_{i})<2 \qquad R>0 \tag{12}$$

$$\partial_{i}W=0,\ R(\varphi_{i})=2 \qquad R=0 \tag{13}$$

$$\partial_{i}W=0,\ R(\varphi_{i})>2 \qquad R<0 \tag{14}$$

Because of the Nelson-Seiberg argument, it is not possible to solve all these equations at the same time. However, it can be possible to solve a subset of these equations. Let's look at two common possibilities:

- In some cases it is possible to solve all the equations with $R\geq 0$ (or $R\leq 0$). In this case it is sufficient to rescale all fields by a factor $\varepsilon^{-R(\varphi)}$ (or $\varepsilon^{R(\varphi)}$) as in (11). Then we send $\varepsilon\to 0$ to solve also the equations with $R<0$ (or $R>0$) and obtain a supersymmetric runaway vacuum: $V\to 0$ as $\varepsilon\to 0$. This is the case of model (1). This case often occurs when there are a few equations with $R=0$.

- In other cases it is not possible to solve the equations with $R = 0$. A simple example is a small modification of the Shih model with another pseudomodulus Y:

$$W = fX + (\lambda X + \eta Y)\phi_{(1)}\phi_{(-1)} + m_1\phi_{(3)}\phi_{(-1)} + \frac{1}{2}m_2\phi_{(1)}^2 \tag{15}$$

  In these cases it is often possible to minimize $V_{R=0} = \sum_{R(\varphi_j)=2} |\partial_j W|^2$ with respect to all the fields and solve the equations with $R > 0$ (or $R < 0$) at the same time. Then the rescaling (11) by a factor $\varepsilon^{-R(\varphi)}$ (or $\varepsilon^{R(\varphi)}$) parametrizes a runaway direction with $V \to V_\infty > 0$ which corresponds to a non-supersymmetric runaway vacuum.

Most of the models with generic R-charges realize one of these two possibilities.

CONCLUSIONS

O'Raifeartaigh models with generic R-charge assignment are interesting for model building because of the possibility of spontaneous R-symmetry breaking, which makes it easier to obtain gaugino masses of the right order of magnitude. The presence of runaway directions in these models is a curious and common feature and can be useful because of the existence of other R-symmetry breaking vacua along these directions.

REFERENCES

1. K. Intriligator, N. Seiberg and D. Shih, JHEP **0604** (2006) 021 [arXiv:hep-th/0602239].
2. A. E. Nelson and N. Seiberg, Nucl. Phys. B **416** (1994) 46 [arXiv:hep-ph/9309299].
3. D. Shih, arXiv:hep-th/0703196.
4. L. Ferretti, arXiv:0705.1959 [hep-th].

Preliminary Results from MINOS on ν_μ Disappearance Based on an Exposure of 2.5×10^{20} 120 GeV Protons on the NuMI Target

Peter J Litchfield

University of Minnesota
for the MINOS collaboration

Abstract. Preliminary neutrino oscillation results are presented on data from the MINOS experiment taken up to March 2007, yeilding an exposure of $2.50x10^{20}$ protons on target. New values of $\Delta m^2_{32} = (2.38^{+0.20}_{-0.16}) \times 10^{-3}\mathrm{eV}^2/c^4$ and $\sin^2 2\theta_{23} = 1.00_{-0.08}$ are obtained.

Keywords: neutrino oscillations, neutrino mixing parameters
PACS: 13.15.+g,14.60.Lm,14.60.Pq

INTRODUCTION

The MINOS Experiment is designed to explore the phenomenon of ν_μ disappearance as observed in experiments studying atmospheric neutrinos [1, 2, 3] and more recently in the K2K accelerator-based experiment [4].

MINOS makes use of a configurable intense neutrino source (NuMI) derived from 120 GeV protons extracted from the Fermilab Main Injector onto a graphite target, and two magnetized-iron and scintillator detectors: a 0.98 kton Near Detector (ND) located on the Fermilab site approximately 1 km downstream of the NuMI target and a 5.4 kton Far Detector (FD) located in the Soudan Underground Laboratory at a distance of 735 km. The NuMI beam line and the MINOS detectors are described in detail elsewhere [5, 6].

In my talk I reported results based on data accumulated during the first period of NuMI operations between May 2005 and February 2006 which have been published in Ref. [7]. We denote this period as 'Run-I'. Far Detector data, corresponding to an exposure of 1.27×10^{20} protons on target (POT), were analyzed in the context of two-flavor $\nu_\mu \rightarrow \nu_\tau$ oscillations.

Following the Fermilab accelerator complex shutdown in spring 2006, NuMI resumed operations in June 2006. The period from then through July 2007 is referred to as 'Run-II'. With some modifications relative to that reported in Ref. [7], we have carried out an analysis of the LE Run-I data plus the portion of the LE Run-II data collected through March 2007 (denoted as Run-IIa), corresponding to a combined exposure of 2.50×10^{20} POT. This analysis was first reported at the Manchester EPS conference, shortly after the PASCOS meeting. Since this supersedes the results I presented, in this report I have updated the data and analysis to these new results. Most aspects of the

CP957, *Particles, Strings, and Cosmology, 13th International Symposium*
edited by A. Rajantie, C. Contaldi, P. Dauncey, and H. Stoica

analysis follow those described in Ref. [7]. However, the new analysis reported here, including updated results from the Run-I data, incorporates several improvements. The most significant involved an upgraded neutrino interaction simulation package [8]; a new track reconstruction algorithm; increased acceptance; and improved rejection of neutral-current (NC) backgrounds. More details of the new data and the differences with the previous analysis can be found in Ref. [9].

As in Ref. [7], we derive an expected FD E_ν spectrum by extrapolating the observed spectrum in the ND. We have employed an extrapolation scheme, denoted the "Beam Matrix" method [10], that is largely insensitive to mis-modeling of the neutrino flux and neutrino interaction cross-sections. The predicted null-oscillation FD E_ν spectrum is shown in Fig. 1, along with the corresponding predicted spectra from three other extrapolation methods used as cross-checks. Agreement between the methods is at the level of 4%.

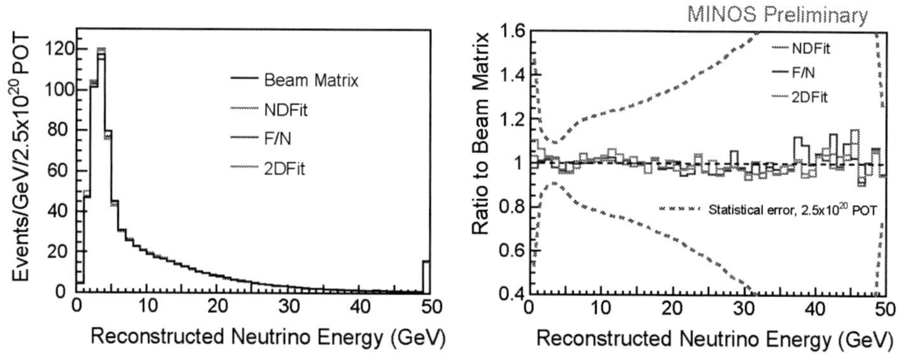

FIGURE 1. Left plot: predicted Far Detector reconstructed E_ν spectra from four extrapolation methods (left) based on ND data; Right plot: ratios of spectra from cross-check extrapolation methods to the Beam Matrix prediction. The dashed line represents the expected statistical uncertainty on FD E_ν spectrum bin contents for the current exposure. In both plots, the final bin at 49 GeV is an overflow bin, including events above 50 GeV as well.

PRELIMINARY RESULTS

>From the Run-I and Run-IIa samples, we select 812 neutrino-like events with a reconstructed track in the FD fiducial volume, of which 563 satisfy ν_μ CC interaction selection (track quality, track charge sign and NC rejection) cuts. Fig. 2 shows the E_ν spectrum for the 563 ν_μ CC interaction candidate events (points). Overlaid is the null-oscillation expectation (black histogram), totaling 750 ± 30 events as inferred from the ND data, where the uncertainty is due to the FD/ND relative normalization systematic error. The expected contamination from the three main background sources is; NC interactions (5.6 events), ν_τ CC interactions (0.8 events) and ν_μ CC interactions in the rock upstream of the FD or within the FD but outside the fiducial volume (1.7 events). The estimates for the latter two sources account for oscillation effects.

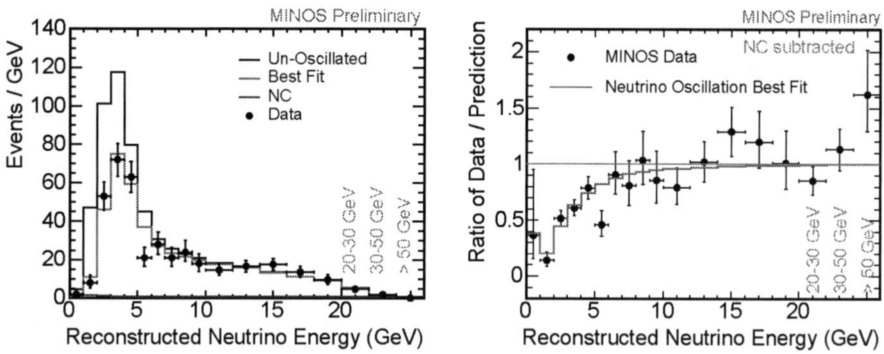

FIGURE 2. Left: the reconstructed ν_μ CC energy spectrum in the combined LE Run-I and IIa Far Detector data (points), with the null oscillation prediction overlaid (black histogram). Also shown are the oscillation-weighted prediction using best-fit parameters (red) and the expected NC background contribution (blue). The numbers of events in the final three bins, corresponding to energy ranges of 20–30, 30–50 and 50–200 GeV, have been scaled according to these bin widths. Right: the ratio of the NC-background subtracted FD spectrum to the null-oscillation prediction (points), with the best-fit oscillation expectation overlaid (red).

We carried out a simultaneous fit of the oscillation-weighted predicted E_ν spectra for Runs I and IIa to the corresponding observed FD spectra.

The plots in Fig. 2 show the predicted FD spectrum weighted according to the best-fit oscillation parameter values (red) overlaid on the observed spectrum (points). The best-fit oscillation parameter values are:

$$\begin{aligned} \left|\Delta m^2_{32}\right| &= (2.38\,^{+0.20}_{-0.16}) \times 10^{-3}\,\mathrm{eV}^2/c^4 \\ \sin^2 2\theta_{23} &= 1.00_{-0.08}\,, \end{aligned}$$

corresponding to $\chi^2 = 41.2$ for 34 degrees of freedom. The uncertainties represent 1-dimensional 68% CL intervals. The 90% CL lower limit on $\sin^2 2\theta$ is found to be 0.84. The 68% and 90% CL contours in oscillation parameter space are shown in Fig. 3.

The new MINOS contours are compared with those from the published analysis of Run-I data [7] in the right plot in Fig. 3. The allowed region is shifted to lower values of Δm^2 in the new analysis. Considering the two running periods separately, we find $\Delta m^2 = (2.50\,^{+0.24}_{-0.20}) \times 10^{-3}\,\mathrm{eV}^2/c^4$ for Run-I and $(2.22\,^{+0.44}_{-0.22}) \times 10^{-3}\,\mathrm{eV}^2/c^4$ for Run-IIa.

We have studied the differences between the current analysis of Run-I and that previously published [7]. Carrying out the Run-I analysis with the new neutrino interaction and event reconstruction software, but applying the same selection criteria as used in the previous analysis, we obtain $\Delta m^2 = 2.46 \times 10^{-3}\,\mathrm{eV}^2/c^4$. We conclude that the differences are largely due to the improved neutrino simulation package and reconstruction algorithms.

We are currently in the process of analyzing the full Run-I and Run-II dataset, with a total exposure of 3.5×10^{20} POT, of which 3.25×10^{20} POT is in the LE beam

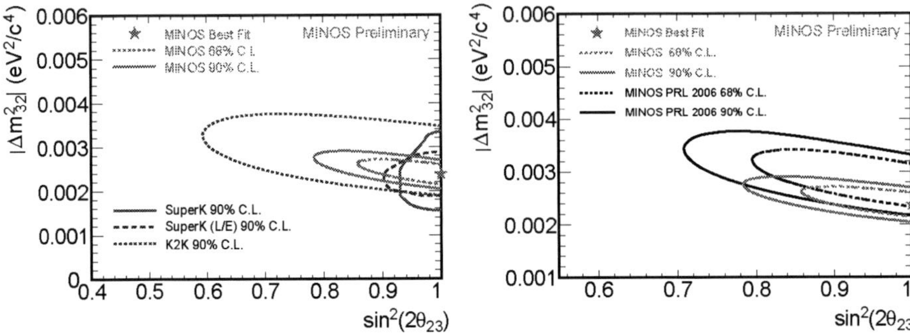

FIGURE 3. Left: The new preliminary MINOS best fit point (star) and the 68% and 90% CL contours (red) as determined according to $\Delta\chi^2 = 2.30$ and 4.61, respectively. Overlaid are the 90% CL contours from the Super-Kamiokande zenith angle [3] and L/E analyses [2], as well as that from the K2K experiment [4]. Right: The new preliminary MINOS contours (red) are compared with the corresponding contours obtained in the original MINOS analysis of Run-I data [7].

configuration and 0.16×10^{20} POT is in the HE configuration.

REFERENCES

1. K. S. Hirata *et al.* (the Kamiokande Collaboration), Phys. Lett. **B280**, 146 (1992);
   R. Becker-Szendy *et al.* (the IMB Collaboration), Phys. Rev. **D 46**, 3720 (1992);
   M. Ambrosio *et al.* (the MACRO Collaboration), Eur. Phys. J. C **36**, 323 (2004);
   W. W. W. Allison *et al.* (the Soudan 2 Collaboration), Phys. Rev. **D 72**, 052005 (2005).
2. Y. Ashie *et al.* (the Super-Kamiokande Collaboration), Phys. Rev. Lett. **93**, 101801 (2004).
3. Y. Ashie *et al.* (the Super-Kamiokande Collaboration), Phys. Rev. **D 71**, 112005 (2005).
4. M. H. Ahn *et al.* (the K2K Collaboration), Phys. Rev. **D 74**, 072003 (2006).
5. S. Kopp, "The NuMI neutrino beam at Fermilab," in the Proc. of 2005 IEEE Part. Accel. Conf. (Knoxville, TN), May, 2005, Fermilab-Conf-05-093-AD and arXiv:physics/0508001.
6. D. G. Michael *et al.*, "The MINOS Detectors," to be submitted to Nucl. Instrum. & Meth.
7. D. G. Michael *et al.* (the MINOS Collaboration), Phys. Rev. Lett. **97**, 191801 (2006); D. G. Michael *et al.* (the MINOS Collaboration), to be submitted to Phys. Rev. D.
8. NEUGEN version 3.5.0. See H. Gallagher, "The NEUGEN Neutrino Event Generator," Nucl.Phys.Proc.Suppl.112:188-194,2002 as a reference for earlier versions.
9. The MINOS collaboration, submitted to the XXIII International Symposium on Lepton and Photon interactions at High Energy, Daegu, Korea, arXiv:0708.1495.
10. M. Szleper and A. Para, hep-ex/0110001 (2001).

Spontaneous breaking of discrete symmetries in QCD on a small volume

B. Lucini[1*], A. Patella[†] and C. Pica[**]

[*]Department of Physics, Swansea University, Singleton Park, Swansea SA2 8PP, UK
[†]Scuola Normale Superiore, Piazza dei Cavalieri 27, 56126 Pisa, Italy
and INFN Pisa, Largo B. Pontecorvo 3 Ed. C, 56127 Pisa, Italy
[**]Physics Department, Brookhaven National Laboratory, Upton, NY 11973, USA

Abstract. In a compact space with non-trivial cycles, for sufficiently small values of the compact dimensions, charge conjugation ($\mathscr{C}$), spatial reflection ($\mathscr{P}$) and time reversal ($\mathscr{T}$) are spontaneously broken in QCD. The order parameter for the symmetry breaking is the trace of the Wilson line wrapping around the compact dimension, which acquires an imaginary part in the broken phase. We show that a physical signature for the symmetry breaking is a persistent baryonic current wrapping in the compact directions. The existence of such a current is derived analytically at first order in perturbation theory and confirmed in the non-perturbative regime by lattice simulations.

Keywords: QCD, Spontaneous Breaking of Discrete Symmetries, Lattice Gauge Theories
PACS: 12.38.Aw, 11.30.Er, 12.38.Bx, 12.38.Gc

BREAKING OF DISCRETE SYMMETRIES

We consider a non-Abelian gauge theory with gauge group SU(N) and N_f families of fundamental Dirac fermions. This theory is described by the Lagrangian

$$\mathscr{L} = -\frac{1}{2g^2}\mathrm{Tr}\left(G_{\mu\nu}(x)G^{\mu\nu}(x)\right) + \sum_{n=1}^{N_f} \bar{\psi}_n(x)\left(i\,\partial\!\!\!/ - A\!\!\!/(x) - m\right)\psi_n(x)\,,$$

where $G_{\mu\nu} = \partial_\mu A_\nu - \partial_\nu A_\mu + [A_\mu, A_\nu]$.
The manifold on which the theory is defined is $T \times L^3$, where T corresponds to time and the compact spatial directions L are equal. Spatial directions are closed with periodic boundary conditions (PBC), while the temporal direction is closed with antiperiodic boundary conditions for fermions and PBC for bosons.
We compute the vacuum expectation value (*vev*) of the Wilson line in a compact direction [1]

$$W_\alpha^{(A)}(x, z^{(\alpha)}) = \mathrm{Pexp}\left(i\int_0^L A_\alpha(x, z^{(\alpha)})\,dz_\alpha\right)\,. \tag{1}$$

The gauge can be fixed in such a way that W is diagonal

$$W_\alpha^{(A)}(x, z^{(\alpha)}) = \left(e^{iv_{1\alpha}^{(A)}(x,z^{(\alpha)})}, \ldots, e^{iv_{N\alpha}^{(A)}(x,z^{(\alpha)})}\right)\,. \tag{2}$$

[1] Presenter at the conference.

CP957, *Particles, Strings, and Cosmology, 13th International Symposium*
edited by A. Rajantie, C. Contaldi, P. Dauncey, and H. Stoica

We focus on the effective potential for those eigenvalues[2]:

$$e^{iTV(v_1,\dots,v_N)} = \int \exp\left\{ i \int \mathscr{L} d^4x \right\} \times \tag{3}$$
$$\times \prod_k \delta\left(v_k - \frac{1}{TL^2} \int v_k^{(A)}(\mathbf{x})\, dt\, dx_\perp \right) \mathscr{D}A \mathscr{D}\bar{\psi} \mathscr{D}\psi \,.$$

The absolute minima of $V\ (v_1,\dots,v_N)$ give the expectation values for the set of the eigenvalues. If $\mathscr{C}$-symmetry is broken, the set of eigenvalues is not invariant under the substitution $v \to -v$. The effective potential for the Wilson line is given by [2]

$$V(\vec{v}_1,\dots,\vec{v}_N) = \left[\sum_{i,j=1}^{N} f(0,\vec{v}_i - \vec{v}_j) - 2N_f \sum_{i=1}^{N} f(m,\vec{v}_i) \right] \,, \tag{4}$$

$$\text{with } f(m,\vec{v}) = \frac{1}{L} \left(\frac{mL}{\pi} \right)^2 \sum_{\vec{k} \neq 0} \frac{K_2(mLk)}{k^2} \sin^2\left(\frac{1}{2} \vec{k} \cdot \vec{v} \right) \,, \tag{5}$$

where K_2 is the modified Bessel function of the second kind of order 2 and the sum runs over three-index integer vectors $\vec{k}$. The first term of Eq. (4) gives rise to an attraction between the eigenvalues. The second term produces an unconstrained absolute minimum at $v_i = \pi$. When the $SU(N)$ constraint is taken into account, the minima are:

$$v_1^* = v_2^* = \cdots = v_N^* = \begin{cases} \pm \frac{N-1}{N} \pi & \text{for } N \text{ odd} \\ \pi & \text{for } N \text{ even} \end{cases} \,. \tag{6}$$

The spontaneous symmetry breaking shows up as a non-zero imaginary part of the Wilson line. Hence, if N is even, there is no symmetry breaking. If N is odd, then $\mathscr{P}$, $\mathscr{C}$, $\mathscr{T}$, $\mathscr{C}\mathscr{P}\mathscr{T}$ are broken. There are 8 vacua. The same calculation can be reproposed for a number n of compact dimensions. We arrive at the same conclusions, with the number of vacua given by 2^n. This result is due to the spacial PBC for the fermions. The validity of the calculation in the non-perturbative regime has been checked on the lattice in [3].

SYMMETRY BREAKING AND PHYSICAL OBSERVABLES

The Wilson line is an order parameter for the symmetry breaking. However, if the theory is not invariant under $\mathscr{P}$, $\mathscr{C}$ and $\mathscr{T}$ we expect this to be reflected by some observable. This observable must be odd under the broken symmetries, but even under the combined action of two of them. The spatial components of the baryon current $j_\alpha = \langle \sum_{i=1}^{N_f} \bar{\psi}_i \gamma_\alpha \psi_i \rangle$ satisfy this property. Why should we expect a non-zero baryon current? A non trivial *vev* of the Wilson line means a non-zero value of the gauge field in that direction. Since the system is translationally invariant along the compact direction, the value of the gauge field must be constant (note that in the presence of toroidal topology a constant field

[2] We drop the subscript α where this does not lead to ambiguities.

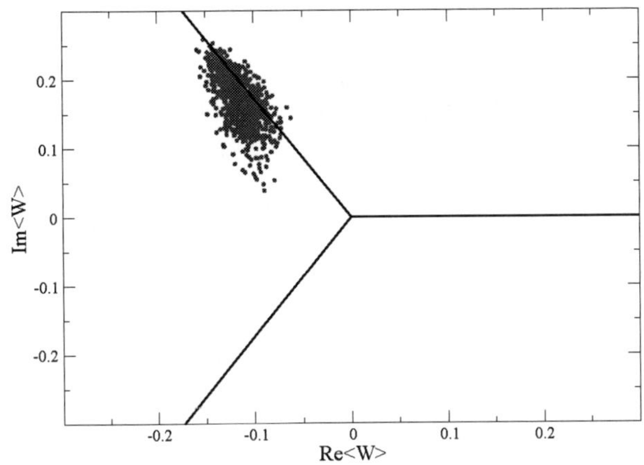

FIGURE 1. Plot of the Wilson loop. Clustering of the phase around $2\pi/3$ means spontaneous breaking of discrete symmetries.

cannot be gauged away). The background gauge field acts as a non-trivial source for the baryon current. Hence, we expect this current to be different from zero.
The expectation value of the baryonic current is generated by the partition function of the system in presence of a generalised "chemical potential", which acts as a source for the current [4]. Using the one-loop value for the effective potential we obtain

$$\langle \vec{j} \rangle = -\frac{N_f N}{L^3}\left(\frac{mL}{\pi}\right)^2 \sum_{\vec{k}\neq 0} \frac{K_2(mLk)}{k^2} \sin\left(\vec{k}\cdot\vec{v}^*\right)\vec{k} \,. \tag{7}$$

$\langle j_\alpha \rangle$ is zero when $v^*_\alpha = 0$ or $v^*_\alpha = \pi$ (i.e. when the symmetry breaking does not occur in direction α), is odd under $v^*_\alpha \to -v^*_\alpha$, goes to zero when $m \to \infty$.

THE LATTICE CALCULATION

Details of the lattice action used in our simulation are provided in [4]. A 24×4^3 lattice at $\beta = 5.5$ was used. This fixes the lattice spacing (determined measuring the Sommer parameter r_0 [5]) to 0.125 fm. This means that $L_t = aN_t = 3$ fm and $L_s = aN_s = 0.5$ fm, in agreement with the assumptions of sufficiently large L_t and L_s below the fermi scale. The breaking of discrete symmetries can be checked by looking at the Wilson line wrapping around a spatial direction. A typical behaviour is plotted in Fig. 1, which shows that the Wilson loop magnetises along $e^{i\frac{2}{3}\pi}$, i.e. $\langle W \rangle$ acquires an imaginary part, as required by the symmetry breaking scenario.
For the Euclidean rotated theory in the broken phase, we expect a non-zero value of the imaginary part of the current. At the given lattice parameters we find $|\mathrm{Im}\langle j_\alpha \rangle| = 0.060 \pm 0.002$, which should be compared with the perturbative prediction $\langle j_\alpha \rangle \simeq 0.037473(4)$:

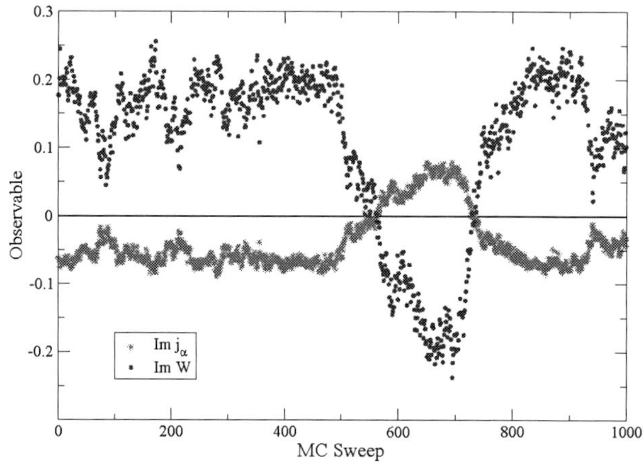

FIGURE 2. Monte Carlo history of the imaginary part of the Wilson line and the current in a compact direction in the broken symmetry phase.

it is remarkable that for compact dimensions of the order of $1/\Lambda_{QCD}$ the perturbative prediction still gives the correct order of magnitude.
The correlation of the current with the phenomenon of the spontaneous symmetry breaking can be understood by looking at the correlations between $\mathrm{Im}j_\alpha$ and $\mathrm{Im}W$. An example is given in Fig. 2. This plot shows that the two observables are closely related. In particular, in the event of a tunneling between the broken vacua (possible in our case because the spatial volume is finite), the current changes sign.

CONCLUSIONS AND OUTLOOK

In the phase in which discrete symmetries are spontaneously broken there is a persistent baryonic current wrapping around the topologically non-trivial compact directions. Unlike persistent currents in superconductors, this current is conserved. The existence of this current is a clear physical signature of the symmetry breaking and could be used to determine the order of the symmetry restoring phase transition that happens at a critical radius of the compact direction, which string-inspired calculations predict to be second order [6]. Another open problem is the interplay between the aforementioned phase transition and the chiral restoring phase transition at finite temperature.

REFERENCES

1. M. Unsal and L. G. Yaffe, *Phys. Rev.* **D74**, 2006, p. 105019 [arXiv:hep-th/0608180].
2. J. L. F. Barbon and C. Hoyos, *Phys. Rev.* **D73**, 2006, p. 126002 [arXiv:hep-th/0602285].
3. T. DeGrand and R. Hoffmann, *JHEP* **0702**, 2007, 022 [arXiv:hep-lat/0612012].
4. B. Lucini, A. Patella and C. Pica, *Phys. Rev.* **D75**, 2007, p. 121701 [arXiv:hep-th/0702167].
5. R. Sommer, *Nucl. Phys.* **B411**, 1994, p. 839 [arXiv:hep-lat/9310022].
6. A. Armoni, *JHEP* **0704**, 2007, 046 [arXiv:hep-th/0703229].

Paraquantum extension of the Wess-Zumino model[1]

L. Maghlaoui*,† and N. Belaloui†

*Université M'hamed Bouguerra, Faculté des Sciences, Département de Physique, Boumerdès, Algeria
†LPMPS, Faculté des Sciences Exactes, Département de Physique, Université Mentouri Constantine, Algeria

Abstract. A most simple paraextension of the Wess-Zumino model is investigated. As a parabosons-parafermions system, this model forms a field theoretical realization of a supersymmetric Poincaré algebra (SPA), where, the parasupercharges satisfy the bilinear commutations relations dictated by these types of systems. The closure of the transformations algebra is established with a bilinear product rule for the fermionic elements. Finally, we verify that these parasupercharges are really the generators of the (para)supersymmetric transformations.

Keywords: Supersymmetry, Parastatistics, Poincaré algebra
PACS: 11.30.Pb, 05.30.Pr, 11.60.Jv

INTRODUCTION

Generalized statistics and supersymmetry are two main interests of theoretical physics in recent years. They may be unified in what is called parasupersymmetry, which is a symmetry between bosons and parafermions. The parasupersymmetry structure will depend on the number P (the order of the paraquantization) of parafermions that can occupy the same state. Recently [1], a most immediate field theoretical realization of the parasupersymmetric Poincaré algebra (PSPA), bosons-parafermions extension of the Wess-Zumino (W-Z) model, without the context of the parasuperspace is investigated. Following the same procedure in [2] , we demonstrated by explicite calculation that the spinor parasupercharges Q of the theory, considered as linear operators in the Fock space, satisfy the cubic (PSPA): $QQQ \sim P_\mu$, where P_μ is the energy-impulsion quadrivector.The distinctive feature of the closure of the transformations is, like in [3], the occurence of a trilinear product rule for the fermionic elements which is the translating of the $P = 2$ parasupersymmetric algebra (PSA). Unlike the ordinary bilinear case, this result imply that the three times repeated applications of the transformations on a field gives rise to a translation of the transformed field. This closure is reinforced by the verification that these parasupercharges are the effective generators of these transformations. The aim of this work is to investigate the parabosons-parafermions extension of the W-Z model. Despite that the parafields satisfy triliner commutation relations, we show that this model form a field theoretical realisation of an ordinary (SPA), where,

[1] This work is supported by the Algerian Ministry of the Higher Education and Scientific Research under contract No.D1602/04/04.

CP957, *Particles, Strings, and Cosmology, 13th International Symposium*
edited by A. Rajantie, C. Contaldi, P. Dauncey, and H. Stoica

as parabosons-parafermions systems, the parasupercharges satisfy the bilinear commutation relations: $QQ \sim P_\mu$ dictated by these types of systems. we verify the closure of the algebra and that the parasupercharges are the generators of the transformations constructed.

THE MODEL

Introducing (para)supersymmetry in the context of a simple four dimensional massive and free field theory : paraquantum version of the W-Z model, let us assume that the model possess one parafermion Ψ^α of order $P = 2$ which is a Majorana spinor, on shell, that is $(i\partial_\mu \gamma^\mu - m)\Psi = 0$. Let us assume that the two degrees of freedom of the parafermionΨ^α impose the introduction of two parabosonic degrees of freedom in order to form a realization of (para)supersymmetry. We introduce then one real scalar parafield A and one real pseudoscalar parafield B subject to $(\Box + m^2)A = (\Box + m^2)B = 0$. Here we assume that the mass dimension of the parafermion Ψ is, like in the ordinary case, always $3/2$ and of course the parafields A and B have the dimension 1. With an adequate symmetrization of the parafermionic and the parabosonic fields, the lagrangian density describing this simple system is given by

$$L = \frac{1}{4}\left[\overline{\Psi},(i\gamma^\mu\partial_\mu - m)\Psi\right] + \frac{1}{4}\{(\partial_\mu A),(\partial^\mu A)\} + \frac{1}{4}\{(\partial_\mu B),(\partial^\mu B)\}$$
$$-\frac{1}{4}m^2\{A,A\} - \frac{1}{4}m^2\{B,B\} \tag{1}$$

where all the parafields have the same mass m and $\overline{\Psi} = \Psi^T\gamma^o$. The plane wave expansions of the Majorana Ψ and the A and B parafields are given by :

$$\Psi^a(x) = \frac{1}{(2\pi)^{3/2}}\sum_s \int d^3p(\frac{m}{\omega_p})^{1/2}(d(\overrightarrow{p},s)\,u_a(\overrightarrow{p},s)\,e^{-ipx} + d^+(\overrightarrow{p},s)\,v_a(\overrightarrow{p},s)\,e^{ipx})) \tag{2}$$

$$A(x) = \frac{1}{(2\pi)^{3/2}}\int d^3p(\frac{1}{2\omega_p})^{1/2}\{a(\overrightarrow{p})\,e^{-ipx} + a^+(\overrightarrow{p})\,e^{ipx}\} \tag{3}$$

$$B(x) = \frac{1}{(2\pi)^{3/2}}\int d^3p(\frac{1}{2\omega_p})^{1/2}\{b(\overrightarrow{p})\,e^{-ipx} + b^+(\overrightarrow{p})\,e^{ipx}\} \tag{4}$$

with the usual relations :

$$\omega_p = \left(\overrightarrow{p}^2 + m^2\right)^{1/2}, \ (\not{p} - m)u(\overrightarrow{p},s) = (\not{p} + m)v(\overrightarrow{p},s) = 0$$

The simplest (para)supersymmetry $P = 2$ theory dealing with para-bosonic and para-fermionic fields, having the same order as required when we want to ensure the validity of the mixed trilinear relations, is concerned with a pair of parafiels characterized by well-defined structure relations which are distributed in two nonequivalent ways leading to the so-called relative para-Bose and relative para-Fermi sets after Greenberg-Messiah [4]. Let us adopt the first way :

$$
\begin{aligned}
\left[\left\{d(\vec{p},s),a(\vec{p'})\right\},a^{+}(\vec{k})\right] &= 2\delta\left(\vec{p'}-\vec{k}\right)d(\vec{p},s) \\
\left[\left\{d(\vec{p},s),a^{+}(\vec{k})\right\},a(\vec{p'})\right] &= -2\delta\left(\vec{p'}-\vec{k}\right)d(\vec{p},s) \\
\left\{\left\{a(\vec{p'}),d^{+}(\vec{p},s)\right\},d\left(\vec{k},l\right)\right\} &= 2\delta\left(\vec{p}-\vec{k}\right)a(\vec{p'}) \\
\left\{\left\{a(\vec{p'}),d\left(\vec{k},l\right)\right\},d^{+}(\vec{p},s)\right\} &= 2\delta\left(\vec{p}-\vec{k}\right)a(\vec{p'}) \\
\left[\left\{d(\vec{p},s),a^{+}(\vec{p'})\right\},a^{+}(\vec{k})\right] &= \left[\left\{d(\vec{p},s),a(\vec{k})\right\},a(\vec{p'})\right]=0 \\
\left\{\left\{a(\vec{p'}),d^{+}(\vec{p},s)\right\},d^{+}\left(\vec{k},l\right)\right\} &= \left\{\left\{a(\vec{p'}),d\left(\vec{k},l\right)\right\},d(\vec{p},s)\right\}=0
\end{aligned} \tag{5}
$$

It is clear that, in this case, the parasupercharge Q_a must check bilinear relations. This implies that the corresponding transformation parameter ε must check the ordinary Grassmann parameter relations and satisfy:

$$
[\varepsilon_a,A]=[\varepsilon_a,B]=\{\varepsilon_a,\Psi_a\}=0 \tag{6}
$$

this leads us to the following transformations

$$
\begin{aligned}
\delta\Psi &= -\left(i\gamma^{\mu}\partial_{\mu}+m\right)(A-i\gamma_5 B)\,\varepsilon \\
\delta\overline{\Psi} &= \overline{\varepsilon}\,(A-i\gamma_5 B)\left(i\gamma^{\mu}\partial_{\mu}-m\right) \\
\delta A &= \overline{\varepsilon}\Psi,\delta B=-i\overline{\varepsilon}\gamma_5\Psi
\end{aligned} \tag{7}
$$

Then, from the zero component of the (para)supersymmetry conserved Noether curent density, one calculates the supercharge and obtains :

$$
\begin{aligned}
Q_a=\frac{i}{2}\lambda\left(\frac{m}{2}\right)^{1/2}\sum_s\int d^3p[\{C_{a\alpha}(\vec{p}),d^{+}(\vec{p},s)\}\,v_{\alpha}(\vec{p},s) \\
-\{D_{a\alpha}(\vec{p}),d(\vec{p},s)\}\,u_{\alpha}(\vec{p},s)]
\end{aligned} \tag{8}
$$

where

$$
C_{a\alpha}(\vec{p})=(a-i\gamma^5 b)_{a\alpha} \text{ and } D_{a\alpha}(\vec{p})=(a^{+}-i\gamma^5 b^{+})_{a\alpha}
$$

SUPERALGEBRA

In analogy with the parasupersymmetry of quantum mechanics, it is easy to see that the parasupercharge of a parabosons - parafermions W-Z model satisfy bilinear commutations relations. Indeed, by using the trilinear commutation relations (5) and choosing $\lambda=2$, one obtains the algebra corresponding to the ordinary case of the W-Z model:

$$
\{Q_a,\overline{Q}_b\}=2P_{\mu}\gamma^{\mu}_{ab} \tag{9}
$$

CLOSURE AND (PARA)SUPERSYMMETRY TRANSFORMATIONS

For the closure, it is clear here that, unlike the first model [1], the infinitesimal transformations close onto an algebra that involves bilinear ordinary relations. Indeed, since the parameters of the transformations even anticommute between them, therefore one can use the usual relations between the γ matrices and these parameters in addition to the equations of the parafield Ψ to obtain:

$$[\delta_1, \delta_2]\Psi_a = -2i(\overline{\varepsilon}_2\gamma^\mu\varepsilon_1)\partial_\mu\Psi \tag{10}$$

Here, ε_1 and ε_2 are the parameters corresponding to the infinitisimal transformations δ_1 and δ_2 respectively. This describes a paraboson-parafermion supersymmetry analogue to the ordinary case ($P = 1$). In the same way, one finds the closure on the parabosons A and B.

Now, as in the first model [1], one can explicitely verify that the parasupercharges (8) are really the generators of the transformations. ε being an ordinary Grassmann parameter, and according to the precedent result, one can describe the transformations in the following way, where, by the use of (2), (5), (8), one can for example obtain the result :

$$\delta\Psi = i\overline{\varepsilon}^a\{Q_a, \Psi\} = -\left(i\gamma^\mu\partial_\mu + m\right)(A - i\gamma_5 B)\,\varepsilon \tag{11}$$

which represents the transformation of the parafield Ψ. In the same way, one finds the transformations of the parabosonic fields A and B.

CONCLUSION

In this work, we have investigated the most simple paraextension of the W-Z model. As a parabosons-parafermions system, this model forms a field theoretical realization of a SPA, where, the parasupercharges satisfy the bilinear commutations relations dictated by these types of systems. In supersymmetry theories, the extensions of the Poincaré algebra are obtained from a "square root" of the transformations, $QQ \sim P_\mu$. In [1,3], new algebra are obtained from yet higher order roots "cubic roots" $QQQ \sim P_\mu$. On the other hand, this model is completely described by parafields of the same order $P = 2$. The results obtained confirm that the parabosons-parafermions supersymmetry of an arbitrary order P is equivalent to the ordinary supersymmetry.

REFERENCES

1. L. Maghlaoui and N. Belaloui: Bosons-Parafermions Wess-Zumino Model, non published.
2. H. J. W. Muller-Kirsten, Supersymmetry: An Introduction with Conceptual and Calculational Details, World Scientific Publishing, 1987.
3. R. Floreanini and L. Vinet: Pararupersymmetry in quantum mechanics and field theory, Phys. Rev. D 44, 3851(1991).
4. O. W. Greenberg and A. M. L. Messiah. Phys. Rev. B 138, 1155 (1965).

Hunting for the top partner in the littlest Higgs model with T parity at the CERN LHC

Shigeki Matsumoto

Tohoku University International Advanced Research and Education Organization, Institute for International Advanced Interdisciplinary Research, Sendai, Miyagi, 980-8578, Japan

Abstract. Little Higgs scenario provides us a new solution for the hierarchy and fine tuning problem of the Standard Model. Furthermore, the Littlest Higgs model with T-parity, which is one of models in this scenario, solves the dark matter problem in our universe. The model predicts existence of a new vector-like quark, in addition of a whole spectrum of heavy particles including the dark matter. In this talk, we discuss the phenomenology of the little Higgs model, focusing on the collider signature of this new vector-like quark at the CERN LHC experiment.

Keywords: Little Higgs, LHC, Dark Matter
PACS: 14.80.Cp, 12.60.Cn, 14.65.Ha

INTRODUCTION

The Higgs sector in the Standard Model (SM) receives large quadratic mass corrections from top and gauge boson loop diagrams. New symmetries involving top-Higgs and gauge-Higgs sectors below 1 TeV are proposed to remove the corrections. One of the physics targets at the LHC is to find new particles predicted by such symmetries. The important new physics candidate is the Minimal Supersymmetric Standard Model (MSSM). Recently, alternative scenarios which do not rely on supersymmetry to cancel the quadratic divergent corrections have been proposed and studied. The little Higgs model [1] is one of these alternatives. In this model, the Higgs boson is regarded as a pseudo Nambu-Goldstone boson, which originates from the spontaneous breaking of a global symmetry at certain high scale, and the global symmetry protects the Higgs mass from the quadratic radiative corrections.

The simplest version of the model is called the littlest Higgs model [2]. The global symmetry of the model is SU(5), which is spontaneously broken into SO(5). The part of the SU(5) symmetry is gauged, and the gauge symmetry is $SU(2)_1 \times SU(2)_2 \times U(1)_1 \times U(1)_2$. The top sector is also extended to respect the part of the global symmetry. This model, however, predicts a large correction to the EW observables because of the direct mixing between heavy and light gauge bosons after the EW symmetry breaking. The precision EW measurements force the masses of heavy gauge bosons and top partners to be 10 TeV, reintroducing the fine tuning problem to the Higgs mass [3]. A

CP957, *Particles, Strings, and Cosmology, 13th International Symposium*
edited by A. Rajantie, C. Contaldi, P. Dauncey, and H. Stoica

solution of the problem is the introduction of T-parity to the model which forbids the mixing [4]. This is the symmetry under the transformations, $SU(2)_1 \leftrightarrow SU(2)_2$ and $U(1)_1 \leftrightarrow U(1)_2$. Almost all heavy gauge bosons are assigned a T-odd charge, while the SM particles have a T-even charge. The matter sector should be extended so that T-odd partners are predicted. The lightest T-odd particle is the heavy photon A_H, which is stable and becomes a candidate for dark matter [5].

In the top sector, the vector-like top quark T and its T-odd partner T_- is introduced in addition to the usual top quark t and its T-odd partner t_-. The T quark plays an important role to cancel quadratic divergent corrections to the Higgs mass term. Among T-odd particles in the model, T_- quark is the lightest colored particle, thus the quark is expected to be copiously produced at the LHC. The purpose of this talk is to show a realistic estimate for the detection of this top partner at the LHC [6].

PROPARTIES OF T_- QUARK

The T-odd quark T_- is the color triplet and singlet under the SM $SU(2)_L$ gauge symmetry. Its electric charge is the same as the SM top quark. Thus, the quark decays simply into the top quark and the heavy photon (dark matter) with almost 100% branching ratio. Its width is about a few GeV when the mass of T_- is 1 TeV and the dark matter mass is 100 GeV. Since this quark has a color charge, it will be produced through the gluon fusion at the LHC. Its production cross section from the pp collision at the LHC is about 0.1 pb when T_- mass is 1 TeV.

We will choose several representative points to study the LHC signature of the T_- production. In order to do that, we have to consider some constraints on the Littlest Higgs model with T-parity. The most severe constraint comes from the electroweak precision measurements at the LEP experiment. The T_- mass is strongly correlated with T quark mass. Once we set the T_- mass is small, then T mass is also small, leading to large contributes to the rho parameter. The detailed study of the electroweak precision constraints shows that the mass of the T_- quark should be larger than 600 GeV. On the other hand, large T_- mass is not favored from the view point of the fine tuning on the Higgs mass term. Thus, the natural region of the T_- mass is between 600 GeV and 1 TeV. In addition to the electroweak precision experiments, the model receives the constraint from the cosmological observation, because the model contains the dark matter. The observation of cosmological parameters, to be more precise, the dark matter abundance from the WMAP experiment gives the strong constraint on the dark matter mass. It is well known that the dark matter abundance is determined by the annihilation cross section of dark matter. In this model, dark matter annihilate into weak gauge bosons through the s-channel Higgs exchange. Once the Higgs mass is set to be around 120 GeV, then it turns out that the dark matter mass should be about 100 GeV. We chose following four representative points to study the LHC signature. All points satisfy both LEP and WMAP constraints.

Point	**Parameters (m_{T_-}, m_{AH})**	**σ(pp $\rightarrow$ T_- \bar{T}$_-$ + X)**
I	(600, 100)	0.940 pb
II	(700, 125)	0.382 pb
II	(800, 150)	0.171 pb
IV	(900, 175)	0.0822 pb

LHC SIGNATURE OF T_- QUARK PRODUCTION

As mentioned in the previous section, the T_- quark decays into the SM top quark and dark matter. Thus, the LHC signal is the SM top pair production with large missing energy due to dark matter emissions. Interestingly, as shown in the table above, the mass of the T_- quark is much larger than the top quark and dark matter. Thus, energetic top quarks are expected in the signal event, which has a large advantage to reduce the SM background as shown in the following discussion.

Main background against the signal comes from the SM top pair production. In fact, the cross section of the background is about 400 pb at the leading order calculation. Thus, it is important to reduce the background to detect the signal. In order to do that, we impose following three cuts. (1) In the background event, the effective mass M_{eff} and the missing energy E_{Tmiss} tend to be small, while these two values can be large in the signal event due to the heavy T_- production and the dark matter emission. Hence, we impose the cut on the effective mass as $2m_{T-} - 200$ (300) GeV $< M_{eff} < 2m_{T-}$ for $2m_{T-} \leq 700$ (≥ 800) GeV, and investigate the event at the large missing energy region. (2) Since the mass of the T_- quark is about 1 TeV, it is expected that the quark is produced almost at rest at the LHC, and it decays into the energetic top quark. Thus, it is convenient to use the hemisphere analysis to reconstruct the signal event. We also use this analysis to reduce the background, which is sometimes called the top cut; the three jets invariant mass in a hemisphere should be less than 200 GeV. We impose the top cut at least one of the hemisphere. (3) Final cut is the lepton veto. Since the large missing energy comes from the leptonic W decay in background events, those are drastically reduced once we impose the cut.

Results of simulations are shown in Figure 1, where missing energy distributions for point I to IV are depicted. The red (dark gray in the black-and-white version) and blue (darkest gray) histograms are for the signal and background distributions, and the green (light gray) histograms are the sum of these. Each distribution shows the clear excess of the signal over the background. Furthermore, it can be seen that the signal event can be detected as a bump in all sample points.

CONCLUSION

Little Higgs scenario is one of interesting scenarios for physics beyond the SM. Among models in this scenario, the Littlest Higgs model with T-parity is attractive, because the model solves the hierarchy problem naturally, and also explains the dark matter abundance observed by the WMAP experiment. In the model, the vector-like T_- quark, which is color triplet and singlet under the SM $SU(2)_L$ gauge symmetry, will play an essential role at the LHC towards the discovery of the Little Higgs mechanism. In this talk, it has been shown that the production cross section of the quark pair is large enough, and it is easy to distinguish the signal from the background. Furthermore, the signal will be observed with a bump structure, it is also possible to extract model parameters by analyzing the structure.

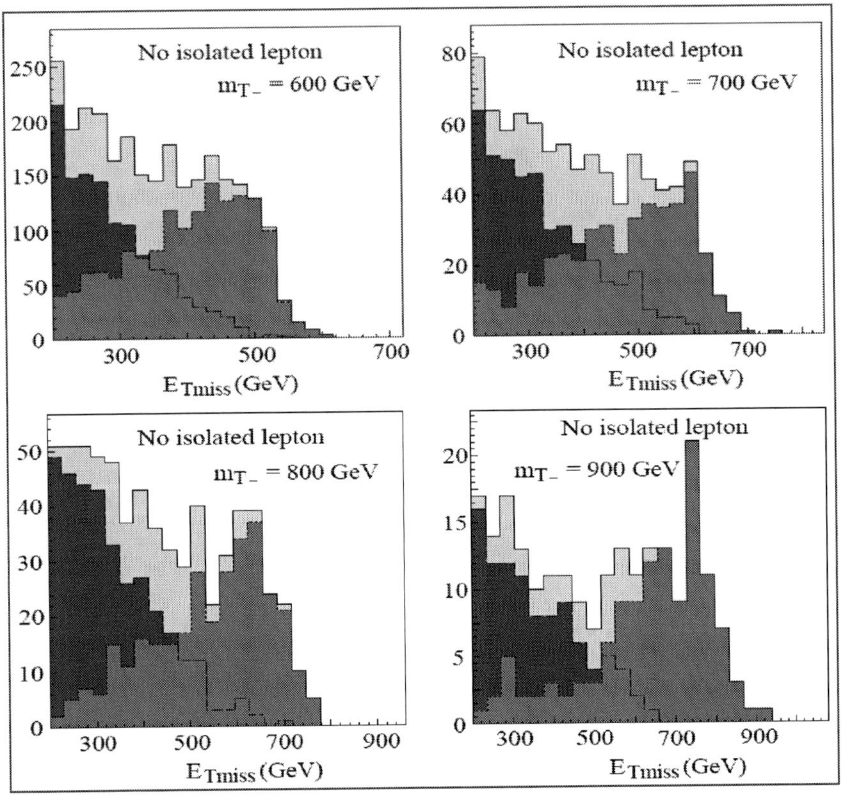

FIGURE 1. Missing energy distributions for point I to IV. The red (dark gray in the black-and-white version) and blue (darkest gray) histograms are for the signal and background distributions, and the green (light gray) histograms are the sum of these.

REFERENCES

1. N. Arkani-Hamed, A. G. Cohen and H. Georgi, *Phys. Letters* B513, 232 (2001); N. Arkani-Hamed, A. G. Cohen, E. Katz, A. E. Nelson, T. Gregoire and J. G. Wacker, *JHEP* 0208, 021 (2002).
2. N. Arkani-Hamed, A. G. Cohen, E. Katz and A. E. Nelson, *JHEP* 0207, 034 (2002).
3. C. Csaki, J. Hubisz, G. D. Kribs, P. Meade and J. Terning, *Phys. Rev.* D67, 115002 (2003); J. L. Hewett, F. J. Petriello and T. G. Rizzo, *JHEP* 0310, 062 (2003); C. Csaki, J. Hubisz, G. D. Kribs, P. Meade and J. Terning, *Phys. Rev.* D68, 035009 (2003); T. Gregoire, D. R. Smith and J. G. Wacker, *Phys. Rev.* D69, 115008 (2004); M. C. Chen and S. Dawson, *Phys. Rev.* D70, 015003 (2004); Z. Han and W. Skiba, *Phys. Rev.* D72, 035005 (2005) 035005; W. Kilian and J. Reuter, *Phys. Rev.* D70, 015004 (2004).
4. H. C. Cheng and I. Low, *JHEP* 0309, 051 (2003); 0408, 061 (2004); I. Low, *JHE*P 0410, 067 (2004).
5. J. Hubisz and P. Meade, *Phys. Rev.* D71 035016 (2005); M. Asano, S. Matsumoto, N. Okada and Y. Okada, Phys. Rev. D75, 063506 (2007); A. Birkedal, A. Noble, M. Perelstein and A. Spray, Phys. Rev. D74 035002 (2006).
6. S. Matsumoto, M. M. Nojiri and D. Nomura, *Phys. Rev.* D75, 055006 (2007).

The dark matter as a light gravitino [1]

Gilbert MOULTAKA

Laboratoire de Physique Théorique et Astroparticules
UMR5207–CNRS, Université Montpellier II
Place E. Bataillon, F–34095 Montpellier Cedex 5, France

Abstract. We address the question of gravitino dark matter in the context of gauge mediated supersymmetry breaking models.

Keywords: supersymmetry, gauge mediation, gravitino, dark matter, entropy dilution
PACS: 12.60.Jv, 04.65.+e, 95.35.+d, 95.30.Cq

INTRODUCTION

In scenarios where supersymmetry breaking is triggered by non-perturbative dynamics of some (secluded) gauge sector and communicated to the MSSM by a messenger sector through perturbative gauge interactions, the susy breaking scale $\sqrt{F}$ and the mass scale Λ of the secluded gauge sector can be well below the Planck scale. Recent developments [1] stressing the existence of metastable susy breaking vacua, have renewed the interest in such gauge-mediated susy breaking (GMSB) scenarios opening new possibilities for the model-building [2], and appear to be very interesting from the early Universe point of view as well [3]. On the other hand, the gravitational interactions which play a minor role for susy breaking in GMSB models remain physically relevant through the coupling to supergravity, at least in order to absorb the unphysical goldstino component, to adjust the cosmological constant to a small value and to avoid a massless R-axion. Moreover, if the above mentioned two scales combine to trigger the electroweak symmetry breaking yielding $G_F^{-1/2} \sim (\alpha/4\pi)kF/\Lambda$, where G_F is Fermi's constant (and $0 < k \leq 1$ measures the secludedness of the secluded sector), then the gravitino mass $m_{3/2} \simeq F/(\sqrt{3}m_{\rm Pl}) \sim (4\pi/\alpha)(\Lambda/\sqrt{3}km_{\rm Pl})\,G_F^{-1/2}$ where $m_{\rm Pl}$ is the reduced Planck mass, is expected to be very small ($\lesssim \mathcal{O}(1)$ GeV) and is the lightest supersymmetric particle (LSP). The question then arises as to which particle can be a good candidate for the cold dark matter (CDM) in this case? To answer this question requires an unconventional treatment as compared to the Neutralino "vanilla" candidate or even to the heavy gravitino candidate in the context of gravity mediated susy breaking models. Indeed, in contrast with the latter where the hidden sector is typically too heavy to be produced at the end of inflation, the secluded and messenger sectors of GMSB provide stable particles that may be present in

---

[1] based on work in collaboration with K. Jedamzik (*LPTA-Montpellier*), M. Lemoine (*IAP-Paris*) [6, 7]; and work in progress, M. Kuroda (Meiji-Gakuin), M. Lemoine (Paris), M. Capdequi-Peyranère (Montpellier).

CP957, *Particles, Strings, and Cosmology, 13th International Symposium*
edited by A. Rajantie, C. Contaldi, P. Dauncey, and H. Stoica

the early Universe for a sufficiently heavy reheat temperature T_{RH}. We consider hereafter such configurations assuming that only the messenger (including the spurion) sector can be produced and illustrate its relevance to the issue of the CDM.

A MESSENGER SOLUTION TO THE GRAVITINO PROBLEM

The mass degeneracy within a supermultiplet of messenger fields is lifted by susy breaking leading to a lighter and a heavier scalar messengers with masses $M_\pm = M_X(1 \pm kF/M_X^2)^{1/2}$ and a fermionic partner with mass M_X (where F and M_X are related to the dynamical scale Λ). Thus $kF/M_X^2 < 1$. Moreover, one has to require $kF/M_X \lesssim 10^5$GeV to ensure an MSSM spectrum $\lesssim \mathcal{O}(1)$TeV. One then expects typically $M_X \gtrsim 10^5$GeV. In typical GMSB models the lightest messenger particle (LMP) with mass M_- is stable due to the conservation of a messenger quantum number. If present in the early Universe the messenger particles are thermalized through their gauge interactions with the thermal bath. The corresponding LMP relic density is calculable similarly to that of the Neutralino LSP. However, it turns out to be typically too large to account for the CDM (albeit fine-tuning) even in the most favorable case of the electrically neutral component of a $\mathbf{5}+\overline{\mathbf{5}}$ representation of $SU(5)$ where it is found to scale as $\Omega_M h^2 \simeq 10^5 \left(M_-/10^3 TeV\right)^2$ with the LMP mass [4]. The situation is even worse in the case of $SO(10)$ where the LMP is an MSSM singlet with a suppressed annihilation cross-section leading to a very large relic density. One possible cure to this messenger overcloser problem, namely to allow the LMP to decay, can actually turn out to be a blessing regarding a solution to the gravitino problem and simultaneously letting the gravitino account for the CDM in the context of GMSB models.[2] The LMP late decay into MSSM particles can release enough entropy to dilute the initial gravitino relic density down to a level which can account for the CDM in the Universe even for very high T_{RH}, [5, 6, 7]. For this to work, though, the LMP should dominate the Universe energy density before it decays, and should decay after the gravitino has freezed-out from the thermal bath. The necessary condition $T_d < T_{MD} < T_{3/2}^f$ [where $T_d, T_{MD}, T_{3/2}^f$ denote respectively the LMP decay and messenger matter domination temperatures, and the gravitino freeze-out temperature] is then determined by the particle properties and annihilation cross-section and decay width of the LMP, delineating the favorable parts of the parameter space. We have studied this scenario in detail for the case of $SU(5)$ [6] and $SO(10)$ [7]. In the next section we concentrate on the latter case with one set of messengers transforming as $\mathbf{16}+\overline{\mathbf{16}}$.

---

[2] One can easily argue for an unstable LMP once the GMSB model is coupled to supergravity, invoking the violation of the messenger number conservation by gravitational interactions akin to discrete accidental symmetries. The resulting messenger number violating operators are then Planck scale suppressed and would not upset the natural suppression of the flavor changing neutral currents in GMSB models.

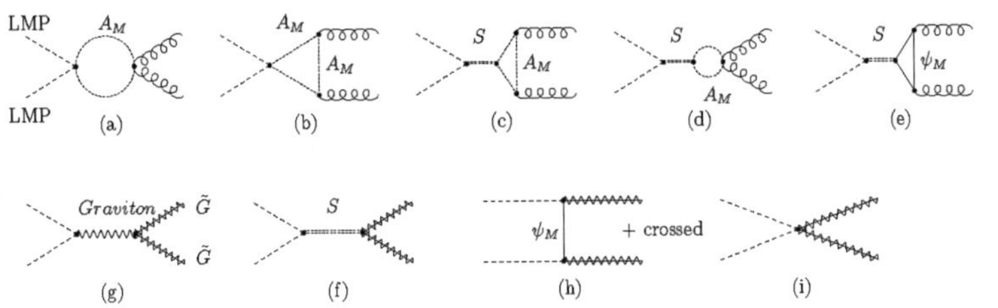

FIGURE 1. Feynman diagrams of the leading LMP annihilation into 2 gluons (a-e) or 2 gravitinos (f-i).

GRAVITINO RELIC ABUNDANCE

The entropy release $\Delta S \equiv S_{\text{after}}/S_{\text{before}}$, diluting the initial gravitino density, is determined by the temperatures before and after LMP decay and can be approximated to T_{MD}/T_d. T_{MD} is given by the LMP yield and mass ($T_{MD} \simeq (4/3)M_- \times Y_{\text{LMP}}$) and T_d is determined by the LMP width ($\Gamma_{\text{LMP}} \simeq H(T_d)$). Y_{LMP} is determined by the LMP annihilation into MSSM particles. Since in our case the LMP is an $SU(5)$ singlet [4, 7], this annihilation proceeds via loop effects of virtual messengers (A_M, ψ_M) and spurion (S) exchange, fig.1. We parameterize the thermally averaged leading annihilation cross-section into 2 gluons as $\langle\sigma_{\text{1loop}}v\rangle \sim f(\alpha_s/4\pi)^2\kappa^4/s$ where κ is the spurion-messenger coupling ($W \supset \kappa\hat{S}\mathbf{16}_M\overline{\mathbf{16}_M}$), α_s the strong coupling constant, $\sqrt{s}$ the C.M. energy and f a form factor depending on the internal masses. The LMP decay is induced by Planck scale suppressed non-renormalizable messenger number violating operators which can originate from the Kähler potential, e.g. $K \supset \mathbf{16}_F\overline{\mathbf{16}}_M^\dagger\mathbf{10}_H/m_{\text{Pl}}$, or from the superpotential, e.g. $W \supset \overline{\mathbf{16}}_M\mathbf{16}_F\mathbf{16}_F\mathbf{10}_H/m_{\text{Pl}}$, leading respectively to 2- and 3-body decays, where $\mathbf{16}_M(\overline{\mathbf{16}_M}), \mathbf{16}_F$ and $\mathbf{10}_H$ denote respectively the messenger, the standard matter and the electroweak Higgs supermultiplets. We assume a typical decay width $\Gamma_{\text{LMP}} = (1/16\pi)f'M_X^3/m_{\text{Pl}}^2$ where f' parameterizes our ignorance of the couplings and possible further phase space suppression. When the necessary temperature conditions are met, the final gravitino relic density is given by $\Omega_{grav} = \Omega^{th}_{grav}/\Delta S + \Omega^{Mess}_{grav} + \Omega^{NLSP}_{grav}$ where the last two contributions denote non-thermal production through late decays or scattering. One should also consider various cosmological constraints (hotness/warmness, BBN, species dilution, etc...). In fig.2 we illustrate the case with $T_{RH} \simeq 10^{12}$ GeV, see also [6]. The horizontal red shading shows the theoretically excluded region where $k > 1$; the other red shading indicates the region excluded by BBN constraints. If the spurion is heavier than the LMP, gravitino cold DM (green region) occurs for relatively light LMPs and $m_{3/2} \sim$ 1kev 10MeV. More generally, in the models of ref. [8] one finds [7] $\Omega_{grav}h^2 \simeq 10^3 f^{0.8}\kappa^{3.2}f'^{1/2}\left(M_-/10^6\,\text{GeV}\right)^{-0.3}\left(m_{3/2}/1\,\text{MeV}\right)$ for non-relativistic LMP freeze-out, putting the gravitino relic abundance in the ballpark of WMAP results, for $\kappa \sim \mathcal{O}(10^{-1})$ and typical ranges for f and f'. The LMP can also annihilate into 2 gravitinos through gravitational interactions, fig.1. For very heavy spurions the annihilation cross-section at rest reads $\langle\sigma v\rangle \simeq (1/24\pi)k^2M_-^2/(m_{3/2}m_{\text{Pl}})^2$. It can dominate the 1-loop annihilation, eventually saturating the unitarity limit (the black dashed line in fig.2), thus

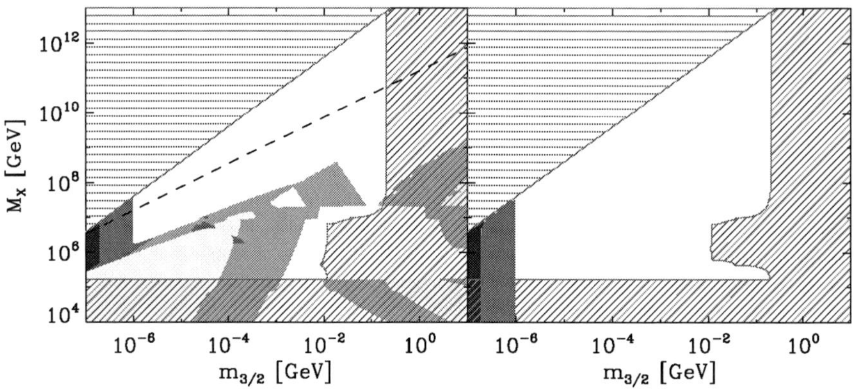

FIGURE 2. Contours of $\Omega_{3/2}$ in the plane $M_X(\equiv M_-) - m_{3/2}$ for one pair of messengers sitting in $\mathbf{16} + \overline{\mathbf{16}}$ representations of $SO(10)$; the LMP is a singlet under $SU(3) \times SU(2) \times U(1)$. We take for illustration $\kappa^2 \simeq \alpha_s/4\pi$, $f \sim \mathcal{O}(1)$ and $f' \simeq 5.10^{-2}$ and a bino NLSP with $M_{\rm NLSP} = 150$GeV, $M_{gluino} = 1$TeV and $kF/M_- \simeq 10^5$GeV; blue (hot), red (warm), green (cold) DM with $0.01 < \Omega_{grav} < 1$; yellow ($\Omega_{grav} < 0.01$), white ($\Omega_{grav} > 1$). In the right (left) panel the spurion is lighter (heavier) than the messenger. (taken from [6].)

disfavouring gravitino CDM solutions for a very heavy LMP.

To summarize, a light gravitino can account for CDM irrespective of T_{RH}, making it a good DM candidate in GMSB: typically if $T_{RH} \lesssim 10^5$GeV then the messengers are not produced and thermal gravitinos with $m_{3/2} \simeq 1$MeV provide the right CDM density, while for $T_{RH} \gtrsim 10^5$GeV the messenger can be present and should be unstable, thus providing a source of entropy production that can reduce a thermally overproduced gravitino to a cosmologically acceptable level. Moreover, various constraints (e.g. on T_{RH}, [9], or on the gravitino mass [10]) simply do not apply in the scenarios we have illustrated, thus escaping possible tension with thermal leptogenesis.

REFERENCES

1. K. Intriligator, N. Seiberg and D. Shih *JHEP 0604:021,2006.*
2. for a review of the earlier GMSB model-building and phenomenology see for insance G.F. Giudice, R. Rattazzi, *Phys.Rept 322: 419, 1999*, and references therein.
3. S.A. Abel *et al.*, *JHEP 0701:089,2007*, *JHEP 0701:015,2007*; N.J. Craig, P.J. Fox, J.G. Wacker *PRD75:085006,2007*; W. Fischler *et al.* *JHEP 0703:107,2007*; V. Kaplunovsky, these proceedings.
4. S. Dimopoulos, G.F. Giudice, A. Pomarol, *PLB 389:37,1996.*
5. M. Fujii, T. Yanagida, *PLB 549:273,2002*; see also E. A. Baltz, H. Murayama, *JHEP 0305:067, 2003.*
6. K. Jedamzik, M. Lemoine, G. Moultaka, *PRD 73:043514,2006*; see also G. Moultaka, *Acta Phys.Polon.B38:645,2007.*
7. M. Lemoine, G. Moultaka, K. Jedamzik, *PLB 645:222,2007*, and M. Kuroda *et al.*, *to appear.*
8. M. Dine, A.E. Nelson, *PRD 48:1277, 1993*; M. Dine, A.E. Nelson, Y. Shirman, *PRD 51:1362, 1995*; M. Dine *et al.* *PRD 53:2658, 1996.*
9. J. Pradler, F.D. Steffen, *PLB 648:224,2007.*
10. M. Viel *et al.* *PRD 71:063534,2005.*

New two-loop contributions to hadronic EDMs in the MSSM

Minoru Nagai

ICRR, University of Tokyo, Kashiwa 277-8582, Japan

Abstract. We discuss the quark EDMs and CEDMs in the MSSM with general flavor-changing terms in the squark mass matrices. In particular, the charged-Higgs mediated contributions to the down-quark EDM and CEDM are evaluated at two-loop level. We point out that these two-loop contributions may dominate over the one-loop induced gluino contribution even when the squark and gluino masses are around few TeV and $\tan\beta$ is moderate. [1]

Keywords: Supersymmetry, CP violation, Flavor
PACS: 12.60.Jv, 11.30.Er, 11.30.Hv

INTRODUCTION

The minimal supersymmetric standard model (MSSM) is the most motivated model beyond the standard model (SM). This model exhibits plenty of new CP-violating phases in addition to the CKM phase. It is known that they constitute a threat for very sensitive CP tests, at least for SUSY particle masses which are within the TeV region. New CP phases may be introduced in both the flavor-conserving and the flavor-changing soft SUSY breaking terms. The former ones are constrained directly by the electron and hadronic EDMs (electric dipole moments). In the latter case, flavor-changing neutral current (FCNC) processes are sensitive to those new phases, and we have to satisfy the stringent bounds arising from the K and B physics low-energy experiments.

In this work, we are interested in the quark EDMs and CEDMs (chromoelectric dipole moments) arising from flavor-changing terms in the squark mass matrices. While the EDMs and CEDMs are highly suppressed within the minimal flavor violation (MFV) framework, as it happens in the SM, those flavor-changing terms allow us to introduce new types of Jarlskog invariants in the MSSM. The Jarlskog invariants are a basis-independent measure of CP violation [2]. Within a general flavor violation (GFV) framework, the invariants are less suppressed by the Yukawa coupling constants compared to the MFV case. This implies that the EDMs are sensitive to non-minimal structures of flavor violation. Especially we will analyze in detail two-loop effects induced by non-holomorphic Yukawa interactions [3, 4] through the effective couplings of down-type quarks/charged-Higgs boson/up-type quarks.

---

[1] This talk is based on work in collaboration with J. Hisano and P. Paradisi in [1].

CP957, *Particles, Strings, and Cosmology, 13th International Symposium*
edited by A. Rajantie, C. Contaldi, P. Dauncey, and H. Stoica

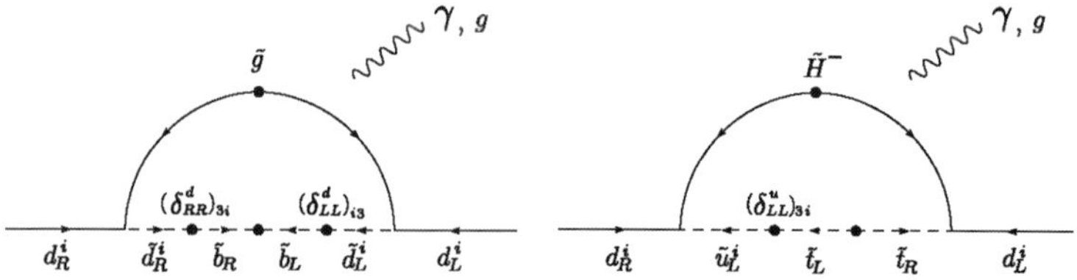

FIGURE 1. Left: Gluino contribution to the i-th down-quark (C)EDM in a case of non-zero values for both left- and right-handed squark mixings. Right: Higgsino contribution to the i-th down-quark (C)EDM in a case of non-vanishing left-handed squark mixing.

NEW JARLSKOG INVARIANTS IN THE MSSM

In this work we focus on the down-type quark contributions to the hadronic EDMs. Relevant CP violating operators that contribute to the hadronic EDMs, up to dimension five operators, are

$$\mathscr{L}_{CP} = -\sum_{q=d,s} i\frac{d_q}{2}\bar{q}(F\cdot\sigma)\gamma_5 q - \sum_{q=d,s} i\frac{d_q^c}{2}\bar{q}(g_s G\cdot\sigma)\gamma_5 q, \tag{1}$$

where $F_{\mu\nu}$ and $G_{\mu\nu}$ are the electromagnetic and the $SU(3)_C$ gauge field strengths respectively [5]. The coefficients of these operators, d_q and d_q^c represent the quark EDM and CEDM respectively.

As explained in the introduction, it is convenient to consider Jarlskog invariants to estimate the CP violating effects by flavor-changing terms. Since there is only one CP violating phase in the CKM matrix in the SM, the corresponding Jarlskog invariant is suppressed by the ninth power of Yukawa couplings, leading to very tiny EDMs. However, there appear new Jarlskog invariants originated from CP phases of squark mass matrices in the MSSM. For down-type quarks, depending on the existence of left-handed and/or right-handed squark mixing, δ_{LL}^q and δ_{RR}^d, three Jarlskog invariants are obtained,

$$J_{LR}^{(d_i)} = \mathrm{Im}\left\{\delta_{LL}^q f_d \delta_{RR}^d\right\}_{ii}, \quad J_{LL}^{(d_i)} = \mathrm{Im}\left\{[f_u f_u^\dagger, \delta_{LL}^q] f_d\right\}_{ii}, \quad J_{RR}^{(d_i)} = \mathrm{Im}\left\{f_u f_u^\dagger f_d \delta_{RR}^d\right\}_{ii}, \tag{2}$$

where f_d $[f_u]$ is the down[up]-type quark Yukawa coupling constant.

The first invariant of Eq.(2) appears at 1-loop gluino-mediated diagrams as shown in Fig. 1 (Left). In spite of the existence of small flavor mixing parameters, this contribution is not so much suppressed due to the enhancement by the bottom Yukawa coupling and some parameters are already constrained by the current experimental bounds [6]. The Higgsino mediated diagrams in Fig. 1 (right) also contribute to the EDMs, being proportional to the $J_{LL}^{d_i}$, which is unfortunately suppressed by the light quark mass m_{d_i} [7]. The Jarlskog invariant $J_{RR}^{d_i}$ first appears at two-loop level since it requires the coupling with both charged and neutral particles. In the next section we see that the relevant diagrams is very important and may dominate over the 1-loop gluino contribution.

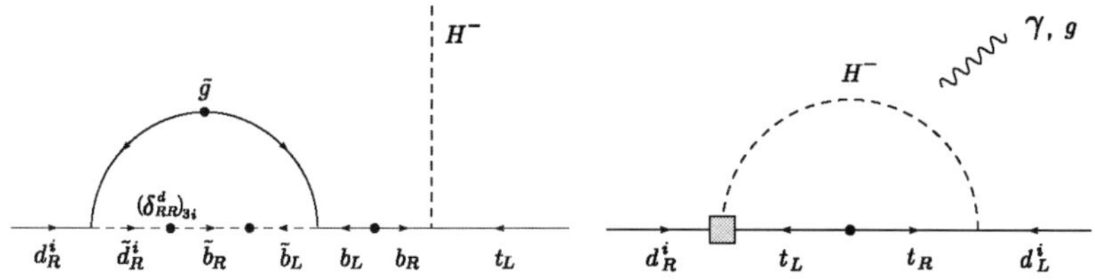

FIGURE 2. Left: Leading contribution to the effective $\overline{t_L} d^i_R H^+$ coupling in Eq. (5) in a case of $(\delta^d_{RR})_{i3} \neq 0$. Right: Charged-Higgs mediated diagrams contributing to the (C)EDMs in a case of $(\delta^d_{RR})_{i3} \neq 0$.

CHARGED HIGGS MEDIATED TWO-LOOP CONTRIBUTIONS

Among the two-loop contributions proportional to $J^{d_i}_{RR}$, a charged Higgs mediated contribution is especially important. Since this contribution are induced by non-holomorphic Yukawa interactions, they do not decouple even when the SUSY particles are very heavy.

The effective interactions of $\overline{t_L} d^i_R H^+$ and $\overline{t_R} d^i_L H^+$ is parametrized as

$$\mathcal{L}^{\rm eff}_{H^\pm} = \frac{g\, m_t}{\sqrt{2} m_W \tan\beta} V_{3i} C^{H^+}_L\, \overline{t_R} d^i_L H^+ + \frac{g\, m_{d_i} \tan\beta}{\sqrt{2} m_W} V_{3i} C^{H^+}_R\, \overline{t_L} d^i_R H^+ + \text{h.c.}. \tag{3}$$

The dominant corrections to the charged-Higgs vertex arise from gluino exchange, and the effective charged-Higgs coupling constants are given by

$$C^{H^+}_L = (1-\varepsilon_s \tan\beta) + \varepsilon_f \frac{(\delta^d_{LL})_{3i}}{V_{3i}} \tan\beta, \tag{4}$$

$$C^{H^+}_R = \frac{1}{(1+\varepsilon_s \tan\beta)} \left(1 - \frac{\varepsilon_f \tan\beta}{(1+\varepsilon_s \tan\beta)} \frac{(\delta^d_{LL})_{3i}}{V_{3i}} + \frac{\varepsilon_f \tan\beta}{(1+\varepsilon_s \tan\beta)} \frac{m_b}{m_{d_i}} \frac{(\delta^d_{RR})_{3i}}{V_{3i}} \right). \tag{5}$$

Here, V refers to the renormalized CKM matrix at one-loop level. and loop factors ε_s and ε_f are given as $\varepsilon_s = \frac{\alpha_s}{3\pi} \frac{\mu}{m_{\tilde{q}}}$ and $\varepsilon_f = \frac{\alpha_s}{9\pi} \frac{\mu}{m_{\tilde{q}}}$ respectively if the squark masses are taken to be equal to the gluino mass. Thanks to the right-handed squark mixing terms, we can always pick up the bottom mass m_b instead of m_d or m_s. We stress that, in the MFV framework, it is not possible to get such a Yukawa enhancement.

The (C)EDMs of down and strange quarks are induced by the one-loop exchange of the charged-Higgs boson and top quark, as it is shown in Fig. 2. In the case of $(\delta^d_{RR})_{i3} \neq 0$ $(i = 1, 2)$, by making use of the effective Lagrangian in Eq. (3), they are evaluated as

$$\left\{ \frac{d_{d_i}}{e}, d^c_{d_i} \right\} = -\frac{\alpha_2}{4\pi} \text{Im} \left[V^\star_{3i} (\delta^d_{RR})_{3i} \right] \frac{m_t^2}{m_W^2} \frac{m_b}{m^2_{H^\pm}} \frac{\varepsilon_f (1-\varepsilon_s \tan\beta) \tan\beta}{(1+\varepsilon_s \tan\beta)^2} \{F_7(y), F_8(y)\}, \tag{6}$$

where $y = m_t^2/m^2_{H^\pm}$ and the loop functions $F_7(y)$ and $F_8(y)$ are defined in [1]

These contributions are not suppressed by the light quark mass and may become large. In fact, as is illustrated on the left of Fig. 3, they can be of order $10^{-(26-27)}$ cm for $\tan\beta = 10$ and $m_{H^\pm} \in (0.2, 2)$ TeV. They are slowly decoupled for $m_{H^\pm} \to \infty$ since the

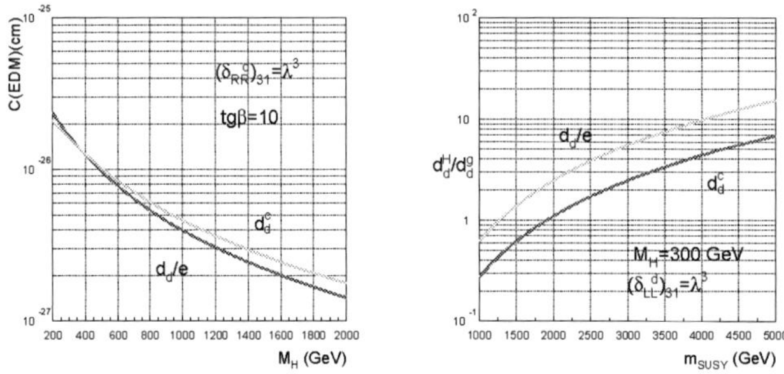

FIGURE 3. Left: Down-quark EDM and CEDM as functions of the charged-Higgs mass $m_{H^\pm}$ for $\tan\beta = 10$. In this figure, a CKM-type mixing is assumed for δ^d_{RR}, *i.e.* $|(\delta^d_{RR})_{31}| = \lambda^3$. Right: Ratio of charged-Higgs and gluino induced down-quark (C)EDM as a function of a common SUSY particle mass m_{SUSY}, setting the Higgs mass to $m_{H^\pm} = 300$GeV. We assume $(\delta^d_{LL})_{13} = -V^\star_{31}$.

loop function contains a large logarithm and it behave as $d_q \sim m_t^2/m^2_{H^\pm}\log(m^2_{H^\pm}/m_t^2)$. In Fig. 3 (right) we show the ratio of charged-Higgs contribution to the gluino one. Since the charged-Higgs contribution is not decoupled for heavy SUSY particle masses, it can become dominant over the gluino one if $m_{\rm SUSY} \gtrsim (1-2)$TeV when $m_{H^\pm} \sim 300$GeV.

CONCLUSIONS

We have pointed out that new two-loop contributions mediated by the charged-Higgs exchange may generate the quark (C)EDMs. In particular, we have evaluated the down- and strange-quark EDMs and CEDMs. Interesting enough, the above contributions do not decouple even for heavy squarks and gluino. These new contributions may dominate over the one-loop contributions in the case of non-vanishing right-handed down-type squark mixing, even when the squark and gluino masses are around the TeV scale. Furthermore, the down-quark EDM and CEDM may be very close to the actual experimental sensitivities and they will be strongly probed with the expected future experimental resolutions.

REFERENCES

1. J. Hisano, M. Nagai and P. Paradisi, Phys. Lett. B **642**, 510 (2006).
2. C. Jarlskog, Phys. Rev. Lett. **55** (1985) 1039.
3. L.J. Hall, R. Rattazzi and U. Sarid, Phys. Rev. D **50** (1994) 7048.
4. C. Hamzaoui, M. Pospelov and M. Toharia, Phys. Rev. D **59** (1999) 59; K. S. Babu and C. F. Kolda, Phys. Rev. Lett. **84** (2000) 228.
5. For a review of EDMs, M. Pospelov and A. Ritz, Annals Phys. **318** (2005) 119.
6. S. Dimopoulos and L. J. Hall, Phys. Lett. B **344** (1995) 185; R. Barbieri, A. Romanino and A. Strumia, Phys. Lett. B **369** (1996) 283; A. Romanino and A. Strumia, Nucl. Phys. B **490** (1997) 3; J. Hisano, M. Kakizaki, M. Nagai and Y. Shimizu, Phys. Lett. B **604** (2004) 216; J. Hisano, M. Kakizaki and M. Nagai, Phys. Lett. B **624** (2005) 239.
7. M. Endo, M. Kakizaki and M. Yamaguchi, Phys. Lett. B **583** (2004) 186.

Field Quantization in Krein space

F. Payandeh*, M. Mehrafarin* and M.V. Takook†

*Department of Physics, Amirkabir University, Tehran 15914 , IRAN
†Department of Physics, Razi University, Kermanshah, IRAN

Abstract. Previously it was shown that the presence of negative-norm states or negative-frequency solutions is indispensable for a fully covariant quantization (Krein space quantization) of the minimally coupled free scalar field in de Sitter spacetime. This new method of quantization has been extended to free boson and spinor fields in Minkowski spacetime. The presence of unphysical negative-frequency states plays the role of an automatic renormalization tool for the theory.

Keywords: Krein space, Negative-norm
PACS: 04.62.+v, 03.70+k, 11.10.Cd, 98.80.H

INTRODUCTION

The origion of this work goes back to the quantization of massless minimally coupled scalar field in de Sitter spacetime which plays an important role in the inflationary model as well as in the linear quantum gravity. It has been shown that negative frequency or norm states necessarily appear in a covariant quantization of the free minimally coupled scalar field in de Sitter spacetime [1]. As proved by Allen [2], a covariant quantization of minimally coupled scalar field cannot be constructed by positive norm states alone. To cope with this problem the Krein space quantization has been utilized [2]. Preserving the covariance principle and ignoring the positivity condition, the field quantization has been performed in Krein space, a combined Hilbert and anti-Hilbert space. This process resembles Gupta-Bleuler quantization of vector field in electrodynamics in Minkowski spacetime. Mathematically, for the minimally coupled scalar field in de Sitter spacetime [1], it has been proven that the use of the two sets of solutions (positive and negative frequency states) is unavoidable for preservation of (i) causality (locality), (ii) covariance, and (iii) elimination of the infrared divergence. This causal approach was generalized further to the calculation of the graviton propagator in de Sitter spacetime, and to the one-loop effective action for the scalar field in a general curved spacetime [3]. In this process ultraviolet and infrared divergences have been automatically eliminated [3]. The new method of quantization has been extended to free boson and spinor fields in Minkowski spacetime [4]. The role of “unphysical” states appears in the method as a natural renormalization tool in the one-loop approximation. In this paper we review the free fields quantization in Krein space.

KREIN QFT CALCULATION IN MINKOWSKI SPACETIME

Generalizing to the Minkowski spacetime, $\eta^{\mu\nu} = diag(1,-1,-1,-1)$, the Krein space method has been successfully implemented in boson field quantization [4], resulting

CP957, *Particles, Strings, and Cosmology, 13th International Symposium*
edited by A. Rajantie, C. Contaldi, P. Dauncey, and H. Stoica

in both positive and negative norms for the auxiliary negative-frequency states. This approach has been applied to the study of the free spinor field, $(s=\frac{1}{2})$, too. In so doing, the auxiliary negative-frequency states, with positive norms, have been utilized, the modes of which do not interact with the physical states or the real physical world. Since the norms of the unphysical states are positive, the Krein space quantization (which deals with negative norms) should be extended to the " generalized krein space quantization". Naturally these modes can not be affected by the physical boundary conditions. Following this scheme, the normal ordering procedure is rendered useless because the ultraviolet divergence in the stress tensor disappears and the vacuum energy remains convergent [1]. The results of the new approach are as the following [4] (K and T stand for quantities in Krein space and the time ordered product, respectively):

(i) Klein-Gordon field

$$(\partial^2+m^2)\phi(x)=0\ ,\ \phi_K(x)=\frac{1}{\sqrt{2}}\left[\phi_p(x)+\phi_n(x)\right]\ ,\ iG_T(x,x')=<0|T\phi_K(x)\phi_K(x')|0>$$

$$u_p(k,x)=\frac{e^{-ik.x}}{\sqrt{(2\pi)^3 2\omega_{\vec{k}}}}\ ,\ u_n(k,x)=\frac{e^{ik.x}}{\sqrt{(2\pi)^3 2\omega_{\vec{k}}}}\ ,\ \omega_{\vec{k}}=\sqrt{\vec{k}.\vec{k}+m^2}$$

$$\phi_K(x)=\frac{1}{\sqrt{2}}\int d^3\vec{k}\left[(a_{\vec{k}}+b^\dagger_{\vec{k}})u_p(k,x)+(a^\dagger_{\vec{k}}+b_{\vec{k}})u_n(k,x)\right]$$

$$[a_{\vec{k}},a_{\vec{k}'}{}^\dagger]=+\delta^3(\vec{k}-\vec{k}')\quad,\quad[b_{\vec{k}},b_{\vec{k}'}{}^\dagger]=-\delta^3(\vec{k}-\vec{k}')$$

$$<1_{\vec{k}'}|1_{\vec{k}}>=+\delta^3(\vec{k}-\vec{k}')\quad,\quad<\bar{1}_{\vec{k}'}|\bar{1}_{\vec{k}}>=-\delta^3(\vec{k}-\vec{k}')$$

$$H=\int d^3\vec{k}\ \omega_{\vec{k}}\ (a_{\vec{k}}{}^\dagger a_{\vec{k}}+b_{\vec{k}}{}^\dagger b_{\vec{k}}+a_{\vec{k}}{}^\dagger b_{\vec{k}}{}^\dagger+a_{\vec{k}}b_{\vec{k}})\quad,\quad<0\,|\,H\,|\,0>=0$$

(ii) Maxwell field

$$\partial^2 A_\mu=0\ ,\ A^K_\mu(x)=\frac{1}{\sqrt{2}}[A^p_\mu(x)+A^n_\mu(x)]\ ,\ D^T_{\mu\nu}(x,x')=-\eta_{\mu\nu}G_T(x,x')$$

$$A^K_\mu(x)=\frac{1}{\sqrt{2}}\int d^3\vec{k}\sum_{\lambda=0}^{3}\varepsilon^\lambda_\mu(\vec{k})[(a^\lambda_{\vec{k}}+b^{\lambda\dagger}_{\vec{k}})\ u_p(k,x)+(a^{\lambda\dagger}_{\vec{k}}+b^\lambda_{\vec{k}})u_n(k,x)]$$

$$[a^\lambda_{\vec{k}},a^{\lambda'\dagger}_{\vec{k}'}]=-\eta^{\lambda\lambda'}\delta^3(\vec{k}-\vec{k}')\quad,\quad[b^\lambda_{\vec{k}},b^{\lambda'\dagger}_{\vec{k}'}]=+\eta^{\lambda\lambda'}\delta^3(\vec{k}-\vec{k}')$$

$$|one\ physical\ state>=|1^\lambda_{\vec{k},a}>\propto a^{\lambda\dagger}_{\vec{k}}|0>\qquad\lambda=1,2$$

$$|one\ unphysical\ state>=|\bar{1}^\lambda_{\vec{k},a}>\propto a^{\lambda\dagger}_{\vec{k}}|0>\qquad\lambda=0,3$$

$$|one\ unphysical\ state>=|\bar{1}^\lambda_{\vec{k},b}>\propto b^{\lambda\dagger}_{\vec{k}}|0>\qquad\lambda=0,1,2,3$$

$$H=\int d^3\vec{k}\ \omega_{\vec{k}}\sum_{\lambda=0}^{3}(-\eta_{\lambda\lambda})(a^{\lambda\dagger}_{\vec{k}}a^\lambda_{\vec{k}}+b^{\lambda\dagger}_{\vec{k}}b^\lambda_{\vec{k}}+a^{\lambda\dagger}_{\vec{k}}b^{\lambda\dagger}_{\vec{k}}+a^\lambda_{\vec{k}}b^\lambda_{\vec{k}})\ ,\ <0\,|\,H\,|\,0>=0$$

Four of the above states have negative norms, and two of them are unphysical positive norm states.

(iii) Dirac field (generalized Krein space)

$$(i\not\partial - m)\psi(x) = 0\ ,\ \psi(x) = \frac{1}{\sqrt{2}}\,[\psi_p(x)+\psi_n(x)]\ ,\ S_T(x,x') = (i\not\partial + m)G_T(x,x')$$

$$\mathscr{U}^s(k,x) = \sqrt{\frac{m}{(2\pi)^3\omega_{\vec{k}}}}\, u^s(\vec{k})e^{-ik.x}\quad ,\quad \mathscr{V}^s(k,x) = \sqrt{\frac{m}{(2\pi)^3\omega_{\vec{k}}}}\, v^s(\vec{k})e^{ik.x}$$

$$\psi_K(x) = \frac{1}{\sqrt{2}}\int d^3\vec{k}\sum_{s=1,2}[(b_{\vec{k}s}+c^\dagger_{\vec{k}s})\mathscr{U}^s(k,x)+(d^\dagger_{\vec{k}s}+a_{\vec{k}s})\mathscr{V}^s(k,x)]$$

$$\{b_{\vec{k}s},b^\dagger_{\vec{k}'s'}\} = \{d_{\vec{k}s},d^\dagger_{\vec{k}'s'}\} = \delta_{ss'}\ \delta^3(\vec{k}-\vec{k}')\ ,\ \{a_{\vec{k}s},a^\dagger_{\vec{k}'s'}\} = \{c_{\vec{k}s},c^\dagger_{\vec{k}'s'}\} = \delta_{ss'}\ \delta^3(\vec{k}-\vec{k}')$$

$$< 1^b_{\vec{k},s} \mid 1^d_{\vec{k}',s} >= \delta_{bd}\ \delta^3(\vec{k}-\vec{k}')\quad ,\quad < \bar{1}^a_{\vec{k},s} \mid \bar{1}^c_{\vec{k}',s} >= \delta_{ac}\ \delta^3(\vec{k}-\vec{k}')$$

$$H = \int d^3\vec{k}\ \omega_{\vec{k}}\sum_{s=1,2}[b^\dagger_{\vec{k}s}b_{\vec{k},s}+b^\dagger_{\vec{k}s}c^\dagger_{\vec{k}s}+c_{\vec{k},s}b_{\vec{k},s}+c_{\vec{k},s}c^\dagger_{\vec{k}s}-d_{\vec{k},s}d^\dagger_{\vec{k}s}-d_{\vec{k},s}a_{\vec{k}s}-a^\dagger_{\vec{k}s}d^\dagger_{\vec{k}s}-a^\dagger_{\vec{k}s}a_{\vec{k}s}]$$

where $< 0 \mid H \mid 0 >= 0$. $b^\dagger_{\vec{k}s}$, $d^\dagger_{\vec{k}s}$ are respectively the creation operators of physical one-particle and one-antiparticle states with positive energy, and $c^\dagger_{\vec{k}s}$, $a^\dagger_{\vec{k}s}$ are called the creation operators of unphysical one-particle and one-antiparticle states respectively with negative energy.

FREE OF DIVERGENCE PROPAGATOR IN KREIN SPACE

The origin of divergences in standard QFT lies in the singularity of the Green's function at short relative distances (ultraviolet divergence) or in the large relative distances (infrared divergence):

$$\begin{aligned} < G^P_F(x,x') > \ &= \ -\frac{1}{8\pi}\delta(\sigma_0)+\frac{m^2}{8\pi}\ \theta(\sigma_0)[\frac{J_1(\sqrt{2m^2\sigma_0})-iN_1(\sqrt{2m^2\sigma_0})}{\sqrt{2m^2\sigma_0}}] \\ &\quad -\frac{im^2}{4\pi^2}\ \theta(-\sigma_0)\frac{K_1(\sqrt{2m^2(-\sigma_0)})}{\sqrt{2m^2(-\sigma_0)}} \end{aligned}$$

The divergence appears in the imaginary part of the Feynman propagator, and the real part is convergent. Consideration of negative frequency states, which are also solutions of the field equation conveniently removes singularity of the Green function with exception of delta function singularity. The latter singularity can be eliminated by the quantum metric fluctuations [5]. It has been shown that the combination of the above methods (*i.e.* Krein space quantization and quantum metric fluctuations) totally removes the divergences of QFT [5]:

$$< G_T(x-x') >= -\frac{1}{8\pi}\sqrt{\frac{\pi}{2<\sigma_1^2>}}\exp(-\frac{\sigma_0^2}{2<\sigma_1^2>})+\frac{m^2}{8\pi}\ \theta(\sigma_0)\frac{J_1(\sqrt{2m^2\sigma_0})}{\sqrt{2m^2\sigma_0}}$$

where $2\sigma_0 = \eta_{\mu\nu}(x^\mu - x'^\mu)(x^\nu - x'^\nu)$, and $< \sigma_1^2 >$ is related to the density of gravitons. When $\sigma_0 = 0$, due to the metric quantum fluctuation $< \sigma_1^2 > \neq 0$, and $< G_T(0) > = -\frac{1}{8\pi}\sqrt{\frac{\pi}{2<\sigma_1^2>}} + \frac{m^2}{8\pi}$. The time-ordered propagator of spinor field in generalized Krein space also reads:

$$S_T(x,x') = \frac{1}{8\pi}\, i\gamma^\mu (x_\mu - x'_\mu)\{\sqrt{\frac{\pi}{2<\sigma_1^2>}}\; e^{-\frac{\sigma_0^2}{2<\sigma_1^2>}}[\frac{\sigma_0}{<\sigma_1^2>} + m^2\frac{J_1(\sqrt{2m^2\sigma_0})}{\sqrt{2m^2\sigma_0}}]$$
$$+\frac{m}{2\sqrt{2}}\,\theta(\sigma_0)[\sqrt{2m^2\sigma_0}\,J_0(\sqrt{2m^2\sigma_0}) - 2J_1(\sqrt{2m^2\sigma_0})]\}$$
$$+\frac{m}{8\pi}\,[-\sqrt{\frac{\pi}{2<\sigma_1^2>}}e^{-\frac{\sigma_0^2}{2<\sigma_1^2>}} + m^2\theta(\sigma_0)\frac{J_1(\sqrt{2m^2\sigma_0})}{\sqrt{2m^2\sigma_0}}]$$

The most interesting result of this construction is the convergence of Green's function at large distances, which means that the infrared divergence is gauge dependent [1, 3].

CONCLUSION

In quantum field theory, to eliminate the divergences that appear in the physical quantities, the normal ordering (renormalization) procedure has been adopted for free (interacting) fields. However, the divergences seem to disappear once the requirement of the positivity of norm and energy is relaxed. The addition of the new unphysical states, thus, leads us to the Krein space quantization. And, considering the quantum metric fluctuations it is found that the theory is automatically renormalized. We hope to apply the new method successfully to reformulate QED in Krein space [4].

ACKNOWLEDGMENTS

The authors would like to thank H. S. Moslemi for his interest in this work, and PASCOS07 organizers at Imperial College.

REFERENCES

1. De Bievre S., Renaud J., Phys. Rev. D **57**, 6230 (1998); Gazeau J. P., Renaud J., Takook M. V., Class. Quant. Grav. **17**, 1415 (2000).
2. Allen B., Phys. Rev. D, **32**, 3136 (1985); M. Mintchev, J. Phys. A: Math. Gen., **13** 1841 (1979).
3. Takook M. V., Proceedings of the 6th International Wigner Symposium (WIGSYM6), Istanbul, Turkey, 16-22 Aug. 1999; Garidi T., Gazeau J. P., Takook M. V., J. Math. Phys. **44**, 3838 (2003); Takook M. V., Int. J. Mod. Phys. E **14**, 219 (2005); Takook M. V., Mod. Phys. Lett. A **16**, 1691 (2001).
4. Payandeh F., Khani S., Takook M. V., Mehrafarin M., "Quantization of free boson fields in Krein space", submitted for publication; Fatahi N., Payandeh F., Rouhani S., Takook M.V., "Quantization of spinor fields in generalized Krein space", submitted for publication; Payandeh F., Mehrafarin M., Takook M. V., "One-loop approximation of Møller scattering in generalized Krein space quantization", in preparation.
5. Ford H. L., gr-qc/9707062 (1997); Rouhani S., Takook M. V., gr-qc/0607027 (2006).

Signatures of pseudo-degenerate WIMP dark matter

Adam Ritz

Department of Physics & Astronomy, University of Victoria, Victoria, BC, V8P 1A1 Canada

Abstract. WIMP models of cold dark matter which exhibit nearly degenerate excited states have a number of interesting signatures that may prove relevant for direct and indirect searches. Such near-degeneracies form a qualitatively distinct regime in the parameter space of many models which, generically due to the little hierarchy problem, now require efficient co-annihilation to ensure the correct relic density. This work discusses the relevance of such scenarios for explaining the observed galactic 511 keV line, and also the impact on direct terrestrial searches which may be complicated by recombination-type processes.

Keywords: Dark Matter, Galactic 511 keV Line
PACS: 12.60.-i, 95.35.+d

Introduction

Given what is now rather compelling evidence for the existence of dark matter on various astrophysical and cosmological scales, finding direct and/or indirect means for discerning its nature stands as one of the most profound problems in particle physics. In this regard, local astrophysical observations are increasingly becoming one of our most powerful indirect probes, and there are a number of galactic observations which can provide useful guidance and constraints on the particle physics of dark matter.

An interesting example of this kind is the 511 keV line from the central region of our galaxy, with a photon flux $\Phi_{\rm exp} = (9.35 \pm 0.54) \times 10^{-4}$ ph cm^{-2} s^{-1}, well-measured using the SPI spectrometer on the INTEGRAL satellite [1]. This source of galactic positrons represents a challenge for conventional astrophysics, and its localization within the inner 1-2 kpc of the galactic bulge invites speculation about an interpretation involving dark matter. Since the injected positrons cannot be highly energetic in order not to violate constraints on the γ-background [2], most scenarios, e.g. those involving annihilation of MeV-scale dark matter [3], fall outside the generic WIMP paradigm. However, a source of low-energy positrons from the interaction of conventional electroweak scale WIMPs would be quite natural if there were MeV-scale splittings in the dark sector. Interestingly enough, near-degeneracies of a less extreme nature are now well-motivated in supersymmetric scenarios as the need for a relatively high LSP mass generically requires co-annihilation to ensure the correct relic abundance.

With this motivation in mind, in the next section we explore three possible mechanisms via which WIMPs could produce the required galactic positron flux and discuss their feasibility; for further details see [4]. In the final section, we also comment on the impact of some of these kinematic scenarios on signals in underground detectors.

CP957, *Particles, Strings, and Cosmology, 13th International Symposium*
edited by A. Rajantie, C. Contaldi, P. Dauncey, and H. Stoica

Pseudo-degenerate WIMPs and Galactic Positrons

We will consider dark sector at or above the electroweak scale composed of a neutral WIMP χ_1^0 and a nearby excited state χ_2, which may be neutral or charged, with a small mass splitting $\Delta m = m_2 - m_1 \sim \mathcal{O}(\mathrm{MeV})$. This *pseudo-degenerate WIMP* framework allows in principle for several possible production mechanisms for galactic positrons that we will now consider in turn.

Collisional excitation in WIMP-WIMP scattering:
A TeV mass WIMP χ_1 moving with velocity $v \sim O(10^{-3})$ has a kinetic energy comparable to the electron mass, and thus collisional excitation $\chi_1 + \chi_1 \to \chi_2^+ + \chi_2^-$ in the galactic core region, where the dark matter density is expected to be enhanced, could provide a source of MeV positrons via the decay $\chi_2^+ \to \chi_1 + e^+$. Rather than build a detailed model to realize this scenario, it will be enough to assume in the relevant kinematic range that such a model attains the maximal inelastic cross-section allowed by unitarity, namely for the lth partial wave, $\sigma_{\mathrm{max}} = \pi(2l+1)/k_{\mathrm{cm}}^2$, where $k_{\mathrm{cm}} = \frac{m_0}{2}(2v)$ is the center-of-mass momentum for a WIMP of mass m_0. With a Maxwellian velocity distribution in the halo, the collisional positron production rate can be computed and compared with that required by the observed photon flux. The required rate of course depends on the dark matter halo profile in the inner region of the galaxy, but taking the existing constraints for the Milky Way into account [5] the flux appears to be maximized with an NFW profile. We obtain

$$\frac{\langle \sigma_{\mathrm{max}} v_{\mathrm{rel}} \rangle}{\langle \sigma v \rangle_{\mathrm{exp}}} \leq 3.4 \times 10^{-3} \times (2l+1) \left(\frac{v_{\mathrm{rms}}}{220\ \mathrm{km/s}} \right)^7, \tag{1}$$

with the bound saturated for a WIMP mass of $m_0 \sim 700$ GeV. The unfortunate conclusion is that the rate is far too small to provide the observed flux for any realistic partial wave l. The rate naturally increases with a much steeper inner dark matter halo profile [6], but this seems difficult to reconcile with astrophysical evidence about the dark matter density in the inner regions of spiral galaxies, and the Milky Way in particular.

Positron emission in WIMP-nucleus recombination:
The excitation of WIMPs through collisions with nuclei in the ISM suffer from the same constraints as in the previous section, as the abundances of very heavy nuclei with masses in excess of 100 GeV are even lower than for dark matter. However, a new mechanism arises involving the 'recombination' of *light* nuclei such as C, N, and O in the ISM with WIMPs which may happen if there are charged states χ_2^+ and χ_2^- near in mass to the WIMP state χ_1^0. In particular, if χ_1 and χ_2 are respectively bosons and fermions or vice versa, electromagnetic processes such as that in Fig. 1 become possible. In particular, the charged states are rapidly depleted in the early universe via processes such as $\chi_2^{\pm} \to \chi_1^0 + e^{\pm}$, while in the current epoch positrons may be produced by 'recombination' of the form $\chi_1^0 + N \to (N\chi_2^-) + e^+$, where N denotes a generic nucleus, and $(N\chi_2^-)$ is a stable bound state. For C, N and O, which have sizable abundances in the ISM, this process is possible if $0.511\ \mathrm{MeV} < m_{\chi_2^-} - m_{\chi_1^0} < \mathcal{O}(2-3)$ MeV The abundance of these nuclei in the galactic bulge is estimated to be similar to that of

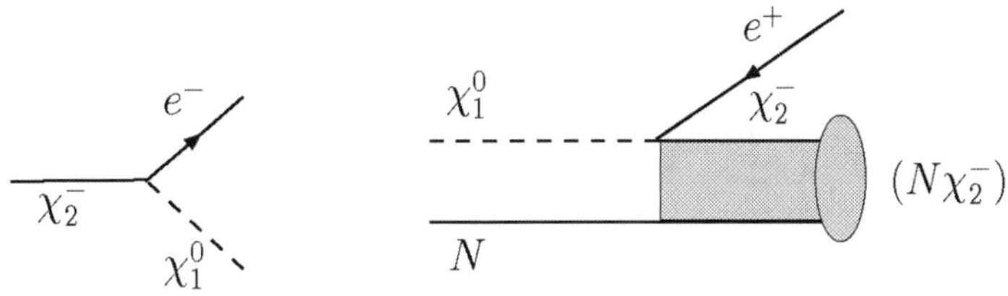

FIGURE 1. One the left, the rapid decay of the excited charged state χ_2^-, and on the right, the capture process of χ_1^0 forming the bound state $(N\chi_2^-)$ and producing a positron.

dark matter and so in this case, since the momentum of the nucleus will be an order of magnitude less than that of the WIMP, the unitarity bound on the cross-section is approximately consistent with that required to supply the observed photon flux.

However, while this appears promising, this scenario runs into severe constraints from limits on the terrestrial abundances of anomalous heavy isotopes. As one example, the relative abundance limit of $f_{\rm exp} \lesssim 10^{-20}$ on heavy isotopes of C [7], e.g. from production of $(N\chi_2^-)$ in the atmosphere, apparently (and unfortunately) restricts the allowed WIMP-nucleon cross-section to be several orders of magnitude below that required to explain the photon flux.

Delayed decays of excited WIMP states:

A third and somewhat different mechanism involves the delayed decay of an excited relic state χ_2. Long-lived charged relics are severely constrained by BBN and so we focus on the decay $\chi_2^0 \to \chi_1^0 + e^+ + e^-$, with once again an MeV-scale mass splitting. It is straightforward to relate the e^+e^- decay width required to produce the observed photon flux to the lifetime of the excited state. Given that $\Gamma_{\rm tot} \geq \Gamma_{e^+e^-}$, one obtains an allowed window on the lifetime of χ_2^0,

$$\text{few} \times 10^9\,\text{yr} \lesssim \tau_{\chi_2^0} \lesssim 10^{13}\,\text{yr}, \tag{2}$$

where the lower limit arises from the requirement that the relic number density of excited states in the present epoch be sufficiently large to fit the observed flux.

Given this window on the allowed lifetime, it is straightforward to construct a simple (albeit tuned) model realizing the required 511 keV flux, assuming the χ_i states interact with the Standard Model at tree-level only through the Higgs sector, $\mathscr{L}_{\rm int} = -\lambda_{ij} v h \chi_i^0 \chi_j^0$, with the λ_{12} coupling mediating a decay to e^+e^- via an off-shell Higgs. The Higgs-mediated decay has the advantage that for an MeV-scale momentum transfer the direct photon production is suppressed relative to e^+e^- by a loop factor, i.e. there is no Yukawa suppression in this momentum range.

For scalar WIMPs, the decay rate, assuming $\Delta m/m_1 \ll 1$ while still producing relativistic e^+e^- pairs, is given by

$$\Gamma_{e^+e^-}^{\rm scalar} = \frac{m_e^2\lambda^2}{384\pi^3 m_2}\left(\frac{\Delta m}{m_h}\right)^4. \tag{3}$$

For the model above this requires $\lambda = \lambda_{12} + \lambda_{21} \sim 10^{-6}$, and the diagonal couplings are fixed to ensure the correct relic abundance, $\lambda_{ii} \sim 10^{-2}$. While the resulting model can explain the required photon flux, one must bear in mind that more detailed features of the line-shape and in particular the highly localized nature of the source in the galactic core may be more problematic within a decaying dark matter scenario [5].

Recombination and Direct Detection

To summarize, we have analyzed a number of possible scenarios within which generic WIMP dark matter with MeV-scale degeneracies could provide the primary source of the galactic 511 keV INTEGRAL/SPI signal. Our conclusions for collisional excitation of WIMPs to a nearby excited state, and radiative capture by nuclei in the galactic interstellar medium, were essentially negative, albeit for different reasons, while the decay of a long-lived relic may be feasible and deserves further study.

Moreover, the general scenario of near-degeneracies in the dark sector is motivated beyond the question of the 511 keV line, and it is also interesting to consider a perhaps more generic scenario with $\Delta m \sim \mathcal{O}(10)$ MeV in which recombination with heavy nuclei would still be possible. Such a scenario arises near the boundary of the WMAP strip region of the CMSSM, and can have important implications for underground detectors searching for the signal of a WIMP recoiling on a heavy nucleus. For pseudo-degenerate WIMPs, the cross-section will likely be dominated by inelastic recombination as considered above, and the large MeV-scale energy release will not easily be detected by existing setups even if the cross-section is above the current bound. Thus, this scenario appears to have qualitatively distinct signatures for direct detection which also appear worthy of further study [8].

ACKNOWLEDGMENTS

I would like to thank the organisers for their invitation to this stimulating meeting, Maxim Pospelov for enjoyable collaboration on the work presented here, and NSERC, Canada, for financial support.

REFERENCES

1. J. Knodlseder *et al.*, Astron. Astrophys. **411**, L457 (2003);
   P. Jean *et al.*, Astron. Astrophys. **407**, L55 (2003).
2. J. F. Beacom and H. Yuksel, Phys. Rev. Lett. **97**, 071102 (2006).
3. C. Boehm, D. Hooper, J. Silk, M. Casse and J. Paul, Phys. Rev. Lett. **92**, 101301 (2004).
4. M. Pospelov and A. Ritz, Phys. Lett. B **651**, 208 (2007).
5. Y. Ascasibar, P. Jean, C. Boehm and J. Knoedlseder, Mon. Not. Roy. Astron. Soc. **368**, 1695 (2006).
6. D. P. Finkbeiner and N. Weiner, arXiv:astro-ph/0702587.
7. T. K. Hemmick *et al.*, Phys. Rev. D **41**, 2074 (1990).
8. M. Pospelov and A. Ritz, to appear.

Beyond the Standard Model Searches at the Tevatron

G. Sajot, on behalf of the CDF and DØ Collaborations

LPSC 53 Avenue des Martyrs 38026 Grenoble France

Abstract. Recent searches for non-SUSY exotics in $p\bar{p}$ collisions at a center-of-mass energy of 1.96 TeV at the Tevatron Run II are reported. The emphasis is put on the results of model-driven analyses which were updated to the full Run IIa datasets corresponding to integrated luminosities of about 1 fb^{-1}.

Keywords: Krein space, Negative-norm
PACS: 04.62.+v, 03.70+k, 11.10.Cd, 98.80.H

INTRODUCTION

Numerous searches for extensions of the Standard Model (SM) are conducted at the Tevatron Run II. In this report we concentrate on the model-driven analyses that were recently updated by CDF and D0 collaborations to the full Run IIa datasets, representing integrated luminosities of about 1 fb^{-1}. Details about these analyses can be found in reference [1]

The SM is constructed with the following ingredients : it's a quantum field theory where the matter fields are replicated into three families of quarks and leptons. This field theory is placed into a four-dimensional space-time. Its Lagrangian is invariant under the Poincaré group and under the $SU_C(3) \times SU_L(2) \times U_Y(1)$ gauge group. The electroweak symmetry breaking is provided by the Higgs mechanism. Up to now, all the experimental tests of the SM have shown no significant deviations with respect to its predictions. However the SM leaves many questions unresolved and is clearly not a full and a satisfactory theory. Therefore many theoretical ideas have been proposed to extend it. Among these ideas is a possible sub-structure of the particles considered as elementary in the SM. This could explain the replication of the quarks and leptons into three families.

Another path is a possible extension of the SM gauge symmetries. This enables to envisage a unification of the three fundamental interactions described by the SM at a very high energy scale, whilst explaining their differences at low energy as results of different symmetry breakings. One can also postulate the existence of extra space dimensions that could explain the hierarchy between the Planck and the electroweak scales as well as the relative weakness of the gravitational interaction with respect to the three other fundamental interactions.

Searches for fermions sub-structure are reported in sections II and III, section IV is devoted to a search for an additional quark generation and sections V and VI contain searches for hints of extended gauge symmetries and of extra space dimensions respec-

CP957, *Particles, Strings, and Cosmology, 13th International Symposium*
edited by A. Rajantie, C. Contaldi, P. Dauncey, and H. Stoica

tively. All the exclusion limits in this note are given at the 95% confidence level.

SEARCH FOR LEPTOQUARKS

As predicted by numerous extensions of the standard model, leptoquarks (LQ) are hypothetical bosons allowing lepton-quark transitions [2]. In hadron collisions, the pair production of scalar leptoquarks (LQ) is a pure QCD process : $q\bar{q} \to LQ\overline{LQ}$. Thus the cross section depends on the LQ mass and not on the unknown coupling (λ) between the LQ and its associated lepton and quark (if one neglects the contribution due to the t-channel lepton exchange which is very small compared with the theoretical uncertainties on the cross section).

D0 performed a search for second generation scalar LQ in the $LQ_2 + LQ_2 \to \mu\ \nu\ q'\ \bar{q}$ channel, in a dataset of $\int Ldt = 1.05 fb^{-1}$. Events containing a high p_T (> 20 GeV) and isolated muon plus jets and large missing transverse energy ($\not{E}_T > 30$ GeV) are selected. The left hand side of Figure 1 shows the LQ_2 reconstructed mass, i.e. the invariant mass of the muon-jet which is closer to the assumed leptoquark mass (and not differing from it by more than 100 GeV). On this plot, the leptoquark sample is generated with $M_{LQ} = 200$ GeV and $Br(LQ \to \mu\ q) = 0.5$. Clearly, the data are background-like. A limit excluding $M_{LQ_2} < 210$ GeV is set in the hypothesis of this semi-leptonic decay of the LQ_2 pairs.

SEARCH FOR EXCITED ELECTRONS

For the hypothesis of quarks and leptons compositeness as detailed in reference [3], the main parameters are the excited fermion mass M_f^* and the compositeness scale Λ. We report a search for excited electrons (e^*) carried by D0 with an integrated luminosity of $1.0\ fb^{-1}$. It is assumed that the excited electrons are produced via the contact interaction process $q\bar{q} \to e^{*\pm}e^{\mp}$ and that the $e^{*\pm}$ subsequently decays into the $\gamma + e^{\pm}$ mode. This leads to $\gamma e^{\pm}e^{\mp}$ final states. The analysis essentially consists in selecting events with two high p_T (> 35 GeV and > 15 GeV) and isolated electrons plus an energetic ($E_T > 15$ GeV) and isolated photon and to search for a resonance in the $M_{\gamma e^{\pm}}$ distribution. Since no excess of data is found with respect to the SM background, exclusion limits are derived in the Λ and $M_{e^{*\pm}}$ plane, as shown on right hand side of Figure 1. For example, for Λ = 1 TeV, e^* masses below 756 GeV are excluded.

SEARCH FOR 4TH GENERATION QUARK b'

CDF searched for new particles that lead to a Z boson plus jets in the final state in a data sample with an integrated luminosity of $1.06\ fb^{-1}$. Z boson decays to ee and $\mu\mu$ are considered. A completely data-based method has been developed to predict the dominant background from standard-model Z+jet events. No significant excess above the background prediction is observed, and a limit is set using a fourth generation quark model to quantify the acceptance. Assuming $BR(b' \to bZ) = 100\%$ and using a leading-

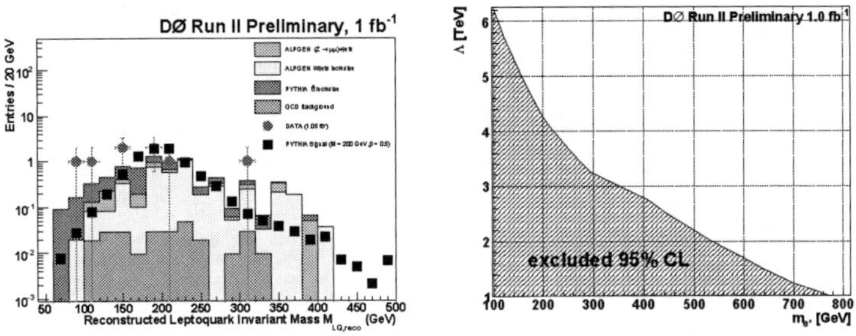

FIGURE 1. D0 collaboration : distribution of the reconstructed LQ_2 invariant mass (M_{LQ_2}) in a $LQ_2 + \bar{LQ}_2 \rightarrow \mu \nu q q'$ search (left) and excluded domain in the plane Λ (compositeness scale) versus m_{e*} by a search for excited electrons (right).

order calculation of the b' cross section, b' quark masses below 268 GeV are excluded (see Figure 2).

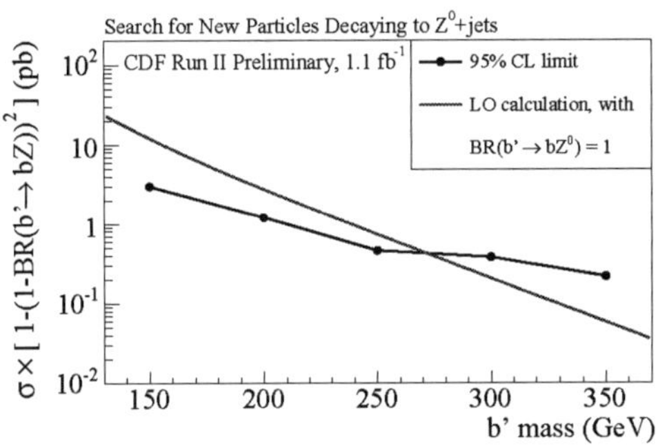

FIGURE 2. CDF search for b' : cross section limit vs b' mass.

EXTENDED GAUGE SYMMETRY

If there exists a grand unification of the strong and the electroweak interactions at a high energy scale, then the breakdown of the corresponding gauge group (i.e. SU(5), SO(10), E6,...) down to the SM gauge group occurs through a cascade of symmetry breakings where extra SU(2) and U(1) factors may appear. Such extra gauge group factors predict the existence of new and heavy W and Z bosons denoted W' and Z'.

We present a D0 search for the $W' \rightarrow e^{\pm} \nu$ process in a dataset of $\int L dt = 0.9\ fb^{-1}$. The events are selected if they contain a high E_T (> 30 GeV) and isolated electron and a significant missing transverse energy ($\not{E}_T > 30$ GeV). The transverse mass $M_T(e^{\pm}, \not{E}_T)$ distribution displayed on the left hand part of Figure 3 is scrutinized especially above 150 GeV. No data excess is found on top of the SM background tail. Therefore a W'

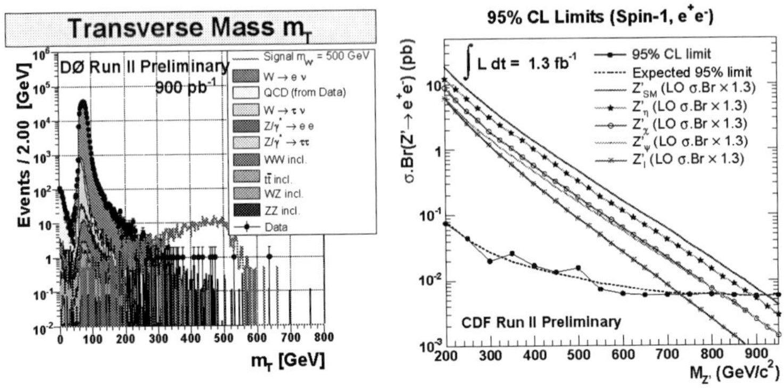

FIGURE 3. Searches for new heavy gauge bosons : W' by D0 (left) and Z' by CDF (right)

with a mass below 965 GeV is excluded.

We present a CDF search for the $Z' \to e^+e^-$ process using a data sample of $\int Ldt = 1.29\ fb^{-1}$. The analysis selects events with two high E_T (> 25 GeV) and isolated electrons with at least one in the central part of the calorimeter and with a matching track. Here the signal region is defined as the tail (above 150 GeV) of the dielectron invariant mass. The data are in good agreement with the SM background causing a Z' with a mass lower than 923 GeV and SM-like couplings to be excluded (see right hand part of Figure 3).

EXTENDED NUMBER OF SPACE DIMENSIONS

The lack of precision measurements of gravity in the sub-millimeter domain leaves some room for possible departures from the Newton's law in this distance range. Such exotic behaviors are obviously predicted in the assumption that space has more than three dimensions. Models postulating the existence of extra spatial dimensions have been proposed to solve the hierarchy problem posed by the large difference between the Planck scale $M_{Planck} \approx 10^{16}$ TeV, at which gravity is expected to become strong, and the scale of electroweak symmetry breaking ≈ 1 TeV.

One such model by Randall and Sundrum (RS) [4] localizes gravity on a (3+1)-dimensional brane, the Planck brane, that is separated from the standard model brane in a fifth dimension with warped metric. Gravity lives on the second brane where it isn't weak, but can propagate along the fifth dimension. The gravitons appear as towers of Kaluza-Klein excitations with masses and widths determined by the parameters of the model. These parameters can be expressed in terms of the mass of the first excited mode of the graviton, M_1, and the dimensionless coupling to the standard model fields $\kappa\sqrt{8\pi}/M_{Planck}$.

The Arkani-Hamed, Dimopoulos and Dvali (ADD) model [5] also localizes the SM fields on one brane and allows gravity to propagate within a bulk possibly made of up to

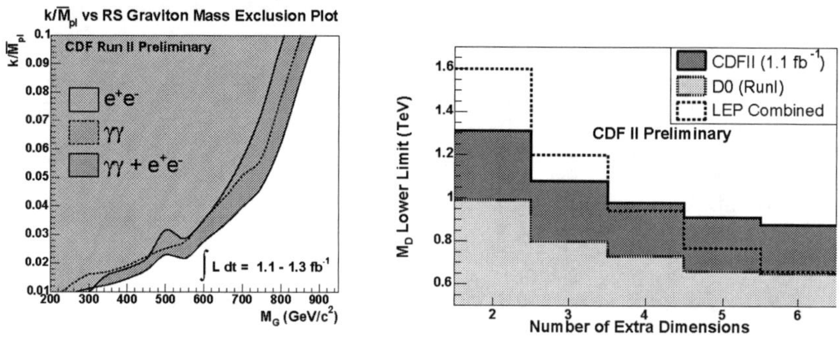

FIGURE 4. CDF searches for extended number of space dimensions in the Randall-Sandrum (left) and the Arkani-Hamed, Dimopoulos and Dvali (right) models.

N = 7 large extra dimensions. Here Kaluza-Klein excitations of the gravitons cannot be resolved. At the Tevatron, gravitons can be produced recoiling against a quark or gluon jet. The large number of kinematically accessible states compensating for the small gravitational coupling to give a sizable cross section. The parameters are the number of extra dimensions N and the effective Planck scale M_D (i.e. the Planck scale in 4+N dimensions).

CDF has searched for RS gravitons in the two decay channels $G \rightarrow e^+e^-$ and $G \rightarrow \gamma\gamma$. The individual exclusion domain and the combination of the two are displayed at the left hand side of Figure 4. We also report on a CDF search for ADD gravitons in a $1.1 fb^{-1}$ dataset. In these LED models, gravitons that are produced directly in processes such as $q\bar{q} \rightarrow gG$, $qg \rightarrow qG$, or $gg \rightarrow gG$ escape undetected in the bulk of extra dimensions, leaving the final state quark or gluon to produce a single jet. Hence the analysis requires events with a very high E_T jet (> 150 GeV) contained in the central part of the calorimeter and confirmed by tracks. In order to allow for a gluon ISR or FSR a second soft jet (E_T jet < 60 GeV) is accepted. The discriminating variable is the $\not{E}_T$, which is compatible with the expected background. Consequently exclusion limits are derived as a function of the number of extra dimensions and the effective Planck scale, M_D, as shown on the right hand side of Figure 4.

CONCLUSION

Many searches covering very different topologies have been studied at the Tevatron Run II by the CDF and D0 collaborations. Despite the recent updates of some of these analyses to the full Run IIa datasets of about $1 fb^{-1}$, no hints of exotic extensions of the SM have been found and more stringent exclusion limits have been derived.

REFERENCES

1. http://www-d0.fnal.gov/Run2Physics/WWW/results/np.html
   http://www-cdf.fnal.gov/physics/exotics/exotic.html
2. J.C. Pati and A. Salam, Phys. Rev. **D10**, 275 (1974)
   H. Georgi and S.L. Glashow, Phys. Rev. Lett. **32**, 438 (1974)
   W. Buchmuller and D. Wyler, Phys. Lett. **B177**, 377 (1986)
   E. Eichten, K.D. Lane, and M.E. Peskin, Phys. Rev. Lett. **50** 811 (1983)
   E. Eichtein, I. Hinchliffe, K.D. Kane, and, C. Quigg, Phys. Rev. **D34**, 1547 (1986)
3. U. Baur et al., Phys. Rev. **D42**, 815 (1990)
4. L. Randall and R; Sundrum, Phys. Rev. Lett. **83**, 3370 (1999)
5. N. Arkani-Hamed, S. Dimopoulos, and D. Dvali, Phys. Lett. **B429**, 23 (1998)

NMSSM neutralino dark matter

Ana M. Teixeira

Laboratoire de Physique Théorique, UMR 8627, Université de Paris-Sud XI, Bâtiment 201, F-91405 Orsay Cedex, France

Abstract. We study the viability of the lightest neutralino as a dark matter candidate in the Next-to-Minimal Supersymmetric Standard Model. Taking into account accelerator constraints as well as bounds on low-energy observables, and imposing consistency with present bounds on the neutralino relic density, we address the prospects for the direct detection of neutralino dark matter. We find regions of the allowed parameter space where the neutralino detection cross section is within the reach of dark matter detectors, essentially owing to the presence of very light singlet-like Higgses, and to either singlino dominated or very light neutralinos.

Keywords: Dark Matter, Supersymmetry Phenomenology, Cosmology of Theories beyond the SM
PACS: 12.60.Jv, 95.35.+d

The Next-to-Minimal Supersymmetric Standard Model (NMSSM) is a well-motivated extension of the Minimal Supersymmetric Standard Model (MSSM) by a singlet superfield $\hat{S}$. In addition to providing an elegant solution to the so-called μ problem of the MSSM, and rendering the Higgs "little fine tuning problem" of the MSSM less severe, the presence of additional fields (CP-even, CP-odd neutral Higgs bosons, and a fifth neutralino) leads to a richer and more complex phenomenology. In the NMSSM, one has possibility of dark matter scenarios that can be very different from those encountered in the MSSM, both regarding the relic density as well as prospects for direct detection. In particular, the exchange of very light Higgses can lead to large direct detection cross sections, within the reach of the present generation of dark matter detectors [1, 2].

The NMSSM superpotential contains new couplings, involving the Higgs doublets and the singlet,

$$W_{\text{NMSSM}} = \varepsilon_{ij} \left(Y_u \hat{H}_2^j \hat{Q}^i \hat{u} + Y_d \hat{H}_1^i \hat{Q}^j \hat{d} + Y_e \hat{H}_1^i \hat{L}^j \hat{e} \right) - \varepsilon_{ij} \lambda \hat{S} \hat{H}_1^i \hat{H}_2^j + \frac{1}{3} \kappa \hat{S}^3 . \quad (1)$$

New terms associated with additional soft supersymmetry (SUSY) breaking trilinear couplings A_λ and A_κ will also appear in the Lagrangian. After the spontaneous breaking of electroweak (EW) symmetry, the neutral Higgs scalars develop vacuum expectation values (VEVs), $\langle H_1^0 \rangle = v_1$, $\langle H_2^0 \rangle = v_2$ and $\langle S \rangle = s$. This leads to the dynamical generation of an effective interaction $\mu \hat{H}_1 \hat{H}_2$, with $\mu \equiv \lambda s$. In the NMSSM spectrum, we now have three CP-even and two CP-odd Higgs states. In particular, the lightest Higgs scalar can be written as $h_1^0 = S_{11} H_1^0 + S_{12} H_2^0 + S_{13} S$, where S is the unitary matrix that diagonalises the 3×3 scalar Higgs mass matrix. In the neutralino sector, the singlino mixes with the bino, wino and Higgsinos. The lightest state can be now expressed as $\tilde{\chi}_1^0 = N_{11} \tilde{B}^0 + N_{12} \tilde{W}_3^0 + N_{13} \tilde{H}_1^0 + N_{14} \tilde{H}_2^0 + N_{15} \tilde{S}$, where N diagonalises the 5×5 neutralino mass matrix.

CP957, *Particles, Strings, and Cosmology, 13th International Symposium*
edited by A. Rajantie, C. Contaldi, P. Dauncey, and H. Stoica

The low-energy NMSSM parameter space can be described in terms of the $\lambda, \kappa, \tan\beta, \mu, A_\lambda, A_\kappa$ degrees of freedom, as well as the soft SUSY-breaking terms, namely gaugino masses, $M_{1,2,3}$. A thorough analysis of the low-energy NMSSM phenomenology (minimisation of the potential, computation of spectrum and compatibility with LEP/Tevatron bounds) can be obtained using the NMHDECAY 2.0 code [3]. Additionally, we have also included in our analysis [2] a more precise computation of the $b \to s\gamma$ decay in the NMSSM [4], taking into account next-to-leading order contributions, and imposing consistency at the 2σ level with the experimental central value [5], $\mathrm{BR}^{\exp}(b \to s\gamma) = (3.55 \pm 0.27) \times 10^{-4}$. Likewise, we have also incorporated the constraints coming from the contribution of a light pseudoscalar a^0 in the NMSSM to the rare B- and K-meson decays [4]. Finally, in our analysis we have also included the constraints coming from the SUSY contributions to the muon anomalous magnetic moment, $a_\mu = (g_\mu - 2)$. At present, the observed excess in $a_\mu^{\exp}$ [6] constrains a possible SUSY contribution to be [7] $a_\mu^{\mathrm{SUSY}} = (27.6 \pm 8) \times 10^{-10}$.

As thoroughly discussed in [1, 2], an extensive part of the low-energy NMSSM parameter space is directly excluded on theoretical grounds, namely the occurrence of tachyons. Moreover, false minima, and Landau poles for the couplings in W_{NMSSM} also render unviable further areas. In regions surviving the latter constraints and the LEP/Tevatron bounds, and that present the most appealing prospects regarding dark matter direct detection, an important role is played by the $b \to s\gamma$ decay, which can in principle exclude important regions of the parameter space (due to very large contributions to $\mathrm{BR}(b \to s\gamma)$ from charged Higgs exchange). Concerning the SUSY contributions to a_μ, in the dark-matter interesting regions, these tend to be in general quite small. A sufficiently large a_μ^{SUSY} can nevertheless be obtained when slepton (and gaugino) masses are decreased, in association with large values of the slepton trilinear couplings.

In order to be a good dark matter candidate, the lightest NMSSM neutralino must also comply with the increasingly stringent bounds on its relic density,

$$0.1 \lesssim \Omega h^2 \lesssim 0.3 \text{ (astrophysical)}, \qquad 0.095 \lesssim \Omega h^2 \lesssim 0.112 \text{ (WMAP)}, \tag{2}$$

respectively arising from astrophysical constraints [8], and from taking into account the recent three years data from the WMAP satellite [9].

Compared to the MSSM, there are several alterations leading to $\Omega_{\tilde{\chi}_1^0} h^2$: first, the possibility of a singlino-like lightest supersymmetric particle (LSP), associated with new couplings in the interaction Lagrangian, may favour the coupling of WIMPs to a singlet-like Higgs, whose mass can be substantially lighter than in the MSSM, given the more relaxed experimental constraints. Secondly, in the NMSSM we have new open channels for neutralino annihilation, e.g. s-channel resonances, new channels with annihilation into Zh_1^0, $h_1^0 h_1^0$, $h_1^0 a_1^0$ and $a_1^0 a_1^0$ (due to light h_1^0 and a_1^0 states), providing important contributions to the annihilation and co-annihilation cross-sections [10]. In our analysis [2], the results for the neutralino relic density, obtained from an NMHDECAY link to MicrOMEGAS [10].

In the regions of the parameter space likely to have large neutralino detection cross sections [1], we have found that, in general, the correct relic density can only be obtained when either the singlino composition of the neutralino is large enough or when the annihilation channels into Z, W, or h_1^0 are kinematically forbidden. Interestingly, some allowed areas were very close to the tachyonic border, which as we verified, can give rise to very large direct detection cross sections.

As pointed out in [1], the existence of a fifth neutralino state, together with the presence of new terms in the Higgs-neutralino-neutralino interaction (which are proportional to λ and κ), trigger new contributions to the spin-independent part of the neutralino-nucleon cross section, $\sigma_{\tilde{\chi}_1^0-p}$. On the one hand, although the term associated with the s-channel squark exchange is formally identical to the MSSM case, it can be significantly reduced if the lightest neutralino has a major singlino composition. On the other hand, and more importantly, the dominant contribution to $\sigma_{\tilde{\chi}_1^0-p}$, associated to the exchange of CP-even Higgs bosons on the t-channel, can be largely enhanced when these are very light. Consequently, large detection cross sections can be obtained, even within the reach of the present generation of dark matter detectors.

As illustrative examples of our analysis, let us consider two cases, which represent the most relevant features of NMSSM dark matter scenarios. Let us begin by considering $M_1 = 160$ GeV, $A_\lambda = 400$ GeV, $A_\kappa = -200$ GeV, and $\mu = 130$ GeV, with $\tan\beta = 5$, leading to a parameter space consistent with bounds on $a_\mu^{\rm SUSY}$ and BR($b \to s\gamma$). In regions of the parameter space where the neutralino is relatively heavy and has a mixed bino-Higgsino composition, the relic density is too small. As the neutralino mass decreases, some annihilation channels become kinematically forbidden, such as annihilation into a pair of Z or W bosons when $m_{\tilde{\chi}_1^0} < M_Z$ or $m_{\tilde{\chi}_1^0} < M_W$, respectively. Below these, the resulting relic density can be large enough to fulfil the WMAP constraint, which occurs for distinct regions in parameter space [2]. One of the allowed regions is close to the tachyonic area and exhibits very light singlet-like Higgses, potentially leading to large detection cross sections. This is indeed the case, as evidenced on the left-hand side of Fig. 1, where the theoretical predictions for $\sigma_{\tilde{\chi}_1^0-p}$ are plotted versus the lightest neutralino mass. The resulting $\sigma_{\tilde{\chi}_1^0-p}$ spans several orders of magnitude, but, remarkably, areas with $\sigma_{\tilde{\chi}_1^0-p} \gtrsim 10^{-7}$ pb are found. These correspond to the above mentioned regions of the parameter space with very light singlet-like Higgses ($25\,\text{GeV} \lesssim m_{h_1^0} \lesssim 50\,\text{GeV}$ with $S_{13}^2 \gtrsim 0.99$). The neutralino is a mixed singlino-Higgsino state ($N_{15}^2 \approx 0.35$) with mass around 75 GeV. The sensitivities of present and projected dark matter experiments are also depicted for comparison. On the right-hand side of Fig. 1 we show the resulting $\sigma_{\tilde{\chi}_1^0-p}$ when the neutralino composition is changed, namely when the Higgsino component is enhanced. In particular, we consider the choice $M_1 = 330$ GeV, $\tan\beta = 5$, $A_\lambda = 570$ GeV, $A_\kappa = -60$ GeV, with $\mu = 160$ GeV. Such neutralinos annihilate more efficiently, thus leading to a reduced $\Omega_{\tilde{\chi}_1^0} h^2$, so that the astrophysical constraint becomes more stringent. On the right-hand side of Fig. 1, the various resonances appear as funnels in the predicted $\sigma_{\tilde{\chi}_1^0-p}$ for the regions with the correct $\Omega_{\tilde{\chi}_1^0} h^2$ at the corresponding values of the neutralino mass ($m_{\tilde{\chi}_1^0} \approx M_Z/2$ and $m_{\tilde{\chi}_1^0} \approx m_{h_1^0}/2$). Below the resonance with the Z boson, light neutralinos are obtained $m_{\tilde{\chi}_1^0} \lesssim M_Z/2$ with a large singlino composition which have the correct relic abundance. The lightest Higgs is also singlet-like and very light, leading to a very large detection cross section, $\sigma_{\tilde{\chi}_1^0-p} \gtrsim 10^{-6}$ pb.

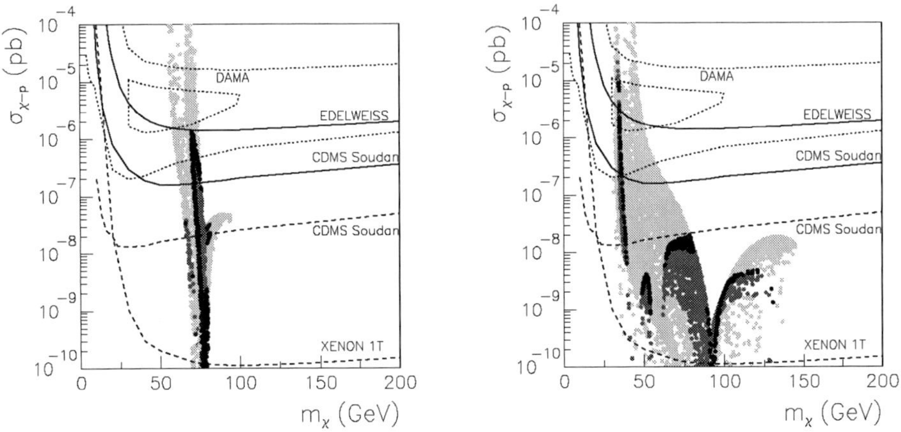

FIGURE 1. Scatter plot of the scalar neutralino-nucleon cross section as a function of the lightest neutralino mass. On the left, $M_1 = 160$ GeV, $\tan\beta = 5$, $A_\lambda = 400$ GeV, $A_\kappa = -200$ GeV, and $\mu = 130$ GeV. All the points represented are in agreement with LEP/Tevatron, $a_\mu^{\rm SUSY}$, and BR($b \to s\gamma$) bounds. Dark grey dots represent points which, in addition, fulfil $0.1 \leq \Omega_{\tilde{\chi}_1^0} h^2 \leq 0.3$, whereas black dots are those in agreement with the WMAP constraint. The sensitivities of present and projected experiments are also depicted, with solid and dashed lines, respectively. On the right, a different example with $M_1 = 330$ GeV, $\tan\beta = 5$, $A_\lambda = 570$ GeV, $A_\kappa = -60$ GeV, with $\mu = 160$ GeV, a case where the resulting $a_\mu^{\rm SUSY}$ is outside the experimental 2σ region.

ACKNOWLEDGMENTS

The work of A. M. Teixeira was supported by the French ANR project PHYS@COL&COS. The work reported here was based in collaborations with D. G. Cerdeño, E. Gabrielli, C. Hugonie, D. E. López-Fogliani and C. Muñoz.

REFERENCES

1. D. G. Cerdeño, C. Hugonie, D. E. López-Fogliani, C. Muñoz and A. M. Teixeira, JHEP **0412** (2004) 048 [arXiv:hep-ph/0408102].
2. D. G. Cerdeño, E. Gabrielli, D. E. López-Fogliani, C. Muñoz and A. M. Teixeira, JCAP **0706** (2007) 008 [arXiv:hep-ph/0701271].
3. U. Ellwanger and C. Hugonie, Comput. Phys. Commun. **175**, 290 (2006) [arXiv:hep-ph/0508022].
4. G. Hiller, Phys. Rev. D **70** (2004) 034018 [arXiv:hep-ph/0404220].
5. The Heavy Flavour Averaging Group, http://www.slac.stanford.edu/xorg/hfag/
6. G. W. Bennett et al. [Muon g-2 Collaboration], Phys. Rev. Lett. **92** (2004) 161802 [arXiv:hep-ex/0401008].
7. See, for example, K. Hagiwara, A. D. Martin, D. Nomura and T. Teubner, Phys. Lett. B **649** (2007) 173 [arXiv:hep-ph/0611102], and references therein.
8. For a review see, C. Muñoz, Int. J. Mod. Phys. A **19** (2004) 3093 [arXiv:hep-ph/0309346].
9. D. N. Spergel et al., arXiv:astro-ph/0603449.
10. G. Belanger, F. Boudjema, C. Hugonie, A. Pukhov and A. Semenov, JCAP **0509** (2005) 001 [arXiv:hep-ph/0505142].

Warped Throats at the LHC

Bret Underwood

Department of Physics, University of Wisconsin, Madison, WI 53706, USA

Abstract. We discuss how the detailed geometry of warped compactification can significantly impact collider data. We show how the details of the cross section for dilepton production (such as the spacing, the width, and the peaks of the KK graviton resonances) for a generic class of warped throats which are closely related to the Randall-Sundrum model depend sensitively on the precise shape of the warp factor.

Keywords: KK modes, LHC, string theory
PACS: 11.25.Wx, 11.10.Kk, 04.50.+h

INTRODUCTION

Extra dimensions show up in many different contexts in theories of physics beyond the standard model. In top down model building, extra dimensions typically appear in attempts at unification, as is seen with the unification of electromagnetism and gravity from Kaluza and Klein, and in a more modern context with string theory. In bottom up model building, extra dimensions typically show up in attempts to solve the hierarchy problem, as in the proposals of ADD [1] and RS [2].

Because extra dimensions show up in different contexts in different ways, it is important to understand how we can distinguish all of these different scenarios. At the LHC, mostly bottom up scenarios have been studied. It is now well understood that the large extra dimensions of ADD will lead to missing energy at colliders, and the warped space of RS will lead to the production of Kaluza-Klein resonances of the graviton or gluon [4]. It is interesting and timely to ask whether top down scenarios of extra dimensions can led to observable signatures, particularly at the LHC. Futher, different top down scenarios may lead to different signatures which may help us distinguish among the scenarios and learn important details about the shape and structure of the extra dimensions.

In this paper, we will discuss how RS and warped string theory models, which appear at first glance to be quite similar, in fact differ in important and observable ways at the LHC through the spectrum and couplings of the resonant KK particles produced. We will also discuss how small differences in the warped geometry, such as may arise in a broad class of models from string theory, can also lead to similar differences. This proceedings is based on work done in [6].

CP957, *Particles, Strings, and Cosmology, 13th International Symposium*
edited by A. Rajantie, C. Contaldi, P. Dauncey, and H. Stoica

WARPED THROATS

The RS warped space solution is given by a slice of AdS_5,

$$ds^2 = \left(\frac{r}{R}\right)^2 \hat{g}_{\mu\nu} dx^\mu dx^\nu + \left(\frac{R}{r}\right)^2 dr^2 \tag{1}$$

with explicit cutoffs in the IR ($r \ll R$) and UV ($r \sim R$) ends of the extra dimension r [2]. Typically the IR cutoff is set so that the fundamental 4D Planck scale in the IR is $\sim$ TeV, e.g. $r_{IR}/R \sim 10^{-16}$, thus providing a solution to the hierarchy problem.

Warped spaces are ubiquitous in type IIB string theory as local sources of energy density backreact on the geometry. In particular, a stack of D3 branes placed at some point in the extra dimensions produces a geometry in the near-horizon limit of $AdS_5 \times X^5$, where X^5 is some angular space (such as S^5 or $T^{(1,1)}$). However, these backgrounds do not quite mimic the RS scenario since, while there is a UV cutoff provided by the string scale, there is no cutoff in the IR, thus these backgrounds are not useful in solving the hierarchy problem.

However, in order to stabilize the moduli of the compact space one must turn on background p-form fluxes. These fluxes can also backreact on the geometry, and if their local density is large then they can also form a warped region of spacetime. The advantage to using fluxes to generate such a background is that they typically give you a smooth IR cutoff, realizing the RS scenario. The most explicit of these solutions is the Klebanov-Strassler solution [8], but other warped geometries with other angular isometries and different behavior near the IR "tip" exist [5].

In general, the metric for a 10D warped space can be written as [7]

$$ds_{10}^2 = f^{-1/2}(r)\hat{g}_{\mu\nu} dx^\mu dx^\nu + f^{1/2}(r)(dr^2 + ds_{X^5}^2). \tag{2}$$

The hierarchy is generated by the small value of the "warp factor" $f(r)$ at the "tip" of the throat where $r \to 0$, $f^{-1/4}(r \to 0) \ll 1$. Notice that if we ignore the angular space X^5 and write $f(r) = R^4/r^4$ we obtain the RS metric. A comparison of the warp factor of the RS and KS solutions is shown in Figure 1. Notice that the warp factors are approximately the same far from the tip, but differ significantly in the IR. In particular, the RS warp factor is cut off abruptly, while the KS warp factor smoothly flattens out. We will see later that this difference in behavior shows up in the masses and couplings of the KK modes.

We can also parameterize the warp factor more generally in the form of a "mass-gap" ansatz [9],

$$f(r) = \frac{R^4}{r_{tip}^4 + f_2 r^2 + r^4} \tag{3}$$

which parameterizes small differences of the behavior of the warp factor in the tip region. These small changes may arise from other types of geometries (see [5]), and we will see below that these small differences also show up as differences in the spectra of the KK graviton resonances.

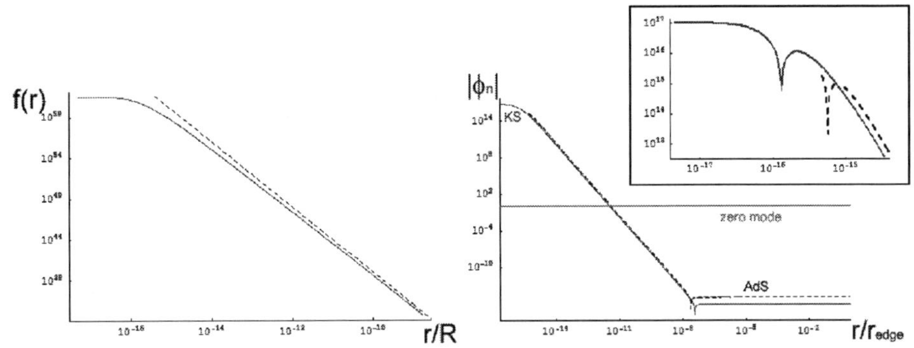

FIGURE 1. [Left] KS (solid blue) and RS (dashed black) warp factors $f(r)$ as a function of the radial coordinate r. [Right] Normalized KK graviton wave functions for the zero mode (solid red) and for the first KK mode of the AdS (dashed black) and KS (solid blue) geometries. Inset: The second KK modes for RS (black) and KS (blue).

WARPED KK SPECTRUM AND COUPLINGS

An important signature of warped spaces at the LHC is the production of resonances of KK gravitons [4]. The masses and eigenfunctions of the gravitons can be found by making an ansatz for the 4D graviton, $h_{\mu\nu}(x,r) = \overline{M}_p^{-1} \sum_n h_{\mu\nu}^{(n)}(x)\phi_n(r)$, and solving the associated Laplace equation on the warped space for the eigenvalues $m_n^2 h_{\mu\nu}^{(n)}(x) = \nabla_4^2 h_{\mu\nu}^{(n)}(x)$ and the eigenfunctions $\phi_n(r)$,

$$\partial_r \left(\sqrt{g_{X^5}}\partial_r\phi_n(r)\right) + m_n^2\sqrt{g_{X^5}}f(r)\phi_n(r) = 0\,. \tag{4}$$

The eigenfunctions and eigenvalues serve to give the masses and couplings of the KK gravitons as seen by an observer in the IR [10]. These differences clearly show up through the resonances in the $q\bar{q}, gg \to KK \to \ell^+\ell^-, t\bar{t}$ cross sections, as shown in Figure 2. In particular, one can see that in the comparison of KS and RS the KS resonances are more closely space, are broader, and the KS cross-section is overall much higher, resulting in an increased chance of observation at the LHC. The differences are even seen in the decay to pairs of tops, which is a more model-independent decay channel.

CONCLUSION

In summary, we have shown how subtle details in the warped geometry, in particular the small differences between the KS and RS models, can lead to large and observable differences in the spectrum and resonance signatures at the LHC. In particular, the KS resonances are broader and more strongly coupled, and should lead to stronger constraints on the warped geometry parameters than in RS. This interplay between the observation of the KK modes and details of the warped geometry is but one example of the broader connection between observational input and understanding the structure

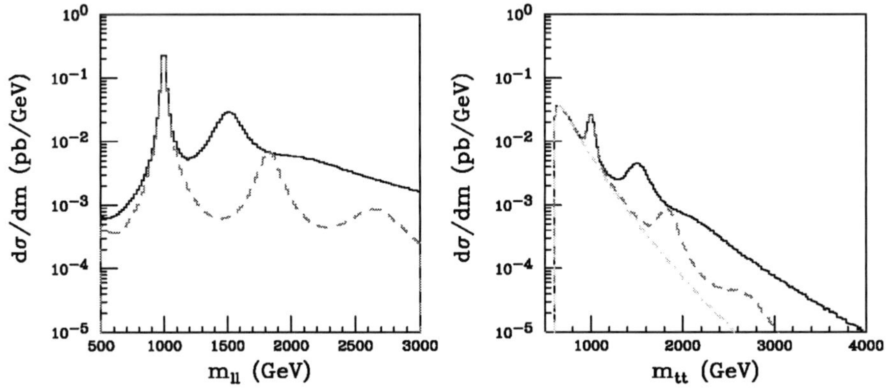

FIGURE 2. $\ell^+\ell^-$ (above) and $t\bar{t}$ (below) invariant mass distributions for KS (black solid) with hierarchy $r_{tip} = 2\times10^{-14}$ (masses and couplings given in Table I) and RS (dashed red) with hierarchy $r_{tip} = 8\times10^{-16}$ (masses given in Table I, with $\Lambda_{KK} = 2$ TeV). The SM backgound (dot-dashed green) is $Z,\gamma \to \ell^+\ell^-$ and gluon $\to t\bar{t}$, respectively. In generating these plots, we have assumed that all Standard Model fields are confined to the IR brane.

of high energy theory and geometry. We hope to investigate other promising directions further in the future.

ACKNOWLEDGMENTS

I would like to thank G. Shiu, D. Walker, and K. Zurek for their important role in this research and for many helpful and enlightening conversations. I would like to thank thank I. Hinchliffe, M. Shapiro, H. Tye for discussions. B.U. is supported in part by NSF CAREER Award No. PHY-0348093, DOE grant DE-FG-02-95ER40896, a Research Innovation Award and a Cottrell Scholar Award from Research Corporation.

REFERENCES

1. N. Arkani-Hamed, S. Dimopoulos and G. R. Dvali, Phys. Lett. B **429**, 263 (1998).
2. L. Randall and R. Sundrum, Phys. Rev. Lett. **83**, 3370 (1999); Phys. Rev. Lett. **83**, 4690 (1999).
3. G. Shiu, B. Underwood, Phys. Rev. Lett. **98** (2007) 051301.
4. For recent work, see e.g., A. L. Fitzpatrick, J. Kaplan, L. Randall, L.-T. Wang, arXiv:hep-ph/0701150; K. Agashe, H. Davoudiasl, G. Perez, A. Soni, arXiv:hep-ph/0701186.
5. A. Butti, M. Graña, R. Minasian, M. Petrini, A. Zaffaroni, JHEP 0503 (2005) **069**.
6. G. Shiu, B. Underwood, K. M. Zurek and D. G. E. Walker, arXiv:0705.4097 [hep-ph].
7. S. B. Giddings, S. Kachru and J. Polchinski, Phys. Rev. D **66**, 106006 (2002).
8. I. R. Klebanov, M. J. Strassler, JHEP **0008**, 052 (2000).
9. S. Kecskemeti, J. Maiden, G. Shiu, B. Underwood, JHEP **0609** (2006) 076.
10. T. Han, J. D. Lykken, R.-J. Zhang, Phys.Rev. **D59** (1999) 105006.

A Natural Nightmare for the LHC?

Thomas E.J. Underwood

Institute for Particle Physics Phenomenology (IPPP), University of Durham, DH1 3LE, UK.

Abstract. A minimal lepton number conserving extension to the Standard Model is considered providing light Dirac neutrinos without resorting to tiny Yukawa couplings. Successful baryogenesis through leptogenesis is not only possible in this case, but even suggests an electroweak scale vacuum expectation value for a gauge singlet scalar in the model. The spectrum contains two massive Higgs bosons and a massless Nambu-Goldstone boson. The existence of the Nambu-Goldstone boson suppresses the Higgs to $b\bar{b}$ branching ratio and instead Higgs bosons will decay mainly into invisible Goldstone bosons. We consider the constraints on the potential and the implications for the LHC.

Nambu-Goldstone bosons, first considered in 1960 [1], have several special properties. In particular, they are massless and they couple to the divergence of the current j^μ associated with a spontaneously broken symmetry [2]. This coupling has a strength inversely proportional to the scale of the symmetry breaking F such that

$$\mathscr{L}_{\text{int}} = \frac{1}{2F} \mathscr{J} \partial_\mu j^\mu , \tag{1}$$

where $\mathscr{J}$ is the Nambu-Goldstone boson (NGB) field. Suzuki and Shrock [3] first proposed that if the Standard Model Higgs boson were to mix with a new scalar field that was charged under a spontaneously broken global symmetry, then it would decay into a pair of NGBs if the scale F was close to the electroweak scale, $F \approx 100$ GeV. Interestingly, such an invisible Higgs decay could actually be searched for at colliders.

This possibility seems less exotic when one considers that mixing between the SM Higgs H and a new complex scalar field Φ, $\mathscr{L}_{\text{int}} = H^\dagger H \Phi^\dagger \Phi$, is one of very few renormalizable operators allowed which could link the SM to a new gauge singlet sector i.e. Φ is charged under a new global symmetry G_P but singlet under the SM gauge group. In this way the Higgs provides a portal into a hidden, or "phantom" sector [4].

Another example of such a portal is the neutrino Yukawa coupling which links SM fields with a gauge singlet right-handed neutrino. This interaction is so commonly invoked that it is now normally considered as part of the SM. In this way, the usual (Majorana) see-saw mechanism is a very simple "phantom sector". The Majorana see-saw mechanism even contains a broken global symmetry (lepton number) which in many extended models is spontaneously broken – leading to a NGB, the Majoron [5].

Stringent bounds can be placed on the coupling of massless particles to matter coming from considerations of energy loss in supernovae, stars and terrestrial collider experiments. For common NGBs such as Majorons, axions [6] and familions [7] these constraints generally mean that $F \gtrsim 10^9$ GeV. The underlying reason for such stringent bounds is that these NGBs couple to a current carried by quarks and/or charged leptons, e.g. lepton number in the case of the Majoron. If the NGB coupled to a current carried by gauge singlet matter (e.g. ν_R) then such stringent constraints on F would not apply.

CP957, *Particles, Strings, and Cosmology, 13th International Symposium*
edited by A. Rajantie, C. Contaldi, P. Dauncey, and H. Stoica

DIRAC NEUTRINO MASSES AND BARYOGENESIS

Neutrinos are not necessarily Majorana particles. As yet, we have no firm evidence of neutrinoless double beta decay, a characteristic signature of the lepton number violation associated with Majorana neutrinos. Along the same lines as the Majorana see-saw mechanism, an operator generating naturally small, Dirac, neutrino masses is

$$\mathscr{L}_\nu = \frac{(\overline{L}\cdot\tilde{H})\,(\Phi\cdot\nu_R)}{\Lambda}\,, \tag{2}$$

where L is the usual $\mathrm{SU}(2)_L$ lepton doublet, $\tilde{H} \equiv i\,\sigma_2 H$ is the SM Higgs doublet, ν_R is a gauge singlet right-handed neutrino and Φ is a new gauge singlet, complex scalar field.

In a model generating eq. (2), some global symmetry G_P carried by only by the gauge singlets Φ and ν_R prevents the neutrinos from acquiring masses via a $\overline{L}\cdot\tilde{H}\nu_R$ term in the Lagrangian. After the spontaneous breakdown of G_P and the electroweak symmetry, eq. (2) results in naturally small Dirac neutrino masses if the vacuum expectation values (VEVs) $\langle\Phi\rangle \approx \langle H\rangle \approx 100$ GeV provided that $\Lambda \sim 10^{16}$ GeV. A model generating the operator (2) was first considered by Roncadelli and Wyler [8].

Although lepton number is conserved in this model, it was recently shown [9] that the model could lead to successful baryogenesis via (Dirac) leptogenesis [10]. This is possible because the model of [8] contains heavy Dirac states, S and $\bar{S}$ much like the heavy, but Majorana, right-handed neutrinos of the usual see-saw. In the early Universe the S particles decay into neutrinos and Higgs scalars. Significantly, CP can be violated in this decay process. After the S have decayed, no reaction can take place quickly enough to bring the left and right-handed neutrinos into equilibrium – this is related to the smallness of the light neutrino masses. Matter/antimatter asymmetries in the left and right-handed neutrino sectors, produced during the CP-violating decays of the S, cannot equilibrate and a net lepton asymmetry remains amongst the left-handed $\mathrm{SU}(2)_L$ doublet leptons. Rapid $B+L$ violating processes in the early Universe, which are insensitive to asymmetries in the gauge singlet ν_R, convert this lepton asymmetry into a baryon asymmetry.

Under reasonable assumptions, one can derive approximate limits on the VEV of the singlet scalar field Φ [9], $0.1\ \mathrm{GeV} \lesssim \langle\Phi\rangle \lesssim \frac{2\ \mathrm{TeV}}{T_{RH}}$, where T_{RH} is the reheating temperature of the Universe after inflation.

HIGGS PHENOMENOLOGY

In models containing a singlet scalar Φ charged under a new global symmetry, e.g. $G_P = U(1)_P$, nothing prevents the term, $\mathscr{L}_{\mathrm{link}} = \eta\, H^\dagger H\, \Phi^*\Phi$, from appearing in the Lagrangian, where η is a new dimensionless coupling. After the spontaneous breakdown of $\mathrm{U}(1)_P$ and electroweak symmetry the two massive Higgs bosons in the model mix. The spectrum also contains a massless NGB $\mathscr{J}$ associated with the breakdown of $\mathrm{U}(1)_P$. The Higgs bosons will couple to this NGB [3] as

$$\mathscr{L}_{\mathrm{int}} = \frac{1}{2\langle\Phi\rangle}\,\mathscr{J}\,\partial_\mu j^\mu \;\to\; \frac{m_{H_i}}{\langle\Phi\rangle}\,O_{2i}\,H_i\,\mathscr{J}\,\mathscr{J}\,, \tag{3}$$

where O_{2i} is a mixing matrix element parameterizing the mixing of the massive H_i. Hence, the massive Higgs bosons H_i will decay into the massless and invisible $\mathscr{J}$ [11]. For Higgs masses $m_{H_i} \lesssim 130$ GeV the dominant decay mode of the Standard Model Higgs is $H \to b\bar{b}$. Comparing the rates $\Gamma(H_1 \to b\bar{b})$ and $\Gamma(H_1 \to \mathscr{J}\,\mathscr{J})$ it can be shown that for 20 GeV $\lesssim m_{H_1} \lesssim 130$ GeV the Higgs will dominantly decay into invisible $\mathscr{J}\,\mathscr{J}$.

LHC (and LEP) Higgs phenomenology is influenced by the number of visible Higgs decay events seen as compared to the SM expected value. This is quantified by the parameter $\mathscr{R}^2$ defined as

$$\mathscr{R}_i^2 \equiv \frac{\sigma(pp \to H_i X)\,\mathrm{Br}(H_i \to YY)}{\sigma(pp \to h_{\mathrm{SM}} X)\,\mathrm{Br}(h_{\mathrm{SM}} \to YY)}, \quad \mathscr{T}_i^2 \equiv \frac{\sigma(pp \to H_i X)}{\sigma(pp \to h_{\mathrm{SM}} X)}\mathrm{Br}(H_i \to \mathscr{J}\,\mathscr{J}), \tag{4}$$

where YY is a visible final state such as $b\bar{b}$ or $\gamma\gamma$, and $\mathscr{T}^2$ is the analogous parameter for invisible decays.

Looking to the future, both the ATLAS and CMS experiments have performed detailed studies exploring the discovery potential of their detectors in cases where Higgs bosons decay to visible final states [12] and also invisible final states [13]. Considering $\mathscr{L} = 30\ \mathrm{fb}^{-1}$ of LHC integrated luminosity, it can be estimated that it would be difficult to discover a visibly decaying Higgs if $\mathscr{R}_i \lesssim 0.2$. Furthermore, ATLAS studies [13] indicate that with the same amount of integrated luminosity, invisible Higgs bosons with $m_{H_i} \lesssim 200$ GeV could be excluded only if $\mathscr{T}_i^2 \gtrsim 0.3$.

Figure 1 (left) shows the areas where either $\mathscr{R}_i^2 \geq 0.3$ or $\mathscr{T}_i^2 \geq 0.3$ in the m_{H_1} vs. m_{H_2} m_{H_1} plane. The plot assumes maximal mixing and $\langle\Phi\rangle = \langle H\rangle$. It is clear that a "nightmare" region remains where no Higgs bosons are accessible to the LHC if experiments do not have the sensitivity to see into areas where $\mathscr{R}_i^2 \leq 0.3$ and $\mathscr{T}_i^2 \leq 0.3$.

TRIVIALITY AND VACUUM STABILITY

The potential of the model being considered reads

$$V = \mu_H^2 H^\dagger H + \mu_\Phi^2 \Phi^\dagger\Phi + \lambda_H (H^\dagger H)^2 + \lambda_\Phi (\Phi^*\Phi)^2 \quad \eta H^\dagger H \Phi^*\Phi. \tag{5}$$

There are two classic constraints regarding this potential; triviality and vacuum stability. The triviality constraint is essentially the requirement that the couplings λ_H, λ_Φ and η stay perturbative up to a certain scale $\Lambda_T \gg \langle H\rangle$. Demanding the vacuum is stable leads to the requirement that the potential is bounded from below, at least up to a scale $\Lambda_V \gg \langle H\rangle$. The vacuum stability bound can be reduced to the requirement that $4\lambda_H(Q)\,\lambda_\Phi(Q) > \eta(Q)^2$, at all scales $Q \lesssim \Lambda_V$.

The running parameters, defined at a scale $Q_0 = M_Z$ can be evolved up to higher scales with 1-loop renormalization group equations [14]

$$16\pi^2 \tfrac{d\lambda_H}{dt} = \eta^2 + 24\lambda_H^2 + 12\lambda Y_t^2 \quad 6Y_t^4 \quad 3\lambda(3g_2^2 + g'^2) + \tfrac{3}{8}\left[2g_2^4 + (g_2^2 + g'^2)^2\right],$$

$$16\pi^2 \tfrac{d\eta}{dt} = \eta\left[12\lambda_H + 8\lambda_\Phi \quad 4\eta + 6Y_t \quad \tfrac{3}{2}(3g_2^2 + g'^2)\right], \quad 16\pi^2 \tfrac{d\lambda_\Phi}{dt} = 2\eta^2 + 20\lambda_\Phi^2,$$

where $t \equiv \ln Q/Q_0$, g' and g_2 are respectively the $\mathrm{U}(1)_Y$ and $\mathrm{SU}(2)_L$ gauge couplings and Y_t is the top quark Yukawa coupling.

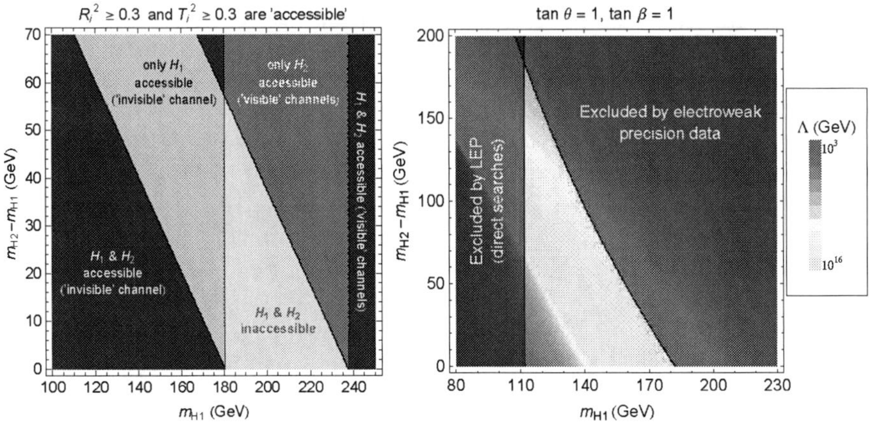

FIGURE 1. *The m_{H_1} vs. $m_{H_2} - m_{H_1}$ plane for $\tan\beta = 1$ and $\tan\theta = 1$. The left panel shows where different Higgs bosons are accessible. We define that a given H_i is accessible if either $\mathscr{R}_i^2 \geq 0.3$ or $\mathscr{T}_i^2 \geq 0.3$. In the dark (blue) regions both Higgs bosons are accessible. In the white (beige) region no Higgs bosons are accessible. The right panel shows the expected cut-off Λ, of the effective theory taking the triviality and positivity of the potential into account (the lower of either Λ_T or Λ_V is shown). The curved line shows the 95% C.L. upper limit on the Higgs masses coming from precision electroweak data (see [9]).*

Figure 1 (right) shows the m_{H1} vs. $m_{H2}\ \ m_{H1}$ plane assuming $\langle\Phi\rangle = \langle H\rangle$ and maximal mixing, where the background colours show the scale of new physics Λ required either by positivity of the potential or triviality (whichever is lower). The plots can be compared to see the correspondence between easily accessible regions at the LHC and regions with a potentially high effective theory cut-off.

I would like to thank Sakis Dedes, Frank Krauss, Terrance Figy and David Cerdeno for collaboration.

REFERENCES

1. Y. Nambu, Phys. Rev. Lett. **4**, 380 (1960); J. Goldstone, Nuovo Cim. **19** (1961) 154.
2. J. Goldstone, A. Salam and S. Weinberg, Phys. Rev. **127**, 965 (1962).
3. R. E. Shrock and M. Suzuki, Phys. Lett. B **110**, 250 (1982).
4. B. Patt and F. Wilczek, arXiv:hep-ph/0605188.
5. Y. Chikashige, R. N. Mohapatra and R. D. Peccei, Phys. Lett. B **98**, 265 (1981); G. B. Gelmini and M. Roncadelli, Phys. Lett. B **99**, 411 (1981).
6. S. Weinberg, Phys. Rev. Lett. **40**, 223 (1978); F. Wilczek, Phys. Rev. Lett. **40**, 279 (1978).
7. F. Wilczek, Phys. Rev. Lett. **49**, 1549 (1982).
8. M. Roncadelli and D. Wyler, Phys. Lett. B **133** (1983) 325.
9. D. G. Cerdeno, A. Dedes and T. E. J. Underwood, JHEP **0609** (2006) 067.
10. K. Dick, M. Lindner, M. Ratz and D. Wright, Phys. Rev. Lett. **84** (2000) 4039.
11. A. S. Joshipura and S. D. Rindani, Phys. Rev. Lett. **69** (1992) 3269; A. S. Joshipura and J. W. F. Valle, Nucl. Phys. B **397** (1993) 105.
12. e.g. F. Gianotti and M. L. Mangano, arXiv:hep-ph/0504221.
13. F. Meisel, M. Dührssen, M. Heldmann, K. Jakobs, ATL-PHYS-PUB-2006-009; L. Neukermans, B. Di Girolamo, ATL-PHYS-2003-006.
14. C. F. Kolda and H. Murayama, JHEP **0007**, 035 (2000); R. Schabinger and J. D. Wells, Phys. Rev. D **72**, 093007 (2005).

Sigma Model Q-balls and Q-Stars

Y. Verbin

Department of Natural Sciences, The Open University of Israel, P.O.B. 808 Raanana 43107, Israel
e-mail: verbin@openu.ac.il

Abstract. A new kind of Q-balls is found: Q-balls in a non-linear sigma model. Their main properties are presented together with those of their self-gravitating generalization, sigma model Q-stars. A special limit of solutions bound by gravity alone ("sigma stars") is also discussed briefly.

Keywords: non-topological solitons, self-gravitating systems, Q-balls, Q-Stars
PACS: 11.27.+d, 04.40.-b, 11.30.Fs

Q-balls [1] occur in a wide variety of (theoretical) physical contexts like the minimal supersymmetric Standard Model [2, 3], as dark matter candidates [4] and as a possible explanation for the baryon asymmetry in the universe [3] and the baryon to dark matter ratio [5]. See e.g. Enqvist&Mazumdar[6] and Dine&Kusenko [7] for further reviews.

The above-mentioned Q-ball studies are based on the "original" flat space Q-balls. However, it is evident that for a large enough mass scale, gravitational effects become important and one needs to study Q-stars [8]. Moreover, unifying theories (typically in higher dimensions) lead frequently to non-linear sigma models, so Q-balls and Q-stars should be studied in these models as well. Although topological solitons in non-linear sigma models [9] have been studied for decades, it seems that non-topological solitons of the same models have received very little attention. This work will be devoted to a special type of those: Q-balls and their self-gravitating counterparts, i.e. sigma model Q-stars. Their properties result from the action

$$S=\int d^4x\sqrt{|g|}\left(\frac{1}{2}\mathscr{E}(|\Phi|)(\nabla_\mu\Phi)^*(\nabla^\mu\Phi)-U(|\Phi|)+\frac{1}{16\pi\mathscr{G}}R\right). \tag{1}$$

A particularly simple system which will be studied here is the O(3) sigma model [9] which corresponds to $\mathscr{E}(|\Phi|)=1/(1+|\Phi|^2/m^2)^2$.

The function $U(|\Phi|)$ is a non-negative potential which will be chosen such as Q-ball solutions will exist, that is solutions of the form $\Phi=f(r)e^{i\omega t}$. Q-stars are the self-gravitating generalization where we take the usual form for the spherically-symmetric line element

$$ds^2=A^2(r)dt^2-dr^2/(1-2\mathscr{G}M(r)/r)-r^2(d\theta^2+\sin^2\theta d\varphi^2). \tag{2}$$

The field equations for spherical solutions are written in a dimensionless form in terms of the dimensionless variables $\mathscr{M}=\mathscr{G}mM$, $x=mr$, $\overline{\omega}=\omega/m$ and $\gamma=4\pi\mathscr{G}m^2$. It is also simpler to revert to an angular field Θ defined by $|\Phi|=m\tan(\Theta/2)$, and to use a

CP957, *Particles, Strings, and Cosmology, 13th International Symposium*
edited by A. Rajantie, C. Contaldi, P. Dauncey, and H. Stoica

dimensionless potential function $u(\Theta)$. In terms of all these, the field equations will be

$$\mathscr{M}' = \gamma x^2 \left[\frac{\overline{\omega}^2 \sin^2(\Theta)}{8A^2} + \left(1 - \frac{2\mathscr{M}}{x}\right) \frac{\Theta'^2}{8} + u(\Theta) \right] \quad (3)$$

$$\left(1 - \frac{2\mathscr{M}}{x}\right) \frac{xA'}{A} = \gamma x^2 \left[\frac{\overline{\omega}^2 \sin^2(\Theta)}{8A^2} + \left(1 - \frac{2\mathscr{M}}{x}\right) \frac{\Theta'^2}{8} - u(\Theta) \right] + \frac{\mathscr{M}}{x} \quad (4)$$

$$\left(1 - \frac{2\mathscr{M}}{x}\right) \Theta'' + 2\left(1 - \frac{\mathscr{M}}{x} - \gamma x^2 u(\Theta)\right) \frac{\Theta'}{x} + \frac{\overline{\omega}^2 \sin(2\Theta)}{2A^2} - 4\frac{du}{d\Theta} = 0. \quad (5)$$

It turns out that Q-balls and Q-stars are easily obtained for a large family of potentials having a global minimum at $\Theta = 0$ and another local one. We will use the simple form $u(\Theta) = \sin^2(\Theta)/8 - \alpha \sin^p(\Theta)/p$ with $p = 3$, 4 so the potentials are:

$$u_{23}(\Theta) = \frac{\sin^2(\Theta)}{8} - \frac{\alpha \sin^3(\Theta)}{3} \quad , \quad u_{24}(\Theta) = \frac{\sin^2(\Theta)}{8} - \frac{\alpha \sin^4(\Theta)}{4} \quad (6)$$

We will see that the difference between the corresponding solutions will be analogous to those between the 2-3-4 and the 2-4-6 potentials of the "linear" system [10]. We choose representative values of $\alpha = 0.35$ for the 2-3 potential and $\alpha = 0.4$ for the 2-4 one. As usual, one may gain some insight from the "effective potential" $u_{eff}(\Theta) = u(\Theta) - \overline{\omega}^2 \sin^2(\Theta)/8$. The parameter $\overline{\omega}$ will take values between $\overline{\omega}_- = \sqrt{1 - 8\alpha/p}$ and $\overline{\omega}_+ = 1$ and the corresponding central fields are $\Theta_*(0) = \pi/2$ and $\Theta(0) \to 0$. These give the "thin wall" and "thick wall" limits.

Due to the "north-south" symmetry which is left in the potential functions there exists of course another family of "mirror" solutions with the boundary condition $\Theta(\infty) = \pi$ instead of the usual $\Theta(\infty) = 0$ that we are using. The second boundary condition will satisfy accordingly $\Theta(0) > \pi/2$. Obviously, our choice does not lead to any loss of generality of the results that will be presented here, but it should be kept in mind that they apply to two different families of solutions.

We will study solutions with both potentials for the three values $\gamma = 0$, 0.02 , 0.2.

It is obvious that there are no flat space solution with $\alpha = 0$, that is a mass term only. Gravity changes the situation and allows solutions which are the sigma model analogues of the boson star solutions, so we may call them "sigma stars". One difference with respect to boson stars is that these sigma stars do not enjoy a scaling symmetry and do not fall on a one-parameter curve in parameter space. Therefore the $\gamma = 0.2$ family of pure self-gravitating sigma model solutions which we present is not universal.

The main results are shown in figures 1-2. The first, figure 1 shows the charge as a function of the central field $\Theta(0)$ in all cases. It covers the region of maximal charge for $\gamma = 0.02$, although the maximum is so narrow that its width cannot be seen it the plots. The plots for the binding energy per particle vs. charge, figure 2 give further information about the masses being larger or smaller than mQ, that is about the stability.

A new characteristic with respect to the "linear" solutions is the limiting value of the central scalar field $\Theta(0)$ which is the same for both potentials: $\Theta_*(0) = \pi/2$. Moreover, now the same limiting $\Theta(0)$ appears also for the self-gravitating solutions and not for

Q-balls only. The reason for this is the special form of the potential functions that we chose, which yields the minimum of the effective potential to be at $\Theta(0) = \pi/2$ for all values of ω. But it is possible that a different choice of potential function will allow larger values of $\Theta_*(0)$.

The "damped oscillations" of Q beyond its maximum exist also in the sigma model case but they are only partially visible in figure 1 due to the $\Theta(0) = \pi/2$ limit. As before these oscillations correspond to unstable solutions which are degenerate (in charge) with the smaller $\Theta(0)$ ones.

Another new characteristic is the existence of "mirror" solutions with different boundary conditions. The possibility of coexistence of solutions of both type and their interaction deserves further study.

We find a close parallelism between the main properties of the sigma Q-balls/Q-stars and the analogous "ordinary" Q-balls/Q-stars [10]. The sigma Q-balls with the 2-3 potential are always stable with ever growing charge. Gravity imposes maximal charge values: for $\gamma = 0.2$ the maximum is at $Q = 382.97$ and $\Theta(0) = 1.56921$ or $2\Theta(0)/\pi = 0.99899$, while for $\gamma = 0.02$ the maximum is at $Q = 32155.3$. The corresponding central scalar field is so close to $\pi/2$ that we give it as $1 - 2\Theta(0)/\pi = 1.5543 \times 10^{-15}$. The sigma star solutions with $\gamma = 0.2$ that appear in figure 1 have a maximum at $Q = 95.07$ and $\Theta(0) = 1.54355$ or $2\Theta(0)/\pi = 0.98265$.

These values define the region of stable solutions. As figure 2 show, all the others are either unbound (with negative binding energy), or may decay into the stable solutions while conserving particle number.

The 2-3 Q-star curve with $\gamma = 0.02$ in figure 1 is very similar to the $\gamma = 0$ Q-balls (below the maximum of $Q = 32155.3$) and it is impossible to distinguish between them. However, the difference shows up in the binding energy which is shown in figure 2.

The solutions with the 2-4 potential exhibit a more involved structure which may be classified according to the central field value. The small $\Theta(0)$ sigma Q-balls (no gravity: $\gamma = 0$) are large and unstable. The stability region starts at $\Theta(0) = 1.11515$ for which $Q = M/m = 23.43$ and extends all the way to $\Theta(0) = \pi/2$ with monotonically increasing charge and mass. For $\gamma = 0.02$ the small $\Theta(0)$ behavior changes completely. The small $\Theta(0)$ solutions have positive binding energies for $\Theta(0) \leq 0.14476$ (for which $Q = 43.37$), passing through a local maximum of charge at $\Theta(0) = 0.06400$ and $Q = 50.17$.

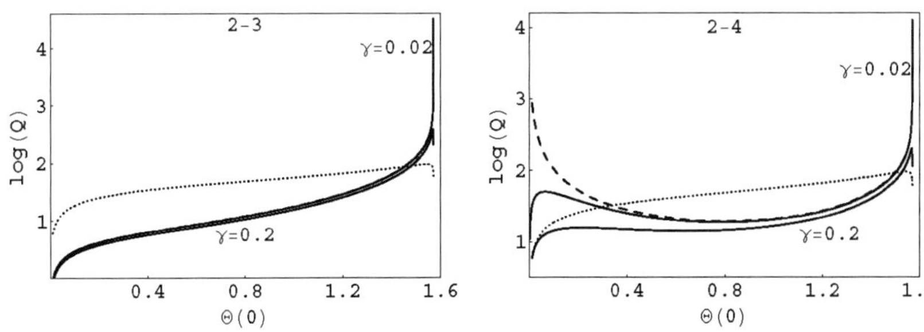

FIGURE 1. Plots of $\log(Q)$ vs. $\Theta(0)$ for $\gamma = 0$ (dashed), $\gamma = 0.02$ and $\gamma = 0.2$. The $\gamma = 0$ line cannot be resolved from the $\gamma = 0.02$ one in the 2-3 potential. The dotted lines correspond to $\gamma = 0.2$ sigma stars.

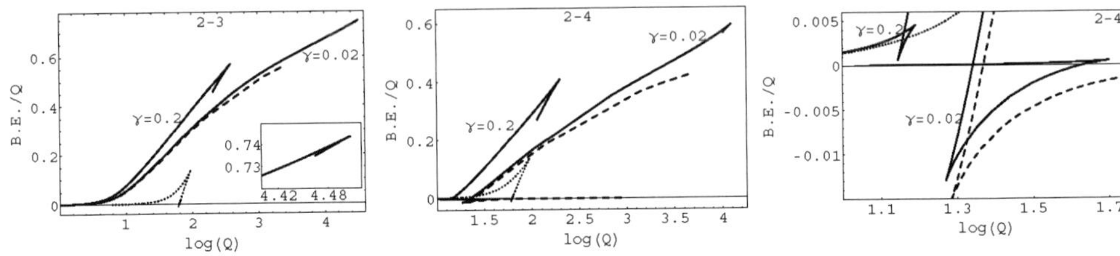

FIGURE 2. Plots of binding energy per particle $(mQ-M)/mQ$ vs. $\log(Q)$ for sigma model Q-stars with $\gamma=0$ (sigma model Q-balls - dashed), $\gamma=0.02$ and $\gamma=0.2$ and for boson stars with $\gamma=0.2$ (dotted). Left: 2-3 potential. Center: 2-4 potential – general view; Right: magnification of the small $B.E./Q$ region.

An instability regions follows for $\Theta(0) \leq 1.08417$. The second range of bound solutions starts at $\Theta(0) = 1.08417$ for which $Q = M/m = 21.99$ and goes toward $\Theta(0) \rightarrow \pi/2$. There is however a global charge maximum of $Q = 12586.03$ very close to $\Theta(0) = \pi/2$. i.e. such that $1-2\Theta(0)/\pi = 1.7686 \times 10^{-13}$. We may therefore conclude that for this 2-4 potential there are stable sigma Q-stars for any charge up to a maximal value of $Q = 12586.03$ for the parameters we chose. Some charge intervals exhibit charge degeneracy, so the higher mass solutions will decay.

The solutions for $\gamma = 0.2$ have similar behavior with two main quantitative differences: the maximal charge is now much smaller and has a value of $Q = 200.18$ and all solutions have in this case positive binding energy due to the stronger gravitational self-attraction.

As far as the flat space limit is concerned, the 2-3 solutions may be divided into two types in accordance with two regions of the central field values $0 < \Theta(0) < \pi/2$. The sigma Q-stars in most of this interval are similar to the flat space ones and may be considered solitonic. The exception is the small region near (and below) $\Theta(0) = \pi/2$ where the solutions are clearly non-solitonic. The 2-4 solutions may be divided into three types: in the region of small $\Theta(0)$ and near $\Theta(0) = \pi/2$ gravity gives rise to solutions with no flat space limit, while in the medium $\Theta(0)$ values the solutions may be considered as solitonic.

REFERENCES

1. S. Coleman, Nucl. Phys. B **262**, 263 (1985) [Erratum-ibid. B **269**, 744 (1986)].
2. A. Kusenko, Phys. Lett. B **405**, 108 (1997).
3. K. Enqvist and J. McDonald, Phys. Lett. B **425**, 309 (1998).
4. A. Kusenko and M. E. Shaposhnikov, Phys. Lett. B **418**, 46 (1998).
5. K. Enqvist and J. McDonald, Nucl. Phys. B **538**, 321 (1999).
6. K. Enqvist and A. Mazumdar, Phys. Rept. **380**, 99 (2003).
7. M. Dine and A. Kusenko, Rev. Mod. Phys. **76**, 1 (2004).
8. B. W. Lynn, Nucl. Phys. B **321**, 465 (1989).
9. R. Rajaraman, *solitons and Instantons*, (North Holland, Amsterdam 1982).
10. Y. Verbin, arXiv: 0708.2673 [gr-qc], to be published in the proceedings of the 11th Marcel Grossmann Meeting, Berlin, July 2006.

A domain-wall-brane-localized standard model

Raymond R. Volkas

School of Physics, Research Centre for High Energy Physics, The University of Melbourne, Victoria 3010, Australia

Abstract.
We sketch an SU(5) bulk model in $4+1$-dimensions that plausibly produces an effective $3+1$-dimensional standard model dynamically-localized on a domain wall.

Keywords: domain wall, brane, standard model, unification
PACS: 04.50.+h, 11.27.+d, 12.10.-g

INTRODUCTION

It has long been speculated that our universe might be a domain wall brane embedded in a background $4+1$-dimensional spacetime [1] (see also [2, 3, 4, 5, 6, 7, 8, 9]). The main challenge in constructing such a model is the simultaneous dynamical localization of fermions, gauge bosons, gravitons and Higgs bosons. In this talk, I sketch how various mechanisms for the localization of these four classes of particle can be assembled into a coherent model realizing the standard model (SM), or something close to it, as the effective $3+1$-dimensional theory for the localized fields [10].

The main aesthetic motivation is to treat the extra dimension on an equal footing to the usual spatial dimensions in the action, with the distinction arising only at the solution level. Since in such a theory the brane is not fundamental but rather a soliton solution, it also has the advantage of answering the question: what is the brane made of? (The answer is scalar fields.) Finally, it seems more satisfying to provide dynamical localization mechanisms rather than positing *ab initio* that certain fields are localized to a fundamental brane.

The theory we shall describe [10] requires an SU(5) gauge symmetry in the bulk, spontaneously broken to SU(3) $\otimes$ SU(2) $\otimes$ U(1) on the wall. It has two immediate advantages over $3+1$-dimensional SU(5) models: absence of the tree-level $m_e = m_d$ mass relations, and the chance to naturally suppress colored-Higgs-induced proton decay.

THE MODEL

Let us start with gauge boson localization. The best understood proposal is that due to Dvali and Shifman [11]: Consider a theory with gauge group G containing an adjoint scalar multiplet χ plus a singlet η with $\eta \to -\eta$ imposed as a discrete symmetry. Arrange the Higgs potential parameters so that a solution of the coupled Klein-Gordon equations has η forming a kink as a function of the extra-dimensional coordinate y, with the relevant component of χ assuming a $1/\cosh$-like configuration. This causes

CP957, *Particles, Strings, and Cosmology, 13th International Symposium*
edited by A. Rajantie, C. Contaldi, P. Dauncey, and H. Stoica

the spontaneous breaking $G \to H$ inside the domain wall, with G rapidly restored in the bulk. Taking the bulk theory to be asymptotically-free and confining, we deduce that the H gauge bosons will be localized to the wall. The point is that these gauge bosons propagate as massless gauge particles (or relatively light H glueballs) inside the wall, but as constituents of massive G glueballs in the bulk. The mass gap plausibly ensures their localization. Alternatively, one may view the bulk as a dual superconductor, and observe that the field lines of a G-charge source in the wall will be repelled from the (smeared out) boundary with the bulk, leading to effective $3+1$-dimensional behavior in the large distance limit (within the wall) [12, 13]. For our application, G is SU(5) and H is the SM subgroup. It is interesting that gauge boson localization immediately motivates grand unification (in the bulk).

It can be readily verified that the Higgs potential

$$V_{\eta\chi} = (c\eta^2 - \mu_\chi^2)\mathrm{Tr}(\chi^2) + \lambda_1 \left[\mathrm{Tr}(\chi^2)\right]^2 + \lambda_2 \mathrm{Tr}(\chi^4) + l(\eta^2 - v^2)^2, \tag{1}$$

admits the required solution-type for a large range of parameters. Generally, the solutions can only be obtained numerically, but the simple analytic form $\eta(y) = v\tanh(ky)$, $\chi_1(y) = A\,\mathrm{sech}(ky)$ follows when certain relations on the Higgs parameters are enforced, with the inverse width k and the amplitude A given in terms of those parameters. The notation χ_1 signifies the component of the SU(5) adjoint that induces breakdown to the SM gauge group. In writing the above quartic potential, we adopt the view that our non-renormalizable $4+1$-d model is but an effective theory, valid below an ultraviolet cutoff, and we analyse here only the dominant, lowest-dimension operators. This Higgs potential, and all subsequent Lagrangians, are truncations.

We now add the fermions $\Psi_5 \sim 5^*$ and $\Psi_{10} \sim 10$, and couple them to η and χ:

$$Y_{DW} = h_{5\eta}\overline{\Psi}_5\Psi_5\eta + h_{5\chi}\overline{\Psi}_5\chi^T\Psi_5 + h_{10\eta}\mathrm{Tr}(\overline{\Psi}_{10}\Psi_{10})\eta - 2h_{10\chi}\mathrm{Tr}(\overline{\Psi}_{10}\chi\Psi_{10}). \tag{2}$$

We solve the Dirac equations,

$$\left[i\Gamma^M\partial_M - h_{n\eta}\eta(y) - \sqrt{\frac{3}{5}\frac{Y}{2}}h_{n\chi}\chi_1(y)\right]\Psi_{nY}(x,y) = 0, \tag{3}$$

in the $\eta(y)$, $\chi_1(y)$ background, where $n = 5, 10$ and Y is the weak-hypercharge of the SM components denoted Ψ_{5Y} and Ψ_{10Y}. We seek separated-variable solutions, $\Psi_{nY}(x,y) = \psi_{nY,L}(x)f_{nY}(y)$, where the ψ's are left-chiral ($\gamma_5\psi_L = -\psi_L$) $3+1$-d fields obeying the massless $3+1$-d Dirac equation. The localization profiles are then

$$f_{nY}(y) \propto e^{-\int^y b_{nY}(y')dy'}, \tag{4}$$

where the zeroes of

$$b_{nY}(y) \equiv h_{n\eta}\eta(y) + \sqrt{\frac{3}{5}\frac{Y}{2}}h_{n\chi}\chi_1(y) \tag{5}$$

are the localization centers. The various SM components are *split* [14] along y due to the different linear combinations of η and χ_1 that they feel. Figure 1 shows examples of such profiles.

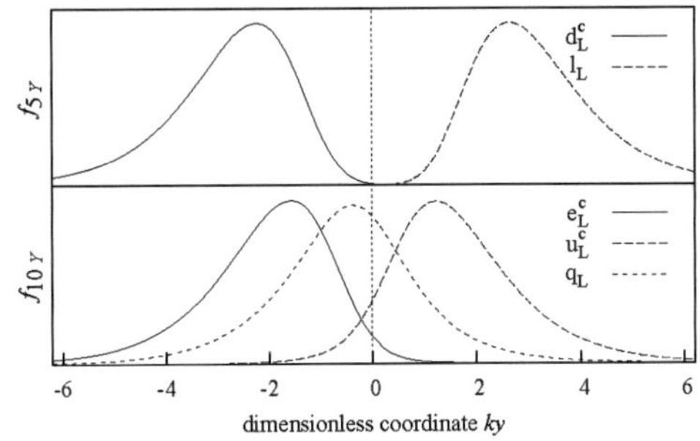

FIGURE 1. Typical fermion localization profiles normalized so they square-integrate to one.

Electroweak symmetry breaking can now be incorporated by introducing the scalar $\Phi \sim 5^*$, which couples as per

$$Y_5 = h_- \overline{(\Psi_5)^c}\Psi_{10}\Phi + h_+ \varepsilon^{ijklm}\overline{(\Psi_{10})^c}_{ij}\Psi_{10kl}\Phi^*_m + h.c. \tag{6}$$

$$\begin{aligned} V_{\text{rest}} &= \mu_\Phi^2 \Phi^\dagger\Phi + \lambda_3(\Phi^\dagger\Phi)^2 + \lambda_4\Phi^\dagger\Phi\eta^2 + 2\lambda_5\Phi^\dagger\Phi\text{Tr}(\chi^2) \\ &+ \lambda_6\Phi^\dagger(\chi^T)^2\Phi + \lambda_7\Phi^\dagger\chi^T\Phi\eta. \end{aligned} \tag{7}$$

The field Φ contains the electroweak Higgs doublet Φ_w and a colored scalar Φ_c. Writing both in separated variable form, $\Phi(x,y) = p(y)\phi(x)$, where the ϕ's are required to satisfy 3 + 1-d massive Klein-Gordon equations, the 4 + 1-d KG equations yield

$$-\frac{d^2}{dy^2}p_{w,c}(y) + W_Y(y)p_{w,c}(y) = m^2_{w,c}p_{w,c}(y), \tag{8}$$

with a weak-hypercharge-dependent effective potential,

$$W_Y(y) = \mu_\Phi^2 + \lambda_4\eta^2 + \lambda_5\chi_1^2 + \frac{3Y^2}{20}\lambda_6\chi_1^2 + \sqrt{\frac{3}{5}}\frac{Y}{2}\lambda_7\eta\chi_1. \tag{9}$$

with $m^2_{w,c}$ being the 3 + 1-d squared masses. The two effective potentials can be arranged to produce $m_c^2 > 0$, $m_w^2 < 0$ and localized ϕ's. The tachyonic m_w^2 triggers spontaneous electroweak symmetry breaking on the brane. Figure 2 displays suitable potentials.

The usual SU(5) mass relations $m_e = m_d$ are *not* obtained from Y_5 because the 3 + 1-d masses depend on overlap integrals of bulk profiles which differ among the SM components. Similarly, one may be able to suppress colored-scalar-induced proton decay because the relevant overlap integrals are small due to the splittings.

The above analysis had gravity switched off. We now include Einstein-Hilbert and bulk cosmological terms. By fine-tuning the latter, one may localize the graviton by the smooth domain-wall-brane version of type-2 Randall-Sundrum [4] (the warp factor exponent behaves something like $-\log(\cosh ky)$ rather than $-|ky|$). The various effective Schrödinger potentials that induce localization now change in character: they get warped-down to zero asymptotically in the bulk, rather than approaching nonzero constants. This means that mode-continua begin at zero mass for all particles, an effect

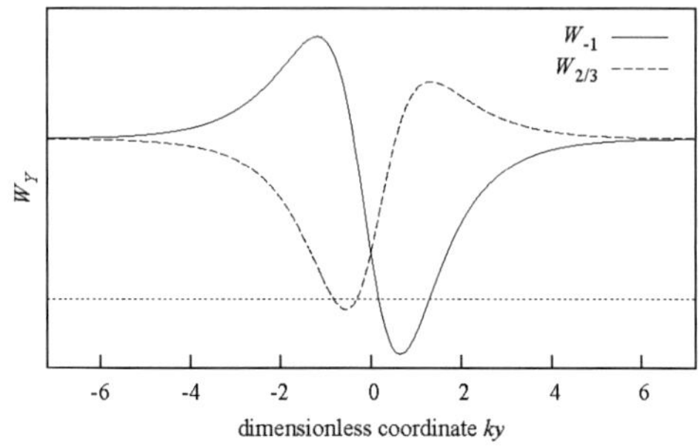

FIGURE 2. Sample Higgs and colored scalar localization potentials. The horizontal line is $W = 0$, and W_{-1} has a negative eigenvalue m_w^2 triggering electroweak symmetry breakdown.

well-known for gravitons in type-2 RS. What were previously excited bound (i.e. localized) states now become *resonances*. The non-resonant modes are greatly suppressed in amplitude on the brane, because the wave-functions have to tunnel through potential barriers. Thus, despite the continua starting at zero mass, an effective $3 + 1$-d theory is still produced (see Ref. [15] for more details), retaining the general character of the flat-space toy model.

We conclude by saying that the above construction plausibly produces an effective wall-localized theory that is close to being the standard model.

ACKNOWLEDGMENTS

This work was performed in collaboration with Damien P. George and Rhys Davies [10]. It was supported by the Australian Research Council.

REFERENCES

1. V. A. Rubakov, and M. E. Shaposhnikov, *Phys. Lett.* **B125**, 136–138 (1983).
2. N. Arkani-Hamed, S. Dimopoulos, and G. R. Dvali, *Phys. Lett.* **B429**, 263–272 (1998).
3. L. Randall, and R. Sundrum, *Phys. Rev. Lett.* **83**, 3370–3373 (1999).
4. L. Randall, and R. Sundrum, *Phys. Rev. Lett.* **83**, 4690–4693 (1999).
5. I. Antoniadis, *Phys. Lett.* **B246**, 377–384 (1990).
6. I. Antoniadis, N. Arkani-Hamed, S. Dimopoulos, and G. Dvali, *Phys. Lett.* **B436**, 257–263 (1998).
7. K. Akama, *Lecture Notes in Physics, Berlin Springer-Verlag* **176**, 136–138 (1983).
8. M. Visser, *Phys. Lett.* **B159**, 22 (1985).
9. G. W. Gibbons, and D. L. Wiltshire, *Nucl. Phys.* **B287**, 717 (1987).
10. R. Davies, D. P. George, and R. R. Volkas (2007), `arXiv:0705.1584`.
11. G. R. Dvali, and M. A. Shifman, *Phys. Lett.* **B396**, 64–69 (1997).
12. N. Arkani-Hamed, and M. Schmaltz, *Phys. Lett.* **B450**, 92 (1999).
13. S. L. Dubovsky, and V. A. Rubakov, *Int. J. Mod. Phys.* **A16**, 4331–4350 (2001).
14. N. Arkani-Hamed, and M. Schmaltz, *Phys. Rev.* **D61**, 033005 (2000).
15. R. Davies, and D. P. George (2007), `arXiv:0705.1391`.

MiniBooNE's First Neutrino Oscillation Result

M.O. Wascko, for the MiniBooNE Collaboration

Imperial College London

Abstract.
The MiniBooNE Collaboration has performed a search for ν_e appearance in a ν_μ beam. No significant excess of events above background was observed in the analysis region. The data are consistent with no oscillation.

Keywords: Neutrino oscillation
PACS: 14.60.Pq, 14.60.St

INTRODUCTION

The discovery of neutrino flavour oscillation, and the associated discovery of neutrino mass, is the first confirmed observation not allowed by the Standard Model of particle physics. Although the Standard Model has been remarkably successful in describing the interactions of matter at low energies, we know it must break down at higher energies. The study of neutrino oscillation, which was first postulated in 1962 [1] but not conclusively observed until 2003 [2], promises to point the way toward the next theory beyond the Standard Model.

For two neutrinos, the probability for a neutrino created as an eigenstate of flavour a to later be observed as flavour b is given by:

$$P(\nu_a \to \nu_b) = \sin^2 2\theta_{12} \sin^2(1.27\Delta m^2_{12}\frac{L}{E}), \tag{1}$$

where the two mass states are m_1 and m_2, θ_{12} is the mixing angle, $\Delta m^2_{12} = m_1^2 - m_2^2$, L is the distance the neutrino has travelled in km and E is the neutrino energy in GeV.

To date, we have observed three distinct neutrino oscillation signals. Atmospheric ν_μ disappearance was first observed by the Super-Kamiokande experiment in 1998, indicating $\Delta m^2 \sim 10^{-3}\text{eV}^2$ [3] with maximal mixing betwenthe states. Oscillation on this scale was confirmed with long-baseline accelerator beams of neutrinos [4, 5]. Neutrinos from the sun were found to oscillate with a large mixing angle and $\Delta m^2 \sim 10^{-5}\text{eV}^2$ [6]. Flavour oscillation at this mass scale was also observed with reactor neutrinos [2]. The third indication of neutrino oscillation comes a from short baseline accelerator neutrino beam experiment, LSND [7], which saw an excess of $\overline{\nu}_e$ in a $\overline{\nu}_\mu$ beam with $\Delta m^2 \sim \text{eV}^2$. The LSND oscillation probability is extremely small, 0.26%.

The LSND signal was interpreted in the context of a simple two neutrino oscillation, which is justified because the mass scale is clearly independent from the other two mass scales. Other experiments, notably KARMEN and BUGEY, were sensitive to oscillation in the LSND allowed mass region but with lower sensitivities; their lack of oscillation signals are thus not able to completely rule out the LSND oscillation hypothesis.

CP957, *Particles, Strings, and Cosmology, 13th International Symposium*
edited by A. Rajantie, C. Contaldi, P. Dauncey, and H. Stoica

Because $\Delta m^2_{LSND} >> \Delta m^2_{atmos} + \Delta m^2_{solar}$, the confirmation of LSND would require at least one more neutrino mass state. The spectacular implications of additional neutrino states made it imperative that the LSND result be checked. MiniBooNE was designed to do exactly that.

MINIBOONE DESCRIPTION

MiniBooNE creates a ν_μ beam at Fermilab which travels 541 m to the MiniBooNE detector. The neutrino beam is comprised of 95% ν_μ, 4.4% $\overline{\nu}_\mu$, 0.6% ν_e and trace amounts of $\overline{\nu}_e$, and has a mean energy of 0.8 GeV. The beamline can also be run in a configuration which produces a beam of antineutrinos. The detector is a 610 cm radius steel sphere filled with $\sim$ 800 tonnes of pure mineral oil. The detector is divided into two regions: a sperical shell *veto* region, 35 cm thick, surrounds the optically isolated *main* detector volume which contains the $\sim$510 cm fiducial volume. Charged particles from the neutrino interactions in the detector cause emission of Cherenkov and scintillation light in the oil, which is detected by 8" photomultiplier tubes (PMTs). The PMT hit patterns are used to reconstruct the particle tracks under the hypotheses that they were created by an e, μ, or two γs from neutral pion decay. Likelihood ratios from these three hypotheses are used to distinguish the ν_e candidate events from the more numerous ν_μ events.

MiniBooNE collected data from February 2002 until January 2005. In total 5.58×10^{20} protons on target (POT) were collected for the neutrino oscillation analysis, yielding 1.7 million neutrino interactions in the detector.

ν_e APPEARANCE SEARCH

The MiniBooNE oscillation search [11] had two aspects: a counting experiment for ν_e appearance in a ν_μ beam and an energy fit for simple two-neutrino appearance-only oscillation. Because the experiment used only one neutrino detector, many consistency checks were applied to the Monte Carlo prediction of the signal and background rates using calibration and neutrino data samples. Systematic uncertainties from neutrino beam prodcution, neutrino interactions in the detector, and the detector modelling were considered. The experiment employed a "closed-box" blind analysis. Before unblinding the data set, the energy limits on the analysis region for the counting experiment and energy fit were set at 475 MeV and 1250 MeV because they maximised the sensitivity of the ν_e appearance search.

If the LSND oscillation hypothesis were correct, MiniBooNE would expect $\sim$163$\pm$21 ν_e events after all analysis cuts. These would appear above a background of 358$\pm$35$\pm$19: 229 from intrinsic ν_es in the beam and 129 from misidentified ν_μ events.

The observed ν_e events are shown as a function of neutrino energy in Figure 1, along with the predicted backgrounda and two representative signals allowed by the LSND result. In the analysis region, 380 ν_e events were observed, yielding an excess of 22$\pm$19$\pm$35 events and a significance of excess of 0.55 σ. The energy fit yielded no significant evidence for $\nu_\mu \rightarrow \nu_e$ oscillation in a simple two-neutrino context. Figure 2

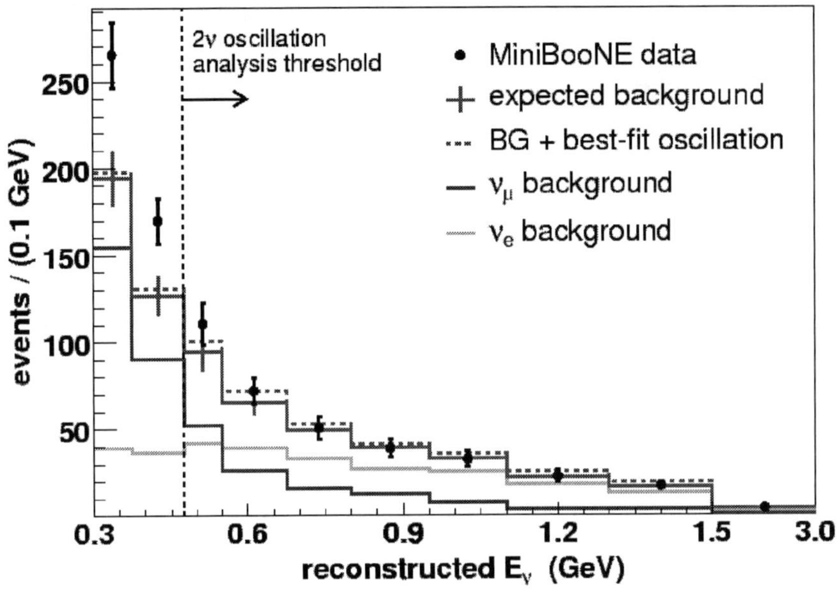

FIGURE 1. The number of candidate ν_e events as a function of neutrino energy . The points represent the data with statistical error, while the histogram is the expected background with systematic errors from all sources. The vertical dashed line indicates the threshold used in the two-neutrino oscillation analysis. Also shown are the best-fit oscillation spectrum (dashed histogram) and the background contributions from ν_μ and ν_e events.

shows the MiniBooNE 90% confidence level (CL) limit curve, which completely covers the LSND 90% CL allowed region. Assuming that neutrinos and antineutrinos oscillate with the same probability, this limit excludes two neutrino appearance-only oscillation as an explanation of the LSND anomaly at 98% CL.

Below 475 MeV, an excess of $96 \pm 17 \pm 20$ ν_e events was observed. While this excess is intriguing, the shape of the excess as a function of energy is not consistent with a two-neutrino appearance signal; this is illustrated in Figure 1. The source of these events is currently under intense investigation.

CONCLUSIONS

Despite the presently unexplained excess of events at low energy, there is excellent agreement between data and Monte Carlo in the oscillation analysis region. If neutrinos and antineutrinos undergo oscillation in the same way, MiniBooNE excludes two neutrino appearance-only oscillation as an explanation of the LSND anomaly at 98%

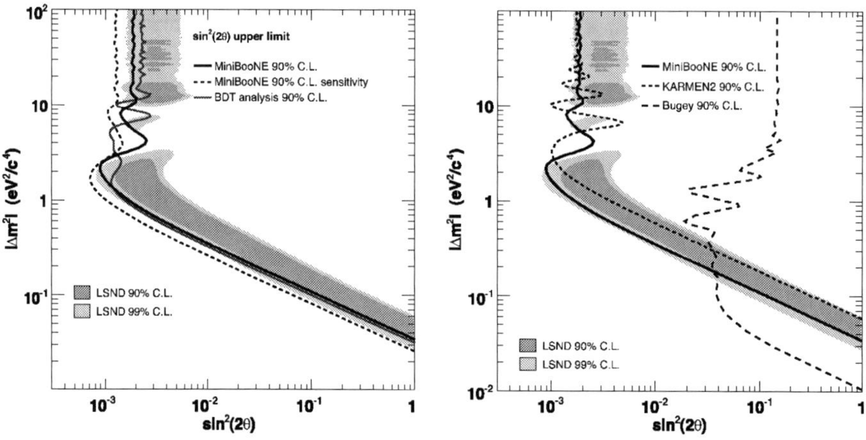

FIGURE 2. The left panel shows the MiniBooNE limit (thick solid curve) and sensitivity (dashed curve) for events in the analysis region within a two neutrino oscillation model. Also shown is the limit from the cross-check analysis (thin solid curve). The right panel compares the MiniBooNE limit with the limits from the KARMEN and Bugey experiments.

CL. MiniBooNE is currently taking data with the beam in antineutrino configuration.

ACKNOWLEDGMENTS

MiniBooNE received support from Fermilab, the US Department of Energy, and the US National Science Foundation. The author gratefully acknowledges support from the conference organisers.

REFERENCES

1. Maki, Nakagawa and Sakata, *Prog. Theor. Phys.*, **28** (1962) 247.
2. Eguchi, *et al.* (KamLAND Collaboration), *Phys.Rev.Lett* **90** (2003) 021802.
3. Hosaka, *et al.* (Super-Kamiokande Collaboration), *Phys.Rev.D* **73** (2006) 112001.
4. Ahn, *et al.* (K2K Collaboration), *Phys.Rev.D* **74** (2006) 0072003.
5. Michael, *et al.* (MINOS Collaboration), *Phys.Rev.Lett* **97** (2006) 191801.
6. Aharmim, *et al.* (SNO Collaboration), *Phys.Rev.D* **75** (2007) 045502.
7. Aguilar-Arevalo, *et al.* (LSND Collaboration), *Phys.Rev.D* **64** (2001) 112007.
8. Armbruster, *et al.* (KARMEN Collaboration), *Phys.Rev.D* **65** (2002) 112001.
9. Church, *et al.*, *Phys.Rev.D* **66**, (2002) 013001.
10. Achkar, *et al.* (BUGEY Collaboration), *Nucl.Phys.* **B434**, (1995) 503.
11. Aguilar-Arevalo *et al.* (MiniBooNE Collaboration), *Phys.Rev.Lett.* **98** (2007) 231801.

STRINGS

A new non-perturbative time-dependent string configuration

Jean Alexandre

Department of Physics, King's College London, WC2R 2LS, UK

Abstract. A time-dependent bosonic string configuration is discussed, in graviton and dilaton backgrounds, which leads to Weyl-symmetry beta-functions which are homogeneous in X^0, to any order in α'. As a consequence, a string reparametrization can always be implemented, such that beta functions can be cancelled, to any order in α'. This non-perturbative conformal invariance is valid for any target space dimension, and leads to a power law expanding Universe, for which the power vanishes if a specific relation between the dimension and dilaton amplitude holds. Finally, $D = 4$ is the minimum dimension (in the case of a spherical world sheet) for which this configuration is consistent with a Wick rotation in a Minkowski target space.

Keywords: non-perturbative string cosmology
PACS: 11.10.Kk, 11.25.Pm, 11.25.Wx

The usual approach in String Cosmology consists in cancelling one-loop (or two-loop) beta functions corresponding to Weyl invariance, and solving the corresponding differential equations in order to find the relevant backgrounds. An alternative approach is discussed here [1], where, instead of cancelling perturbative orders of the beta functions, we look for a configuration of the bosonic string leading to beta functions which are homogeneous functions of X^0, to all orders in α'. This property will be enough to recover Weyl invariance, as the two-loop and higher orders in α' of the beta functions are not fixed in a unique way, but depend on the renormalization scheme which is used when computing the trace of the energy-momentum tensor of the string [2]. Changing renormalization scheme corresponds to reparametrizing the string couplings, which in our case will not affect the time-dependence of the configuration, but will only rescale the metric by a constant and add another constant to the dilaton.

In order to find a hint on what configuration can lead to homogeneous beta functions, we first present an exact functional method, dealing with the sigma model provided by the string action, where the evolution of the quantum theory with the amplitude of α' is derived. We will see that, if the target space metric is conformally flat and proportional to the second derivative of the dilaton (in the string frame), the latter is independent of the amplitude of quantum fluctuations. Homogeneous beta functions will then be obtained for a specific case, where the graviton is a power law of X^0.

We consider a bosonic string on a spherical world sheet, with bare action

$$S = \frac{1}{4\pi}\int d^2\xi\sqrt{\gamma}\left\{\gamma^{ab}\lambda\eta_{\mu\nu}\partial_a X^\mu\partial_b X^\nu + R^{(2)}\phi_{bare}(X^0)\right\}, \tag{1}$$

where $\lambda = 1/\alpha'$ is left as a free parameter which controls the amplitude of quantum

CP957, *Particles, Strings, and Cosmology, 13th International Symposium*
edited by A. Rajantie, C. Contaldi, P. Dauncey, and H. Stoica

fluctuations. $\phi_{bare}(X^0)$ is a bare dilaton, whose precise form is not relevant, but it should in principle contain at least a cubic term in X^0, in order to generate quantum fluctuations. The quantum theory is obtained by defining the connected graphs generator functional and its Legendre transform Γ, for a given value of λ. We obtain then a family of quantum theories, labelled by λ and described by Γ_λ. The exact evolution equation for Γ with λ was derived in [1], and leads to the following evolution for the dilaton in the quantum theory (no summation on i: g_{ii} is any of the space components of the metric):

$$\begin{aligned} g_{00}\dot{\phi} &= -\frac{\Lambda^2}{8\pi R^{(2)}}\left(1+(D-1)\frac{g_{00}}{g_{ii}}\right) \\ &\quad +\frac{\phi''}{4g_{00}}\ln\left(1+\frac{2\Lambda^2 g_{00}}{R^{(2)}\phi''}\right), \end{aligned} \tag{2}$$

where a dot represents a derivative with respect to λ and a prime a derivative with respect to X^0. Λ is a *fixed* world sheet cut off, and we stress here the difference with a Wilsonian approach, where $\lambda = 1/\alpha'$ would be fixed and the world sheet cut off would be running. As a consistency check, one can easily see that a linear dilaton and flat metric [3] lead to $\dot{\phi} = 0$ (after a rescaling of the space coordinates $X^i \to X^i\sqrt{D-1}$), which is expected, as such a configuration does not generate quantum fluctuations. There is another possibility, thought, to obtain an α'-independent dilaton: from the evolution equation (2), if g_{00} is proportional to g_{ii} and to ϕ'', it is always possible to obtain $\dot{\phi} = 0$ by rescaling the space coordinates X^i (this rescaling absorbs the world sheet cut off). As a consequence, a necessary condition for the non-trivial solution to be independent of the amplitude of quantum fluctuations is that

$$g_{\mu\nu}(X^0) \propto \phi''(X^0)\ \eta_{\mu\nu}. \tag{3}$$

We now turn to the conformal properties of a configuration satisfying the condition (3). We first note that the one-loop beta functions corresponding to Weyl invariance cannot be cancelled by a configuration satisfying the condition (3). But, for the following specific case

$$\begin{aligned} g_{\mu\nu}(X^0) &= \frac{\kappa}{(X^0)^2}\eta_{\mu\nu} \\ \phi(X^0) &= \phi_0 \ln X^0, \end{aligned} \tag{4}$$

where κ and ϕ_0 are constants, the different terms in the one-loop beta functions happen to be homogeneous to the same power of X^0. We checked that the next order in α' again contains homogeneous functions of X^0, which are also homogeneous to the first order. Finally, this property is valid to any order in α': whatever power of the Ricci or Riemann tensor is considered, and multiplied by covariant derivatives of the dilaton, contracting the indices with the metric or the inverse metric will always lead to the same power of X^0. As a consequence, we have for the configuration (4)

$$\beta^g_{00} = \frac{1}{(X^0)^2}\sum_{m=1}^{\infty}\xi_m\left(\frac{\alpha'}{\kappa}\right)^m,$$

$$\beta^g_{ij} = \frac{\delta_{ij}}{(X^0)^2}\sum_{m=1}^{\infty}\zeta_m\left(\frac{\alpha'}{\kappa}\right)^m,$$

$$\beta^\phi = \frac{1}{\alpha'}\sum_{m=1}^{\infty}\eta_m\left(\frac{\alpha'}{\kappa}\right)^m, \tag{5}$$

where ξ_n, ζ_n, η_n are coefficients independent of α'.
The next step is to argue that, from two-loops and above, the coefficients ξ_n, ζ_n, η_n are not unique but depend on the renormalization scheme which is used, in order to calculate the trace of the energy momentum tensor of the string [2]. A change of renormalization scheme corresponds to a reparametrization of the string, which leaves the S matrix invariant, and reads, at two loops

$$\tilde{g}_{\mu\nu} = g_{\mu\nu} + \alpha' g_{\mu\nu}\left(b_1 R + b_2\partial^\rho\phi\partial_\rho\phi + b_3\nabla^2\phi\right),$$

$$\tilde{\phi} = \phi + \alpha'\left(c_1 R + c_2\partial^\rho\phi\partial_\rho\phi + c_3\nabla^2\phi\right), \tag{6}$$

where $b_1,...,c_1,...$ are any constants. In the specific case of the configuration (4), this reparametrization just rescales the metric by a constant factor and adds another constant to the dilaton. But the important point is that the reparametrization (6) changes the beta functions in the following way [2]

$$\tilde{\beta}^i = \beta^i + (\tilde{g}^j - g^j)\frac{\partial\beta^i}{\partial g^j} - \beta^j\frac{\partial}{\partial g^j}(\tilde{g}^i - g^i), \tag{7}$$

such that it is always possible to choose the constants $b_1,...,c_1,...$ to cancel the two-loop beta functions $\tilde{\beta}^i$, what was done explicitly in [1]. We then conjecture that the configuration (4) satisfies Weyl invariance to any order in α', since the reparametrization (6) can be extended to any order, and the modification (7) is always valid. This cancellation obtained after a reparametrization of the string is possible for any target space dimension, and for any dilaton amplitude.

Concerning Wilsonian properties of the configuration (4), it was shown in [1] that the latter is an infrared fixed point of momentum flows defined on the world sheet. To show this, an exact renormalization equation was derived, following the approach given in [4], where a sharp cut off was used (this is indeed enough if one considers the evolution for the potential part only of the Wilsonian action). As a consequence, the α'-fixed point solution of the equation (2) was identified with a world sheet Wilsonian IR fixed point.

In order to find the cosmological properties of the configuration (4), we remind that the metrics in the string frame and the Einstein frame are related by

$$dt^2 - a^2(t)(d\vec{x})^2 = \exp\left(\frac{-4\phi(x^0)}{D-2}\right) g_{\mu\nu}(x^0)dx^\mu dx^\nu, \tag{8}$$

where t is the cosmic time and $a(t)$ is the scale factor of the corresponding spatially flat FRW Universe. Plugging the configuration (4) into the identity (8), we obtain the cosmic time as a power law of x^0, and the scale factor is

$$a(t) = a_0 t^{1+\frac{D-2}{2\phi_0}}, \tag{9}$$

where a_0 is a constant. We obtain then a power-law expanding Universe, whose power vanishes if the following relation holds

$$D-2+2\phi_0 = 0, \tag{10}$$

which is therefore the condition to have a Minkowski target space. Note that, if expressed in terms of the cosmic time, the dilaton is independent of ϕ_0 and reads

$$\phi = -\frac{D-2}{2}\ln t. \tag{11}$$

To conclude with the configuration (4), we discuss its properties under a target space Wick rotation. Up to now, the world sheet metric had a Euclidean signature, but the target space metric had a Minkowski signature. Because of the logarithmic dilaton, the analytic continuation $X^0 \to iX^0$ generates an imaginary part in the action. As a consequence, the partition function acquires a phase and becomes, in the case of a spherical world sheet,

$$Z \to Z_E \times \exp(i\pi\phi_0), \tag{12}$$

where Z_E is the Euclidean partition function (for both world sheet and target space metrics). In order to have a real partition function, we need ϕ_0 to be an integer. Taking in addition the condition (10) to have a Minkowski target space, we find that the allowed dimensions are all the even integers, starting from 4:

$$D = 4, 6, 8, \cdots \tag{13}$$

$D = 4$ is thus the minimum dimension where the Wick rotation in target space is consistent with the configuration (4).

As a last remark, one can notice that the logarithmic dilaton, in X^0, leads to a diverging Euclidean partition function, for late times, in the case $D = 4$ and $\phi_0 = -1$, for which the target space is flat and static. This feature shows that a UV cut off in time is necessary for the consistency of the whole picture.

Finally, the extension of this work to a superstring would not change the results, as the homogeneity of the beta functions corresponding to Weyl invariance would not be affected.

REFERENCES

1. J. Alexandre, J. Ellis and N. E. Mavromatos, JHEP **0612** (2006) 071 [arXiv:hep-th/0610072].
2. R. R. Metsaev and A. A. Tseytlin, Nucl. Phys. B **293** (1987) 385.
3. I. Antoniadis, C. Bachas, J. R. Ellis and D. V. Nanopoulos, Nucl. Phys. B **328** (1989) 117.
4. F. J. Wegner and A. Houghton, Phys. Rev. A **8** (1973) 401.

Projective superspace and hyperkähler sigma models on cotangent bundles of Hermitian symmetric spaces

Masato Arai*, Sergei M. Kuzenko† and Ulf Lindström**,‡

*High Energy Physics Division, Department of Physical Sciences
University of Helsinki and Helsinki Institute of Physics
P.O. Box 64, FIN-00014, Finland
†School of Physics M013, The University of Western Australia
35 Stirling Highway, Crawley W.A. 6009, Australia
**Department of Theoretical Physics, Uppsala University
Box 803, SE-751 08 Uppsala, Sweden
‡Helsinki Institute of Physics,
P.O. Box 64, FIN-00014, Finland

Abstract. We review the projective-superspace construction of four-dimensional $\mathscr{N}=2$ supersymmetric sigma models on (co)tangent bundles of the classical Hermitian symmetric spaces.

Keywords: hyperkähler manifold, sigma model, supersymmetry, projective superspace
PACS: 11.10.Lm, 11.30.Pb

Supersymmetry in sigma models is closely related to the geometry of target space [1]. In $\mathscr{N}=1$ models in four space-time dimensions, the target space manifold must be Kähler [1]. General $\mathscr{N}=1$ sigma models can be written in terms of a Kähler potential depending on $\mathscr{N}=1$ chiral superfields and their conjugates. 4D $\mathscr{N}=2$ models require the target space geometry to be hyperkähler [2]. Hyperkähler metrics are difficult to construct explicitly. Thus, manifest $\mathscr{N}=2$ formulations are needed in order to generate $\mathscr{N}=2$ sigma models and, therefore, hyperkähler metrics. Projective superspace [3, 4] provides such a formulation, and it is thus suitable for the construction.

Projective superspace formulation has led to the discovery of several multiplets that can be used to construct new hyperkähler metrics. One of the most interesting projective multiplets is the so-called polar multiplet [5, 6] (see also [7]), for it can be used to describe a charged $U(1)$ hypermultiplet coupled to a vector multiplet [6], and therefore is analogous to the $\mathscr{N}=1$ chiral superfield. The polar multiplet is described by an arctic superfield $\Upsilon(\zeta)$ and its complex conjugate composed with the antipodal map $\zeta\to-1/\zeta$, the antarctic superfield $\breve{\Upsilon}(\zeta)$. It is required to possess certain holomorphy properties on a punctured two-plane parametrized by the complex variable ζ (the latter may be interpreted as a projective coordinate on $\mathbf{C}P^1$). When realized in ordinary $\mathscr{N}=1$ superspace, $\Upsilon(\zeta)$ and $\breve{\Upsilon}(\zeta)$ are generated by an infinite set of ordinary superfields:

$$\Upsilon(\zeta)=\sum_{n=0}^{\infty}\Upsilon_n\zeta^n=\Phi+\Sigma\zeta+O(\zeta^2)\,,\qquad \breve{\Upsilon}(\zeta)=\sum_{n=0}^{\infty}\bar{\Upsilon}_n(-\zeta)^{-n}\,. \tag{1}$$

CP957, *Particles, Strings, and Cosmology, 13th International Symposium*
edited by A. Rajantie, C. Contaldi, P. Dauncey, and H. Stoica

Here Φ is chiral, Σ complex linear,

$$\bar{D}_{\dot{\alpha}}\Phi = 0 \ , \qquad\qquad \bar{D}^2\Sigma = 0 \ , \tag{2}$$

and the remaining component superfields are unconstrained complex superfields. Sigma model couplings of several polar multiplets are described by an action of the form [5]

$$S[\Upsilon, \breve{\Upsilon}] = \frac{1}{2\pi \mathrm{i}} \oint \frac{\mathrm{d}\zeta}{\zeta} \int \mathrm{d}^8 z \, \mathscr{K}\big(\Upsilon(\zeta), \breve{\Upsilon}(\zeta), \zeta\big) \ . \tag{3}$$

It was observed in [8, 9] that there exists a large subclass in the family of models (3) with interesting geometric properties. It corresponds to the case when the Lagrangian $\mathscr{K}$ in (3) has no explicit ζ-dependence,

$$S[\Upsilon, \breve{\Upsilon}] = \frac{1}{2\pi \mathrm{i}} \oint \frac{\mathrm{d}\zeta}{\zeta} \int \mathrm{d}^8 z \, K\big(\Upsilon^I(\zeta), \breve{\Upsilon}^{\bar{J}}(\zeta)\big) \ , \tag{4}$$

and then K can be interpreted as the Kähler potential of a Kähler manifold $\mathscr{M}$. Such a theory occurs as a minimal $\mathscr{N} = 2$ extension of the general four-dimensional $\mathscr{N} = 1$ supersymmetric nonlinear sigma model [1]

$$S[\Phi, \bar{\Phi}] = \int \mathrm{d}^8 z \, K(\Phi^I, \bar{\Phi}^{\bar{J}}) \ . \tag{5}$$

The target space of the $\mathscr{N} = 2$ sigma model (4) turns out to be (an open domain of the zero section) of the cotangent bundle of $\mathscr{M}$ [9]. By construction, the model (4) involves an infinite set of auxiliary superfields. The hard technical problem is to eliminate these auxiliaries, and then dualize the complex linear superfields Σ's into chiral ones, in accordance with the generalized Legendre transform [5]. This problem was solved in [9, 10] for the case $\mathscr{M} = \mathbf{C}P^n$, the complex projective space, and incomplete results were also obtained in [10] for the complex quadric $SO(n+2)/SO(n) \times SO(2)$. Recently we have solved the problem for a large class of classical Hermitian symmetric spaces [11]. In what follows, a brief review of our construction is presented.

The extended supersymmetric sigma model (4) inherits all the geometric features of its $\mathscr{N} = 1$ predecessor (5). The Kähler invariance of the latter, $K(\Phi, \bar{\Phi}) \to K(\Phi, \bar{\Phi}) + \Lambda(\Phi) + \bar{\Lambda}(\bar{\Phi})$ turns into $K(\Upsilon, \breve{\Upsilon}) \to K(\Upsilon, \breve{\Upsilon}) + \Lambda(\Upsilon) + \bar{\Lambda}(\breve{\Upsilon})$ for the model (4). A holomorphic reparametrization of the Kähler manifold, $\Phi^I \to f^I(\Phi)$, has the following counterpart $\Upsilon^I(\zeta) \to f^I\big(\Upsilon(\zeta)\big)$ in the $\mathscr{N} = 2$ case. Therefore, the physical superfields of the $\mathscr{N} = 2$ theory

$$\Upsilon^I(\zeta)\Big|_{\zeta=0} = \Phi^I \ , \qquad \frac{\mathrm{d}\Upsilon^I(\zeta)}{\mathrm{d}\zeta}\Big|_{\zeta=0} = \Sigma^I \ , \tag{6}$$

should be regarded, respectively, as coordinates of a point in the Kähler manifold and a tangent vector at the same point. Thus the variables (Φ^I, Σ^J) parametrize the tangent bundle $T\mathscr{M}$ of the Kähler manifold $\mathscr{M}$ [8].

The presence of auxiliary superfields Υ_2, $\Upsilon_3 \ldots$ in (1) makes $\mathscr{N} = 2$ supersymmetry manifest, but the physical content of the theory is hidden. To describe the theory in terms of the physical superfields Φ and Σ only, all the auxiliary superfields have to be eliminated with the aid of the corresponding algebraic equations of motion

$$\oint \frac{d\zeta}{\zeta} \zeta^n \frac{\partial K(\Upsilon, \breve{\Upsilon})}{\partial \Upsilon^I} = \oint \frac{d\zeta}{\zeta} \zeta^{-n} \frac{\partial K(\Upsilon, \breve{\Upsilon})}{\partial \breve{\Upsilon}^{\bar{I}}} = 0 , \qquad n \geq 2 . \tag{7}$$

In general, these equations can be solved only pertubatively. However, as outlined in [9] and elaborated in detail in [11], the auxiliary fields may be eliminated exactly for any Hermitian symmetric space $\mathcal{M}$. Let us sketch the procedure of constructing such a solution $\Upsilon_*(\zeta) \equiv \Upsilon_*(\zeta;\Phi,\bar{\Phi},\Sigma,\bar{\Sigma})$ under the initial conditions (6).

Given an arbitrary point $p_0 \in \mathcal{M}$, there exists a Kähler normal coordinate frame with origin at p_0 such that

$$K_{I_1..I_m \bar{J}_1..\bar{J}_n}\Big|_{\Phi=0} = 0 , \qquad m \neq n \tag{8}$$

and therefore $K(\Phi,\bar{\Phi}) = F(\Phi\bar{\Phi})$, with $F(\Phi\bar{\Phi})$ a real analytic function. Now, for any tangent vector Σ_0 at the origin, one can see that

$$\Upsilon_0(\zeta) = \zeta \Sigma_0, \qquad \breve{\Upsilon}_0(\zeta) = -\frac{\bar{\Sigma}_0}{\zeta} \tag{9}$$

solve the equations (7) at $\Phi = 0$. We set $\Upsilon_*(\zeta;\Phi = 0,\bar{\Phi} = 0,\Sigma_0,\bar{\Sigma}_0) = \zeta\Sigma_0$. A next step is to distribute this solution to any point Φ of the manifold $\mathcal{M}$, that is to make use of $\Upsilon_*(\zeta;\Phi = 0,\bar{\Phi} = 0,\Sigma_0,\bar{\Sigma}_0)$ in order to obtain $\Upsilon_*(\zeta;\Phi,\bar{\Phi},\Sigma,\bar{\Sigma})$.

Let G be the isometry group of $\mathcal{M}$. It acts transitively on $\mathcal{M}$ by holomorphic transformations. Let U be the open domain on which the normal coordinate system is defined. It can always be chosen such that we can construct a coset representative, $\mathcal{S}\colon U \to G$, where $\mathcal{S}(p)\colon \mathcal{M} \to \mathcal{M}$ is a holomorphic isometry transformation with the property

$$\mathcal{S}(p)\, p_0 = p , \qquad \mathcal{S}(p) \in G , \qquad \forall p \in U .$$

In other words, $\mathcal{S}(p)$ maps the origin to p. In local coordinates, $\mathcal{S}(p) = \mathcal{S}(\Phi,\bar{\Phi})$, and it acts on a generic point $q \in U$ parametrized by complex variables $(\Psi^I, \bar{\Psi}^{\bar{J}})$ as follows:

$$\Psi \to \Psi' = f(\Psi;\Phi,\bar{\Phi}) , \qquad f(0;\Phi,\bar{\Phi}) = \Phi . \tag{10}$$

It is crucial that the holomorphic isometry transformations leave the equations (7) invariant. This means that applying $\mathcal{S}(\Phi,\bar{\Phi})$ to $\Upsilon_0(\zeta)$, eq. (9), gives

$$\Upsilon_0(\zeta) \;\to\; \Upsilon_*(\zeta) = f(\Upsilon_0(\zeta);\Phi,\bar{\Phi}) = f(\Sigma_0\,\zeta;\Phi,\bar{\Phi}) , \qquad \Upsilon_*(0) = \Phi . \tag{11}$$

Imposing the second initial condition in (6),

$$\Sigma^I = \Sigma_0^J \frac{\partial}{\partial \Psi^J} f^I(\Psi;\Phi,\bar{\Phi})\Big|_{\Psi=0} , \tag{12}$$

we are in a position to uniquely express Σ_0 in terms of Σ and Φ, $\bar{\Phi}$. By construction, Σ is a complex linear superfield constrained as in (2). As to Σ_0, it obeys a generalized linear constraint that follows from (12) by requiring $\bar{D}^2\Sigma = 0$.

Once the solution $\Upsilon_*(\zeta) \equiv \Upsilon_*(\zeta;\Phi,\bar{\Phi},\Sigma,\bar{\Sigma})$ is given, we are in a position in principle to compute the tangent bundle action

$$S_{\rm tb}[\Phi,\bar{\Phi},\Sigma,\bar{\Sigma}] = \frac{1}{2\pi {\rm i}} \oint \frac{{\rm d}\zeta}{\zeta} \int {\rm d}^8 z K\big(\Upsilon_*(\zeta),\breve{\Upsilon}_*(\zeta)\big) . \tag{13}$$

But it is extremely important to choose a simplest coset representative (using the natural freedom in its choice $\mathscr{S}(\Phi,\bar{\Phi}) \to \mathscr{S}(\Phi,\bar{\Phi})\,h(\Phi,\bar{\Phi})$, with $h(\Phi,\bar{\Phi})$ taking its values in the isotropy subgroup H at p_0). With a complicated coset representative chosen, it will be practically impossible to do the contour integral on the right of (13). Finally, after having evaluated the contour integral in (13), it only remains to dualize the complex linear tangent variables Σ's into chiral one-forms, in accordance with the generalized Legendre transform [5]. The target space for the model obtained is $T^*\mathscr{M}$ in the compact case, and a part of $T^*\mathscr{M}$ in the non-compact case. In our work [11], the above procedure was carried out for the following Hermitian symmetric spaces (HSS):

compact HSS	$\frac{U(n+m)}{U(n)\times U(m)}$	$\frac{SO(2n)}{U(n)}$	$\frac{Sp(n)}{U(n)}$	$\frac{SO(n+2)}{SO(n)\times SO(2)}$
non-compact HSS	$\frac{U(n,m)}{U(n)\times U(m)}$	$\frac{SO^*(2n)}{U(n)}$	$\frac{Sp(n,\mathbf{R})}{U(n)}$	$\frac{SO_0(n,2)}{SO(n)\times SO(2)}$

As an example, consider the Grassmannian $\mathscr{M} = G_{m,n+m}$. Its Kähler potential is

$$K(\Phi,\Phi^\dagger) = \ln\det(\mathbf{1}_m + \Phi^\dagger\Phi) = \ln\det(\mathbf{1}_n + \Phi\Phi^\dagger) . \tag{14}$$

where $\Phi = (\Phi^{i\alpha})(i = 1,\dots,n,\ \alpha = 1,\dots,m)$. The useful coset representative is

$$\mathscr{S}(\Phi,\bar{\Phi}) = \begin{pmatrix} \underline{s} & \Phi s \\ -\Phi^\dagger \underline{s} & s \end{pmatrix} , \qquad s^{-2} = \Phi^\dagger\Phi + \mathbf{1}_m , \quad \underline{s}^{-2} = \Phi\Phi^\dagger + \mathbf{1}_n . \tag{15}$$

Its use leads to the tangent bundle action [11]

$$S_{\rm tb} = \int {\rm d}^8 z \Big\{ K(\Phi,\Phi^\dagger) + \ln\det\Big(\mathbf{1}_m - (\mathbf{1}_m + \Phi^\dagger\Phi)^{-1}\Sigma^\dagger(\mathbf{1}_n + \Phi\Phi^\dagger)^{-1}\Sigma\Big)\Big\} . \tag{16}$$

Finally, applying the generalized Legendre transform to (16) gives the $\mathscr{N} = 2$ sigma model originally constructed in [12].

REFERENCES

1. B. Zumino, Phys. Lett. B **87**, 203 (1979).
2. L. Alvarez-Gaumé and D. Z. Freedman, Commun. Math. Phys. **80**, 443 (1981).
3. A. Karlhede, U. Lindström and M. Roček, Phys. Lett. B **147**, 297 (1984).
4. N. J. Hitchin, A. Karlhede, U. Lindström and M. Roček, Commun. Math. Phys. **108**, 535 (1987).
5. U. Lindström and M. Roček, Commun. Math. Phys. **115**, 21 (1988).
6. U. Lindström and M. Roček, Commun. Math. Phys. **128**, 191 (1990).
7. F. Gonzalez-Rey, M. Roček, S. Wiles, U. Lindström and R. von Unge, Nucl. Phys. B **516**, 426 (1998).
8. S. M. Kuzenko, Int. J. Mod. Phys. A **14**, 1737 (1999).
9. S. J. Gates Jr. and S. M. Kuzenko, Nucl. Phys. B **543**, 122 (1999); Fortsch. Phys. **48**, 115 (2000).
10. M. Arai and M. Nitta, Nucl. Phys. B **745**, 208 (2006).
11. M. Arai, S. M. Kuzenko and U. Lindström, JHEP **0702** (2007) 100.
12. U. Lindström and M. Roček, Nucl. Phys. B **222** (1983) 285.

Stringy Model of Cosmological Dark Energy

Irina Ya. Aref'eva[1]

Steklov Mathematical Institute of Russian Academy of Sciences,
Gubkin st., 8, 119991, Moscow, Russia

Abstract. A string field theory (SFT) nonlocal model of the cosmological dark energy providing $w < -1$ is briefly surveyed. We summarize recent developments and open problems, as well as point out some theoretical issues related with others applications of the SFT nonlocal models in cosmology, in particular, in inflation and cosmological singularity.

Keywords: Strings, cosmology, dark energy
PACS: 11.25.-w, 11.25.uv, 11.10Lm

INTRODUCTION

The origin of the dark energy (DE) is still a fascinating puzzle. Present cosmological observations do not exclude an evolving DE state parameter w. According recent data (U.Seljak, Pascos, July 07)

$$w = -1.04 \pm 0.06.$$

There are two questions to experimental data:

- Can we rule out a dynamical DE?
- Can we rule out $w < -1$?

Recent data are not enough to answer these two questions and moreover, within the next few years answers on these two questions will not accessible as it has previously expected. However one might wonder whether there is a room for $w < -1$ in a theory.

Dark energy models with the state parameter $w < -1$ violate the null energy condition (NEC). All local models realizing the NEC violation are unstable and violate usual physical requirements. To provide $w < -1$ a string field theory (SFT) nonlocal DE model has been proposed [1].

In this SFT nonlocal model our Universe is considered as a D3 non-BPS brane embedded in the 10 dimensional space-time. The role of the dark energy plays the Neveu-Schwarz (NS) string tachyon leaving in GSO$-$ sector. The tachyon action is dictated by the cubic fermionic SFT [2, 3] and it is nonlocal due to string effects [4].

We postulate a minimal form of the tachyon interaction with gravity. In the spatially flat FRW metric the model is described by a system of two nonlinear nonlocal equations for the tachyon field and the Hubble parameter. The corresponding potential has perturbative and nonperturbative minima. A transition from a perturbative vacuum to a non-perturbative one is interpreted as D-brane decay. It happens that this model under

[1] Email: arefeva@mi.ras.ru

CP957, *Particles, Strings, and Cosmology, 13th International Symposium*
edited by A. Rajantie, C. Contaldi, P. Dauncey, and H. Stoica

some conditions displays a phantom behaviour [5]. Note that unlike phenomenological phantom models here phantom appears in an effective theory. Since SFT is a consistent theory this approach does not suffer from usual problems which are inevitable for phenomenological phantom models. The UV completion is supposed to be solved by extending the one mode (tachyon) approximation.

SFT in the flat background dictates a particular value of the D-brane tension. It can be found from the requirement that the total energy of the system in the true non-perturbative vacuum is zero. In cosmology this total energy of the system in the true non-perturbative vacuum can be interpreted as the cosmological constant. It has been conjectured that an existence of a rolling solution describing a smooth transition to the true vacuum does define the value of the cosmological constant [1]. We cannot prove this conjecture but arguments to it favor can be given using the local approximation [8]. A recent breakthrough in solving numerically the full nonlinear and nonlocal system of equations [7] also supports this expectation.

MODEL

Our model is given by the following action [1]

$$S=\int d^4x\sqrt{-g}\left(\frac{R}{2\kappa^2}+\frac{1}{\lambda_4^2}\left(-\frac{\xi^2\alpha'}{2}g^{\mu\nu}\partial_\mu\phi(x)\partial_\nu\phi(x)+\frac{1}{2}\phi^2(x)-\frac{1}{4}\Phi^4(x)-T'\right)\right)$$

Here $g_{\mu\nu}$, κ and λ_4 are the four-dimensional metric, gravitational coupling constant and scalar field coupling constant, respectively; $\frac{1}{\lambda_4^2}=\frac{v_6M_s^4}{g_o}(\frac{M_s}{M_c})^6$, g_o is the open string dimensionless coupling constant, M_s is the string scale $M_s=1/\sqrt{\alpha'}$ and M_c is a scale of the compactification, v_6 is a number related with a volume of the 6-dimensional compact space. $T'=1/4+\Lambda'$, where Λ' is a dimensionless cosmological constant. ϕ is a tachyon field and Φ is related with ϕ by the following relation $\Phi=e^{\frac{\alpha'}{8}\Box_g}\phi$, where $\Box_g=\frac{1}{\sqrt{-g}}\partial_\mu\sqrt{-g}g^{\mu\nu}\partial_\nu$. This form of the nonlocal interaction is defined by the cubic fermionic SFT (CFSFT) [3]. More precisely, the CFSFT brings a more complicated form of the interaction, but by an analogy with the flat case [6, 5, 7] we believe that this approximation catches essential physical properties of the model and $\xi^2\approx 0.9556$ is a constant defined by the CFSFT.

In the spatially flat FRW metric with a scale factor $a(t)$ the equation for a space homogenous tachyon field Φ and the Friedmann equations have the form [1]

$$\left(\xi^2\mathscr{D}+1\right)e^{-\frac{1}{4}\mathscr{D}}\Phi = \Phi^3,\quad \mathscr{D}=-\partial_t^2-3H(t)\partial_t,\ \ H=\partial_t a/a \tag{1}$$

$$3H^2 = \frac{\kappa^2}{\lambda_4^2}\left(\frac{\xi^2}{2}\partial_t\phi^2-\frac{1}{2}\phi^2+\frac{1}{4}\Phi^4+\mathscr{E}_1+\mathscr{E}_2+T'\right), \tag{2}$$

$$\mathscr{E}_1 = -\frac{1}{8}\int_0^1 ds\left(\left(\xi^2\mathscr{D}+1\right)e^{\frac{s-2}{8}\mathscr{D}}\Phi\right)\cdot\left(\mathscr{D}\,e^{-\frac{1}{8}s\mathscr{D}}\Phi\right), \tag{3}$$

$$\mathscr{E}_2 = -\frac{1}{8}\int_0^1 ds\left(\partial_t(\xi^2\mathscr{D}+1)\,e^{\frac{s-2}{8}\mathscr{D}}\Phi\right)\cdot\left(\partial_t e^{-\frac{1}{8}s\mathscr{D}}\Phi\right). \tag{4}$$

The non-local energy $\mathscr{E}_1$ plays the role of an extra potential term and $\mathscr{E}_2$ the role of the kinetic term. Note that here we use a dimensionless time $t \to t\sqrt{\alpha'}$.

HOW WE STUDY OUR MODEL AND WHAT WE GET

Equations (1) and (2) form a rather complicated system of nonlinear nonlocal equations for functions Φ and $H(t)$ because of the presence of an infinite number of derivatives and a non-flat metric. Before to discuss the methods of study this model let us mention the known methods of study equation (1) in the flat background, $H = 0$,

$$(-\xi^2\partial_t^2 + 1)\, e^{\frac{1}{4}\partial_t^2}\Phi(t) = \Phi(t)^3. \tag{5}$$

Equation (2) in the flat case describes the energy conservation [5].

A boundary problem $\Phi(\pm\infty) = \pm 1$ for (5) has been studied using:

- a numerical method [6] based on an integral representation of (5); it is related with a diffusion equation method [10, 11] which uses an auxiliary function of two variables $\Psi(r,t)$ that is the subject of a linear equation and $\Psi(\frac{1}{4},t) = \Phi(t)$;
- a decomposition on local fields [12, 13, 14]; this method works well for linear equations and has been used to study solutions to (5) near vacuum ± 1;
- existence theorems [15, 16, 17] and almost exact solution methods [18, 19].

The following two characteristic properties of (5) have been obtained

- an existence of a critical point $\xi_{\text{cr}}^2 \approx 1.38$ such that for $\xi^2 < \xi_{\text{cr}}^2$ eq. (5) has a rolling solution [6] interpolating between ± 1;
- an existence of a dominance of an extra non-local kinetic term $\mathscr{E}_2$ over the local kinetic one [5] and as a result, an appearance of a phantom behavior providing $w < -1$.

These result have been obtained using numerical calculations. It is very interesting to study the problem analytically and also try to find approximate models admitting explicit solutions and having above mentioned properties. They could be two or more components local models.

An investigation of non-flat eqs. (1) and (2) is essentially more complicated. The following methods are used:

- A decomposition on local fields and a modification of the potential have been used in [8, 19, 9, 14]. A simplest one phantom mode approximation with an explicit form of the solution $\phi(t) = \tanh(t)$ is realized for a six-order potential [8] and gives $H_0 = 1/3m_p^2$. Assuming that $M_c \sim M_p$ and $M_s \sim 10^{-6.6}M_p$ we get

$$H_0 \sim 10^{-60}M_p.$$

- A numerical study has been performed in [7], where a diffusion equation method has been used to define $\exp\mathscr{D}$ and a double-step iteration procedure has been proposed.

The following physical effects are found in [7]

- For $\xi^2 < \xi^2_{cr} \approx 1.18$ and a suitable Λ the system (1), (2) has a rolling solution.
- For $\xi^2 < \xi^2_{shape}$ and $t > 0$ the Hubble function $H(t)$ is a monotonic function with an asymptotic H_0 and for $\xi^2_{shape} < \xi^2 < \xi^2_{cr}$ it has two maximum. To realize this shape by local fields one needs at least two fields [9].

As in the flat case it would be very interesting to find approximate analytical solutions which exhibit these properties. Note that two maximum shape regime for $H(t)$ is interesting in a context of building an unified cosmological evolution. Let us also note that there are applications of non-local SFT models to inflation [21, 22] and cosmological singularity (see refs. in [14]).

ACKNOWLEDGMENTS

I thank the organizers of PASCOS for the invitation to a very stimulating conference. I am grateful to L.V. Joukovskaya, A.S. Koshelev, S.Yu. Vernov and I.V. Volovich for collaboration and useful discussions. The work is supported in part by RFBR grant 05-01-00758, INTAS grant 03-51-6346 and Russian President's grant NSh-2052.2003.1.

REFERENCES

1. I.Ya. Aref'eva, AIP Conf. Proc. **826** (2006) 301–311, astro-ph/0410443; in: Contents and Structures of the Universe, Proc. of the XLIst Rencontres de Moriond, 2006, pp. 131-135.
2. I.Ya. Aref'eva, P.B. Medvedev, A.P. Zubarev, Phys.Let. **B** (1990); C.R. Preitschopf, C.B. Thorn, S.A. Yost, Nucl. Phys. **B337** (1990) 363; I.Ya. Aref'eva, P.B. Medvedev, A.P. Zubarev, Nucl. Phys. **B341** (1990) 464.
3. I.Ya. Arefeva, D.M. Belov, A.S. Koshelev, P.B. Medvedev, Nucl. Phys **B638**(2002) 3.
4. K. Ohmori, hep-th/0102085; I.Ya. Aref'eva, D.M. Belov, A.A. Giryavets, A.S. Koshelev, P.B. Medvedev, hep-th/0111208; W. Taylor, hep-th/0301094.
5. I.Ya. Aref'eva, L.V. Joukovskaya, A.S. Koshelev, JHEP **0309** (2003) 012.
6. Ya.I. Volovich, J. Phys. **A36** (2003) 8685.
7. L. Joukovskaya, arXiv:0707.1545; arXiv:0710.0404
8. I.Ya. Aref'eva, A.S. Koshelev, S.Yu. Vernov, astro-ph/0412619; Phys. Lett. **B628** (2005) 1.
9. I.Ya. Aref'eva, A.S. Koshelev, S.Yu. Vernov, Phys. Rev. **D72** (2005) 064017; S.Yu. Vernov, astro-ph/0612487.
10. V.S. Vladimirov, math-ph/0507018.
11. Calcagni, M.Montobbio and G.Nardelli, arXiv: 0705.3043.
12. I.Ya. Aref'eva and I.V. Volovich, hep-th/0612098; Int. J. of Geom. Meth. Mod. Phys. V.4 (2007) 881-895, hep-th/0701284.
13. I.Ya. Aref'eva, A.S. Koshelev, JHEP 0702 (2007) 041; A.S. Koshelev, JHEP 0704 (2007) 029.
14. I.Ya. Aref'eva, L.V. Joukovskaya and S.Yu.Vernov, JHEP 0707 (2007) 087.
15. V.S. Vladimirov, Ya.I. Volovich, math-ph/0306018.
16. L.V. Joukovskaya, arXiv:0708.0642.
17. D.V. Prokhorenko, math-ph/0611068.
18. V. Forini, G. Grignani, G. Nardelli, JHEP 0503 (2005) 079.
19. I.Ya. Aref'eva and L.V. Joukovskaya, JHEP 0510:087,2005.
20. G. Calcagni, JHEP **05** (2006) 012.
21. J.Lidsey, hep-th/0703007.
22. N. Barnaby, T. Biswas, J.M. Cline, hep-th/0612230; N. Barnaby, J.M. Cline, arXiv:0704.3426.

Almost the supersymmetric Standard Model from the $\mathbb{Z}_6'$ orientifold

David Bailin and Alex Love

Department of Physics & Astronomy, University of Sussex, Brighton BN1 9QH, UK

Abstract. Intersecting stacks of supersymmetric fractional branes on the $\mathbb{Z}_6'$ orientifold may be used to construct the supersymmetric Standard Model. We construct a four-stack model, which has the matter spectrum of the supersymmetric Standard Model, including a single pair of Higgs doublets, plus three neutrino-singlet states. However, the gauge group is $SU(3)_{\text{colour}} \times SU(2)_L \times U(1)_Y \times U(1)$, and there is no overall cancellation of Ramond-Ramond tadpoles

Keywords: Intersecting D6-branes, supersymmetric Standard Model, orientifold
PACS: 11.25.Wx, 12.10.Dm, 12.60.Jv

Open strings that begin and end on a stack a of N_a D6-branes wrapping a 3-cycle of $T^6 = T_1^2 \otimes T_2^2 \otimes T_3^2$ give the (massless) gauge bosons of $U(N_a) = U(1)_a \otimes SU(N_a)$. At the intersections of two stacks a and b there is chiral matter in the bi-fundamental $(\mathbf{N}_a, \overline{\mathbf{N}}_b)$ representation of $U(N_a) \otimes U(N_b)$, where $\mathbf{N}_a$ and $\overline{\mathbf{N}}_b$ respectively have charges $Q_a = +1$ and $Q_b = -1$ with respect to $U(1)_a$ and $U(1)_b$. To get the standard model, we start with $N_a = 3$ and $N_b = 2$ and then arrange that there are $a \cap b = 3$ intersections. This will give three quark doublets $Q_L = (\mathbf{3}, \overline{\mathbf{2}})$. However, $3(\mathbf{3}, \overline{\mathbf{2}})$ has $Q_b = -9$. The only other (chiral) doublets in the standard model are the three lepton doublets $L = (\mathbf{1}, \mathbf{2})$. Such states occur at intersections of the stack b with a single D6-brane c. If we arrange that there are $b \cap c = 3$ such intersections, there would remain $Q_b = 6$ units of doublet charge to be found from other sources to cancel Q_b overall. This requires additional (vector-like) non-standard model matter. If instead we wrap an "orientifold" T^6/Ω, where Ω is the world-sheet parity operator that acts on the world-sheet coordinates (τ, σ) as $\Omega(\tau, \sigma) = (\tau, \pi - \sigma)$, then at the intersections of a and the orientifold image b' of b the chiral matter is in the $(\mathbf{N}_a, \mathbf{N}_b) = (\mathbf{3}, \mathbf{2})$ representation. Then $2(\mathbf{3}, \overline{\mathbf{2}}) + 1(\mathbf{3}, \mathbf{2})$ has $Q_b = -3$ which *can* be cancelled by three lepton doublets. To get *just* the standard model [1] we require that $(a \cap b, a \cap b') = (2, 1)$ or $(1, 2)$.

D6-branes wrapping 3-cycles of T^6/Ω generally give a non-supersymmetric spectrum. This requires a low, (TeV-scale) unification/string scale to avoid the hierarchy problem. Such low-scale models have unacceptable levels of flavour changing neutral currents induced by world-sheet instantons [2]. Non-supersymmetric theories generally also have uncancelled NSNS tadpoles, which reflect the instability of the complex structure moduli of T^6. We can stabilise (some of) these moduli using an orbifold T^6/P. That is, we identify points on the torus that are related by the action of the point group P. The surviving states are then those that are invariant under the action of P. Also there are twisted sectors associated with the fixed points and planes of P. We only consider the point group $P = \mathbb{Z}_6'$. We study orientifolds in which the orbifold is quotiented with the world-sheet parity operator Ω. A supersymmetric theory allows a high string scale.

CP957, *Particles, Strings, and Cosmology, 13th International Symposium*
edited by A. Rajantie, C. Contaldi, P. Dauncey, and H. Stoica

Also, RR tadpole cancellation, which ensures anomaly cancellation in the associated quantum field theory, ensures NSNS tadpole cancellation too [3, 4]. The question we address is whether we can get *just* the supersymmetric standard model spectrum from the $\mathbb{Z}_6'$ orientifold.

We use a complex coordinate z_k $(k = 1,2,3)$ in each torus T_k^2. The generator θ of the point group P acts on these as $\theta z_k = e^{2\pi i v_k} z_k$. A supersymmetric embedding has $\sum_k v_k = 0$ mod 1. For the $\mathbb{Z}_6'$ orbifold, $\mathbf{v} = \frac{1}{6}(1,2,-3)$. The action of θ must be an automorphism of the lattice, so we may use an $SU(3)$ root lattice in the tori $T_{1,2}^2$. Then their complex structures $U_{1,2} = e^{i\pi/3}$. On T_3^2 the point group acts as a reflection, so U_3 is unconstrained. The embedding $\mathscr{R}$ of Ω acts as $\mathscr{R} z_k = \bar{z}_k$ and for this too to be an automorphism each lattice must be in either the **A** or **B** configuration. This fixes Re $U_3 = 0$ or $\frac{1}{2}$, but Im U_3 remains unconstrained. There are four independent (invariant, untwisted) bulk 3-cycles $\rho_{1,3,4,6}$. For example, $\rho_1 \equiv 2(\pi_{1,3,5} + \pi_{2,3,5} + \pi_{1,4,5} - 2\pi_{2,4,5})$, where $\pi_{1,3,5} \equiv \pi_1 \times \pi_3 \times \pi_5$ etc. The definition of the other ρ_p may be found in [5]. The general bulk 3-cycle Π_κ^{bulk}then has the form

$$\Pi_\kappa^{\text{bulk}} = \sum_{p=1,3,4,6} A_p^\kappa \rho_p$$

with the "bulk coefficients" $A_p^\kappa(n_k^\kappa, m_k^\kappa)$ determined by the wrapping numbers (n_k^κ, m_k^κ), $k = 1,2,3$ on the three tori. For example,

$$A_1^\kappa = (n_1^\kappa n_2^\kappa + n_1^\kappa m_2^\kappa + m_1^\kappa n_2^\kappa) n_3^\kappa$$

The formulae for the other A_p^κ are given in [5]. The only non-zero intersections are *even*: $\rho_1 \cap \rho_4 = 4 = \rho_3 \cap \rho_6$ and $\rho_1 \cap \rho_6 = -2 = \rho_3 \cap \rho_4$.

Supersymmetry constrains the bulk coefficients A_p^κ. For example, on the **AAA**-lattice

$$\sqrt{3} A_3^\kappa = (A_6^\kappa - 2A_4^\kappa) \text{Im } U_3$$
$$2A_1^\kappa - A_3^\kappa - \sqrt{3} A_6^\kappa \text{Im } U_3 > 0$$

To get odd intersection numbers you have to use "fractional" branes

$$\kappa = \frac{1}{2}\left[\Pi_\kappa^{\text{bulk}} + \Pi_\kappa^{\text{exceptional}}\right]$$

where, in the θ^3 twisted sector, $\Pi_\kappa^{\text{exceptional}}$ consists of a collapsed 2-cycle stuck at a ($\mathbb{Z}_2$) fixed point in $T_1^2 \times T_3^2$ times a 1-cycle in T_2^2. Exceptional cycles also occur in the $\theta^{2,4}$ twisted sectors. The former generates 8 independent cycles $\varepsilon_i, \tilde{\varepsilon}_i$, $(i = 1,4,5,6)$ with $\varepsilon_i \cap \tilde{\varepsilon}_j = -2\delta_{ij}$. Only this sector is rich enough to generate the required intersection numbers. The pairs of fixed points to be used are determined by the wrapping numbers in $T_{1,3}^2$ of the bulk part

In orientifolds there is also chiral matter in the $\mathbf{N}_\kappa \times \mathbf{N}_\kappa$ representation of $SU(N_\kappa)$ on the branes (as well as gauge particles). For the $SU(3)$ stack a this gives $\mathbf{3} \times \mathbf{3} = \mathbf{6} + \bar{\mathbf{3}}$, and we must exclude the symmetric representation $\mathbf{S}_a = \mathbf{6}$. Likewise, for the $SU(2)$ stack

b this gives $\mathbf{2} \times \mathbf{2} = \mathbf{3} + \mathbf{1}$, and we must exclude the symmetric representation $\mathbf{S}_b = \mathbf{3}$. The antisymmetric pieces are $\mathbf{A}_a = \overline{\mathbf{3}} \sim q_L^c$ and $\mathbf{A}_b = \mathbf{1} \sim \nu_L^c, \ell_L^c$, so we also require that $\#(\mathbf{A}_a) \leq 3$ and $\#(\mathbf{A}_b) \leq 3$. We have shown that we *can* find supersymmetric stacks a and b satisfying these constraints and with $(a \cap b, a \cap b') = (2,1)$ or $(1,2)$. In all cases there is neither symmetric nor antisymmetric matter on one of the stacks (usually the $SU(2)_L$ stack b). Whether or not there is any antisymmetric matter on the other stack depends upon the lattice.

For example[1], on the **AAA** lattice, take $(n_k^a, m_k^a) = (1,-1;1,0;1,-3)$. Then

$$a = \frac{1}{2}(-\rho_3 + 3\rho_6 + \varepsilon_1 + \varepsilon_6 + 2\tilde{\varepsilon}_1 + 2\tilde{\varepsilon}_6)$$

This gives $\#(\mathbf{S}_a) = 0$ and $\#(\mathbf{A}_a) = 0$. Supersymmetry fixes $\mathrm{Im}\, U_3 = -\frac{1}{\sqrt{3}}$. Also take $(n_k^b, m_k^b) = (1,0;1,0;1,0)$ so that

$$b = \frac{1}{2}(\rho_1 + \tilde{\varepsilon}_1 + \tilde{\varepsilon}_4)$$

and $\#(\mathbf{S}_b) = 0 = \#(\mathbf{A}_b)$. Then $(a \cap b, a \cap b') = (1,2)$ which gives the required quark doublets $3Q_L$.

To get the full standard-model spectrum we have to introduce further stacks $c, d, ..$ with just a single brane in each $N_{c,d,..} = 1$. The weak hypercharge Y is a linear combination of the $U(1)$ charges Q_κ associated with each stack $\kappa = a, b, c, d, ...$

$$Y = \sum_\kappa y_\kappa Q_\kappa$$

Since all 3 quark doublets Q_L have $Y = \frac{1}{6}$, and since they arise as $2(\mathbf{3}, \overline{\mathbf{2}}) + 1(\mathbf{3}, \mathbf{2})$, we must have $y_a = \frac{1}{6}$ and $y_b = 0$. The $U(1)$ stacks $c, d, ..$ are needed to make quark singlets u_L^c having $Y = -2/3$ and d_L^c having $Y = 1/3$ at intersections with a, and lepton and Higgs doublets having $Y = \pm\frac{1}{2}$ at intersections with b. Then $y_{c,d,...} = \pm\frac{1}{2}$. We have two further stacks. We take $(n_k^c, m_k^c) = (0,-1;-1,-1;0,1)$ so that

$$c = \frac{1}{2}(\rho_4 + 2\rho_6 - 2\varepsilon_1 - 2\varepsilon_5 - \tilde{\varepsilon}_1 - \tilde{\varepsilon}_5)$$

Then $(a \cap c, a \cap c') = (-3,-3)$, so taking $y_c = 1/2$ this gives the required $3u_L^c$ and $3d_L^c$ quark singlets. Similarly, $(b \cap c, b \cap c') = (1,1)$, which gives one Higgs doublet H with $Y = \frac{1}{2}$ and one lepton (or Higgs) doublets L (or $\bar{H}$) having $Y = -\frac{1}{2}$. Evidently we require a fourth stack d to generate the remaining doublets with $Y = -\frac{1}{2}$. We take $(n_k^d, m_k^d) = (1,-1;1,1;1,-1)$ so that

$$d = \frac{1}{2}(\rho_1 - \rho_3 - \rho_4 + \rho_6 - \varepsilon_1 - \varepsilon_6 + \tilde{\varepsilon}_1 + \tilde{\varepsilon}_6)$$

[1] The following example is *not* that presented at the PASCOS-07 meeting. The errors reported in the erratum to [5] make that example unviable.

Then $(a \cap d, a \cap d') = (0,0)$, so there are no (unwanted) further quark singlet states. Also, $(b \cap d, b \cap d') = (-2,1)$, which gives precisely the required three L or H doublets provided that $y_d = -\frac{1}{2}$. It is easy to verify that $\#(\mathbf{S}_c) = 0 = \#(\mathbf{S}_d)$, so that there are no lepton singlet states on c or d. Finally, $(d \cap c, d \cap c') = (-3,-3)$, so that there are three charged lepton $3\ell^c$ and three neutral lepton $3\nu_L^c$ singlet states at these intersections .

In total we have precisely the spectrum of the supersymmetric standard model, plus three neutrino singlet states ν_L^c. In the first instance the gauge group from these four stacks is

$$\begin{aligned} G &= U(3)_a \times U(2)_b \times U(1)_c \times U(1)_d \\ &= SU(3)_{\text{colour}} \times SU(2)_L \times U(1)_Y \times U(1)^3 \end{aligned}$$

The last 3 $U(1)$ factors are unwanted and unobserved. The Green-Schwarz mechanism gives any anomalous $U(1)$ gauge boson a string scale mass, $O(10^{17})$GeV in a supersymmetric theory, and the $U(1)$ survives only as a global symmetry. However, a non-anomalous $U(1)$ might remain massless, and therefore might be observable. The gauge boson associated with the general linear combination

$$X = \sum_{\kappa} x_\kappa Q_\kappa$$

does *not* acquire a string scale mass (via the GS mechanism) if

$$\sum_{\kappa} x_\kappa N_\kappa (\kappa - \kappa') = 0$$

Then $U(1)_Y$ *is* massless, as required. Unfortunately, $U(1)_c$ is also massless (on the string scale), so might be observable. In addition, for this to be a consistent string theory realisation there must be cancellation of RR tadpoles, which requires that the overall homology class of the D6-branes and the O6-plane must vanish

$$\sum_{\kappa=a,b,c,d} N_\kappa(\kappa + \kappa') = 4\Pi_{O6}$$

It is easy to see that this is *not* satisfied. This is consistent with the conclusions of Gmeiner and Honecker [6].

REFERENCES

1. L. E. Ibáñez, F. Marchesano and R. Rabadán, Getting just the standard model at intersecting branes, JHEP **0111** (2001) 002 [hep-th/0105155].
2. S. A. Abel, O. Lebedev and J. Santiago, Flavour in intersecting brane models and bounds on the string scale, Nucl. Phys. B **696** (2004) 141 [hep-ph/0312157].
3. M. Cvetič, G. Shiu and A. M. Uranga, Three-family supersymmetric standard like models from intersecting brane worlds, Phys. Rev. Lett. **87** (2001) 201801 [hep-th/0107143].
4. M. Cvetič, G. Shiu and A. M. Uranga, Chiral four-dimensional N = 1 supersymmetric type IIA orientifolds from intersecting D6-branes, Nucl. Phys. B **615** (2001) 3 [hep-th/0107166].
5. D. Bailin and A. Love, "Towards the supersymmetric standard model from intersecting D6-branes on the Z'(6) orientifold," Nucl. Phys. B **755** (2006) 79 [arXiv:hep-th/0603172] Erratum: Nucl. Phys. B **783** (2007) 176.
6. F. Gmeiner and G. Honecker, "Mapping an Island in the Landscape," arXiv:0708.2285 [hep-th].

Magnetogenesis from Rotating Cosmic String Loops

Thorsten Battefeld

DAMTP, Center for Mathematical Sciences, University of Cambridge, Wilberforce Road, Cambridge,CB3 0WA, UK

Abstract. We present a mechanism to create vortices in a plasma via gravitational dragging behind rotating cosmic string loops. The vortical motions create magnetic fields by means of the Harrison-Rees mechanism; the fields are further enhanced through galactic collapse and dynamo amplification. Employing the Velocity dependent One Scale model (VOS) for the string network and incorporating loop dynamics, we compute the magnetic fields generated around the time of decoupling: these are just strong and coherent enough to account for presently observed magnetic fields in spiral galaxies if efficient dynamos with $\Gamma_{dy}^{-1} \approx 0.3\,\text{Gyr}$ are present.

Keywords: cosmic strings, loops, magnetogenesis
PACS: 98.80.Cq,98.54.Kt

INTRODUCTION

Magnetic fields in the μG range have been observed in galaxies and clusters [1], though their origin remains a mystery. In the case of spiral galaxies, fields can be amplified considerably during the galaxy's numerous rotations by means of a dynamo; however, the difficulty of the problem still lies in the coherent and reasonably strong seed field requirement.

Over the years, several scenarios generating seed fields have been proposed, but very few compelling explanations exist: astrophysical mechanisms operating within the horizon are not optimal for producing large coherence lengths. Naturally, inflationary perturbations are present on the needed scales, but magnetogenesis is difficult to achieve due to conformal invariance in electromagnetism [1].

In this note we summarize the proposal of [2], explaining a possible mechanism to create plasma vortices via gravitational dragging in the vicinity of rotating cosmic string loops near the time of decoupling. These vortices subsequently cause magnetic fields through the Harrison-Rees mechanism [3, 4], which in turn are amplified during galaxy collapse and dynamo amplification. We try to be as conservative as possible regarding the generation mechanism: we use string network parameters well within observational limits and do not assume turbulence, inverse cascades or large scale averaging.

Previous work on magnetogenesis from string networks include [5, 6, 7, 8], which is extended in this study. We refer the reader to [1, 9, 10, 11] for general reviews on magnetogenesis, and to [12, 13] for string networks. We set $c \equiv 1$ throughout.

CP957, *Particles, Strings, and Cosmology, 13th International Symposium*
edited by A. Rajantie, C. Contaldi, P. Dauncey, and H. Stoica

THE MECHANISM

If a network of cosmic strings is present in the early universe, it has the intriguing feature to cause motions in the plasma on very large scales. This motion is caused by the gravitational interaction of strings with the plasma [14, 15, 16]: if one averages over small scale wiggles on strings and considers only mildly relativistic velocities, a string acts as if it were a classical string with mass per unit length λ in Newtonian gravity [14, 16] ($\lambda \approx 0.56\mu_0$ in the matter era, where μ_0 is the bare string mass density). Hence, a straight string moving with velocity v_s causes a wake with infalling velocity $v_i \sim G\lambda/v_s$ towards the sheet overswept by the string, and dragging velocity $v_d \sim v_i^2/v_s$. The effect increases due to the contribution of the string's deficit angle for relativistic velocities [14, 15], but we will be satisfied with the Newtonian estimate.

Vortices with $v_{rot} \sim v_i$ will arise on small scales due to turbulence in the wake, but it is rather questionable if these are large enough for magnetogenesis; on inter-string distances and without invoking turbulence, we expect vortices with $v_{rot} \sim v_d$ at most [1].

However, stronger vortices arise naturally behind rotating loops due to gravitational dragging (still disregarding turbulence). Furthermore, loops are constantly produced in a network due to string intersection. Thus, the loop population will naturally lead to a wide variety of vortices, up to inter-string distances. As an illustration, consider a circular loop of length ℓ, translational velocity v_t and rotational velocity $v_r \perp v_t$; we can then estimate the magnitude of the angular velocity after the loop traversed to [2]

$$\omega_{pl} \sim \frac{v_d}{\ell} \sim \frac{v_i^2}{\ell v_r} \sim \frac{(2\pi)^2\lambda^2 G^2}{7\,\ell v_t^2 v_r}\,. \tag{1}$$

This expression is valid for $v_t < v_r$, so that the plasma in the vicinity of the loop is exposed to a few rotations before the loop moves away. As expected, ω_{pl} is second order in λG, since it is due to gravitational dragging. Although not considered here, the angular velocity of the plasma might be further amplified by turbulent effects.

Once a vortex is produced in the plasma, it can cause a magnetic field via the Harrison-Rees (HR) effect: Compton scattering on the CMB affects electrons and protons differently [4], leading to the build-up of a current and ultimately to a magnetic field of magnitude [6]

$$B = \frac{2m}{e}\omega_{pl} \approx 10^{-4}\omega_{pl}\,, \tag{2}$$

where B is in Gauss and ω_{pl} in s^{-1}. The HR effect ceases to be efficient around the time of decoupling t_{dec}. As a result, vortices/magnetic fields produced during the matter era just before t_{dec}, which have the largest possible size, are of prime interest to us. After field generation, it redshifts as $B \propto a^{-2}$ due to flux conservation up until the protogalactic cloud collapses at around $z_{gf} \sim 10$.

At this redshift, one needs at least $B \propto 10^{-29}$ G on $0.5-5$ kpc scales to account for magnetic fields in spiral Galaxies: during the collapse of the protogalactic cloud the

[1] It was assumed in [7, 8] that vortices with $v_{rot} \sim v_i$ arise on inter-string distances, but no compelling argument was given for their presence – see [2] for a more detailed analysis.

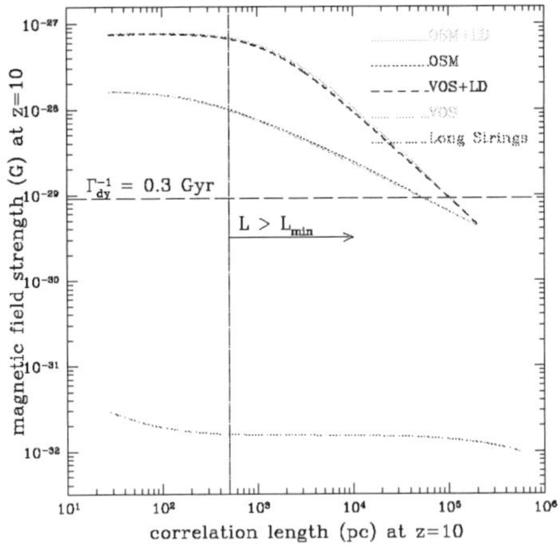

FIGURE 1. Magnetic field strength at $z_{gf} = 10$ for different string network models: OSM = One Scale Model, VOS = Velocity dependent One Scale model, with or without Loop Dynamics (LD); results are compared with the prediction from long string encounters, where the vortex is created in the region between two oppositely moving long strings. $L_{min} = 0.5\,\mathrm{kpc}$ covers only about 10% of the protogalactic cloud and is the minimal coherence length needed. $\Gamma_{dy}^{-1} = 0.3\,\mathrm{Gyr}$ indicates the dynamo needed to amplify B up to micro-Gauss field strength today – dynamos below that line are strongly disfavored. In the numerical simulation we use $G\mu_0 = 2\times10^{-7}$, $\alpha = 0.01$, $v_r(t_F) = 0.4$ and $v_t(t_F) = 0.1$.

field strength amplifies by a factor of order 8×10^3 [1]. If a very efficient dynamo amplification follows, for instance with a rate of $\Gamma_{dy}^{-1} = 0.3\,\mathrm{Gyr}$ [17], the field amplifies further by a factor of $\exp(\Gamma_{dy}^{-1}(t_0 - t_{gf})) \approx 1.4\times10^{19}$ up to today's micro-Gauss field strength. If dynamos are less efficient, as argued in [1], stronger seed fields are needed.

How do magnetic fields behind loops compare to these needed seed fields? Simulating a cosmic string network by means of the velocity dependent one scale model [18, 19], as well as incorporating the loop population's evolution due to redshifting, emission of gravitational waves and dragging effects, leads to field strength of order $B \sim 5\times10^{-28}\,\mathrm{G}$ on kpc scales at a redshift of $z_{gf} = 10$, see Figure 1. Here, we use a conservative choice of network parameters [2]: $G\mu_0 = 2\times10^{-7}$, initial loop length/horizon of $\alpha = 0.01$, as well as initial loop velocities of $v_r = 0.4$ and $v_t = 0.1$. Even if we feed only 10% of the energy transferred from the long string network into suitable large loops, fields cover already up to 10% of the universe. Further, since loops accrete matter and are attracted to existing overdense regions as well, a 10% coverage should be enough to cover most spiral galaxies. In addition, the coverage increases drastically if v_t or α are increased mildly. For variations of parameters we refer the interested reader to [2].

CONCLUSION AND OUTLOOK

In this note, we demonstrated a concrete mechanism to create vortices due to gravitational dragging behind rotating string loops. Subsequently, magnetic fields are generated by the Harrison-Rees mechanism. The resulting fields are just strong and coherent enough to account for magnetic fields in spiral galaxies, if an efficient dynamo is present. However, magnetic fields in elliptical galaxies or clusters seem to be out of reach.

In case dynamos with rates $\Gamma^{-1} \approx 0.3\,\text{Gyr}$ are ruled out, or if $G\mu_0 < 10^{-8}$ in future observations, our mechanism would be refuted. However, there is still the possibility that turbulent effects, which have been ignored entirely in this study, strengthen the existing vortices behind rotating loops. For instance, it is conceivable that turbulence redirects the infalling plasma into the existing vortex, increasing the angular velocity and correspondingly the generated magnetic field.

Thus, a numerical investigation into the flow patterns behind rotating loops is an interesting topic for future studies, especially if cosmic strings are observed in the near future.

ACKNOWLEDGMENTS

I would like to thank S. Alexander for early discussions motivating this project as well as A. C. Davis and K. Dimopoulos for feedback. Special thanks belong to my collaborators D. Battefeld, M. Wyman and D. Wesley. The work of T. B. is supported by PPARC grant PP/D507366/1.

REFERENCES

1. L. M. Widrow, Rev. Mod. Phys. **74**, 775 (2003) [arXiv:astro-ph/0207240].
2. D. Battefeld, T. Battefeld, D. H. Wesley and M. Wyman arXiv:0708.2901.
3. E. R. Harrison, Mon. Not. astr. Soc **147**, 279 (1970).
4. Martin J. Rees, Q. Jl R. astr. Soc **28**, 197 (1987).
5. D. N. Vollick, Phys. Rev. D **48**, 3585 (1993).
6. P. P. Avelino and E. P. S. Shellard, Phys. Rev. D **51**, 5946 (1995).
7. K. Dimopoulos, Phys. Rev. D **57**, 4629 (1998) [arXiv:hep-ph/9706513].
8. A. C. Davis and K. Dimopoulos, Phys. Rev. D **72**, 043517 (2005) [arXiv:hep-ph/0505242].
9. D. Grasso and H. R. Rubinstein, Phys. Rept. **348**, 163 (2001) [arXiv:astro-ph/0009061].
10. V. B. Semikoz and D. D. Sokoloff, Int. J. Mod. Phys. D **14**, 1839 (2005).
11. M. Giovannini, arXiv:astro-ph/0612378.
12. M. B. Hindmarsh and T. W. B. Kibble, Rept. Prog. Phys. **58**, 477 (1995) [arXiv:hep-ph/9411342].
13. Vilenkin, A. & Shellard, E.P.S. 2000, Cosmic Strings and Other Topological Defects (Cambridge:CUP)
14. T. Vachaspati and A. Vilenkin, Phys. Rev. Lett. **67**, 1057 (1991).
15. T. Vachaspati, Phys. Rev. D **45**, 3487 (1992).
16. D. N. Vollick, Phys. Rev. D **45**, 1884 (1992).
17. A.A. Ruzmaikin, D.D. Sokolov and V.I. Turchaninov, Astron. Zh. **57**, 311 (1980) [Sov. Astron. **24**, 182 (1980)].
18. C. J. A. P. Martins and E. P. S. Shellard, Phys. Rev. **D54** 2535 (1996), hep-ph/9602271; Phys. Rev. **D53** R575 (1996), hep-ph/9507335; Phys. Rev. **D65** 043514 (2002), hep-ph/0003298.
19. S. H. Tye, I. Wasserman and M. Wyman, Phys. Rev. D **71**, 103508 (2005) [Erratum-ibid. D **71**, 129906 (2005)] [arXiv:astro-ph/0503506].

Classical Renormalization of Codimension-two Brane Couplings

Claudia de Rham

Dept. of Physics & Astronomy, McMaster University, Hamilton ON, Canada
Perimeter Institute for Theoretical Physics, Waterloo, ON, Canada

Abstract. The curvature on codimension-two and higher branes is not regular for arbitrary matter sources. Nevertheless, the low-energy theory for an observer on such a brane should be well-defined and independent to any regularization procedure. This is achieved via appropriate *classical* renormalization of the brane couplings, and leads to a natural hierarchy between standard model couplings and couplings to gravity.

Keywords: large extra dimensions, six-dimensional gravity, brane couplings, renormalization
PACS: 11.10.Gh, 11.10.Kk, 11.25.-w

General matter sources on codimension-two branes

Codimension-two braneworld models have recently enjoyed an increased interest in their potential resolution of the cosmological constant problem, [1]. However, understanding the cosmology on such objects, requires the introduction of matter sources other than pure tension, which leads to metric divergences on the defect, [2]. As a consequence, it is necessary to regularize codimension-two branes before drawing any physical conclusion. A priori this is not a problem since any codimension-two object arises from an underlying theory which naturally provides a regularization scheme. However, we still expect from field theory that the low-energy theory on the brane is independent from such a regularization. Or in other words, at low-energy, one should be able to integrate out the fields responsible for the object regularization, and thus end up with a low-energy physics independent to the high-energy regularization scale. To this end, the same technique as that usually employed in field theory should be followed and the couplings of the theory should be renormalized, [3].

Origin of the Problem

To understand the origin of the problem, let us consider a free massless scalar field ϕ (symbolizing the graviton) living in a flat six-dimensional space-time

$$ds^2 = \eta_{\mu\nu}dx^\mu dx^\nu + dr^2 + r^2 d\theta^2, \tag{1}$$

with $0 \leq \theta < 2\pi\alpha$, and where $(1-\alpha)$ is the deficit angle ($\alpha \leq 1$). Its propagator $D(x^a, x'^a)$, satisfying $r\Box_x^{(6d)} D(x^a, x'^a) = i\delta^{(6)}(a^a - x'^a)$ is finite as long as one of the end point is taken away from $r = 0$, but it contains a logarithmic singularity when

CP957, *Particles, Strings, and Cosmology, 13th International Symposium*
edited by A. Rajantie, C. Contaldi, P. Dauncey, and H. Stoica

evaluating both points on the tip of the cone. Introducing a momentum cutoff scale Λ in the evaluation of the propagator, one has (see ref. [3]),

$$D_k(0,0) = -\int_0^\Lambda \frac{\mathrm{d}qq}{2\pi\alpha}\frac{i}{k^2+q^2} = \frac{-i}{2\pi\alpha}\log\frac{\Lambda}{k}, \tag{2}$$

where k^μ is the four-dimensional momentum. In principle, this should not be a problem, unless a physical source is introduced at $r=0$. In that case, gravity diverges on the brane. We can therefore wonder what observers on such a codimension-two brane located at $r=0$ and coupled to gravity will feel.

Observer on the brane

To mimic the effect on observers on the brane, we consider a second scalar field χ, that we call the Standard Model (SM) field, localized on the tip of the cone and coupled to "gravity" (the field ϕ):

$$S = -\int \mathrm{d}^4x\mathrm{d}\theta\mathrm{d}r\, r\left(\frac{1}{2}(\partial_a\phi)^2 + \delta^2(y)\left[\frac{1}{2}(\partial_\mu\chi)^2 + \frac{m^2}{2}\chi^2 + \lambda\chi\phi\right]\right). \tag{3}$$

In the absence of the coupling λ between the SM field χ and the graviton ϕ, the SM propagator would simply be $G_k^{\chi\chi} = -i(k^2+m^2)^{-1}$. The coupling with gravity will however induce divergences to the SM propagator, leading to the following "dressed" two-point function, [4]

$$G_k^{\chi\chi} = -\frac{i}{(k^2+m^2) - i\lambda^2 D_k(0,0)}. \tag{4}$$

This clearly shows how the coupling of gravity to the SM field introduces a singular term $D_k(0,0)$ in the expression of the SM propagator. As it stands, one can consider arbitrarily small energies, the propagator will still depend on the cutoff scale Λ, through $D_k(0,0) = -i/2\pi\alpha\log\Lambda/k$. To make sense of the theory at low energies, $k \gg \Lambda$, this divergence ought to be absorbed in the coupling constants. In particular the divergence disappears if the SM mass m is renormalized in the following way

$$m^2(\mu) = m^2(\Lambda) - \frac{1}{2\pi\alpha}\lambda^2(\Lambda)\log\frac{\Lambda}{\mu}, \tag{5}$$

leading to the following Renormalization Group (RG) flows

$$\mu\frac{\mathrm{d}m^2(\mu)}{\mathrm{d}\mu} = \frac{\lambda^2(\mu)}{2\pi\alpha}, \qquad \mu\frac{\mathrm{d}\lambda(\mu)}{\mathrm{d}\mu} = 0, \tag{6}$$

where μ is the physical scale. This mass renormalization is sufficient to ensure that the propagators for both fields are finite and independent of the cutoff scale. The Green's function of the matter field on the brane is thus well-defined despite being coupled with the graviton field which is itself not well-defined on the brane.

Further interactions

To extend this setup, further brane-bulk interactions can also be considered, also affecting the brane observables. In particular, restricting ourselves to the relevant and marginal operators, the most general brane interactions are then

$$S = -\int d^6x \left[\frac{1}{2}(\partial\phi)^2 + \frac{\delta(r)}{2r\pi\alpha}\mathscr{L}_{\chi\phi}\right], \tag{7}$$

$$\mathscr{L}_{\chi\phi} = \left(\frac{1}{2}(\partial\chi)^2 + \frac{1}{2}m^2\chi^2 + \beta_3\chi^3 + \beta_4\chi^4\right) + (\lambda\chi\phi + \lambda_2\chi^2\phi)\,. \tag{8}$$

At this point two independent kinds of divergences can be distinguished. Interactions of the form $\beta_n\chi^n$ with $n > 2$, induce loop corrections to the standard Green's functions which diverge in the ultra-violet. These divergences, standard in field theory, can be absorbed by renormalization of the mass m, the coupling β_4 as well as the wave-function. However, such divergences can be treated in a completely independent way to that arising at the tree-level in our codimension-two scenario. Interactions of the form $\lambda_3\phi\chi^2$, for instance will typically induce tree-level divergences which can be absorbed by appropriate renormalization of the couplings.

At tree-level, the divergent parts of the brane field three and four-point function are proportional to

$$\langle\chi^3\rangle_{\rm div} \propto (\beta_3 - i\lambda_3\lambda D_k(0,0)) \tag{9}$$

$$\langle\chi^4\rangle_{\rm div} \propto \left(\beta_4 - \frac{i}{2}\lambda_3^2 D_k(0,0)\right), \tag{10}$$

which again can be absorbed by appropriate renormalization of the couplings β_3 and β_4:

$$\mu\frac{d\beta_3(\mu)}{d\mu} = \frac{\lambda(\mu)\lambda_3(\mu)}{2\pi\alpha}, \qquad \mu\frac{d\beta_4(\mu)}{d\mu} = \frac{\lambda_3(\mu)^2}{4\pi\alpha}, \tag{11}$$

while the coupling λ_3 remains constant $\partial_\mu\lambda_3 = 0$, similarly as λ. Furthermore, these coupling flows is actually sufficient to ensure that every n-point function (with $n \geq 2$) is finite as the codimension-two cutoff scale Λ is sent to infinity, [4]. With respect to this divergence, the theory is thus completely renormalizable. We have therefore achieved to derive a low-energy effective field theory on the codimension-two brane, independent to any regularization procedure.

Hierarchy problem

Already at the level of this simple toy-model, one can distinguish between two different kinds of couplings:

- Bulk-Brane couplings λ_B, ($\lambda_B = \lambda, \lambda_3, \cdots$), symbolizing interaction with gravity, that do not flow in this specific example, $\partial_\mu\lambda = \partial_\mu\lambda_3 = 0$,

- Pure brane couplings β_b, ($\beta_b = m, \beta_3, \beta_4, \cdots$), symbolizing interactions from the SM, which typically flow as $\mu \partial_\mu \beta \propto \lambda \lambda_B$.

Couplings from the SM thus behave very differently from couplings with bulk fields (*i.e.* gravity), leading to the important conclusion that interactions with bulk fields are naturally suppressed compared to interactions of brane fields between themselves! Although this suppression is not sufficient to solve the hierarchy problem by itself, it can act constructively to the ADD scenario, [5]. Moreover, in higher-codimension scenarios, the suppression could be much more significant, since the divergence in this case is that of a power law rather than logarithmic. This could potentially lead to a potential hierarchy between SM forces and gravity in braneworld scenarios with codimension greater than two.

Discussion

So far, distributional sources on codimension-two branes only made sense as regularized objects, since bulk fields typically diverge logarithmically when evaluated on the brane. We have presented here how classical regularization of brane couplings lead to well-defined observables on the brane, and this independently to any regularization procedure. This provides a well defined low-energy effective theory on the brane which resulting RG flows can offer interesting physical implications. A hierarchy between the SM forces and gravity appears for instance as a natural consequence, and could be amplified in higher-codimension setups.

ACKNOWLEDGMENTS

Research at McMaster is supported by the Natural Sciences and Engineering Research Council of Canada. Research at Perimeter Institute for Theoretical Physics is supported in part by the Government of Canada through NSERC and by the Province of Ontario through MRI.

REFERENCES

1. V. A. Rubakov and M. E. Shaposhnikov, Phys. Lett. B **125**, 139 (1983); Y. Aghababaie, C. P. Burgess, S. L. Parameswaran and F. Quevedo, Nucl. Phys. B **680**, 389 (2004) [arXiv:hep-th/0304256]; A. Albrecht, C. P. Burgess, F. Ravndal and C. Skordis, Phys. Rev. D **65**, 123507 (2002) [arXiv:astro-ph/0107573];
2. R. Geroch and J. H. Traschen, Phys. Rev. D **36**, 1017 (1987).
3. W. D. Goldberger and M. B. Wise, Phys. Rev. D **65**, 025011 (2002) [arXiv:hep-th/0104170].
4. C. de Rham, arXiv:0707.0884 [hep-th].
5. N. Arkani-Hamed, S. Dimopoulos and G. R. Dvali, [arXiv:hep-ph/9803315].

Spinflation and Cycling Branes in Warped Throats

Damien A. Easson[1]

Centre for Particle Theory, Department of Mathematical Sciences, Durham University, Science Laboratories, South Road, Durham, DH1 3LE, UK

Abstract. The implications of brane motion in angular directions of Calabi-Yau flux compactifications is discussed from the point of view of an observer living on the worldvolume of such a brane and from the point of view of an observer living elsewhere in the three non-compact dimensions. The brane observer experiences cosmological bounces and cyclic behavior induced by centrifugal angular momentum barriers. Observers living elsewhere in the compactification experience marginally prolonged periods of inflation due to large angular momentum.

Keywords: D-branes, Flux compactifications, String theory and cosmology, Inflation
PACS: 98.80.Cq, 11.25.-w, 11.25.Uv, 11.25.Mj

INTRODUCTION

The most promising constructions of brane inflation in string theory are within the context of orientifolds of Calabi-Yau flux compactifications of Type IIB superstring theory. [2] The fluxes can produce a large warped throat region providing a relatively stable and concrete setting for the study of brane dynamics. We consider the motion of a D3-brane in this throat both from the point of view of an observer on the brane and from the point of view of an observer living elsewhere (*i.e.* somewhere in the compactification where the Standard Model can be embedded). In addition to examining radial motion of the brane we allow the brane to have angular momentum in the extra dimensions.

CYCLING BRANES AND ANGULAR MOMENTUM

The throat region is taken to be the nonsingular warped deformed conifold (or Klebanov-Strassler (KS) geometry) [7]. [3] Far from the tip the space is approximated by $T^{(1,1)}$ which is topologically an S^2 fibered over an S^3. The fluxes stabilize the dilaton and complex structure moduli but do not lift the closed string Kähler moduli associated with the volumes of four-cycles in the Calabi-Yau space [8]. This stabilization is accomplished by nonperturbative effects such as D7-branes wrapped around the four-cycles $\Sigma_4^{(i)}$ [9].

Consider the dynamics of a D3-brane probe in this background (see Figure 1). The D3 brane fills four-dimensional spacetime $\{t,x^1,x^2,x^3\}$ and appears pointlike in the internal

---

[1] Email: damien.easson@durham.ac.uk

[2] Some early constructions of brane inflation are [1, 2, 3, 4, 5, 6].

[3] Other possible geometry such as pure AdS_5 and the singular warped conifold are considered in [10].

CP957, *Particles, Strings, and Cosmology, 13th International Symposium*
edited by A. Rajantie, C. Contaldi, P. Dauncey, and H. Stoica

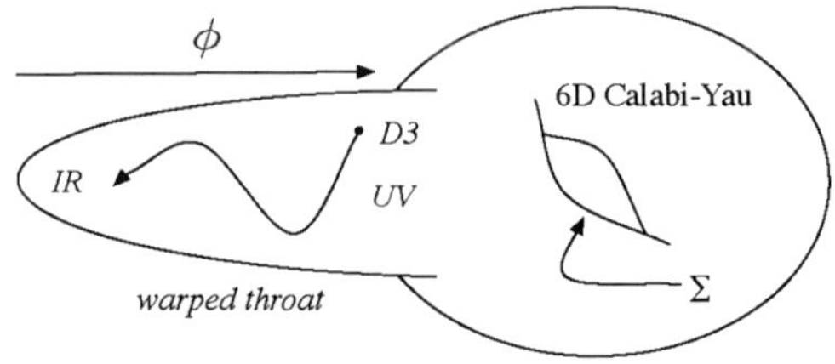

FIGURE 1. Sketch of Calabi-Yau flux compactification with warped throat region. The position of the D3-brane in the throat is parameterized by ϕ and Σ is a four-cycle. There are wrapped D-branes and orientifold planes not included in this diagram.

space. The location of the brane in the extra dimensions is parameterized by a field ϕ related to the radial throat coordinate r. In principle the brane is free to move in the five angular coordinates $y^s = \{\psi, \phi_1, \phi_2, \theta_1, \theta_2\}$. It is the motion in these angular coordinates that is usually ignored and that we focus on here.

The brane motion is described by the Dirac-Born-Infeld (DBI) action plus Wess-Zumino term:

$$S = -T_3 g_s^{-1} \int d^4x h^{-1} \left[e^{-3\phi} \sqrt{1 - hv^2} - 1 \right], \tag{1}$$

where T_3 is the brane tension, g_s is the string coupling, ϕ is the dilaton, $v^2 = g_{mn}\dot{y}^m\dot{y}^n$ and h is the warp factor of the KS geometry:

$$h(\eta) = (g_s M\alpha')^2 2^{2/3} \zeta^{-8/3} \int_\eta^\infty dx \frac{x \coth x - 1}{\sinh^2 x} (\sinh(2x) - 2x)^{1/3}. \tag{2}$$

Here $\alpha' = \ell_s^2$ is the square of the string length, ζ is a free parameter and η is the dimensionless coordinate parameterizing the throat. For an illustrative example, we allow the brane to move in the radial η direction and in the angular ψ and θ_1 directions: $v^2 = A(\eta)\left[\dot{\eta}^2 + \dot{\psi}^2\right] + B(\eta)\,\dot{\theta}_1^2$ where A and B are complicated functions of η defined in [10]. The two conserved angular momenta along ψ and θ_1 are $l_\psi = \frac{A\dot{\psi}}{\sqrt{1-hv^2}}$, $l_\theta = \frac{B\dot{\theta}_1}{\sqrt{1-hv^2}}$, and the conserved energy is

$$\varepsilon = \frac{1}{h}\left[\sqrt{\frac{1 + h\ell^2(\eta)}{1 - hg_{\eta\eta}\,\dot{\eta}^2}} - q\right] \tag{3}$$

where $\ell^2(\eta) = \frac{l_\psi^2}{A} + \frac{l_{\theta_1}^2}{B}$. The time evolution of the radial coordinate η is

$$\dot{\eta}^2 = \frac{\left[\varepsilon(h\varepsilon + 2) - \ell^2(\eta)\right]}{A(h\varepsilon + 1)^2} = Q. \tag{4}$$

the zeros of Q determine turning (bounce) points in the brane trajectory. We plot $Q(\eta)$ for typical values of the parameters in Figure 2.

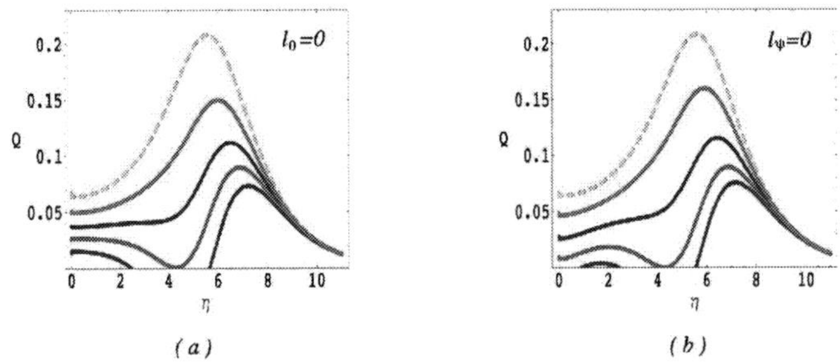

FIGURE 2. The function Q for a brane moving in the KS background as the angular momenta is increased from 0 (top, dashed line) to higher values. In (a) we keep $l_{\theta_1}=0$, and allow l_ψ to vary, whereas in (b) we keep $l_\psi=0$ and change l_{θ_1}.

Induced bouncing and cyclic cosmologies

An observer living on the D3-brane experiences expansion and contraction depending on the motion of the brane in the warped throat. The projected metric in four dimensions may be written in Friedmann-Robertson-Walker (FRW) form:

$$ds^2 = h^{-1/2}\left(-(1-hv^2)\,dt^2 + dx_i dx^i\right) = -d\tau^2 + a^2(\tau) dx_i dx^i\,, \tag{5}$$

with scale factor $a(\tau)=h^{-1/4}(\tau)$ and with the brane cosmic time related to the bulk time coordinate by $d\tau = h^{-1/4}\sqrt{(1-hv^2)}\,dt$. The resulting "cosmology" on the brane is induced by the warping of the background as in [11, 12]. From Figure 2 it is clear that the brane observer can experience bounces in the scale factor $a(\tau)$ either when the brane reaches the end of the smooth KS geometry or when a bounce occurs due to large angular momentum (zeros of Q). When the brane moves down the throat the brane universe contracts and when the brane moves up the throat it expands. From the lowermost solution in Figure 2 (b) we see a bound state where the brane bounces back and forth between two zeros of Q. This corresponds to a cyclic cosmology from the brane observer's point of view.

Spinflation

Alternatively one may consider dynamics from the perspective of an observer living elsewhere in the Calabi-Yau (*e.g.* on a stack of $\mathcal{N}$ D3-branes). In this case the motion of the D3-brane in the throat gives rise to an inflaton field ϕ (related to η) in the four-dimensional effective theory of the elsewhere observer. It is natural to write the effective theory obtained by integrating out the internal coordinates in terms of fields ϕ^m:

$$\begin{aligned} S &= \frac{M_{Pl}^2}{2}\int d^4x\sqrt{-g}R \\ &\quad - g_s^{-1}\int d^4x\sqrt{-g}\left[h^{-1}\sqrt{1+hg_{mn}g^{\mu\nu}\partial_\mu\phi^m\partial_\nu\phi^n} - qh^{-1} + V(\phi^m)\right]. \end{aligned} \tag{6}$$

The ordinary four-dimensional Einstein-Hilbert action arises from dimensional reduction of the closed string sector of the ten-dimensional action [13]. The Planck mass, M_{Pl} is relate to the internal six-dimensional volume by $M_{Pl}^2 = V_6/\kappa_{10}^2$, where $\kappa_{10}^2 = (2\pi)^7 g_s^2 \alpha'^4$ and V_6 is the volume of the internal six-dimensional manifold.

Assuming an FRW ansatz for the metric, the equations of motion can be written

$$h g_{\phi\phi}\dot{\phi}^2 = 1-\left(1+\frac{h\ell^2(\phi)}{a^6}\right)\cdot\left(1+h\left(\frac{H^2}{\beta}-V\right)\right)^{-2} \tag{7}$$

$$\dot{H} = -\frac{3\beta}{2}\left[2\left(\frac{H^2}{\beta}-V\right)+h\left(\frac{H^2}{\beta}-V\right)^2\right]\cdot\left(1+h\left(\frac{H^2}{\beta}-V\right)\right)^{-1}, \tag{8}$$

where we have defined the Hubble parameter $H \equiv \dot{a}/a$ and angular momentum $\ell^2(\phi) = g^{rs} l_r l_s$, $l_m \equiv a^3 g_{mn}\dot{\theta}^n\gamma$ and γ is the "lorentz" factor generated by the non-standard kinetic term for the fields in (6):

$$\gamma = \sqrt{\frac{1+h\ell^2(\phi)/a^6}{1-hg_{\phi\phi}\dot{\phi}^2}}. \tag{9}$$

From the point of view of the elsewhere observer and in contrast to the brane-bound observer the D3-brane inflaton *always* drives expansion. The brane probe undergoes damped bounces up and down through the tip of the KS throat leading to inflation sourced by the DBI action (6). In [14] we show that the presence of angular momentum ("spinflaton" fields) can help a given inflationary scenario by providing a small number of extra e-foldings of expansion at early times.

ACKNOWLEDGMENTS

It is a pleasure to thank my collaborators R. Gregory, D. Mota, G. Tasinato and I. Zavala and the organizers of PASCOS 07. This work is supported in part by PPARC and by the EU 6th Framework Marie Curie Research and Training network "UniverseNet" (MRTN-CT-2006-035863). DCPT-07/51.

REFERENCES

1. G. R. Dvali and S. H. H. Tye, Phys. Lett. B **450**, 72 (1999) [arXiv:hep-ph/9812483].
2. S. H. S. Alexander, Phys. Rev. D **65**, 023507 (2002) [arXiv:hep-th/0105032].
3. G. R. Dvali, Q. Shafi and S. Solganik, arXiv:hep-th/0105203.
4. C. P. Burgess, M. Majumdar, D. Nolte, F. Quevedo, G. Rajesh and R. J. Zhang, JHEP **0107**, 047 (2001) [arXiv:hep-th/0105204].
5. J. H. Brodie and D. A. Easson, JCAP **0312**, 004 (2003) [arXiv:hep-th/0301138].
6. S. Kachru, R. Kallosh, A. Linde, J. M. Maldacena, L. McAllister and S. P. Trivedi, JCAP **0310**, 013 (2003) [arXiv:hep-th/0308055].
7. I. R. Klebanov and M. J. Strassler, JHEP **0008**, 052 (2000) [arXiv:hep-th/0007191].
8. S. B. Giddings, S. Kachru and J. Polchinski, Phys. Rev. D **66**, 106006 (2002) [arXiv:hep-th/0105097].
9. S. Kachru, R. Kallosh, A. Linde and S. P. Trivedi, Phys. Rev. D **68**, 046005 (2003) [arXiv:hep-th/0301240].
10. D. Easson, R. Gregory, G. Tasinato and I. Zavala, JHEP **0704**, 026 (2007) [arXiv:hep-th/0701252].
11. A. Kehagias and E. Kiritsis, JHEP **9911**, 022 (1999) [arXiv:hep-th/9910174].
12. S. Kachru and L. McAllister, JHEP **0303**, 018 (2003) [arXiv:hep-th/0205209].
13. E. Silverstein and D. Tong, Phys. Rev. D **70**, 103505 (2004) [arXiv:hep-th/0310221].
14. D. Easson, R. Gregory, D. Mota, G. Tasinato and I. Zavala, to appear.

Consistent Kaluza-Klein reductions and supersymmetric AdS solutions

Jerome P. Gauntlett*,†, Eoin Ó Colgáin*,† and Oscar Varela*,†

*Theoretical Physics Group, Blackett Laboratory, Imperial College London, London SW7 2AZ, U.K.
†The Institute for Mathematical Sciences, Imperial College London, London SW7 2PE, U.K.

Abstract.
We discuss the conjecture that for any supersymmetric solution of $D = 10$ or $D = 11$ supergravity consisting of a warped product of a $d+1$-dimensional of anti-de-Sitter space with a Riemannian manifold M, there is a consistent Kaluza-Klein truncation on M to a gauged supergravity theory in $d+1$-dimensions for which the fields are dual to those in the superconformal current multiplet of the d-dimensional dual SCFT. This means that any solution of the gauged supergravity theory can be uplifted on M to obtain an exact solution of $D = 10$ or $D = 11$ supergravity. We verify this conjecture for the most general class of supersymmetric AdS_5 solutions of $D = 11$ supergravity by constructing an explicit KK reduction ansatz to minimal gauged supergravity in $D = 5$, at the level of the bosonic fields.

Keywords: M-Theory, Kaluza-Klein theory, AdS/CFT
PACS: 11.25.Yb 11.25.Tq 04.50.+h

INTRODUCTION

An interesting problem in Kaluza-Klein (KK) theory is to determine whether a lower dimensional theory can be embedded into a higher dimensional one, typically supergravity in ten or eleven dimensions. This in turn provides a useful method to construct solutions in $D = 10, 11$ by uplifting solutions of the lower dimensional theory. For this to work it is necessary that there is a *consistent* KK truncation on some internal manifold M from $D = 10$ or $D = 11$ down to the lower-dimensional theory. In general, a KK expansion on M leads to a lower dimensional theory involving an infinite tower of fields. Splitting these fields into a finite number of "light" fields and an infinite tower of "heavy" fields, the KK reduction is called consistent if it is in fact consistent to set all of the heavy fields to zero in the equations of motion, leaving equations of motion for the light fields only. Clearly this is only possible if the on-shell light fields do not source the heavy fields.

Only in certain cases, like toroidal or group manifold reductions, there are group theoretical arguments ensuring the consistency of the truncation. In fact, apart from these cases, consistent KK reductions are the exception rather than the rule. For example, it is only in very special circumstances that there is a consistent KK reduction on a sphere (see [1]). An interesting class of examples are those associated with the maximally supersymmetric solutions of $D = 10, 11$ supergravity that consist of products of AdS spaces and spheres. Corresponding to the $AdS_4 \times S^7$ and $AdS_7 \times S^4$ solutions of $D = 11$ supergravity, there are consistent KK reductions on S^7 [2] and S^4 [3, 4] to $D = 4$ $SO(8)$

CP957, *Particles, Strings, and Cosmology, 13th International Symposium*
edited by A. Rajantie, C. Contaldi, P. Dauncey, and H. Stoica

gauged supergravity and $D = 7$ $SO(5)$ gauged supergravity, respectively. Similarly, associated to the $AdS_5 \times S^5$ solution of type IIB supergravity there is expected to be a consistent KK reduction to $SO(6)$ gauged supergravity [5, 6, 7, 8].

These examples can be viewed as special cases of the following conjecture [9]:

> *For any supersymmetric solution of $D = 10$ or $D = 11$ supergravity that consists of a warped product of $d+1$ dimensional anti-de-Sitter space with a Riemannian manifold M, $AdS_{d+1} \times_w M$, there is a consistent Kaluza-Klein truncation on M to a gauged supergravity theory in $d+1$-dimensions for which the fields are dual to those in the superconformal current multiplet of the d-dimensional dual SCFT.*

Equivalently, one can characterise the fields of the gauged supergravity as those that contain the $d+1$-dimensional graviton and fill out an irreducible representation of the superisometry algebra of the $D = 10$ or $D = 11$ supergravity solution. This conjecture is essentially a restricted version of the one in [10], for which general arguments were put forward in [11].

Here, following [12], we will verify the conjecture, at the level of the bosonic fields, for the most general class of AdS_5 solutions of $D = 11$ supergravity [13] which are dual to $N = 1$ SCFTs in $d = 4$ dimensions. For this case, the bosonic fields in the superconformal current multiplet are the energy momentum tensor and the abelian R-symmetry current. Thus we seek a consistent truncation to minimal $D = 5$ gauged supergravity whose bosonic fields are the metric (dual to the energy momentum tensor of the SCFT) and an abelian gauge field (dual to the R-symmetry current). We do this by constructing an explicit consistent KK reduction ansatz, but before doing that, we review the undeformed geometry.

SUPERSYMMETRIC AdS_5 SOLUTIONS OF $D = 11$ SUGRA

The most general solution of $D = 11$ supergravity of the form $AdS_5 \times_w M_6$ was first analysed in [13] using G-structure techniques [14, 15]. The metric is a warped product

$$ds_{11}^2 = e^{2\lambda}[ds^2(AdS_5) + ds^2(M_6)], \tag{1}$$

where the metric $ds^2(AdS_5)$ is of unit radius and the warp factor λ is a function on M_6. Supersymmetry imposes an $SU(2)$ structure on M_6, specified by two real one-forms K^1, K^2, a real two-form J, a complex two-form Ω and a real scalar $\cos\zeta$, all of them constructed as bilinears of the Killing spinor. These forms are subject to a set of differential and algebraic constrains that restrict the $SU(2)$ structure. Also, the vector dual to $\tilde{K}^2 \equiv \cos\zeta K^2$ is a Killing vector (related to the R-symmetry of the dual CFT). The metric on M_6 can be written

$$ds^2(M_6) = e^i e^i + (K^1)^2 + (K^2)^2, \tag{2}$$

with $i = 1, \ldots, 4$, $J = e^1 \wedge e^2 + e^3 \wedge e^4$ and $\Omega = (e^1 + ie^2) \wedge (e^3 + ie^4)$.

As for the four-form G_4, it is restricted to lie entirely on M_6 to preserve the isometries of AdS_5. Supersymmetry further constrains it to be determined by the bilinear forms defining the $SU(2)$ structure, but its explicit expression will not be needed here.

REDUCTION OF $D = 11$ SUPERGRAVITY ON M_6

The KK reduction ansatz must express the $D = 11$ fields ds^2_{11}, $G_4 = dA_3$ in terms of the $D = 5$ fields ds^2_5, $F_2 = dA$. It is natural to think of the $D = 5$ $U(1)$ gauge field A as arising from the $U(1)$ isometry of M_6 generated by $\tilde{K}^2$. We can, accordingly, replace the metric on AdS_5 in (1) by the arbitrary $D = 5$ metric ds^2_5, and shift the Killing direction K^2 by the (normalised for convenience) $U(1)$ gauge field A,

$$K^2 \quad \longrightarrow \quad \hat{K}^2 \equiv K^2 + \tfrac{1}{3}\cos\zeta A\,, \tag{3}$$

to obtain the usual KK ansatz for the metric

$$ds^2_{11} = e^{2\lambda}\left[ds^2_5 + ds^2(\hat{M}_6)\right]\,, \tag{4}$$

with $ds^2(\hat{M}_6)$ obtained by making the shift (3) in (2).

The ansatz G_4' for the four-form is given by

$$G_4' = \hat{G}_4 + F_2 \wedge \hat{\beta}_2 + *_5 F_2 \wedge \hat{\beta}_1\,, \tag{5}$$

where hatted forms are defined on $\hat{M}_6$ and are obtained from their unhatted counterparts on M_6 through the prescription (3). In particular, $\hat{G}_4$ is obtained from the four-form on M_6 corresponding to the undeformed background $AdS_5 \times_w M_6$ and β_1, β_2 are forms on M_6 to be determined.

Now, direct substitution shows that the $D = 11$ fields (4), (5) satisfy the equations of motion and Bianchi identities of $D = 11$ supergravity provided that 1) the $D = 5$ fields ds^2_5, F_2 satisfy the equations of motion and Bianchi identities of gauged minimal supergravity in $D = 5$, and 2) the unknown forms β_1, β_2 on M_6 satisfy certain algebraic and differential relations. These relations are, precisely, of the same form as those satisfied by some of the bilinears defining the $SU(2)$ structure on M_6. Indeed, the forms β_1, β_2 have a solution in terms of the spinor bilinears as

$$\beta_1 = -\tfrac{1}{3}e^{3\lambda}\cos\zeta K^1 \tag{6}$$

$$\beta_2 = \tfrac{e^{3\lambda}}{3}(-\sin\zeta J + K^1 \wedge K^2) \tag{7}$$

This proves, at the level of the bosonic fields, that there is a consistent Kaluza-Klein truncation of $D = 11$ supergravity down to $D = 5$ minimal gauged supergravity [12]. Any solution of the latter theory can be uplifted, using the expressions (4), (5) with (6), (7), to solutions of the former theory. The truncation is also consistent at the level of the supersymmetry variations of the gravitino [12].

DISCUSSION

A consistent KK reduction was constructed in [16] (see also [17]) for the class of IIB solutions of the form $AdS_5 \times SE_5$, where SE_5 is a $D = 5$ Sasaki-Einstein manifold, and only the self-dual five-form is non-vanishing. This result was extended in [9] to the most general $AdS_5 \times_w M_5$ supersymmetric IIB solutions with all of the fluxes active, that were analysed in [18]. These results, together with those of those of [12] reviewed here, cover all AdS_5 solutions in $D = 10$ and $D = 11$ dimensions, considering that IIA solutions are special cases of $D = 11$ ones, and assuming that there are no AdS_5 solutions in type I supergravity. The KK reductions associated to general classes of $D = 11$ supergravity solutions containing AdS_4 factors considered in [19] were also analysed in [9]. It would be interesting to cast the known sphere compactifications in this language.

The consistency of the KK truncation makes it manifest from the gravity side that SCFTs with a type IIB or $D = 11$ dual share common sectors. For example, if we consider such SCFTs in $d = 4$, the black hole solutions of minimal gauged supergravity constructed in [20] should be relevant for any of the SCFTs. In fact, it would be interesting to try and prove the conjecture of [9] from the SCFT point of view.

ACKNOWLEDGMENTS

JPG is supported by an EPSRC Senior Fellowship and a Royal Society Wolfson Award. OV is supported by the Spanish Ministry of Science and Education through a postdoctoral fellowship and partially through the research grant FIS2005-02761.

REFERENCES

1. M. Cvetic, H. Lu and C. N. Pope, Phys. Rev. D **62** (2000) 064028 [arXiv:hep-th/0003286].
2. B. de Wit and H. Nicolai, Nucl. Phys. B **281** (1987) 211.
3. H. Nastase, D. Vaman and P. van Nieuwenhuizen, Phys. Lett. B **469** (1999) 96 [arXiv:hep-th/9905075].
4. H. Nastase, D. Vaman and P. van Nieuwenhuizen, Nucl. Phys. B **581** (2000) 179 [arXiv:hep-th/9911238].
5. M. Cvetic *et al.*, Nucl. Phys. B **558** (1999) 96 [arXiv:hep-th/9903214].
6. H. Lu, C. N. Pope and T. A. Tran, Phys. Lett. B **475** (2000) 261 [arXiv:hep-th/9909203].
7. M. Cvetic, H. Lu, C. N. Pope, A. Sadrzadeh and T. A. Tran, Nucl. Phys. B **586** (2000) 275 [arXiv:hep-th/0003103].
8. A. Khavaev, K. Pilch and N. P. Warner, Phys. Lett. B **487** (2000) 14 [arXiv:hep-th/9812035].
9. J. P. Gauntlett and O. Varela, arXiv:0707.2315 [hep-th].
10. M. J. Duff and C. N. Pope, Nucl. Phys. B **255** (1985) 355.
11. C. N. Pope and K. S. Stelle, Phys. Lett. B **198** (1987) 151.
12. J. P. Gauntlett, E. O. Colgain and O. Varela, JHEP **0702** (2007) 049 [arXiv:hep-th/0611219].
13. J. P. Gauntlett, D. Martelli, J. Sparks and D. Waldram, Class. Quant. Grav. **21** (2004) 4335 [arXiv:hep-th/0402153].
14. J. P. Gauntlett, D. Martelli, S. Pakis and D. Waldram, Commun. Math. Phys. **247** (2004) 421 [arXiv:hep-th/0205050].
15. J. P. Gauntlett and S. Pakis, JHEP **0304** (2003) 039 [arXiv:hep-th/0212008].
16. A. Buchel and J. T. Liu, Nucl. Phys. B **771** (2007) 93 [arXiv:hep-th/0608002].
17. T. T. Tsikas, Class. Quant. Grav. **3** (1986) 733.
18. J. P. Gauntlett, D. Martelli, J. Sparks and D. Waldram, Class. Quant. Grav. **23** (2006) 4693 [arXiv:hep-th/0510125].
19. J. P. Gauntlett, O. A. P. Mac Conamhna, T. Mateos and D. Waldram, JHEP **0611** (2006) 053 [arXiv:hep-th/0605146].
20. J. B. Gutowski and H. S. Reall, JHEP **0402** (2004) 006 [arXiv:hep-th/0401042].

Renormalisation of $\mathcal{N}=\frac{1}{2}$ Supersymmetry

I. Jack, D.R.T. Jones, R. Purdy and L. Worthy

Dept. of Mathematical Sciences, University of Liverpool, L69 3BX, UK

Abstract. We investigate the one-loop renormalisation of $\mathcal{N}=\frac{1}{2}$ supersymmetry in two and four dimensions.

Keywords: non-anticommutative, supersymmetry
PACS: 11.10.Gh,12.60.Jv,11.10.Kk

$\mathcal{N}=\frac{1}{2}$ SUPERSYMMETRIC GAUGE THEORY IN 4D

$\mathcal{N}=1$ supersymmetry can be deformed [1] by making the superspace co-ordinates non-anticommuting, e.g.

$$\{\theta_\alpha,\theta_\beta\}=C_{\alpha\beta},\quad \{\overline{\theta}_{\dot\alpha},\overline{\theta}_{\dot\beta}\}=\{\theta_\alpha,\overline{\theta}_{\dot\beta}\}=0. \tag{1}$$

This leads to a supersymmetry algebra

$$\{Q_\alpha,Q_\beta\}=0,\quad \{\overline{Q}_{\dot\alpha},Q_\beta\}=2i\sigma^\mu_{\alpha\dot\alpha}\frac{\partial}{\partial y^\mu},\qquad \{\overline{Q}_{\dot\alpha},\overline{Q}_{\dot\beta}\}=-4C^{\alpha\beta}\sigma^\mu_{\alpha\dot\alpha}\sigma^\nu_{\beta\dot\beta}\frac{\partial^2}{\partial y^\mu\partial y^\nu},$$
$$\text{where}\quad y^\mu = x^\mu+i\theta^\alpha\sigma^\mu_{\alpha\dot\alpha}\overline{\theta}^{\dot\alpha}. \tag{2}$$

The algebra of the Qs is unchanged, i.e. half the supersymmetry is unbroken–hence the name "$\mathcal{N}=\frac{1}{2}$".

The action for an ordinary $\mathcal{N}=1$ supersymmetric gauge theory is written as an integral of vector superfields $V(x,\theta,\overline{\theta})$ over superspace. $V(x,\theta,\overline{\theta})$ can be written as a power series in θ, $\overline{\theta}$ in terms of ordinary fields $A_\mu(x)$, $\lambda(x)$ (gaugino), $D(x)$ (auxiliary field). In the deformed case, we obtain the action for pure $N=\frac{1}{2}$ supersymmetric gauge theory[2]:

$$S=\int d^4x\,\mathrm{tr}\left[-\tfrac{1}{2}F^{\mu\nu}F_{\mu\nu}-2i\overline{\lambda}\overline{\sigma}^\mu(D_\mu\lambda)+D^2-2igC^{\mu\nu}F_{\mu\nu}\overline{\lambda}\overline{\lambda}+g^2\mathrm{C}^2(\overline{\lambda}\overline{\lambda})(\overline{\lambda}\overline{\lambda})\right], \tag{3}$$

$$\text{where}\quad C^{\mu\nu}=C^{\alpha\beta}\varepsilon_{\beta\gamma}\sigma^{\mu\nu\gamma}_{\alpha},\quad \sigma^{\mu\nu}=\tfrac{1}{4}[\sigma^\mu,\overline{\sigma}^\nu],\quad \mathrm{C}^2=C^{\mu\nu}C_{\mu\nu}. \tag{4}$$

This action is invariant under

$$\begin{aligned}\delta A_\mu &= -i\overline{\lambda}\overline{\sigma}_\mu\varepsilon,\quad \delta D=-\varepsilon\sigma^\mu D_\mu\overline{\lambda}\\ \delta\lambda &= i\varepsilon D+(\sigma^{\mu\nu}\varepsilon)\left[F_{\mu\nu}+iC_{\mu\nu}\overline{\lambda}\overline{\lambda}\right],\quad \delta\overline{\lambda}=0,\end{aligned} \tag{5}$$

which, apart from the $C^{\mu\nu}$ term, are simply the $\mathcal{N}=1$ transformations with $\overline{\varepsilon}=0$. The natural gauge group is $U(N)$ since the transformations fail to close in $SU(N)$. Recall that

CP957, *Particles, Strings, and Cosmology, 13th International Symposium*
edited by A. Rajantie, C. Contaldi, P. Dauncey, and H. Stoica

for $SU(N)$

$$[R^a, R^b] = i f^{abc} R^c, \quad \{R^a, R^b\} = d^{abc} R^c + \frac{1}{N}\delta^{ab}, \tag{6}$$

whereas for $U(N)$

$$[R^A, R^B] = i f^{ABC} R^C, \quad \{R^A, R^B\} = d^{ABC} R^C \quad \text{where} \quad R^0 = \sqrt{\frac{1}{2N}}1, \quad d^{ab0} = \sqrt{\frac{2}{N}}\delta^{ab}, \tag{7}$$

where $A,B,C\ldots$ run over the $SU(N)$ indices $a,b,c\ldots$ and the $U(1)$ index 0. The problem is that the $U(N)$ gauge theory is not renormalisable; the $SU(N)$ and $U(1)$ couplings renormalise differently. For the pure gauge theory we can specialise to $SU(N)$ just by striking out $U(1)$ pieces from action and transformations[3]-[5]. (This doesn't work for chiral matter, as we shall see shortly.) We can write the action as

$$S = \tfrac{1}{2}\int d^4x \Big[-\tfrac{1}{2}F^{\mu\nu a}F^a_{\mu\nu} - 2i\bar{\lambda}^a\bar{\sigma}^\mu (D_\mu\lambda)^a + D^aD^a - (T_1 - T_2) + \tfrac{1}{4}T_3 \Big] \tag{8}$$

$$\text{where} \quad T_1 = i d^{abc} g C^{\mu\nu}\partial_\mu A^a_\nu \bar{\lambda}^b\bar{\lambda}^c, \qquad T_2 = i d^{abe} f^{cde} g^2 C^{\mu\nu} A^c_\mu A^d_\nu \bar{\lambda}^a\bar{\lambda}^b,$$
$$T_3 = d^{abe} d^{cde} g^2 \mathrm{C}^2 (\bar{\lambda}^a\bar{\lambda}^b)(\bar{\lambda}^c\bar{\lambda}^d). \tag{9}$$

If we introduce bare fields and couplings according to

$$\lambda_B = Z_\lambda^{\frac{1}{2}}\lambda, \quad \bar{\lambda}_B = Z_\lambda^{\frac{1}{2}}\bar{\lambda}, \qquad A_{B\mu} = Z_A^{\frac{1}{2}} A_\mu, \quad D_B = Z_D^{\frac{1}{2}} D,$$
$$g_B = Z_g g, \quad C_B = Z_C C, \quad \mathrm{C}^2_B = Z_{\mathrm{C}^2}\mathrm{C}^2, \tag{10}$$

the renormalisation constants for the fields and for the gauge coupling g are the same as in the ordinary $N = 1$ supersymmetric theory. We find that the divergent part of the one-loop effective action is given by

$$\Gamma^{(1)\text{pole}}_{1\text{PI}} = \Gamma^{(1)\text{pole}}_{1\text{PI}(\mathcal{N}=1)} + \int d^4x \Big\{ -Z_g Z_A^{\frac{1}{2}} Z_\lambda T_1 + \tfrac{1}{2} Z_g^2 Z_A Z_\lambda T_2 + \tfrac{1}{8} Z_g^2 Z_\lambda^2 T_3$$
$$+ NL\left[-\tfrac{1}{4}(T_1 - T_2) + A_1 - \tfrac{1}{2}A_2\right]\Big\}, \tag{11}$$

$$\text{where}\, L = \frac{g^2}{16\pi^2\varepsilon}, A_1 = i d^{abc} g C^\mu{}_\nu A^a_\mu \bar{\lambda}^b \bar{\sigma}^{\nu\rho}\partial_\rho\bar{\lambda}^c, A_2 = i d^{cde} f^{abe} g^2 C^{\mu\rho} A^c_\mu A^d_\nu \bar{\lambda}^a \bar{\sigma}^{\nu\rho}\bar{\lambda}^b. \tag{12}$$

The last four terms conflict with renormalisability and gauge invariance. We find that the unwanted terms can all be removed by a divergent redefinition

$$\delta\lambda^a = -\tfrac{1}{2} NLgC^{\mu\nu} d^{abc}\sigma_\mu \bar{\lambda}^c A^b_\nu. \tag{13}$$

The theory is therefore renormalisable with $Z_C = Z_{\mathrm{C}^2} = 1$. Upon including matter the $SU(N)$ transformations fail to close. However, it turns out that we can define an

$SU(N)\times U(1)$ invariant action (with separate couplings g and g_0 for $SU(N)$ and $U(1)$ respectively):

$$\begin{aligned} S &= \int d^4x\Big[-\tfrac{1}{4}F^{\mu\nu A}F^A_{\mu\nu} - i\bar{\lambda}^A\bar{\sigma}^\mu(D_\mu\lambda)^A + \tfrac{1}{2}D^AD^A - \tfrac{1}{2}iC^{\mu\nu}d^{ABC}e^{ABC}F^A_{\mu\nu}\bar{\lambda}^B\bar{\lambda}^C \\ &+ \tfrac{1}{8}g^2\mathrm{C}^2 d^{abe}d^{cde}(\bar{\lambda}^a\bar{\lambda}^b)(\bar{\lambda}^c\bar{\lambda}^d) + \tfrac{1}{4N}\tfrac{g^4}{g_0^2}\mathrm{C}^2(\bar{\lambda}^a\bar{\lambda}^a)(\bar{\lambda}^b\bar{\lambda}^b) \\ &+ \tfrac{1}{N}\gamma g_0^2\mathrm{C}^2(\bar{\lambda}^a\bar{\lambda}^a)(\bar{\lambda}^0\bar{\lambda}^0), \end{aligned} \tag{14}$$

where

$$e^{abc} = g, \quad e^{a0b} = e^{ab0} = e^{000} = g_0, \quad e^{0ab} = \frac{g^2}{g_0}. \tag{15}$$

(We omit the matter terms for conciseness.) The γ term is not present classically but required for renormalisability. It is separately $\mathcal{N}=\frac{1}{2}$ invariant.

This action is invariant under

$$\begin{aligned} \delta A^A_\mu &= -i\bar{\lambda}^A\bar{\sigma}_\mu\varepsilon, \quad \delta D^A = -\varepsilon\sigma^\mu D_\mu\bar{\lambda}^A \\ \delta\lambda^A &= i\varepsilon D^A + (\sigma^{\mu\nu}\varepsilon)\left[F^A_{\mu\nu} + \tfrac{1}{2}iC_{\mu\nu}e^{ABC}d^{ABC}\bar{\lambda}^B\bar{\lambda}^C\right], \quad \delta\bar{\lambda}^A = 0. \end{aligned} \tag{16}$$

We find that the theory is renormalisable, now with[4] $Z_C = Z_{C^2} = 1, \quad \gamma = -3NLg^2$. The analysis can also be extended to a theory with superpotential[5].

$\mathcal{N}=\frac{1}{2}$ SUPERSYMMETRY IN TWO DIMENSIONS

In two dimensions it is convenient to use $x^\pm$, $\theta^\pm$, $\bar{\theta}^\pm$. We now consider a theory of chiral superfields $\Phi(x^\pm,\theta^\pm,\bar{\theta}^\pm)$ (with components ϕ,ψ,F). The simplest model is the 2-d $\mathcal{N}=2$ non-linear σ model which is in (undeformed) superspace

$$L_{\mathcal{N}=2} = \int d^2\theta d^2\bar{\theta} K(\Phi,\bar{\Phi}) \tag{17}$$

where K is the Kähler potential.

For the deformed version we take

$$(\theta^\pm)^2 = (\bar{\theta}^\pm)^2 = 0, \quad \{\bar{\theta}^+,\bar{\theta}^-\} = 0, \quad \{\theta^+,\theta^-\} = \frac{1}{M}. \tag{18}$$

The $\mathcal{N}=\frac{1}{2}$ algebra is then

$$Q_+^2 = Q_-^2 = \bar{Q}_+^2 = \bar{Q}_-^2 = \{Q_+,Q_-\} = 0, \{\bar{Q}_+,\bar{Q}_-\} = -\frac{4}{M}\frac{\partial^2}{\partial y^+\partial y^-}, \quad y^\pm = x^\pm - i\theta^\pm\bar{\theta}^\pm. \tag{19}$$

In components the deformed lagrangian is[6]-[8]

$$\mathcal{L} = \partial_{\bar{j}}K_0\partial_+\partial_-\bar{\phi}^{\bar{j}} + \partial_{\bar{j}}\partial_{\bar{k}}K_0\partial_+\bar{\phi}^{\bar{j}}\partial_-\bar{\phi}^{\bar{k}} + \partial_i\partial_{\bar{j}}K_0\left(i\psi^i_+\partial_-\bar{\psi}^{\bar{j}}_+ + i\psi^i_-\partial_+\bar{\psi}^{\bar{j}}_- + F^i\bar{F}^{\bar{j}}\right)$$

$$
\begin{aligned}
- & \partial_i \partial_k \partial_{\bar{j}} K_0 \psi_+^i \psi_-^k \overline{F}^{\bar{j}} - \partial_{\bar{i}} \partial_{\bar{k}} \partial_j K_0 \overline{\psi}_+^{\bar{i}} \overline{\psi}_-^{\bar{k}} F^j + i \partial_i \partial_{\bar{j}} \partial_{\bar{k}} K_0 \left(\psi_+^i \overline{\psi}_+^{\bar{j}} \partial_- \overline{\phi}^{\bar{k}} + \psi_-^i \overline{\psi}_-^{\bar{j}} \partial_+ \overline{\phi}^{\bar{k}} \right) \\
+ & \partial_i \partial_j \partial_{\bar{i}} \partial_{\bar{j}} K_0 \psi_+^i \psi_-^j \overline{\psi}_+^{\bar{j}} \overline{\psi}_-^{\bar{j}} + \frac{1}{M} \Big(\partial_i \partial_{\bar{j}} K_1 F^i \partial_+ \partial_- \overline{\phi}^{\bar{j}} - \partial_i \partial_k \partial_{\bar{j}} K_1 \psi_+^i \psi_-^k \partial_+ \partial_- \overline{\phi}^{\bar{j}} \\
+ & \partial_i \partial_{\bar{j}} \partial_{\bar{k}} K_1 F^i \partial_+ \overline{\phi}^{\bar{j}} \partial_- \overline{\phi}^{\bar{k}} - \partial_i \partial_k \partial_{\bar{j}} \partial_{\bar{k}} K_1 \psi_+^i \psi_-^k \partial_+ \overline{\phi}^{\bar{j}} \partial_- \overline{\phi}^{\bar{k}} \Big),
\end{aligned}
\tag{20}
$$

where

$$
K_m(\phi, F, \overline{\phi}) = \int_{\frac{1}{2}}^{-\frac{1}{2}} d\xi \xi^m K \left(\phi + \tfrac{\xi}{M} F, \overline{\phi} \right). \tag{21}
$$

The operators

$$
Q_{\pm} = \psi_{\pm} \frac{\partial}{\partial \phi} \mp F \frac{\partial}{\partial \psi_{\mp}} - i \partial_{\pm} \overline{\phi} \frac{\partial}{\partial \overline{\psi}_{\pm}} + i \partial_{\pm} \overline{\psi}_{\mp} \frac{\partial}{\partial \overline{F}} \tag{22}
$$

implement the unbroken SUSY algebra acting on the fields, so that e.g. $\delta\phi = \varepsilon^- Q_- \phi + \varepsilon^+ Q_+ \phi$. They satisfy the algebra given earlier. So anything of the form $Q_+ Q_- X$ is $\mathcal{N} = \frac{1}{2}$ invariant. Our hope is that the deformed one-loop diagrams may also be written as $Q_+ Q_-$ acting on a "kernel". So far we have succeeded in writing the full set of one-loop diagrams to leading order in $\frac{1}{M}$ in this form; also some sets of the NLO one-loop contributions.

CONCLUSIONS

The one-loop corrections for the four-dimensional theory have revealed some surprising features: the non-anti-commutativity parameter C is unrenormalised; a divergent non-linear field redefinition is required to make gauge-invariance and renormalisability manifest; $SU(N) \times U(1)$ is more appropriate as a gauge group than $SU(N)$ or $U(N)$; extra terms are required for renormalisability which are not in classical action but separately preserve $\mathcal{N} = \frac{1}{2}$ invariance.

Moreover in the two-dimensional theory, to leading order in $\frac{1}{M}$ the one-loop corrections can be written in a succinct form. Work is in progress to see if we can do this to all orders in $\frac{1}{M}$, and if the result be can interpreted as a renormalisation of the classical action (maybe with some kind of "smearing" of the undeformed geometry).

REFERENCES

1. N. Seiberg, JHEP 0306 (2003) 010
2. T. Araki, K. Ito and A. Ohtsuka, Phys. Lett. B573 (2003) 209
3. I. Jack, D.R.T. Jones and L.A. Worthy, Phys. Lett. B611 (2005) 199
4. I. Jack, D.R.T. Jones and L.A. Worthy, Phys. Rev. D72 (2005) 065002
5. I. Jack, D.R.T. Jones and L.A. Worthy, Phys. Rev. D75 (2007) 045014
6. B. Chandrasekhar and A. Kumar, JHEP03 (2004) 103
7. B. Chandrasekhar, Phys. Rev. D70 (2004) 125003
8. L. Álvarez-Gaumé and M.A. Vázquez-Mozo, JHEP 0504 (2005) 007

Rolling Tachyon in Nonlocal Cosmology

L. Joukovskaya

DAMTP, Centre for Mathematical Sciences, University of Cambridge, Wilberforce Road, Cambridge CB3 0WA, UK

Abstract. Nonlocal cosmological models derived from String Field Theory are considered. A new method for constructing rolling tachyon solutions in the FRW metric in two field configuration is proposed and solutions of the Friedman equations with nonlocal operator are presented. The cosmological properties of these solutions are discussed.

Keywords: Rolling Tachyon Solutions, String Field Theory, Nonlocal cosmology
PACS: 11.25.-w, 11.25.uv, 11.10Lm

INTRODUCTION

In recent works there appears interest in nonlocal cosmological models derived from String Field Theory in connection with problem of describing cosmological inflation or accelerating expansion of the Universe.

Modern cosmological data indicates that expansion of the Universe is accelerating. It may be owing to a component of the Universe with negative pressure, Dark Energy. Recent results of WMAP [1] together with Ia supernovae data give us the following range for the dark energy state parameter $w = -0.97^{+0.07}_{-0.09}$. From theoretical point of view it is natural to distinguish three essential different cases, $w > -1$, $w = -1$ and $w < -1$. The third one is more challenging one because it violates all natural energy conditions and there might appear problems of an instability at classical and quantum level. Recently it was proposed to consider D-brane decay in cubic superstring field theory as a model of cosmological Dark Energy [2]. At the same time there have been a number of attempts to realize description of the early Universe via nonlocal cosmological models (for more details see [3] and refs therein).

In this work we will continue research [2, 3, 4, 5] of properties of nonlocal cosmological models derived from string field theory. More precisely, we will consider a cosmological model based on gravity interacting with tachyon matter governed by the tachyon action in CSSFT when fields up to zero mass are taken into account [6].

SETUP

To describe open string states living on a single non-BPS D-brane one has to consider GSO$\pm$ states [8]. GSO$-$ states are Grassmann even, while GSO$+$ states are Grassmann

CP957, *Particles, Strings, and Cosmology, 13th International Symposium*
edited by A. Rajantie, C. Contaldi, P. Dauncey, and H. Stoica

odd. The low level action [6] contains the tachyon field ϕ and one auxiliary field u

$$S[u,\phi]=\frac{1}{g_o^2\alpha'^{\frac{(p+1)}{2}}}\int d^{p+1}x\left[u^2(x)-\frac{\alpha'}{2}\partial_\mu\phi(x)\partial^\mu\phi(x)+\frac{1}{4}\phi^2(x)-\frac{1}{3\gamma^2}U(x)\Phi^2(x)\right], \tag{1}$$

where the Φ and U are defined as $\Phi=e^{k\Box}\phi$, $U=e^{k\Box}u$, $k=-\alpha'\ln\gamma$, $\gamma=\frac{4}{3\sqrt{3}}$.

Covariant generalization of action (1) after rescaling of fields and coordinates in dimensionless space-time variables has the form

$$S=\int d^4x\sqrt{-g}\left(\frac{m_p^2}{2}R+\frac{\xi^2}{2}\psi\Box_g\psi+\frac{1}{2}\psi^2+\frac{1}{4}\upsilon^2-\frac{1}{2}\Upsilon\Psi^2-\Lambda'-T\right), \tag{2}$$

where g is the metric, $\Box_g=\frac{1}{\sqrt{-g}}\partial_\mu\sqrt{-g}g^{\mu\nu}\partial_\nu$, ψ is a dimensionless scalar tachyon field, $\Psi=e^{\frac{1}{8}\Box_g}\phi$, υ is a dimensionless auxiliary scalar field, $\Upsilon=e^{\frac{1}{8}\Box_g}\upsilon$, $m_p^2=g_4\frac{M_p^2}{M_s^2}$, M_p is a Planck mass, M_s is a characteristic string scale, g_4 is a dimensionless four dimensional effective coupling constant, Λ' is an effective cosmological constant, T is the brane tension. In this work we will be interested in the case of the FRW metric $ds^2=-dt^2+a^2(t)\left(dx_1^2+dx_2^2+dx_3^2\right)$. We will consider spatially homogeneous configurations for which Beltrami-Laplace operator takes the form $\Box_g=-\partial_t^2-3H(t)\partial_t$. For the convenience of numerical calculations let us introduce the following notation $\mathcal{D}_{\mathrm{H}}^2=\partial_t^2+3H(t)\partial_t$. The corresponding equations of motion have the form

$$e^{\frac{1}{4}\mathcal{D}_{\mathrm{H}}^2}\Upsilon(t)=\Psi^2(t), \tag{3a}$$

$$\left(-\xi^2\mathcal{D}_{\mathrm{H}}^2+1\right)e^{\frac{1}{4}\mathcal{D}_{\mathrm{H}}^2}\Psi(t)=\Upsilon\Psi(t), \tag{3b}$$

and corresponding Friedman equations have the following form

$$3H^2=\frac{1}{m_p^2}\left(\frac{\xi^2}{2}(\partial\psi)^2-\frac{1}{4}\upsilon^2-\frac{1}{2}\psi^2+\frac{1}{2}\Upsilon\Psi^2+E_{nl_1}+E_{nl_2}+E_{nl_3}+E_{nl_4}+\Lambda'+T\right), \tag{4a}$$

$$\dot{H}=-\frac{1}{m_p^2}\left(\frac{\xi^2}{2}(\partial\psi)^2+E_{nl_2}+E_{nl_4}\right), \tag{4b}$$

where

$$E_{nl1}(t) = \frac{1}{8}\int_0^1 d\rho\left(e^{\frac{2-\rho}{8}\mathcal{D}_{\mathrm{H}}^2}(-\xi^2\mathcal{D}_{\mathrm{H}}^2+1)\Psi\right)\left(\mathcal{D}_{\mathrm{H}}^2e^{\frac{\rho}{8}\mathcal{D}_{\mathrm{H}}^2}\Psi\right), \tag{5a}$$

$$E_{nl2}(t) = -\frac{1}{8}\int_0^1 d\rho\left(\partial e^{\frac{2-\rho}{8}\mathcal{D}_{\mathrm{H}}^2}(-\xi^2\mathcal{D}_{\mathrm{H}}^2+1)\Psi\right)\left(\partial e^{\frac{\rho}{8}\mathcal{D}_{\mathrm{H}}^2}\Psi\right), \tag{5b}$$

$$E_{nl3}(t) = \frac{1}{16}\int_0^1 d\rho\left(e^{\frac{2-\rho}{8}\mathcal{D}_{\mathrm{H}}^2}\Upsilon\right)\left(\mathcal{D}_{\mathrm{H}}^2e^{\frac{\rho}{8}\mathcal{D}_{\mathrm{H}}^2}\Upsilon\right), \tag{5c}$$

$$E_{nl4}(t) = -\frac{1}{16}\int_0^1 d\rho\left(\partial e^{\frac{2-\rho}{8}\mathcal{D}_{\mathrm{H}}^2}\Upsilon\right)\left(\partial e^{\frac{\rho}{8}\mathcal{D}_{\mathrm{H}}^2}\Upsilon\right). \tag{5d}$$

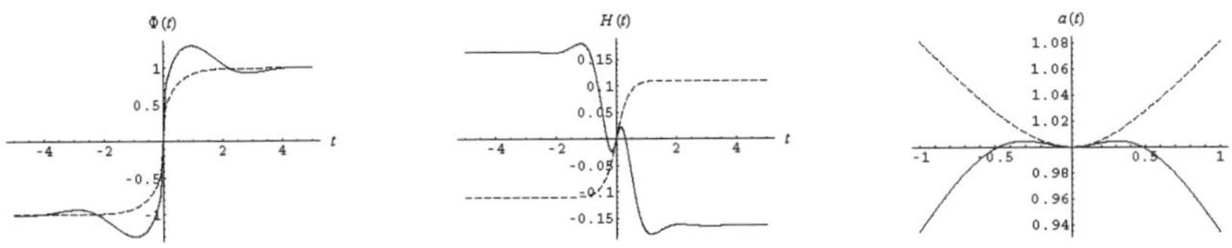

FIGURE 1. Solution for approximation $u \approx e^{k\Box}u$ with $\xi^2 = 0$ (dashed line) and $\xi^2 = -\frac{1}{4\ln\gamma}$ (solid line).

ROLLING TACHYON IN THE FRW UNIVERSE

The approximation $u \approx e^{k\Box}u$ in the action (1) for Minkowski case was considered in [2, 3] as a model example since it considerably simplifies scalar field equation and with $\xi = 0$ reproduces p-adic model for $p = 3$ which is also interesting for the applications. Recently for example it was proposed to consider p-adic theory as a theory of inflation [10]. This approximation was studied in Minkowski space in [6, 7] and recently in FRW space in [3]. Here we will study the model with and without this approximation focusing on two physically interesting cases $\xi^2 = 0$ and $\xi^2 = -\frac{1}{4\ln\gamma} \approx 0.96$ [6, 3]. The obtained results in the approximate case are presented on Fig. 1. It might be more convincible to consider approximation indicated above at the stage of equations of motion. Then in terms of redefined fields Ψ and Υ it is equivalent to approximation $\Psi \approx e^{-\frac{1}{4}\mathcal{D}_{\mathrm{H}}^2}\Psi$ and the system of equations (3) transforms into one $\left(-\xi^2\mathcal{D}_{\mathrm{H}}^2+1\right)e^{-\frac{1}{4}\mathcal{D}_{\mathrm{H}}^2}\Psi = \Psi^3$, simplification of Friedmann equations also takes place (see [3] for details). Figure 1 presents solutions ψ, H and a for $\xi^2 = 0$ and $\xi^2 = -\frac{1}{4\ln\gamma}$.

Let us return to the case of two field configuration [6, 9] Ψ and Υ which is more interesting. To find solutions of the system of equations (3) and integrating equation (4b), $H(t) = -\frac{1}{m_p^2}\int_0^t d\tau\, \dot{H}(\tau)$, we construct the following iterative process

$$\Psi = \lim_{n\to\infty}\Psi_n, \quad \Upsilon = \lim_{n\to\infty}\Upsilon_n, \quad H = \lim_{n\to\infty}H_n,$$

where iterations Ψ_n, Υ_n, and H_n are in turn obtained as limits of sub-iterations $\Psi_{n+1} = \lim_{m\to\infty}\Psi_{n,m}$, $\Upsilon_{n+1} = \lim_{m\to\infty}\Upsilon_{n,m}$, $H_{n+1} = \lim_{m\to\infty}H_{n,m}$, which are recursively defined as ($m \geqslant 0$)

$$\Upsilon_{n,m+1} = \frac{(-\xi^2\mathcal{D}_{\mathrm{H}_n}^2+1)e^{\frac{1}{4}\mathcal{D}_{\mathrm{H}_n}^2}\Psi_{n,m}}{\Psi_{n,m}}, \qquad \Psi_{n,m+1} = \mathrm{sign}(t)\sqrt{e^{\frac{1}{4}\mathcal{D}_{\mathrm{H}_n}^2}\Upsilon_{n,m+1}},$$

$$H_{n,m+1} = -\frac{1}{m_p^2}\int_0^t d\tau\left[\frac{\xi^2}{2}(\partial e^{k\mathcal{D}_{\mathrm{H}_{n,m}}^2}\Psi_{n+1})^2 - \frac{1}{16}\int_0^1 d\rho\left(\partial e^{\frac{2-\rho}{8}\mathcal{D}_{\mathrm{H}_{n,m}}^2}\Upsilon_{n+1}\right)\times\right.$$
$$\left.\left(\partial e^{\frac{\rho}{8}\mathcal{D}_{\mathrm{H}_{n,m}}^2}\Upsilon_{n+1}\right) - \frac{1}{8}\int_0^1 d\rho\left(\partial(-\xi^2\mathcal{D}_{\mathrm{H}_{n,m}}^2+1)e^{\frac{2-\rho}{8}\mathcal{D}_{\mathrm{H}_{n,m}}^2}\Psi_{n+1}\right)\left(\partial e^{\frac{\rho}{8}\mathcal{D}_{\mathrm{H}_{n,m}}^2}\Psi_{n+1}\right)\right].$$

The initial iterations in m is taken as $\Phi_{n,0} = \Phi_n$, $H_{n,0} = H_n$; $\Psi_0(t) = \tanh(t)$, $H_0(t) = 0$ (see [3] for the details of analogous iterative process). Obtained solutions are presented

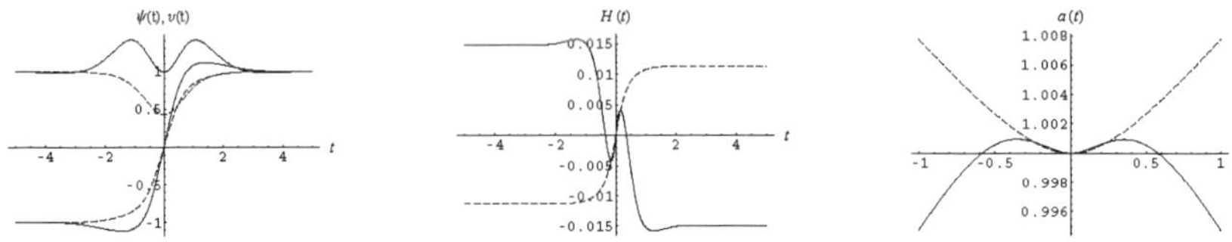

FIGURE 2. Solution of equations (3), (4) with $\xi^2 = 0$ (dashed line) and $\xi^2 = -\frac{1}{4\ln\gamma}$ (solid line). On the most left figure ψ has odd kink type shape, while υ has even wavelet type shape.

on the Fig. 2, from which we see that for $\xi^2 = 0$ we obtained bouncing solutions. Note that Λ' does not enter equations (3), (4b), but can be determined from (4a). Following Sen's conjecture we put D-brane's tension to $T = -V(\Psi = \pm 1, \Upsilon = 1) = \frac{1}{4})$ [2, 6] and thus Λ' is determined uniquely for each field configuration in the same way as in [3].

In this work we have studied the properties of nonlocal cosmological models derived from String Field Theory in the Friedmann space-time. We have construct new method for solving partial differential equations with infinitely many time derivatives, which constitute a new class of mathematical physics equations and are recently discussed in the literature [7, 10, 11]. We obtained classical solutions of the corresponding Friedmann equations which can be considered as a first approximation to the quantum solutions and might be useful for the study of ways to avoid the cosmological singularity problem [12].

ACKNOWLEDGMENTS

The author is grateful to Neil Turok for his support. The author also would like to thank I. Aref'eva, G. Calcagni, D. Mulryne, Ya. Volovich for fruitful discussions. The author gratefully acknowledge the use of the UK National Supercomputer, COSMOS, funded by PPARC, HEFCE and Silicon Graphics. This work is supported by the Center for Theoretical Cosmology, in Cambridge and in part by RFBR grant 05-01-00758, INTAS grant 03-51-6346 and Russian President's grant NSh–672.2006.1.

REFERENCES

1. D.N. Spergel et al., [WMAP collaboration], astro-ph/0603449
2. I. Ya. Aref'eva, AIP Conf.Proc. 826, 301-311, 2006, astro-ph/0410443.
3. L. Joukovskaya, arxiv:0707.1545[hep-th].
4. I.Ya. Aref'eva, L.V. Joukovskaya, *JHEP* **0510** (2005) 087, hep-th/0504200.
5. I.Ya. Aref'eva, L.V. Joukovskaya, S.V. Vernov, JHEP **0707** (2007) 087.
6. I.Ya. Aref'eva, L.V. Joukovskaya and A.S. Koshelev, *JHEP* **0309** (2003) 012.
7. Ya.I. Volovich, *J. Phys.* **A 36** (2003) 8685–8702.
8. A. Sen, Int.J.Mod.Phys. A20 (2005) 5513-5656.
9. L. Joukovskaya, Ya. Volovich, math-ph/0308034; S. Yu. Vernov, astro-ph/0612487.
10. N. Barnaby, T. Biswas, J. M. Cline JHEP **0704** (2007) 056.
11. L. Joukovskaya, arxiv: 0708.0642[math-ph]; V.S. Vladimirov, math-ph/0507018, arXiv: 0705.4600[math-ph]; D. Prokhorenko, math-ph/0611068.
12. J. Khoury, B.A. Ovrut, P.J. Steinhardt and N. Turok, *Phys. Rev.* **D 64** (2001) 123522; *Phys. Rev.* **D 66** (2002) 046005.

Motion in Brane World Models: The Bazanski Approach

M.E.Kahil

Mathematics Department, Modern Sciences and Arts University, Giza, EGYPT
e.mail: kahil@aucegypt.edu

Abstract. Recently, path equations have been obtained for charged and spinning objects in brane world models, using a modified Bazanski Lagrangian. In this study, path deviation equations of extended objects are derived. The significance of moving extended objects in brane world models is examined. Motion in non- symmetric brane world models is also considered.

Keywords: equations of motion, strings and branes phenomology, Klauza-Klein theories, higher dimensional gravity
PACS: 04-25.-g, 11.25.wx, 40.50.cd ,40.50.-h

The Bazanski Approach in Brane World Models

Geodesic and geodesic deviation equations can be obtained simultaneously by applying the action principle on the Bazanski Lagrangian [1]. It is worth mentioning that the Bazanski approach has been successfully applied in geometries different from the Riemannian successfully applied in geometries different from the Riemannian one [2] and [3]. Also, the Bazanski Lagrangian can be amended to describe path and path deviation equations of spinning charged particles in Kaluza-Klein theories [4].

In the brane world scenario, it is well known that our universe can be described in terms of a 4+N dimensional, with $N \geq 1$ and the 4D space-time part of it is embedded in a 4+N manifold [5]. Accordingly, the bulk geodesic motion is observed by a four dimensional observer to reproduce the physics of 4D space-time [6]. Consequently, we are going to apply the Bazanski approach [7] on the following Lagrangian

$$L = g_{\mu\nu}(x^\rho, y)U^\mu \frac{D\Psi^\nu}{Ds} + \hat{F}_{\mu\nu}U^\nu + f_\mu)\Psi^\mu, \tag{1}$$

to obtain path and path deviation equations for a spinning charged particle on a brane , respectively

$$\frac{dU^\alpha}{ds} + \left\{ {\alpha \atop \mu\nu} \right\} U^\mu U^\nu = \hat{F}^\alpha_\nu U^\nu + 2(\frac{1}{2}U^\alpha U^\rho - g^{\alpha\rho})\Omega_{\rho\delta}\frac{dy}{ds}U^\delta \tag{2}$$

and

$$\frac{D^2\Psi^\alpha}{Ds^2} = R^\alpha_{.\mu\nu\rho}U^\mu U^\nu \Psi^\rho + \hat{F}^\alpha_{\nu;\rho}U^\nu\Psi^\rho + \hat{F}^\alpha_\rho \frac{D\Psi^\rho}{Ds} + f^\alpha_{;\rho}\Psi^\rho + 2(\frac{1}{2}U^\alpha U^\sigma - g^{\alpha\sigma})\Omega_{\sigma\nu}\frac{dy}{ds}\frac{D\Psi^\nu}{Ds}, \tag{3}$$

where $g_{\mu\nu}(x^\rho, y)$ is the induced metric , $\hat{F}^\mu_\nu = \frac{e}{m}F^\mu_\nu + \frac{1}{2m}R^\mu_{\nu\rho\delta}S^{\rho\delta}$ and $f_\mu =$

CP957, *Particles, Strings, and Cosmology, 13th International Symposium*
edited by A. Rajantie, C. Contaldi, P. Dauncey, and H. Stoica

$\frac{1}{2}U^{\rho}U^{\sigma}\frac{\partial g_{\rho\sigma}}{\partial y}\frac{dy}{ds}U_{\mu}$ describes a parallel force due to the effect of non-compactified extra dimension [9]. In brane world model one can find the extrinsic curvature $\Omega_{\mu\nu} = \frac{1}{2}\frac{\partial g_{\mu\nu}}{\partial y}$ [8].

Path & Path Deviation Equations of Non-Symmetric Geometries in Brane World Models

Path equation and Path deviation equations in Brane World Models defined in non-symmetric geometries can be obtained by suggesting the following Lagrangian:

$$L = \mathbf{g}_{\mu\nu}U^{\mu}\frac{D\Psi^{\nu}}{Ds} + \lambda f_{[\mu\nu]}U^{\mu}\Psi^{\nu} + \frac{1}{2}U^{\alpha}U^{\beta}U_{\rho}\frac{\partial \mathbf{g}_{\alpha\beta}}{\partial y}\frac{dy}{ds}\Psi^{\rho}, \tag{4}$$

where $\mathbf{g}_{\mu\nu} = g_{(\mu\nu)} + g_{[\mu\nu]}$,$\lambda$ is a parameter and, $f_{[\mu\nu]} = \hat{A}_{\mu,\nu} - \hat{A}_{\nu,\mu}$ is a skew symmetric tensor related to the Yukawa force [10]. Applying the Bazanski approach we obtain the path equation

$$\frac{dU^{\alpha}}{ds} + \left\{ {\alpha \atop \mu\nu} \right\} U^{\mu}U^{\nu} = \lambda \mathbf{g}^{\alpha\mu} f_{[\mu\nu]}U^{\nu} + \mathbf{g}^{\alpha\sigma} g_{[\nu\sigma];\rho}U^{\nu}U^{\rho} + (\frac{1}{2}U^{\rho}U^{\sigma} - \mathbf{g}^{\rho\sigma})\frac{\partial \mathbf{g}_{\rho\sigma}}{\partial y}\frac{dy}{ds}U^{\mu}. \tag{5}$$

and its path deviation equation:

$$\frac{D^2\Psi^{\alpha}}{Ds^2} = R^{\alpha}_{.\mu\nu\rho}U^{\mu}U^{\nu}\Psi^{\rho} + 2\mathbf{g}^{\sigma\alpha}(g_{[\nu[\sigma];\rho]})\frac{D\Psi^{\nu}}{Ds}U^{\rho} + \lambda(f^{\alpha}_{.\nu}\frac{D\Psi^{\nu}}{Ds} + f^{\alpha}_{.\nu;\rho}U^{\nu}\Psi^{\rho}).$$

$$+(\frac{1}{2}U^{\alpha}U^{\rho}\frac{\partial \mathbf{g}_{\delta\rho}}{dy}U^{\delta}\frac{dy}{ds})_{;\nu}\Psi^{\nu} + ((\frac{1}{2}U^{\alpha}U^{\mu} - g^{\alpha\mu})\frac{\partial \mathbf{g}_{\mu\nu}}{dy}\frac{dy}{ds})\frac{D\Psi^{\nu}}{Ds} \tag{6}$$

The Bazanski Approach in Curved Clifford Space

It is well known that extended objects can be expressed by p-branes [11]. This type of representation is defined in curved Clifford Space. The advantage of this space is to show that extended objects are purely poly-geodesics satisfying a metamorphic transformation [12].

Thus, we suggest the following Lagrangian:

$$L = g_{AB}P^{\alpha}\frac{\nabla\phi^{\beta}}{\nabla\tau} + \frac{1}{2}S_{\alpha\beta}\frac{\nabla\phi^{\alpha\beta}}{\nabla\tau}, \tag{7}$$

such that

$$\frac{\nabla\phi^{\alpha}}{\nabla\tau} = \frac{d\phi^{\alpha}}{d\tau} + (\Gamma^{\alpha}_{\mu\nu}\phi^{\mu} + \frac{1}{2}\Xi^{\alpha}_{.\mu\rho\nu}\phi^{\mu\rho})U^{\nu} + \frac{1}{2m}(R_{\gamma\delta.\mu}^{\ \ \alpha}\phi^{\mu} + \frac{1}{2}\Omega_{\gamma\delta.\mu\nu}^{\ \ \alpha}\phi^{\mu\nu})S^{\gamma\delta}, \tag{8}$$

where $\Gamma^{\alpha}_{\beta\rho}$ is the Cartan connection, $\hat{R}^{\alpha}_{.\beta\gamma\delta}$ is the Cartan curvature and the two other metamorphic connections $\Xi_{\beta\rho}^{\alpha\delta}\&\Omega_{\beta\rho\gamma}^{\alpha\delta}$ existed due to the presence of bivector quantities. By taking variation with respect to its deviation vector ϕ^{δ} we get

$$\frac{\nabla U^{\alpha}}{\nabla \tau} = 0, \tag{9}$$

and taking the variation with respect to its deviation bivector $\phi^{\delta\rho}$ we obtain

$$\frac{\nabla S^{\alpha\beta}}{\nabla \tau} = 0. \tag{10}$$

In the case of a charged particle described by a polyvector we can find that the corresponding linear momentum equation becomes [12]

$$\frac{\nabla U^{\alpha}}{\nabla \tau} = \frac{e}{m} F^{\alpha}_{\beta} U^{\beta}, \tag{11}$$

and its angular momentum equation takes the following form

$$\frac{\nabla S^{\alpha\beta}}{\nabla \tau} = F^{\alpha}_{\nu} S^{\nu\beta} - F^{\beta}_{\nu} S^{\nu\alpha}. \tag{12}$$

We suggest the following Lagrangian to derive (11) and (12)

$$L = g_{\mu\nu} P^{\mu} \frac{\nabla \phi^{\nu}}{\nabla \tau} + \frac{1}{2} S_{\mu\nu} \frac{\nabla \phi^{\mu\nu}}{\nabla \tau} + F_{\mu\nu} U^{\nu} \phi^{\mu} + \frac{1}{2} (F_{\mu\rho} S^{\rho}_{\nu} - F_{\nu}{}^{\rho} S_{\rho\mu}) \phi^{\mu\nu}. \tag{13}$$

If we put the flavor of Kaluza-Klein in curved Clifford space for combining gravity and electromagnetism, i.e, increasing the spatial dimension by a compacted one, then equations (11) and (12) can be derived from the following Lagrangian:

$$L = g_{AB} P^{A} \frac{\nabla \phi^{B}}{\nabla \hat{\tau}} + \frac{1}{2} S_{AB} \frac{\nabla \phi^{AB}}{\nabla \hat{\tau}}, \tag{14}$$

where $A = 1,2,3,4,5$ to become

$$\frac{\nabla U^{A}}{\nabla \hat{\tau}} = 0 \quad \text{and} \quad \frac{\nabla S^{AB}}{\nabla \hat{\tau}} = 0, \tag{15}$$

which means that the path of a charged spinning particle defined in Riemannain geometry behave like as a test particle in 5D curved Clifford space.

Discussion and Concluding Remarks

In this study, we have derived path and path deviation equations for spinning charged particles in Brane World Models from one single Lagrangian. Also, the procedure has been used to derive path and path deviation equations for a test particle existing in non-symmetric theory of gravity and examined how the extrinsic curvature term would be amended due to the existence of non-symmetric terms of gravitational field. Thus, we can find that the added term appeared in the extrinsic curvature of non-symmetric part may be connected to the spin. Finally, we have dealt with extended objects as p-branes defined in curved Clifford space [13]. This type of space has a deeper understanding of physics, i.e., quantities in nature could be defined by polyvectors. From this perspective we have developed its corresponding Bazanski Lagrangian to include deviation bi-vectors together with deviation vectors. The importance of this approach is that we can derive from one Lagrangian(7) two simultaneous quantities responsible for conservation of momentum (8) and angular momentum (9) respectively. Using this mechanism, we have found that the Dixon-Souriau equations in Riemannian geometry could be seen like equations for charged object and their angular momentum part defined in curved clifford spaces. In addition, if we have the flavor of Kaluza-Klein of unifying gravity and electromagnetism in curved Clifford space, we must increase the dimension of this space by an extra compacted dimension to preserve the conservation of charges. Consequently, from equations of (15) we can find that charged spinning particles behave like test particles defined in higher dimensional curved clifford space. To conclude this study, we must take into consideration that unification processes can be achieved if we increase the number of dimensions and extend the geometries to geometrize all quantities that appeared in our approach.

ACKNOWLEDGMENTS

The author would like to thank Professors M.I. Wanas, M. Abdel-Megied , G.S. Hall, T. Harko, G. De Young , A. Rajantie, Dr. F. H.Stoica and Mr. W.S. El-Hanafy for their useful comments and support to participate in PASCOS07. A word of thankfulness goes to Dr. N. El-Degwi and Professor K. Abdel Hamid for helping me to represent MSA University at this confenence.

REFERENCES

1. Bazanski, S.L. (1989) J. Math. Phys., **30**, 1018.
2. Wanas, M.I., Melek, M. and Kahil, M.E.(1995) Astrophys. Space Sci.,**228**, 273.; gr-qc/0207113.
3. Wanas, M.I. and Kahil, M.E.(1999) Gen. Rel. Grav., **31**, 1921. ; gr-qc/9912007
4. Kahil, M.E. (2006), J. Math. Physics **47**,052501.
5. Liu, H. and Mashhoon, B. (2000), Phys. Lett. A, **272**,26 ; gr-qc/0005079
6. Youm, D. (2000) Phys.Rev.D**62**, 084002; hep-th/0004144
7. Kahil, M.E. (2007) ;gr-qc/0701015
8. Ponce de Leon, J. (2001) Phys Lett B, **523**, 311 ;gr-qc/0110063
9. Dick, R. (2001) Class. Quant. Grav., **18**, R1.
10. Legaré, J. and Moffat, J.W. (1996), Gen. Rel. Grav., **28**, 1221.
11. Castro, C. and Pavsic, M. (2002) Phys. Lett B **539**, 133; hep-th/0110079
12. Pezzaglia, W.M. (1999) gr-qc/9912025
13. Castro,C. (2002) physics/0011040

Inflation from Intersecting Branes

Louis Leblond

Department of Physics, Texas A&M University
College Station, TX 77843, USA

Abstract. We propose a new scenario for D-term inflation which appears quite straightforwardly in the open string sector of intersecting brane models. We take the inflaton to be a chiral field in a bifundamental representation of the hidden sector and we argue that a sufficiently flat potential can be brane engineered. This type of model generically predicts a near gaussian red spectrum with negligible tensor modes.

Keywords: String Cosmology, Branes, Inflation
PACS: 11.25.Uv, 11.25.Wx, 98.80.Cq

SET-UP AND MOTIVATIONS

In this talk, we would like to propose a model of inflation where the inflaton appear as non-vectorlike (bifundamental) matter arising from strings stretching between different hidden sector branes [1]. Unlike the usual open string modes used in brane inflation [2] we will show that gauge invariance can help restrict the superpotential and lessen fine-tunings. In particular, we describe a scenario for D-term inflation [3, 4] which seems to appear straightforwardly in the context of Type IIA/B intersecting brane models (IBM – see [5] for a recent review) where a flat direction can be engineered by a discrete choice of intersection number of branes.

We assumed that we have a fixed closed string background where all the closed string moduli are fixed at high scale. We then add branes (with their respective gauge theories) wrapping various cycles all intersecting with each other. Living at each intersection between two stacks of branes, we have a bifundamental field charged under each of the two gauge groups.

The low energy effective field theory is assumed to be a softly broken $N=1$ supegravity theory where the D-term dominates over the F-term, $F \ll D$. In the following, we will work in the regime where the inflaton vev is always smaller than the string/Planck-scale, as well as below the masses of the moduli. In our context of Type IIA/B IBM's, we naturally tend to find a large number of D-flat directions. Indeed $N\,U(1)$ brane stacks give N D-term equations but since each brane stack generically intersect with any other one, we have $O(N^2)$ scalar fields in the D-term equations and a solution can always be found.

Our basic idea will be to separate out one $U(1)$, whose D-term potential will give us the inflationary potential

$$V_{\rm inf}^{D} = g^2(|\phi_+|^2 - |\phi_-|^2 - \xi)^2\,, \tag{1}$$

CP957, *Particles, Strings, and Cosmology, 13th International Symposium*
edited by A. Rajantie, C. Contaldi, P. Dauncey, and H. Stoica

where ξ is a Fayet-Illioupoulos term that will give us the necessary vacuum energy during inflation when $\phi_\pm = 0$. The remaining D-terms will form

$$V^D_{\text{rest}} = \sum_j g_j^2 \left(|S|^2 + \sum_i q_i |\rho_i|^2 - \xi_j \right)^2 + \cdots \tag{2}$$

where the dots represents other terms of the same form as the first one, ρ_i are various chiral matter with charge q_i under this particular $U(1)$ gauge group. In order to achieve inflation, we seek a flat direction of V^D_{rest} on which all tachyonic direction of V^D_{inf} are lifted (via Yukawa couplings of the type $\lambda S\phi_+\phi_-$). For large value of S, the mass squared of $\phi_\pm$ are positive and the fields are stabilized at zero and the potential (1) is constant.

As we mentioned already, we can always ensure the existence of a flat direction in V_{inf} by having enough ρ_i. At tree level, the potential is then completely flat but since supersymmetry is broken by the positive vacuum energy, loop corrections will generate a non-trivial potential of the Coleman-Weinberg type. This potential will depend on the scalar field S since it is this field that gives a mass to $\phi_\pm$. At some point, as the inflaton rolls, the Yukawa coupling will no longer lift the waterfall fields sufficiently. A tachyonic direction will then appear causing inflation to end.

Now, one might worry that Yukawa coupling in the F-terms could also lift the flat direction in the D-terms. As we will show, it is possible in these kind of models to ensure that this does not happen via some discrete fine-tuning of intersection numbers. Indeed gauge invariance strongly constrained the form of the superpotential for bifundamental fields and it is possible to brane engineer a set-up such that the contribution from the F-terms are of higher order in the inflaton field S.

This type of slow-roll inflation generically predicts a near gaussian red spectra ($n \sim 0.98$) with an inflationary scale of order of the GUT scale. The inflaton vev is always lower than the Planck scale (which is necessary for a consistent compactification) and therefore it predicts a negligible amount of tensor fluctuations.

A SIMPLE EXAMPLE

Consider six stacks of hidden sector D6 branes a, b, c, d, e and f. These branes each fill spacetime and wrap different 3-cycles on the compact 6D space. The topological intersection numbers of our toy model are $I_{ab} = I_{bc} = I_{ca} = I_{be} = I_{ef} = I_{fa} = I_{db} = I_{ad} = 1$, with all other intersection numbers being zero. Each brane stack yields a $U(1)_{a,b,c,d,e,f}$ gauge group, where $g_{a,b,c,d,e,f}$ are the gauge couplings and $\xi_{a,b,c,d,e,f}$ are the Fayet-Iliopoulos terms. We denote by S, ϕ_+ and ϕ_- the chiral multiplets stretching between a and b, b and c, and c and a respectively (see Fig. 1).

The brane stack 'd' in the picture is there simply for the anomalies to cancel and it plays no role during inflation and for simplicity we set $\xi_d = \delta_1 = \delta_2 = 0$. We choose the c brane stack as our inflationary potential, and we have

$$\begin{aligned} V^D_{\text{inf}} &= g_c^2(|\phi_+|^2 - |\phi_-|^2 - \xi_c)^2 \,, \\ V^D_{\text{rest}} &= g_a^2(|\phi_-|^2 - |S|^2 + |\rho_3|^2 - \xi_a)^2 + g_b^2(|S|^2 - |\phi_+|^2 - |\rho_1|^2 - \xi_b)^2 \\ &\quad + g_e^2(|\rho_1|^2 - |\rho_2|^2 - \xi_e)^2 + g_f^2(|\rho_2|^2 - |\rho_3|^2 - \xi_f)^2 \,. \end{aligned} \tag{3}$$

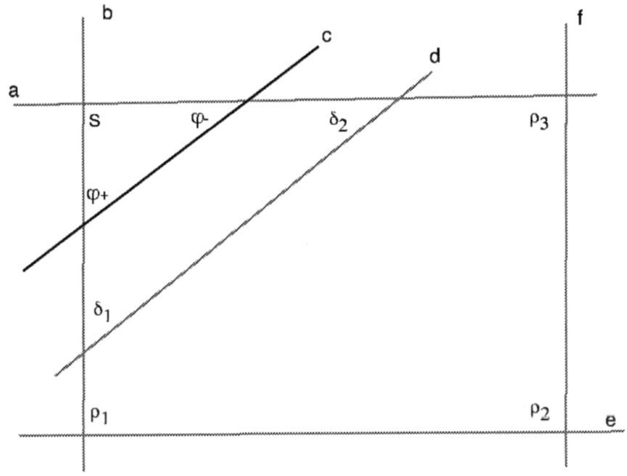

FIGURE 1. This shows a setup of intersecting branes in type IIA that can yield D-term inflation. We have a total of 6 branes. Brane c is our inflationary brane. Branes a and b give rises to the necessary Yukawa coupling between the inflaton and the waterfall field. Branes e and f are there to ensure that the D-term potential has a flat direction. This flat direction is broken by the F-term potential but this is planckian suppressed. Finally, brane d is necessary for the anomalies to cancel, it plays no other role during the inflationary era. S, $\phi_\pm$, $\delta_{1,2}$ and $\rho_{1,2,3}$ are the various bifundamental fields at each intersections. Also note that this picture is a two dimensional depiction of a configuration that actually lives in six curved dimensions.

The flat direction is the path in field space along which the variation $\delta(|S|^2) = \delta(|\rho_{1,2,3}|^2) = r^2$ and V^D_{rest} is constant. A Yukawa coupling will arise from a worldsheet instanton stretching over the triangle generated by the three intersecting branes a, b and c or d:

$$W_{\text{yuk}} = \lambda_\phi S\phi_+\phi_- + \lambda_\delta S\delta_1\delta_2 \,. \tag{4}$$

Due to gauge invariance, these are the only two renormalizable terms in the superpotential containing these fields. This superpopential insures that when $|S| > S_C = \frac{g\sqrt{2\xi_c}}{\lambda_\phi}$, the mass of $\phi_\pm$ is positive and $\phi_\pm = 0$ during inflation. In this case the potential is constant $V_{inf} = g_c^2\xi_c^2$ up to logarithmic correction from susy breaking. This potential gives rise to the standard D-term inflation (see [6] for a review).

FINE-TUNINGS AND FUTURE DIRECTIONS

The inflationary potential receives many planckian corrections that can spoil the flatness of the potential. The only Yukawa coupling relevant to the flatness of our inflaton direction is

$$W_\rho = \frac{\lambda_\rho}{M_{pl}} S\rho_1\rho_2\rho_3 \,. \tag{5}$$

Note that this coupling is Planck-suppressed because it arises from a "square" in the quiver diagram, rather than a triangle. During inflation, we must have $\rho_i \sim S$ to ensure the existence of a flat direction and so this superpotential will generate a term in the potential of the order

$$V \sim \frac{\lambda_\rho^2 S^6}{M_p^2} \tag{6}$$

Which can be made small enough by tuning λ_ρ. But since gauge invariance strongly constrains the coupling between S and the ρ_i we can also insure the flatness of the potential by a discrete tuning of intersection numbers. Indeed, higher suppressions are easily obtained by simply choosing more hidden sectors branes such that the flat direction is described by a polygon with more sides. For a pentagon with four ρ's, we get that the first term in the potential goes like S^8/M_p^4 and so forth for even higher polygons. This is where we can see the full power of gauge invariance and we believe this is the main advantage of this type of models over the usual brane inflation ones. It is simple to show that Planckian corrections to the Kähler potential affect the mass of $\phi_\pm$ but *not* the flatness of the potential.

There are two main working assumptions in this work that remain to be worked out in detail in order to achieve a consistent string theory embedding of this scenario. One assumption is that there exists a consistent moduli stabilization scheme where the D-term energy dominates over the F-term. Secondly, we have not included any quantum corrections to the FI term which may depend on the inflaton field. One might expect that the quantum corrections are suppressed since the inflaton field is a bifundamental living at an intersection separated from the brane that generates the dominant FI term during inflation.

Since this scenario is set in the context of IBM's, it offers a rich phenomenology and it is possible to tie in various phenomenological processes with the inflationary phase. In particular, we have shown [1] that this scenario naturally admits a type of hidden sector baryogenesis, fed by tachyonic preheating at the end of inflation. We also have also noted that it is possible to generate the neutrino masses via a see-saw mechanism at the end of inflation. Finally we should point out that a different D-term inflationary senario in IBM's was proposed in [7].

I have benefitted from my collaborators B. Dutta and J. Kumar and I want to thank the organizers of Pascos 07 for organizing such a nice conference. This work is supported by an NSF grant PHY-0505757.

REFERENCES

1. B. Dutta, J. Kumar, and L. Leblond, *JHEP* **07**, 045 (2007), hep-th/0703278.
2. G. R. Dvali, and S.-H. H. Tye, *Phys. Lett.* **B450**, 72 (1999), hep-ph/9812483.
3. P. Binetruy, and G. R. Dvali, *Phys. Lett.* **B388**, 241–246 (1996), hep-ph/9606342.
4. E. Halyo, *Phys. Lett.* **B387**, 43–47 (1996), hep-ph/9606423.
5. F. Marchesano, *Fortsch. Phys.* **55**, 491–518 (2007), hep-th/0702094.
6. D. H. Lyth, and A. Riotto, *Phys. Rept.* **314**, 1–146 (1999), hep-ph/9807278.
7. R. Blumenhagen, B. Kors, D. Lust, and T. Ott, *Nucl. Phys.* **B641**, 235–255 (2002), hep-th/0202124.

Bouncing and Colliding Branes

Jean-Luc Lehners

DAMTP, CMS, Wilberforce Road, CB3 0WA, Cambridge, UK

Abstract. In a braneworld description of our universe, we must allow for the possibility of having dynamical branes around the time of the big bang. Some properties of such domain walls in motion are discussed here, for example the ability of negative-tension domain walls to bounce off spacetime singularities and the consequences for cosmological perturbations. In this context, we will also review a colliding branes solution of heterotic M-theory that has been proposed as a model for early universe cosmology.

Keywords: Branes, String cosmology
PACS: 98.80.Cq, 04.50.+h

INTRODUCTION

The duality between M-theory and the $E_8 \times E_8$ heterotic string theory [1] provides us with a braneworld picture of our universe, in which one spatial dimension is a line segment whose two boundaries are 10-dimensional (10d) branes containing gauge fields. If 6 spatial dimensions are compactified on a Calabi-Yau (CY) manifold, then we obtain a 5d picture, known as heterotic M-theory, of a bulk spacetime bounded by two (3+1)-dimensional boundary branes [2, 3]. The action is given by

$$S = \int_{5d} \sqrt{-g}[R - \frac{1}{2}(\partial\phi)^2 - 6\alpha^2 e^{-2\phi}] \pm 12\alpha \int_{4d,y=\mp 1} \sqrt{-g}e^{-\phi}, \tag{1}$$

where the scalar field ϕ parameterises the volume of the CY manifold, and α the flux on it. Since the boundary branes contain Standard Model-type gauge fields, we identify one of these branes with our currently visible universe. If we take this braneworld picture seriously, then we know from the constancy of the coupling constants that the distance between the branes, as well as the size of the internal CY manifold, must have been varying exceedingly slowly, if at all, since shortly after the big bang. However, we must allow for the possibility that the branes were truly dynamic around the time of the big bang. Such dynamic branes can have important consequences for cosmology: for example, the basis of the ekpyrotic and cyclic models is the identification of the big bang with a collision of the two boundary branes [4, 5].

Apart from a purely cosmological motivation, it is in any case of theoretical interest to study time-dependent branes regarding their behaviour with respect to singularities. In 5d the branes are domain walls, and the geometry can therefore be described in terms of a linear harmonic function $h(y)$, where y denotes the line segment coordinate. At the location of the branes, the harmonic function contains kinks whose magnitudes are related to the brane tensions. In a time-dependent setting, one can thus imagine the slope as well as the height of this harmonic function to be varying. Moreover, a zero of

CP957, *Particles, Strings, and Cosmology, 13th International Symposium*
edited by A. Rajantie, C. Contaldi, P. Dauncey, and H. Stoica

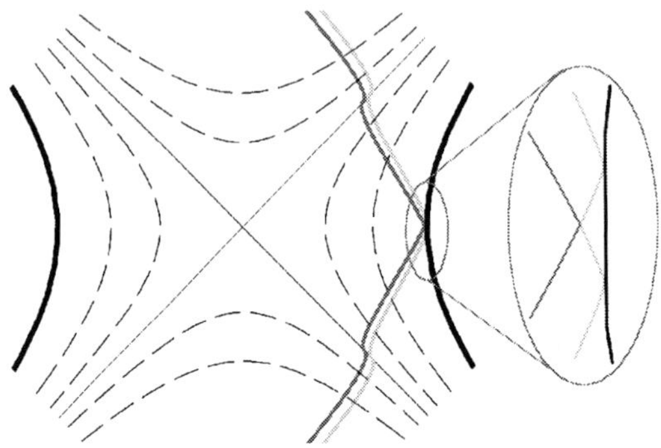

FIGURE 1. A Kruskal plot of the colliding branes solution. The timelike naked singularity is denoted by thick black lines. Here the trajectory of the positive-tension brane is shown in red and that of the negative-tension brane in green. The collision of the branes, as well as the two bounces of the negative-tension brane off the naked singularity, are shown at a magnified scale in the inset.

the harmonic function corresponds to a timelike naked singularity. One might therefore be worried that the negative-tension brane, at whose location the harmonic function is always lowest, could crush into this singularity [6]. A separate worry is that naively one would expect a collision of the two boundary branes to be accompanied by the internal manifold shrinking to zero size and hence the couplings and curvature invariants to be blowing up [7]. Here we will show how both of these potential catastrophes can be avoided, and how avoiding them can lead to unexpected benefits.

COLLIDING BRANES - BOUNCING BRANES

We are looking for a solution in which the collision of the boundary branes is the least singular possible. Thus we impose the boundary conditions that at the collision, the scale factors on the branes and the CY size approach a finite and non-zero constant [8]. Then the spacetime tends to compactified Milne $\times \mathbb{R}^3$ (plus a static CY) at the collision, which means that curvature invariants are small close to the collision. Hence all eventual higher-derivative corrections will also be small, and we can trust our solution right up to the collision (As an aside, note that the brane tensions go to zero at the collision. Also, close to the collision, one expects all relevant modes to be winding modes, which are well-behaved at small orbifold size, see [9] for details.). It turns out that imposing the boundary conditions enunciated above determines the solution almost uniquely [8]: there is just one free parameter, namely the velocity of the branes at the collision.

In describing time-dependent branes, there exist two coordinate systems which are particularly useful. In the first, which has been used implicitly up to now, the branes are kept at fixed coordinate locations and the bulk is dynamical, while in the second, the branes are moving in a static bulk spacetime. The latter description turns out to be more convenient in presenting the colliding branes solution. The bulk metric and scalar field

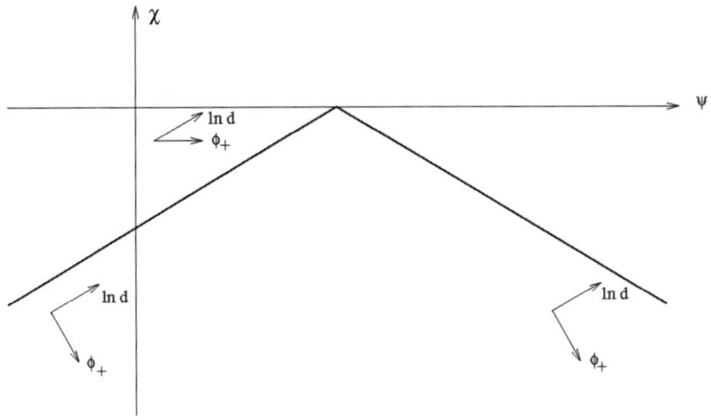

FIGURE 2. The trajectory of the colliding branes solution as seen in the ψ - χ plane. The brane collision occurs as $\psi, \chi \rightarrow -\infty$. Also shown are the directions of increasing distance between the boundary branes ($\ln d$) and increasing CY volume at the location of the positive-tension brane (ϕ_+).

are given by

$$ds^2 = -(\alpha^2 r^2 - \mu r^4)\,dt^2 + (\alpha^2 r^2 - \mu r^4)^{-1} r^{12}\,dr^2 + r^2 d\vec{x}^2, \qquad e^{\phi} = r^6. \tag{2}$$

Here r denotes the coordinate transverse to the branes, and μ is related to the collision velocity. At $r = 0$ there is a timelike naked singularity, which, after changing coordinate systems, would correspond to a zero of the linear harmonic function. The branes' motion in the static bulk is determined by solving their Israel matching conditions. From Figure 1 it can be seen that the negative-tension brane grazes the singularity twice. In the absence of matter on the branes, the brane actually touches the singularity, leading to a catastrophic crunch. However, if matter is added to the branes, then the negative-tension brane can bounce back before reaching $r = 0$, thus shielding the positive-tension brane as well as the bulk from the singularity [10]. The precise conditions for a bounce to occur depend on the coupling of the brane-bound matter to the CY volume scalar, and have been presented in [10]. Typically, for a given matter type, there is a continuous, semi-infinite range of couplings that lead to a bounce. Also, for one specific value of the coupling, only the brane trajectory is altered, and the bulk remains unchanged.

It is instructive to see what the bouncing and colliding branes described above look like from the 4d point of view. The surprisingly simple 4d effective theory is given by

$$S_{\text{moduli}} = 6 \int_{4d} [-\dot{a}_4{}^2 + a_4^2(\dot{\psi}^2 + \dot{\chi}^2)], \tag{3}$$

where a_4 is the effective 4d scale factor, and the two scalars ψ and χ are related to the inter-brane distance and the size of the CY manifold [11]. While the brane collision corresponds to a big crunch/big bang ($a_4 \rightarrow 0$), it turns out that in the $\psi - \chi$ plane, the bounce corresponds to a reflection off a boundary of moduli space, see Figure 2. This boundary can be inferred from the requirement of positivity of the volume of the internal manifold, and it corresponds to imposing $\chi \leq 0$. The reflection has important consequences for cosmological perturbations around the background solution

[12]. Indeed, it is well-known that a non-geodesic motion of the background trajectory has the effect of converting entropy perturbations into curvature perturbations [13]. Here this conversion happens very rapidly, and is thus spectrum-preserving. However, in order to have a working model of early universe cosmology, we must ensure that we can obtain the right spectrum of entropy perturbations in the first place. A nearly scale-invariant spectrum is generated if we add two negative nearly-exponential potentials for the scalar fields ψ and χ in the region where perturbations are produced [12] (It has so far been impossible to directly compute the forces between the boundary branes, and so the best we can do at the moment is to simply assume a certain form for the scalar field potentials.). Then the background trajectory falls off a ridge of the potential, and is thus inherently unstable (This means that one will need a theory of initial conditions. For a proposed solution to this problem in a similar setting, see [14].). The reflection of the background trajectory off $\chi = 0$ then converts the entropy perturbations into nearly scale-invariant curvature perturbations shortly before the big crunch/big bang transition, and in the ensuing expanding phase, the curvature perturbations grow.

CONCLUSION

Perhaps the main conclusion to be drawn from this work is that in the well-motivated setting of heterotic M-theory, the singularities threatening to plague a dynamical braneworld model of the early universe can be dealt with consistently and can even have unanticipated and beneficial side effects, such as the bounce of the negative-tension brane, which not only stabilises the braneworld, but at the same time solves the problem of how to obtain growing-mode curvature perturbations from a contracting phase.

ACKNOWLEDGMENTS

It is a pleasure to thank Paul McFadden, Paul Steinhardt and Neil Turok for collaboration on this work. My research is supported by PPARC.

REFERENCES

1. P. Horava, and E. Witten, *Nucl. Phys.* **B460**, 506–524 (1996).
2. A. Lukas, B. A. Ovrut, K. S. Stelle, and D. Waldram, *Phys. Rev.* **D59**, 086001 (1999).
3. A. Lukas, B. A. Ovrut, K. S. Stelle, and D. Waldram, *Nucl. Phys.* **B552**, 246–290 (1999).
4. J. Khoury, B. A. Ovrut, P. J. Steinhardt, and N. Turok, *Phys. Rev.* **D64**, 123522 (2001).
5. P. J. Steinhardt, and N. Turok, *Phys. Rev.* **D65**, 126003 (2002).
6. W. Chen, Z. W. Chong, G. W. Gibbons, H. Lu, and C. N. Pope, *Nucl. Phys.* **B732**, 118–135 (2006).
7. J.-L. Lehners, and K. S. Stelle, *Nucl. Phys.* **B661**, 273–288 (2003).
8. J.-L. Lehners, P. McFadden, and N. Turok, *Phys. Rev.* **D75**, 103510 (2007).
9. N. Turok, M. Perry, and P. J. Steinhardt, *Phys. Rev.* **D70**, 106004 (2004).
10. J.-L. Lehners, and N. Turok (2007), `arXiv:0708.0743 [hep-th]`.
11. J.-L. Lehners, P. McFadden, and N. Turok, *Phys. Rev.* **D76**, 023501 (2007).
12. J.-L. Lehners, P. McFadden, N. Turok, and P. J. Steinhardt (2007), `hep-th/0702153`.
13. C. Gordon, D. Wands, B. A. Bassett, and R. Maartens, *Phys. Rev.* **D63**, 023506 (2001).
14. E. I. Buchbinder, J. Khoury, and B. A. Ovrut (2007), `arXiv:0706.3903 [hep-th]`.

Short distance modifications to Newton's law in SUSY braneworld scenarios

Gonzalo A. Palma

Deutsches Elektronen-Synchrotron DESY, Theory Group, Notkestrasse 85, D-22603 Hamburg, Germany

Abstract. In braneworld models coming from string theory one generally encounters massless scalar degrees of freedom –moduli– parameterizing the volume of small compact extra-dimensions. Here we discuss the effects of such moduli on Newtont's law for a fairly general 5-D supersymmetric braneworld scenario with a bulk scalar field ϕ. We show that the Newtonian potential describing the gravitational interaction between two bodies localized on the visible brane picks up a non-trivial contribution at short distances that depends on the shape of the superpotential $W(\phi)$ of the theory. In particular, we compute this contribution for dilatonic braneworld scenarios $W(\phi) \propto e^{\alpha\phi}$ (where α is a constant) and discuss the particular case of 5-D Heterotic M-theory.

Keywords: Extra-dimensions, branes, gravity
PACS: 04.50.-h, 04.50.Kd, 11.25.-w

INTRODUCTION

In theories where matter confines to a 4-D brane and gravity is the only massless field able to propagate along the extra dimensional volume, one generally expects short distance corrections to the usual 4-D Newtonian potential. The shape and distance at which these corrections become relevant generally depend on the geometry and size of the extra dimensional volume, thus allowing for distinctive signals dependent of the particular content of the theory. For instance, in the single-brane Randall-Sundrum scenario [1], where a 4-D brane of constant tension $\propto k$ is immersed in an infinitely large AdS_5 volume, a zero mode graviton $g_{\mu\nu}$ localizes about the brane. This zero mode is exponentially suppressed away from the brane with a warp factor $\propto e^{-kz}$, where z is the distance from the brane along the fifth extra-dimensional direction. The Newtonian potential describing the gravitational interaction between two bodies of masses m_1 and m_2 in the brane, and separated by a distance r, is then found to be [1, 2, 3]

$$V(r) = -G_{\rm N}\frac{m_1 m_2}{r}\left(1+\frac{2}{3k^2r^2}\right), \tag{1}$$

where $G_{\rm N}$ is Newton's constant. The correction $2/3k^2r^2$ springs out directly from the way in which gravitons propagate in an AdS_5 spacetime. If the tension k is small enough as compared to the Planck mass $M_{\rm Pl} = (8\pi G_{\rm N})^{-1/2}$, then it would be possible to distinguish this type of scenario from other extra-dimensional models in short distance tests of gravity. Present tests [4, 5] give the robust constraint $1/k < 11\mu$m.

It is therefore sensible to ask how other braneworld scenarios may differ from the Randall-Sundrum case at short distances, especially within the context of more realis-

CP957, *Particles, Strings, and Cosmology, 13th International Symposium*
edited by A. Rajantie, C. Contaldi, P. Dauncey, and H. Stoica

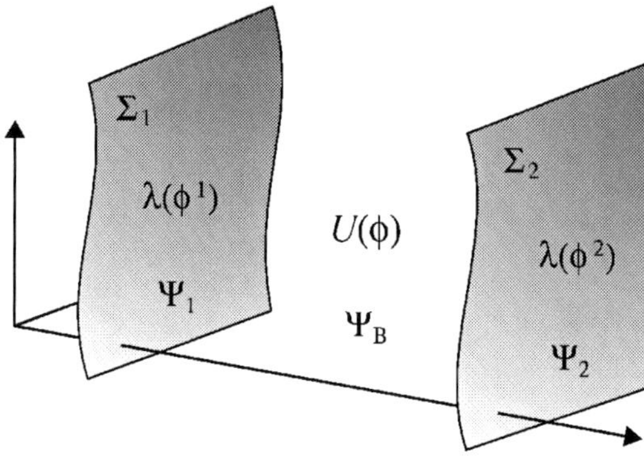

FIGURE 1. In the bulk there is a scalar field ϕ with a bulk potential $U(\phi)$. Additionally, the bulk space is bounded by branes Σ_1 and Σ_2 located at the orbifold fixed points. The branes are characterized by tensions λ_1 and λ_2, and may contain matter fields Ψ_1 and Ψ_2 respectively.

tic models. In what follows we address this question for a fairly general class of supersymmetric braneworld scenarios with a bulk scalar field, where the geometry of the extra-dimensional space differs from the usual AdS profile. We show that the Newtonian potential for this type of models picks up a non-trivial correction at scales comparable to the tension of the brane [6], that differs dramatically from the one shown in Eq. (1).

SUSY BRANEWORLDS

Let us consider a 5-D spacetime $M = \mathrm{R}^4 \times S^1/\mathrm{Z}_2$, where R^4 is a fixed 4-D Lorentzian manifold and S^1/Z_2 is the orbifold constructed from a circle with points identified through a Z_2-symmetry. M is bounded by two 3-branes, Σ_1 and Σ_2, located at the fixed points of S^1/Z_2. There is a bulk scalar field ϕ with a bulk potential $U(\phi)$ and boundary values ϕ^1 and ϕ^2 at the branes. Additionally, the branes have tensions λ_1 and λ_2 which are given functions of ϕ^1 and ϕ^2 (see FIG.1). The total action of the system is

$$S = \frac{M_5^3}{8}\int_M \left[4R^{(5)} - 3(\partial\phi)^2 - 3U(\phi)\right] - \frac{3M_5^3}{2}\int_{\Sigma_1}\lambda_1(\phi^1) - \frac{3M_5^3}{2}\int_{\Sigma_2}\lambda_2(\phi^2). \tag{2}$$

Here $\int_M$ is the short notation for $\int d^5x\sqrt{-g_5}$, where g_5 is the determinant of the 5-D metric g_{AB} of signature $(-++++)$ (a similar convention follows for $\int_\Sigma$). M_5 is the 5-D fundamental mass scale and $R^{(5)}$ is the 5-D Ricci scalar. Additionally, $\lambda_1(\phi^1)$ and $\lambda_2(\phi^2)$ are the brane tensions.

Our interest is focused on a class of models embedded in supergravity, where the bulk potential $U(\phi)$ and the brane tensions $\lambda_1(\phi^1)$ and $\lambda_2(\phi^2)$ satisfy a special relation so as to preserve half of the local supersymmetry near the branes [7]. This is

$$U = (\partial_\phi W)^2 - W^2, \qquad \lambda_1 = W(\phi^1), \qquad \text{and} \qquad \lambda_2 = -W(\phi^2), \tag{3}$$

where $W = W(\phi)$ is the superpotential of the system. Under these conditions the system presents an important property: There is a BPS vacuum state consisting of a static bulk background in which the branes can be allocated anywhere, without obstruction. Indeed, suppose a metric $ds^2 = dz^2 + g_{\mu\nu}dx^\mu dx^\nu$, where z parameterizes the extra-dimension and $g_{\mu\nu}$ is the induced metric on the 4-D foliations of M parallel to the branes. If the bulk fields depend only on z and $g_{\mu\nu} = \omega^2(z)\eta_{\mu\nu}$ (with $\eta_{\mu\nu}$ the Minkowski metric) then one finds that the entire system is solved by functions $\omega(z)$ and $\phi(z)$ satisfying

$$\omega'/\omega = -W/4 \qquad \text{and} \qquad \phi' = \partial_\phi W, \tag{4}$$

where $' = \partial_z$. Remarkably, boundary conditions at the fixed points are also given by these two equations. Thus, the presence of the branes forces the system to acquire a domain-wall-like vacuum background, instead of a flat 5-D Minkowski background.

Let us mention here that in order to have the right phenomenology in this type of scenarios, it is important to have the moduli ϕ_1 and ϕ_2 stabilized (in the 4-D effective low energy theory, they are found to be massless and dangerously coupled to matter [8, 9]). This can be done in a simple way by breaking supersymmetry on the branes in an appropriate manner [6].

THE NEWTONIAN POTENTIAL

It is possible to compute the modifications to Newton's law arising from the way in which bulk gravitons propagate in a background given by Eq. (4). In general, we can write the Newtonian potential describing the gravitational interaction between two masses m_1 and m_2 on the visible brane as $V(r) = -G_{\rm N}\frac{m_1 m_2}{r}[1 + f(r)]$, where $f(r)$ is a function of r whose shape is dictated by the form of $W(\phi)$. For concreteness, let us consider the case of dilatonic braneworlds $W(\phi) = \Lambda e^{\alpha\phi}$ where $\Lambda > 0$ is some fundamental mass scale. Then, the background geometry of the system is

$$\phi(z) = \phi_1 - \frac{1}{\alpha}\ln\left[1 - \alpha^2 W_0 z\right], \qquad \text{and} \qquad \omega(z) = \left[1 - \alpha^2 W_0 z\right]^{1/4\alpha^2}. \tag{5}$$

Notice the presence of a singularity $\omega = 0$ at $z = 1/\alpha^2 W_0$. Without loss of generality, one may take the position of Σ_1 at $z = 0$ (since $\Lambda > 0$, this is a positive tension brane). Then, Σ_2 can be anywhere between $z = 0$ and $z = 1/\alpha^2 W_0$. Interestingly, the relevant case of 5-D Heterotic M-theory [10, 11] corresponds to $\alpha^2 = 3/2$.

If, for simplicity, we further assume that the visible brane is Σ_1 while the second brane Σ_2 is very close to the bulk singularity, then it is possible to find three different solutions for $f(r)$, depending on the value of α (see [6] for a detailed analysis of this)

$$f(r) = \begin{cases} \frac{8}{3\pi^2}\frac{1-4\alpha^2}{1+2\alpha^2}\int_0^\infty \frac{dm}{m}\frac{e^{-mr}}{J^2_{\nu-1}[m/k] + Y^2_{\nu-1}[m/k]} & \text{if } \alpha^2 < 1/4 \\ \frac{32}{9\pi W_0}\int_{\sqrt{b}}^\infty dm\frac{\sqrt{m^2-b}}{m}e^{-mr} & \text{if } \alpha^2 = 1/4 \\ \frac{4}{3}\frac{4\alpha^2-1}{1+2\alpha^2}\sum_n e^{-ku_n^{\mu+1}r} & \text{if } \alpha^2 > 1/4 \end{cases} \tag{6}$$

In the previous expression we have defined $b = (3W_0/8)^2$, $k \equiv |1 - 4\alpha^2|W_0/4$, $\nu \equiv \frac{3}{2}(1 - 4\alpha^2)^{-1} + \frac{1}{2}$ and $\mu \equiv \frac{3}{2}(4\alpha^2 - 1)^{-1} - \frac{1}{2}$, where $W_0 = \Lambda e^{\alpha\phi_1}$ with ϕ_1 the value of ϕ at the positive tension brane. Additionally, $u_n^{\mu+1}$ is the n-th zero of the Bessel function $J_{\mu+1}[x]$, that is $J_{\mu+1}[u_n^{\mu+1}] = 0$. Observe that Eq. (1) is recovered for $\alpha = 0$.

DISCUSSION

We have shown the corrections to Newton's law for SUSY braneworld models arising from the way in which gravitons propagate in a bulk with a geometry that differs from the more commonly studied AdS_5. A sensible question regarding this type of models is whether there are any chances of observing short distance modifications of general relativity in the near future. To explore this, notice that the relevant energy scale at which the corrections become significant is $W_0 = \Lambda e^{\alpha\phi_1}$, instead of the more fundamental mass scale Λ. Typically one would expect $\Lambda \simeq M_5$ which has to be above TeV scales to agree with particle physics constraints. Nevertheless, the factor $e^{\alpha\phi_1}$ leaves open the possibility of bringing $\lambda = W_0^{-1}$ up to micron scales.

In the case of 5-D Heterotic M-theory ($\alpha^2 = 3/2$) one has $e^{\alpha\phi_1} = 1/\mathscr{V}$, where $\mathscr{V}$ is the volume of the Calabi-Yau 3-fold in units of M_5. In order to have an accessible scale $\lambda \simeq 10\mu\text{m}$, it would be required $\mathscr{V}\frac{M_{\text{Pl}}}{M_5} \simeq 10^{29}$, where we assumed $\Lambda \simeq M_5$. On the other hand, Newton's constant is given by $G_{\text{N}}^{-1} = \frac{32\pi}{1+2\alpha^2} M_5^2 W_0^{-1}$, which implies $M_{\text{Pl}}^2 \simeq M_5^2 \mathscr{V}$. Thus, to achieve $\mathscr{V}\frac{M_{\text{Pl}}}{M_5} \simeq 10^{29}$ one requires the following values for M_5 and $\mathscr{V}$

$$M_5 \simeq 10^{-10} M_{\text{Pl}}, \qquad \text{and} \qquad \mathscr{V}^{1/6} \simeq 10^3, \tag{7}$$

which are in no conflict with present phenomenological constraints coming from high energy physics. In particular, non-zero Kaluza-Klein modes coming from the compactified volume $\mathscr{V}$ would have masses of order 10^6GeV. On the other hand, if M_5 is of the order of the grand unification scale $M_{\text{GUT}} \sim 10^{16}$GeV, then corrections to the Newtonian potential would be present at the non-accessible scale $\lambda \sim 10^{-20}\mu\text{m}$.

ACKNOWLEDGMENTS

This work was supported by the Collaborative Research Centre 676 (Sonderforschungsbereich 676) Hamburg, Germany.

REFERENCES

1. L. Randall and R. Sundrum, Phys. Rev. Lett. **83**, 4690 (1999).
2. J. Garriga and T. Tanaka, Phys. Rev. Lett. **84**, 2778 (2000).
3. P. Callin and F. Ravndal, Phys. Rev. D **70**, 104009 (2004).
4. D. J. Kapner, T. S. Cook, E. G. Adelberger, J. H. Gundlach, B. R. Heckel, C. D. Hoyle and H. E. Swanson, Phys. Rev. Lett. **98**, 021101 (2007).
5. E. G. Adelberger, B. R. Heckel, S. Hoedl, C. D. Hoyle, D. J. Kapner and A. Upadhye, Phys. Rev. Lett. **98**, 131104 (2007).
6. G. A. Palma, JHEP **0009**, 091 (2007).
7. E. Bergshoeff, R. Kallosh and A. Van Proeyen, JHEP **0010**, 033 (2000).
8. G. A. Palma and A. C. Davis, Phys. Rev. D **70**, 064021 (2004); Phys. Rev. D **70**, 106003 (2004).
9. G. A. Palma, Phys. Rev. D **73**, 044010 (2006) [Erratum-ibid. D **74**, 089902 (2006)].
10. P. Horava and E. Witten, Nucl. Phys. B **475**, 94 (1996).
11. A. Lukas, B. A. Ovrut, K. S. Stelle and D. Waldram, Nucl. Phys. B **552**, 246 (1999).

Field Theoretical Approach to the Formation of Junctions of Cosmic Strings

Petja Salmi

Instituut-Lorentz, Universiteit Leiden, P. O. Box 9506, 2300 RA Leiden, The Netherlands

Abstract. Superstring theory predicts the potential formation of string networks with junctions. Kinematic constraints for junction formation were derived in [1], based on Nambu-Goto action. Here we test these constraints numerically within the framework of Abelian-Higgs model and report on good agreement with the analytical predictions.

Keywords: cosmic strings, string junctions, Abelian-Higgs model
PACS: 11.27.+d, 98.80.Cq

INTRODUCTION

Ever since it was argued that cosmic strings could have formed in the early universe [2], they have been a potential observational window to the physics at very high energy scales. Even though to date there has been no success on the observational side, theoretically cosmic strings are currently more topical than ever; many models of inflation motivated by fundamental string theory suggest the formation of defect networks. The first study of, so called cosmic superstrings pointed out severe problems related to the formation and growth of superstrings to cosmologically relevant length scales [3]. Advances in fundamental string theory have since changed this picture and all these issues can be circumvented; for instance, in the scenarios based on warped compactifications the warp-factor can lower the string tension and models of brane inflation predict copious production of strings in brane collisions (see e.g. [4]).

The appeal of cosmic strings in comparison with other topological defects, like domain walls, is their relative harmlessness: string networks lose energy efficiently and a scaling string network contributes only a fixed, tiny fraction of the total energy density of the Universe. Considerations inspired by superstring theory have suggested the formation of cosmic string networks that consists of fundamental F-strings, Dirichlet D-strings and bound states of these two, (p,q)-strings [5, 6]. Where strings meet to form a bound state, a junction will form. The effect of junctions on the evolution of the networks is of profound interest: instead of reaching a scaling regime the network could end in a frozen-out state. The (p,q)-strings have attracted lots of attention recently, see for instance [7, 8, 9, 10, 11].

The observational potential of cosmic strings is based on various effects that strings at cosmic length scales would cause generally independently of their origin, fundamental or solitonic. Firstly, strings could leave an imprint in the Cosmic Microwave Background. Furthermore, cosmic strings are likely to emit gravitational radiation that could be detectable with present or forthcoming experiments. Finally, strings cause gravita-

CP957, *Particles, Strings, and Cosmology, 13th International Symposium*
edited by A. Rajantie, C. Contaldi, P. Dauncey, and H. Stoica

tional lensing (for a recent proposal to observe string loops via gravitational lensing in future surveys at radio frequencies see [12]).

STRING JUNCTIONS

Collision of two strings that can interchange partners was studied in [1, 13]. Approach based on the Nambu-Goto action showed that strong kinematic constraints apply to the formation of the bound state.

Consider a straight string making an angle α with the positive x-axis and another one with an angle $\ \alpha$ (the total angle between the strings being thus 2α). Both strings are on the xy-plane and have a velocity v along the z-axis with opposite directions. Once intersecting, these can potentially form a third string along the x-axis, an x-link. The kinematic constraints follow from the necessary requirement that this third string must grow in length. The allowed parameter region for the link formation can be determined at least numerically for any combination of string tensions [13]. However, when the colliding strings have the same tension, denoted hereafter by μ_1, it is possible to express this requirement in a simple closed form

$$\alpha < \arccos\left(\frac{\gamma\mu_3}{2\mu_1}\right), \tag{1}$$

where μ_3 is the tension of the third string (necessarily $\mu_3 < 2\mu_1$) and $\gamma = 1/\sqrt{1\ \ v^2}$.

Here we want to test this result numerically in a field theory where strings are solitonic objects with a *non-zero* width.

ABELIAN-HIGGS MODEL

The Abelian-Higgs model is governed by the Lagrangean

$$L = (\partial_\mu + iqA_\mu)\,\phi\,(\partial^\mu - iqA^\mu)\,\phi^\dagger - \frac{1}{4}F_{\mu\nu}F^{\mu\nu} - \frac{\lambda}{4}(|\phi|^2 - \eta^2)^2, \tag{2}$$

and has vortex solutions [14], which are topologically stable for any value of the scaled coupling $\beta = \lambda/2q$. When $\beta < 1$ a string with winding number $n = 2$ has lower energy than two strings of winding number $n = 1$. In other words using the previous notation $\mu_3 < 2\mu_1$, and two $n = 1$ strings could collide to form an $n = 2$ string. Thus this regime provides a testing ground for the formation of links.

RESULTS

Local cosmic strings always intercommute when intersecting. However, after intercommutation, strings can come together and merge to form a string of winding number $n = 2$. Figure 1 illustrates a typical case when collision occurs at relatively small velocity and angle. The left panel shows the two strings well before the collision, while on the right

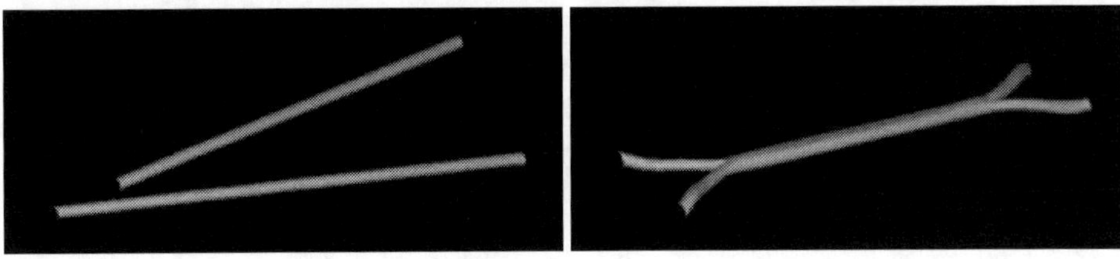

FIGURE 1. Snapshots showing the constant energy density isosurfaces in a simulation of two strings at $\beta = 0.125$ colliding with velocity $v = 0.1$ and angle $\alpha = 15°$.

they have joined later on to form an x-link. Alternatively strings intercommute and never meet again.

The main achievement of this study is presented in Figure 2, where the results from the simulations, standard intercommutation versus link-formation, are shown together with the prediction of (1) for the allowed parameter region in the (α, v)-plane. A decrease in the scalar coupling λ reduces the tension of a string with winding number $n = 2$ relatively more than that of a string with $n = 1$. This allows one to vary the ratio μ_3/μ_1 to a certain extent; at $\beta = 0.36$ we obtain $\mu_3/\mu_1 \simeq 1.88$, whereas $\beta = 0.125$ yields $\mu_3/\mu_1 \simeq 1.77$. This already leads to a considerable difference when the limiting curve of the equality in (1) is plotted for $\beta = 0.36$ (left) and $\beta = 0.125$ (right) in Figure 2. Events of standard intercommutation are indicated by crosses and the formation of links by circles. At $\beta = 0.36$ the agreement between the simulations and the analytical prediction is perfect within the precision of the grid used in the (α, v)-parameter space. There is a discrepancy at the largest angles when $\beta = 0.125$; while the formation of an x-link should be allowed, this is not observed in the simulations.

In addition, at $\beta = 0.125$ there is a band just above the highest velocity allowing x-links to form suggested by (1) where a link does not form, but strings come together again and intercommute for a second time, events denoted by triangles in Figure 2. The end result is thus two strings consisting of the same segments as initially and indistinguishable from the situation where strings had passed through each others. A similar type of effective non-intercommutation has been reported when strings collide at very high velocities [15]. There is no sign of this kind of phenomenon at $\beta = 0.36$.

DISCUSSION

We have reported on numerical simulations of string collisions in the Abelian-Higgs model. The objective was to monitor the kinematic parameter space (α, v) and compare the outcome to the analytical prediction. We do not observe x-links outside the area where they are not expected to be kinematically allowed. On the other hand, x-links generically form whenever allowed to appear. The agreement to analytical results based on the Nambu-Goto action is remarkably good.

It is not surprising that the observed discrepancy occurs when strings intersect at large angles. As strings in the Abelian-Higgs model always intercommute, effectively the original strings break up, and when the new strings straighten after this reconnection,

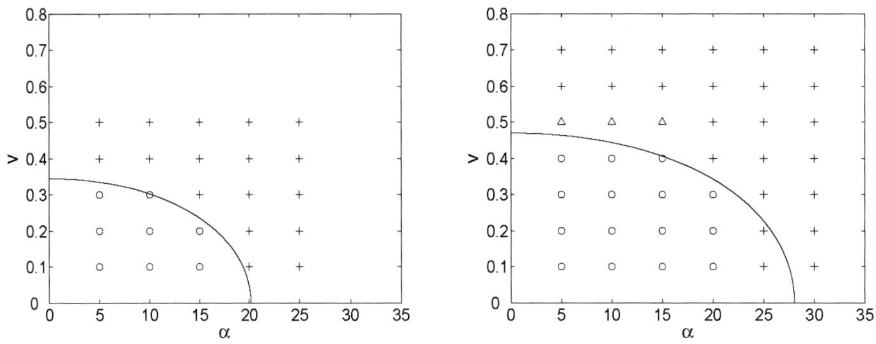

FIGURE 2. Kinematic constraints in the (α, v)-plane for $\beta = 0.36$ (left) and $\beta = 0.125$ (right). The solid line shows the curve of equality in (1), + marks the events of single intercommutation, whereas o indicates strings merging together to form an x-link, a string with winding number $n = 2$.

they do not come together and merge at large angles. It is plausible to think that (p,q)-strings that would not exchange segments (two different U(1) gauge fields have been used to model (p,q)-strings e.g. in [7]) could form an x-link at larger angles than we observe here and this effect could explain the discrepancy.

ACKNOWLEDGMENTS

The author is greatly indebted to Tom Kibble for invaluable communications during this work. Roland de Putter is acknowledged for the code reported in [15]. Furthermore, the author is grateful to Ana Achúcarro and especially to Daniele Steer for carefully reading this manuscript and to Serena Bertone for the help in visualisation. The simulations have been performed at the Lisa cluster the use permitted by NCF grant.

REFERENCES

1. E. J. Copeland, T. W. B. Kibble and D. A. Steer, Phys. Rev. Lett. **97** (2006) 021602.
2. T. W. B. Kibble, J. Phys. A **9** (1976) 1387.
3. E. Witten, Phys. Lett. B **153** (1985) 243.
4. S. Sarangi and S. H. H. Tye, Phys. Lett. B **536** (2002) 185.
5. E. J. Copeland, R. C. Myers and J. Polchinski, JHEP **0406** (2004) 013.
6. H. Firouzjahi, L. Leblond and S. H. H. Tye, JHEP **0605** (2006) 047.
7. P. M. Saffin, JHEP **0509** (2005) 011.
8. M. Hindmarsh and P. M. Saffin, JHEP **0608** (2006) 066.
9. L. Leblond and M. Wyman, Phys. Rev. D **75** (2007) 123522.
10. A. Avgoustidis and E. P. S. Shellard, arXiv:0705.3395 [astro-ph].
11. A. Rajantie, M. Sakellariadou and H. Stoica, arXiv:0706.3662 [hep-th].
12. K. J. Mack, D. H. Wesley and L. J. King, arXiv:astro-ph/0702648.
13. E. J. Copeland, T. W. B. Kibble and D. A. Steer, Phys. Rev. D **75** (2007) 065024.
14. H. B. Nielsen and P. Olesen, Nucl. Phys. B **61** (1973) 45.
15. A. Achúcarro and R. de Putter, Phys. Rev. D **74** (2006) 121701.

Gauge-Higgs Unification from the Heterotic String

Jonas Schmidt

Deutsches Elektronen-Synchrotron DESY, Notkestrasse 85, 22603 Hamburg, Germany

Abstract. We present a 6D orbifold model on $T^2/\mathbb{Z}_2$ which emerges as intermediate step in the compactification of the heterotic string to the supersymmetric standard model in four dimensions. It has SU(6) gauge symmetry in the bulk and two pairs of inequivalent fixed points with unbroken SU(5) and SU(2) × SU(4) symmetry, respectively. All anomalies are cancelled by the Green-Schwarz mechanism. Two quark-lepton generations are located at the SU(5) branes, the third family is composed of split bulk hypermultiplets. The model has vacua with partial or full gauge-Higgs unification and non-vanishing Yukawa couplings which generically avoid the unsuccessful SU(5) mass relations.

Keywords: Grand Unification, Extra Dimensions, Heterotic String Vacua
PACS: 11.25.Wx,11.25.Mj,12.10.-g,12.60.-i

INTRODUCTION

A completion of the standard model in terms of string theory may reveal an intermediate stage of grand unification. This idea is supported by the symmetries and the matter content of the standard model as well as the unification of gauge couplings at the scale $M_{\mathrm{GUT}} \simeq 2 \times 10^{16}$ GeV for the minimal supersymmetric standard model (MSSM). Here we present the construction of such an intermediate grand unifying theory (GUT) in six dimensions [1] and discuss some interesting aspects of its local GUT structure.

The model is based on a $\mathbb{Z}_{6-\mathrm{II}}$ orbifold compactification of the $\mathrm{E}_8 \times \mathrm{E}_8$ heterotic string which leads to the MSSM in four dimensions [2, 3]. Four extra dimensions are compactified on $T^4/\mathbb{Z}_3$, where T^4 is the Lie algebra lattice of $\mathrm{G}_2 \times \mathrm{SU}(3)$ and the SU(3) torus is supplemented by a Wilson line. This leads to a six-dimensional (6D) supergravity theory with unbroken gauge group

$$\mathrm{G}_6 = \mathrm{SU}(6) \times \mathrm{U}(1)^3 \times [\mathrm{SU}(3) \times \mathrm{SO}(8) \times \mathrm{U}(1)^2], \qquad (1)$$

where the brackets denote the subgroup of the second E_8. Additionally to gravity, dilaton and vector multiplets it has $\mathcal{N} = 2$ hypermultiplets[1]

$$(\mathbf{20}; 1, 1) + (1; 1, \mathbf{8}) + (1; 1, \mathbf{8}_s) + (1; 1, \mathbf{8}_c) + 4 \times (1; 1, 1) \qquad (2)$$

from the untwisted sector. Furthermore, there are three fixed points in the SU(3) plane. They all have unbroken SO(14) × U(1) × [SO(14) × U(1)] symmetry, differently em-

[1] All U(1) charges are surpressed in the following. The full list is given in [1].

CP957, *Particles, Strings, and Cosmology, 13th International Symposium*
edited by A. Rajantie, C. Contaldi, P. Dauncey, and H. Stoica

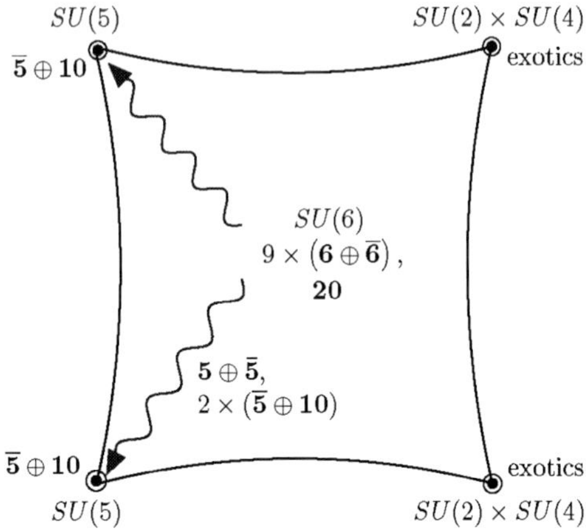

FIGURE 1. The orbifold $T^2/\mathbb{Z}_2$. The arrows indicate the local reduction of bulk states at the two equivalent SU(5)-fixed points ($n_2 = 0$) to one pair of Higgs fields ($\mathbf{5} \oplus \overline{\mathbf{5}}$) and two bulk quark-lepton families ($\overline{\mathbf{5}} \oplus \mathbf{10}$), due to chiral orbifold projections and decoupling. All exotics located at the remaining two fixed points ($n_2 = 1$) can be decoupled.

bedded into $E_8 \times E_8$. Each fixed point contributes three copies of

$$(\mathbf{14},1;1,1) = (\mathbf{6};1,1) + (\overline{\mathbf{6}};1,1) + 2 \times (1;1,1)\,, \tag{3a}$$

$$(1,1;\mathbf{14},1) = (1;\mathbf{3},1) + (1;\overline{\mathbf{3}},1) + (1;1,\hat{\mathbf{8}}) \tag{3b}$$

to the 6D theory. At the three SU(3) fixed points, $(1;1,\hat{\mathbf{8}})$ corresponds to $(1;1,\mathbf{8})$, $(1;1,\mathbf{8}_s)$ and $(1;1,\mathbf{8}_c)$, respectively. Finally, there are oscillator states for the two small compact planes which yield two non-Abelian singlet hypermultiplets for each fixed point. Compactification from six to four dimensions occurs on the orbifold $T^2/\mathbb{Z}_2$, where T^2 is the Lie algebra lattice of SO(4) (cf. Figure 1). Due to the presence of one Wilson line in this torus there are two inequivalent pairs of fixed points, labelled by $n_2 = 0, 1$, with local gauge groups

$$n_2 = 0: \; G_0 = \mathrm{SU}(5) \times \mathrm{U}(1)^4 \times [\mathrm{SU}(3) \times \mathrm{SO}(8) \times \mathrm{U}(1)^2]\,, \tag{4}$$

$$n_2 = 1: \; G_1 = \mathrm{SU}(2) \times \mathrm{SU}(4) \times \mathrm{U}(1)^4 \times [\mathrm{SU}(2)' \times \mathrm{SU}(4)' \times \mathrm{U}(1)^4]\,. \tag{5}$$

The four dimensional gauge group is given by the intersection of the local symmetries. At the fixed points we find localized states which transform under the non-Abelian symmetries as

$$n_2 = 0: (\overline{\mathbf{5}};1,1) + (\mathbf{10};1,1) + (1;1,\mathbf{8_c}) + 2 \times (1;\mathbf{3},1) + 2 \times (1;\overline{\mathbf{3}},1) + 9 \times (1;1,1),$$
$$n_2 = 1: 4 \times (\mathbf{2},1;1,1) + 16 \times (1,1;1,1)\,.$$

All exotics are localized at $n_2 = 1$ and can be decoupled from the zero mode spectrum. Phenomenologically interesting vacua and the four dimensional limit of the model were studied in [3]. Here we focus on the anisotropic limit of the compactification in which the SO(4) torus is considerably larger than the other two tori. This introduces an intermediate field theoretical orbifold GUT in six dimensions. Finally, at string scale energies, all ten dimensions can be resolved and full heterotic string theory has to be considered.

ANOMALIES

The cancellation of anomalies in string derived models follows by application of the Green-Schwarz (GS) mechanism [4]. For the bulk and brane anomalies of our 6D heterotic orbifold model [5, 6, 7] this relies on gauge and Lorentz variations of the bulk two-form B_2. All contributions from box diagrams in the bulk and local triangle diagrams at the fixed points add to reducible anomaly polynomials

$$I_8^{\text{bulk}} \sim X_4 \wedge Y_4 \,, \qquad\qquad I_6^f \sim X_4^f \wedge Y_2^f \,. \tag{6}$$

Here a subscript n denotes a (formal) n-form and f a projection onto the fixed point f. Note that the factorizations (6) invoke ~ 500 conditions in the case of our model which imposes a highly non-trivial check for the spectrum presented before. Furthermore, the explicit calculation of the two-forms Y_2^f allow the identification of local anomalous U(1)'s at the fixed points. They lead to the following coefficients ξ_{n_2} of local Fayet-Iliopoulos (FI) D-terms:

$$\xi_0 = 2\,\frac{gM_{\rm P}^2}{384\pi^2}\,, \qquad\qquad \xi_1 = \frac{gM_{\rm P}^2}{384\pi^2}\,. \tag{7}$$

It is known that localized FI terms lead to an instability of bulk fields and spontaneous localization towards the fixed points [6]. It is intriguing that the mass scale of the FI terms is of the order of the grand unification scale, $M_{\rm P}/\sqrt{384\pi^2} \sim M_{\rm GUT}$.

LOCAL SU(5) UNIFICATION

All bulk multiplets (2), (3) are $\mathcal{N}=2$ hypermultiplets $H=(H_L,H_R)$, where H_L and H_R are $\mathcal{N}=1$ left and right chiral multiplets, respectively. The multiplets (3) come in three copies, due to three fixed points in the G_2 plane. Thus even after the chiral projection onto $n_2=0$ it is still possible to form singlets $H_L H_R^c$. A large number of fields can in this way easily be decoupled by giving a vacuum expectation value (VEV) to only three singlet fields, two of which are localized at $n_2=0$. This yields localized $\mathcal{N}=1$ chiral multiplets $3\times\mathbf{5}\oplus 5\times\bar{\mathbf{5}}\oplus 2\times\mathbf{10}$ from the bulk, including a $\mathbf{5}\oplus\bar{\mathbf{5}}$ pair originating from the SU(6) gauge multiplet. In a second step, one can again form two singlets $\bar{\mathbf{5}}\mathbf{5}$ and specify VEVs which induce masses for the involved fields. Thus finally one $\mathbf{5}\oplus\bar{\mathbf{5}}$ pair of Higgs fields and two SU(5) families $\bar{\mathbf{5}}\oplus\mathbf{10}$ remain light after decoupling (cf. Figure 1).

Depending on the origin of the Higgs multiplets the model can have full, partial or no gauge-Higgs unification. As an example, we realized partial gauge-Higgs unification and the decoupling of all exotic doublets by giving VEVs to 19 singlet fields. To $\mathcal{O}(8)$ in the fields the Yukawa couplings of the zero modes, $W_{\text{Yuk}} = Y^{(u)}_{ij} u^c_i q_j H_u + Y^{(d)}_{ij} d^c_i q_j H_d + Y^{(l)}_{ij} l_i e^c_j H_d$, where $i, j = 1, 2, 3$ is a family index, then take the form

$$Y^{(u)}_{ij} = \begin{pmatrix} a_1 & 0 & a_3 \\ 0 & a_1 & a_3 \\ a_2 & a_2 & g \end{pmatrix}, \quad Y^{(d)}_{ij} = \begin{pmatrix} 0 & 0 & b_2 \\ 0 & 0 & b_2 \\ b_5 & b_5 & b_7 \end{pmatrix}, \quad Y^{(l)}_{ij} = \begin{pmatrix} 0 & 0 & b_1 \\ 0 & 0 & b_1 \\ b_3 & b_3 & b_4 \end{pmatrix}.$$

Here g denotes the $\mathrm{SU}(6)$ gauge coupling and is responsible for the large top mass. The a_i and b_i represent VEVs of distinct combinations of singlets. This example illustrates how the mainly unsuccessfull mass relations of standard $\mathrm{SU}(5)$ GUTs are avoided in our model. It is due to the fact that the third quark-lepton generation forms two distinct split bulk matter families which consequently have distinct Yukawa couplings.

CONCLUSIONS

We have constructed a 6D orbifold GUT from the heterotic string. Two of its four fixed points realize local $\mathrm{SU}(5)$ GUTs with localized quark-lepton families. The third family follows from two split bulk generations with distinct couplings to the Higgs doublets. This avoids the standard $\mathrm{SU}(5)$ equality of lepton and down-type quark Yukawa couplings. It is possible to find vacua with partial or full gauge-Higgs unification, ensuring a large top Yukawa coupling. However, the implementation of a matter parity into the model needs further investigation. We calculated the local FI-terms at the fixed points which have crucial influence on the dynamics of bulk fields. Open issues concern their rôle in a blow-up to a smooth manifold and the stabilization of the GUT scale. Having in mind the goal of grand unification and the large number of similar compactifications [8, 9], one may hope that a local GUT structure as investigated here can help to identify realistic string vacua.

ACKNOWLEDGMENTS

I would like to thank my collaborators W. Buchmüller and C. Lüdeling.

REFERENCES

1. W. Buchmuller, C. Ludeling, and J. Schmidt (2007), arXiv:0707.1651 [hep-ph].
2. W. Buchmuller, K. Hamaguchi, O. Lebedev, and M. Ratz, *Phys. Rev. Lett.* **96**, 121602 (2006), hep-ph/0511035.
3. W. Buchmuller, K. Hamaguchi, O. Lebedev, and M. Ratz (2006), hep-th/0606187.
4. M. B. Green, and J. H. Schwarz, *Phys. Lett.* **B149**, 117–122 (1984).
5. J. Erler, *J. Math. Phys.* **35**, 1819–1833 (1994), hep-th/9304104.
6. H. M. Lee, H. P. Nilles, and M. Zucker, *Nucl. Phys.* **B680**, 177–198 (2004), hep-th/0309195.
7. T. Asaka, W. Buchmuller, and L. Covi, *Nucl. Phys.* **B648**, 231–253 (2003), hep-ph/0209144.
8. O. Lebedev, et al., *Phys. Lett.* **B645**, 88–94 (2007), hep-th/0611095.
9. O. Lebedev, et al. (2007), arXiv:0708.2691 [hep-th].

Brane Decay and Defect Formation

Horace Stoica

Theoretical Physics, Blackett Laboratory, Imperial College, London SW7 2AZ U.K.

Abstract. Topological defects are generically expected to form in models of brane inflation. Brane-anti-brane annihilation provides a way to gracefully end inflation, and the dynamics of the tachyon field results in defect formation. The formation of defects has been studied mainly from the brane world-volume point of view, but the defects are themselves lower-dimensional branes, and as a result they couple to bulk fields. To investigate the impact of bulk fields on brane defect formation, we construct a toy model that captures the essential features of the tachyon condensation with bulk fields. In this toy model, we study the structure of defects and simulate their formation and evolution on a lattice. We find that while bulk fields do not have a significant effect on defect formation, they drastically influence the subsequent evolution of the defects, as they re-introduce long-range interactions between them.

Keywords: brane, topological defect, bulk field
PACS: 11.27.+d, 11.15.Ha, 11.25.-w, 98.80.Cq

INTRODUCTION

Among the successes of the brane world model is the realization of the inflationary universe [1]. The brane-anti-brane separation plays the role of the inflaton and the non-zero attractive interaction generates the inflaton potential. Moduli stabilisation has been recently added to the brane-world model [2], leading to even more realistic brane inflation models.

The brane inflation model contains a natural mechanism for ending inflation via the brane-anti-brane annihilation process [3]. When the inter-brane separation decreases below a critical value, the tachyon field which corresponds to the open string stretching between the brane and the anti-brane develops an instability (Fig. 1), and the rolling of the tachyon field signals the decay of the brane-anti-brane pair. The tachyon field for a brane-anti-brane pair is a complex field and it has a non-trivial vacuum manifold which leads to the formation of topologically stable vortex configurations. These vortices are themselves lower-dimensional branes [4, 5] and their formation [6] and stability [7] have been studied for a variety of brane inflation models. These branes couple with bulk fields and would appear as cosmic strings to a 4-dimensional observer, therefore they are expected to provide a direct observational window into String Theory.

Brane-Anti-Brane Annihilation

The action for the tachyon field of a coincident brane-anti-brane pair has been computed in [4, 5]. Neglecting the gauge fields present in the brane and the anti-brane, the

CP957, *Particles, Strings, and Cosmology, 13th International Symposium*
edited by A. Rajantie, C. Contaldi, P. Dauncey, and H. Stoica

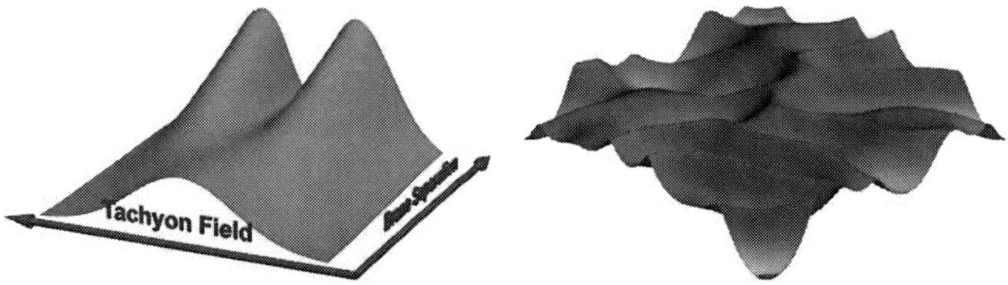

FIGURE 1. Left: The potential of the tachyon field as a function of the inter-brane separation. Right: The configuration of the rolling tachyon field during the brane-anti-brane annihilation process.

action has the form:

$$S = \int d^{p+1}x\sqrt{-g}e^{-T\overline{T}/2M_s^2}\left(1+\partial_\mu T\partial^\mu \overline{T}\right) \tag{1}$$

Performing a lattice evolution of the tachyon field [8], one notices that the field develops singularities [Fig. 1, right panel]. These singularities are the locations of the lower-dimensional branes formed in the brane annihilation process. In [4, 5] it was shown that a linear spatial tachyon profile reproduces the correct tension of the lower dimensional branes in the limit of infinite slope, so the occurrence of singularities was expected. One can use the lattice evolution of the tachyon field to estimate the density of vortices formed in brane annihilation, and the results suggest a much larger density than the one estimated via the Kibble mechanism [8]. However, the occurrence of singularities does not allow one to study the evolution of the network of vortices.

Coupling to Bulk fields

To obtain a more complete picture of the tachyon vortex formation we have to include the effects of all the fields involved. The tachyon field is charged under a linear combination of brane world-volume gauge fields and this same linear combination couples to a bulk antisymmetric tensor fields via a Chern-Simons (CS) term. The relevant terms in the action are [4, 5]:

$$S = \frac{1}{2\kappa_{10}^2}\int\sqrt{-G}d^{10}x\left[e^{-2\phi}\left(R+2\nabla\phi^2\right)-\frac{F_p^2}{2p!}\right]+2\pi\alpha' T_{Dp}\int C_{p-1}\wedge\left(F^+-F^-\right)$$
$$-2T_{Dp}\int d^{p+1}xe^{-\phi}\,e^{-2\pi\alpha' T\overline{T}}\left[1+8\pi\alpha'\ln(2)D^\mu\overline{T}D_\mu T+\frac{\gamma\alpha'^2}{8}\left(F^+_{\mu\nu}-F^-_{\mu\nu}\right)^2\right]$$

We want to study the formation and evolution of the tachyon vortices using the full non-linear equations of motion. Since the above action is 10-dimensional, we will first build a lower-dimensional toy model that captures the main features of the full 10-dimensional model and is amenable to study via a lattice simulation. We study the formation of vortices, so the minimal dimensionality of a brane that allows vortices

TABLE 1. The correspondence between the full 10-dimensional model and the toy model

	10 D Model	Toy Model
brane dim.	$p+1$	$2+1$
bulk dim.	$9+1$	$3+1$
scalar field	tachyon, T	Higgs, ϕ
vector field	$A_\mu^+ - A_\mu^-$	A_μ
tensor field	$C_{M_1 \ldots M_{p-1}}$	C_M
gravity	G_{MN}	flat space
coupling strength	dilaton, ϕ	fixed coupling
potential	$e^{-\lvert T\rvert^2}$	$\lambda\left(\lvert\phi\rvert^2 - \eta^2/2\right)^2$

to form is 2. We choose the bulk to also have the minimum possible dimensionality, 3, and replace the tachyon with a Higgs field in order to avoid singularities. The rest of the correspondence between the full and the toy model is presented in Table 1. The action for the toy model follows that of the 10-dimensional model:

$$\mathscr{L} = -\int_{\mathscr{M}_3} dx^2 dt \left[\frac{F^2}{4g^2_{\text{brane}}} + D_\mu\phi D^\mu\phi^* + V(\phi)\right] - \frac{c_{CS}}{2}\int_{\mathscr{M}_3} F\wedge C - \int_{\mathscr{M}_4} dx^3 dt \frac{H^2}{4g^2_{\text{bulk}}} \tag{2}$$

The details of the lattice regularisation of the above action are presented in [9]. The important results are presented in Fig. 2 and 3. Reducing the size of the extra dimension forces the bulk field to have a larger gradient along the extra dimension, therefore increasing the total energy of the defect-anti-defect pair. The most dramatic effect is that of changing the strength of the CS coupling. Increasing the coupling reduces the quantum of brane gauge field magnetic flux trapped at the core of the defect, at large values of the coupling the flux being almost completely eliminated. This can be understood by first performing a dimensional reduction to 2 dimensions and then writing the equation of motion for the bulk field. The bulk field is sourced by the magnetic flux of the brane gauge field and this magnetic flux is localised at the defect core.

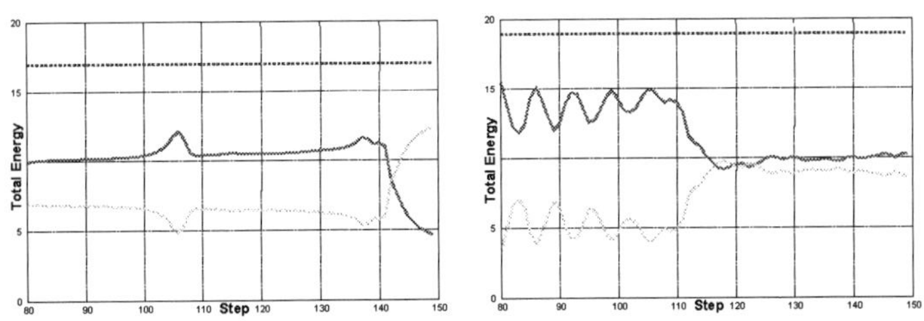

FIGURE 2. Effect of the size of the extra dimension on the energy of a defect pair. Left panel: the extra dimension is large and the bulk is effectively 3-dimensional. Right panel: the extra dimension is comparable to the size of the defect core, making the model effectively 2-dimensional. This also changes the strength of the interaction between the vortices, which annihilate faster when the extra dimension is small.

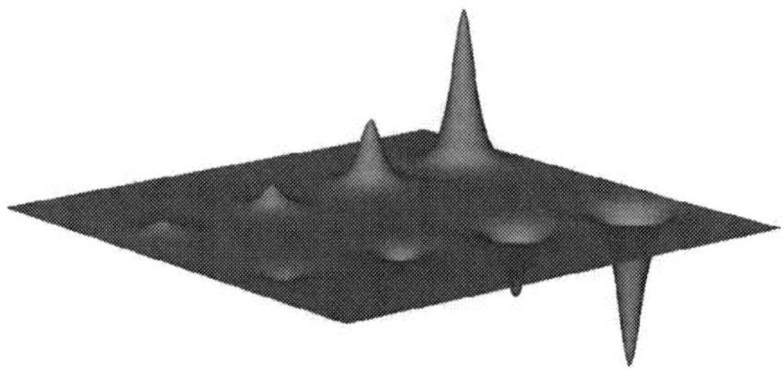

FIGURE 3. The magnetic flux at the core of the defect as a function of the CS coupling. The larger the coupling, the smaller the flux.

$$\nabla^2 C^0 = c_{cs} g_{bulk}^2 F_{12}/2\pi R \tag{3}$$

We can use the solution for C^0 to find the energy density of the fields away from the defect core.

$$\mathscr{E} = |D_\theta \phi|^2 + \frac{2\pi R}{2g_{bulk}^2}|\nabla C^0|^2 \tag{4}$$

where $D_\theta \phi = (in/\rho - iA_\theta)\phi$, $\phi^*\phi = v^2/2$, $A_\theta = \Phi/2\pi\rho$ and $C^0(\rho) = \frac{c_{cs} g_{bulk}^2}{2\pi R}\frac{\Phi}{2\pi}\ln(\rho/\rho_0)$. Φ is the brane gauge field magnetic flux. Minimising the energy density with respect to Φ gives the results:

$$\mathscr{E} = \frac{v^2 n^2}{2\rho^2}\frac{a}{1+a}, \qquad \Phi = \frac{2\pi n}{1+a}, \qquad a \equiv \frac{c_{cs}^2 g_{bulk}^2}{2\pi R v^2}. \tag{5}$$

As the CS coupling is increased the magnetic flux unit at the defect core decreases and the energy of the defect interpolates smoothly between that of a *local* and that of a *global* defect. The bulk field also re-introduces long-range interactions between the local vortices.

ACKNOWLEDGMENTS

I would like to thank N. Barnaby, A. Berndsen, J. M. Cline and G. D. Moore for collaboration on the work presented here.

REFERENCES

1. G. R. Dvali and S. H. H. Tye, *Phys. Lett. B* **450** (1999) 72
2. S. Kachru, *et. al. JCAP* **0310**, 013 (2003)
3. A. Sen, *JHEP* **9808**, 012 (1998)
4. P. Kraus and F. Larsen, *Phys. Rev. D* **63**, 106004 (2001),
5. T. Takayanagi, S. Terashima and T. Uesugi, *JHEP* **0103**, 019 (2001)
6. N. T. Jones, H. Stoica and S. H. H. Tye, *Phys. Lett. B* **563**, 6 (2003)
7. E. J. Copeland, R. C. Myers and J. Polchinski, *JHEP* **0406**, 013 (2004)
8. N. Barnaby, A. Berndsen, J. M. Cline and H. Stoica, *JHEP* **0506**, 075 (2005)
9. G. D. Moore and H. Stoica, *Phys. Rev. D* **74**, 065003 (2006)

COSMOLOGY

Child Universe Production

S. Ansoldi*,† and E. I. Guendelman**

*Department of Physics, University of Kyoto, Kyoto, Japan
†International Center for Relativistic Astrophysics (ICRA) - Pescara - Italy
**Department of Physics, Ben Gurion University, Beer Sheva, Israel

Abstract. There are solutions of Israel junction conditions that describe the detachment of a region of spacetime, a *bubble*, from what can be called the *parent spacetime*. In the past these solutions have been mostly considered in the context of inflationary cosmology (the so called *baby universes*). We will i) briefly review these ideas, ii) show that they can be of interest under more general settings and iii) present solutions where it is possible to create a universe at arbitrarily small energy cost. This could have interesting applications for quantum gravity and particle physics, besides cosmology.

Keywords: Baby Universes, Child Universes
PACS: 98.80.Bp, 98.80.Cq, 98.80.-k, 04.60.Kz, 98.80.Jk, 04.40.-b

The dynamics of spacetime often shows unexpected and astonishing phenomena and it is usually very interesting to move on from the mathematical properties of the solutions to analyze in detail their physical consequences. In what follows we will consider solutions of Einstein equations when spacetime can be divided in two domains separated by an infinitesimally thin layer of matter and energy. We will restrict ourself to the case in which these configurations can be described by the Israel junction formalism and we will analyze a particular case, where one part of spacetime is Minkowskii space and the other is Schwarzschild space, for a not yet considered shell matter content.

Shell-based models have, in fact, been studied in the context of inflationary cosmology [1, 2, 3], where an isolated vacuum bubble in asymptotically flat space (*external space*) can evolve in what is called a *baby universe* [4, 5, 6, 7, 8, 9, 10, 11] (classical as well as quantum processes [10, 12, 13] and related issues [14] have been considered). The bubble energy density supports the expansion, which is counteracted by the pressure difference (negative inside, zero outside): although this implies that the bubble cannot expand into external space, for a big enough bubble of false vacuum its growth to an arbitrarily large size is not only possible, but inevitable (provided the bubble can become larger than the de Sitter space [15, 16] horizon). This counterintuitive result is made possible by the particular structure of the Kruskal extension [17], which describes the external space: indeed, the presence of a wormhole allows the bubble to expand by making its own space if its evolution takes place in the wormhole region of Schwarzschild space [18]. This then implies that the bubble, during its evolution, disconnects from ambient space, i.e. it is a *baby universe*. It is noteworthy to realize that the mechanical force that accelerates the wall always pushes from outside to inside. It is the presence of the wormhole that makes this direction also the direction of increasing radius; thus all the forces that usually would make the bubble shrink, now make it expand. In the cases in which the above scenario has been considered, there is a critical value for the total mass–energy in the asymptotically flat region, $M_{\rm cr}$, such that only for values bigger than $M_{\rm cr}$

CP957, *Particles, Strings, and Cosmology, 13th International Symposium*
edited by A. Rajantie, C. Contaldi, P. Dauncey, and H. Stoica

the bubble can expand up to infinity [8] (i.e. it is "kicked" hard enough to become bigger than the de Sitter horizon). A possibility to lower the mass threshold is to introduce a *hedgehog* [19], i.e. a multiplet of scalar fields living on the bubble in the configuration of a global monopole of big enough strength (for the idea of *topological inflation* see also [20] and [21], where the gauged case is also analyzed). Another common feature (unfortunately, *unwanted*) of the classical models discussed above is the presence of an initial singularity (as we will also do, spherical symmetry is assumed for simplicity; in view of general theorems [22], this is likely to hold in more general cases). In this respect magnetic monopoles might also be of help, and mechanisms have been proposed [23] in which two magnetic monopoles with *below critical* magnetic charge merge into a *supercritical* one giving rise to a baby universe. Other approaches, in which baby universe formation is achieved by collapsing a surrounding membrane onto the monopole [24], can also be singularity free in the causal past of the newly formed baby universe (but not globally, since trapped regions *do* exist [22]).

As we have seen, in the existing literature baby universes have been considered in the context of inflationary cosmology. We will now show that they can be of interest also outside this specific domain, and use the terminology *child universe* to indicate *an arbitrary region of spacetime that expands disconnecting from an ambient space*, irrespectively from the energy-matter content and the structure of both the disconnecting and ambient domains (moreover, exponential expansion is *not* required). An example of child universe that satisfies the slightly more general definition introduced above, can be obtained with a (technically) very simple model, which, as we anticipated at the end of the first paragraph, consists of a spherically symmetric, *thin*, *relativistic shell* Σ joining a domain of Minkowskii spacetime $\mathscr{M}_-$ with a domain of Schwarzschild spacetime $\mathscr{M}_+$. The dynamics of the system can be studied using Israel junction conditions[1] [25, 26]

$$\varepsilon\ \sqrt{\dot{R}^2+1}-\varepsilon_+\sqrt{\dot{R}^2+1-2GM/R}=Gm(R)/R, \qquad (1)$$

provided we choose the matter-energy content of the shell, which is described by the function $m(R)$, the non-trivial content of the shell stress–energy tensor in spherical symmetry after the equation of state relating the pressure p and the surface energy density ρ has been specified. Let us now choose $p=-\rho/2$, p being a uniform pressure and ρ a uniform energy density. Since a string gas in n spatial dimensions satisfies $p=-\rho/n$, the two dimensional shell Σ that we are considering is a *sphere of strings*. Then $\rho=\rho_0/R$, where ρ_0 is a constant, implies $m(R)=cR$, with $c=4\pi\rho_0$. With this choice, and using the fact that solving (1) is equivalent to solving an effective classical-looking problem [8, 27] as detailed in table 1, we conclude that a child universes which disconnects from the ambient Schwarzschild space is produced if the strength of the string shell, c, is increased to above $2/G$. Moreover, this is *independent from the value*

[1] In presence of spherical symmetry only one equation survives. We also set the following notation: G is the gravitational constant; $R(\tau)$ is the radius of the spherical shell expressed as a function of the proper time τ of an observer co-moving with the shell and is the only remaining degree of freedom; an overdot indicates a derivative with respect to τ; $\varepsilon_\pm=\mathrm{sgn}(n^\mu\partial_\mu r)\rceil_{\mathscr{M}_\pm}$ are signs, expressing the fact that the radial coordinate r can increase ($\varepsilon_\pm=+1$) or decrease ($\varepsilon_\pm=-1$) along the normal direction, defined by $n^\mu\rceil_{\mathscr{M}_\pm}$ in $\mathscr{M}_\pm$, respectively (our convention is that the normal is pointing from $\mathscr{M}_-$ to $\mathscr{M}_+$).

TABLE 1. Global spacetime structure and classical formation of a child universe for the junction described in the text. The junction condition (1) is equivalent to the classical looking equation $\dot{R}^2 + V(R) = 0$ with $V(R) = 1 - (2GM/R + G^2c^2)^2/(4G^2c^2)$, with the signs determined as $\varepsilon_- = +1$ and $\varepsilon_+ = \mathrm{sgn}(2GM/R - G^2c^2)$. From these results, from the graph in panel (a) and from the additional result $\lim_{R\to\infty} V(R) = 1 - G^2c^2/4$ it is seen that we can have unbounded trajectories only if $c \geq 2/G$: this is independent from the choice of $M > 0$; moreover, from the result for ε_+, we have that on all the unbounded trajectories ε_+ changes sign (being positive for small enough R and negative for large enough R: the general property that this change of sign must happen behind an horizon, is also obviously satisfied since $c > 2/G$, so that at $R = 2GM$ the sign ε_+ is already negative although for small enough R it is positive). In view of the global spacetime structure associated with the above properties (see panel (b), where boundaries are named following [30]) we see that unbounded solutions realize child universe formation.

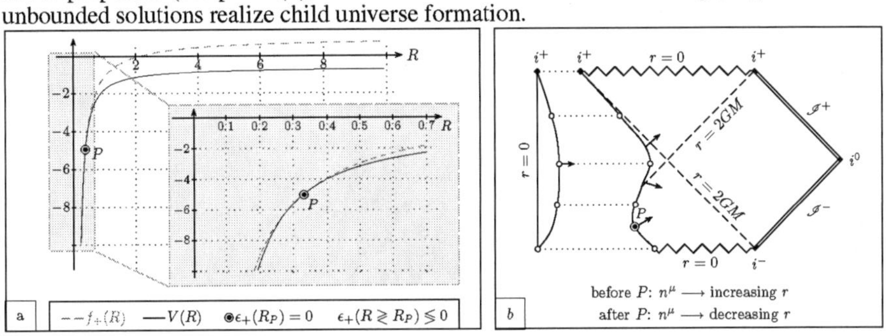

of M, i.e. the critical mass M_{cr} for child universe production is arbitrarily small: thus, in this model, child universes can be produced at an *arbitrarily small energy price* [28].

Despite its simplicity, this model is capable of generalizations and it will be very interesting to consider under which conditions these generalizations maintain the feature of having child universe production using arbitrarily small energy. It has, in fact, already been shown that a similar property also holds under more traditional settings, i.e. in presence of a non-zero energy density inside the child universe and of a uniform surface tension on the shell. In this case crucial parameters are, apart from the total mass–energy of the asymptotically flat region, also the false vacuum energy density and the shell surface tension. In [29] the limits of infinite surface tension and infinite false vacuum energy density have been considered in connection with the corresponding value for the critical mass M_{cr}. Indeed, in the above limits, it was found $M_{\mathrm{cr}} \to 0$. This effectively enhances the situation described in [10]. In the model considered in [29], surface tension and false vacuum energy density are constants (so that taking the limits of infinite surface tension and infinite false vacuum energy density does not correspond to a dynamical process); in the perspective of [29] it would have been, of course, very interesting if the limit of infinite surface tension could have corresponded to a physical process in the context of a model where the tension would become a dynamical variable. This possibility has been addressed in this contribution, where we have seen that the condition $M_{\mathrm{cr}} \to 0$ can be *classically* realized with the introduction of a dynamical surface tension, which here is due to a gas of strings being located on the shell; this strongly affects the critical mass threshold M_{cr} for child universe production. In fact as the strength of this string shell, parametrized by the constant c is increased above $2/G$, not only child universe production becomes possible but the critical mass for the process becomes arbitrarily small. This means that even a minimal (in terms of energy) quantum fluctuation could trigger the creation of a child universe. The condition $c > 2/G$ means that the string shells are transplanckian. This then suggests a very fundamental role of child universes for transplanckian physics: the fact that $M_{\rm cr} \to 0$, i.e. that there is no barrier for child universe creation, could then have important consequences for both particle physics and quantum gravity, besides the standard field of application of early universe cosmology. In this respect, the fact that smaller and smaller length scales are becoming available to observation, like the transplanckian scales, in conjunction with the result that creation of child universes could be unsuppressed at these scales, might imply that in a near future these observations might be able to provide us with the relevant information which is necessary to understand the birth of our universe and tell us if it was, in fact, disconnecting from a pre-existing universe. This and other interesting issues will be discussed elsewhere.

REFERENCES

1. A. H. Guth, *Phys. Rev. D* **23**, 347 (1981).
2. A. Linde, *Phys. Lett.* **108B**, 389 (1982).
3. A. Albrecht, and P. J. Steinhardt, *Phys. Rev. Lett.* **48**, 1220 (1982).
4. S. Coleman, *Phys. Rev. D* **15**, 2929 (1977); errata *ibid.* **16**, 1248(E) (1977).
5. J. C. G. Callan, and S. Coleman, *Phys. Rev. D* **16**, 1762 (1977).
6. S. Coleman, and F. D. Luccia, *Phys. Rev. D* **21**, 3305 (1980).
7. K. Sato, M. Sasaki, H. Kodama, and K.-I. Maeda, *Progr. Theor. Phys.* **65**, 1443 (1981).
8. S. K. Blau, E. I. Guendelman, and A. H. Guth, *Phys. Rev. D* **35**, 1747 (1987).
9. V. A. Berezin, V. A. Kuzmin, and I. I. Tkachev, *Phys. Rev. D* **36**, 2919 (1987).
10. E. Farhi, A. H. Guth, and J. Guven, *Nucl. Phys.* **B339**, 417 (1990).
11. W. Fishler, D. Morgan, and J. Polchinski, *Phys. Rev. D* **41**, 2638 (1990).
12. E. I. Guendelman, and J. Portnoy, *Class. Quantum Grav.* **16**, 3315 (1999).
13. E. I. Guendelman, and J. Portnoy, *Mod. Phys. Lett. A* **16**, 1079 (2001).
14. S. Ansoldi (2007), gr-qc/0701117, in the proceedings of the *16th Workshop on General Relativity and Gravitation (JGRG16)*, Niigata, Japan, 27 Nov - 1 Dec 2006.
15. W. De Sitter, *Proc. Kon. Ned. Akad. Wet.* **19**, 1217 (1917), 9 pages.
16. W. De Sitter, *Proc. Kon. Ned. Akad. Wet.* **20**, 229 (1917).
17. M. D. Kruskal, *Phys. Rev.* **119**, 1743 (1960).
18. K. Schwarzschild, *Sitzber. Deut. Akad. Wiss. Berlin, Kl. Math.-Phys. Tech.* , 189 (1916).
19. E. I. Guendelman, and A. Rabinowitz, *Phys. Rev. D* **44**, 3152 (1991).
20. A. Linde, *Phys. Lett. B* **327**, 208 (1994).
21. A. Vilenkin, *Phys. Rev. Lett.* **72**, 3137 (1994).
22. R. Penrose, *Phys. Rev. Lett.* **14**, 57 (1965).
23. A. Borde, M. Trodden, and T. Vachaspati, *Phys. Rev. D* **59**, 043513 (1999).
24. N. Sakai, K. Nakao, H. Ishihara, and M. Kobayashi, *Phys. Rev. D* **74**, 024026 (2006).
25. W. Israel, *Nuovo Cimento* **B44**, 1 (1966); errata *ibid.* **B48**, 463 (1967).
26. C. Barrabes, and W. Israel, *Phys. Rev. D* **43**, 1129 (1991).
27. A. Aurilia, M. Palmer, and E. Spallucci, *Phys. Rev. D* **40**, 2511 (1989).
28. S. Ansoldi, and E. I. Guendelman, arXiv:0706.1233[gr-qc], (2007).
29. E. I. Guendelman, gr-qc/0703105, *Honorable mention at the Gravity Research Foundation Essay Competion*, (2007).
30. S. W. Hawking, and G. F. R. Ellis, "*The Large Scale Structure of Space and Time*" (Cambridge University Press, Cambridge, 1973).

Flavour-dependent leptogenesis with reheating

Stefan Antusch

Max-Planck-Institut für Physik (Werner-Heisenberg-Institut)
Föhringer Ring 6, D-80805 München, Germany

Abstract. Upper bounds on the reheat temperature of the early universe, as they appear for example in classes of supergravity models, impose severe constraints on the thermal leptogenesis mechanism. To analyse these constraints, we extend the flavour-dependent treatment of leptogenesis to include reheating. We solve the flavour-dependent Boltzmann equations to obtain the leptogenesis efficiency as a function of the flavour dependent washout parameter $\tilde{m}_{1,\alpha}$ and of m_{N_1}/T_{RH}, the ratio of the mass of the lightest right-handed neutrino over the reheat temperature, and calculate the minimal values of the reheat temperature compatible with thermal leptogenesis in type I and type II seesaw scenarios.

Keywords: Early universe; Supersymmetry; Supergravity; Neutrinos; Leptogenesis
PACS: 98.80.Cq,14.60.St,11.30.Pb,04.65.+e

INTRODUCTION

The mechanism is leptogenesis [1] is one of the most attractive and minimal ways to explain the observed baryon asymmetry of the universe $n_{\mathrm{B}}/n_\gamma \approx (6.0965 \pm 0.2055) \times 10^{-10}$ [2]. In thermal leptogenesis [3], the population of right-handed neutrinos N_1 is produced from scattering processes in the thermal bath. To generate the observed baryon asymmetry comparatively high temperatures of the early Universe are required, and these should not lie much below the mass of the lightest right-handed neutrino. High temperatures compatible with thermal leptogenesis can arise in the process of reheating which takes place after cosmic inflation. However, particularly in locally supersymmetric theories, the reheat temperature T_{RH} is often constrained. In the presence of such bounds on T_{RH}, the requirement of successful thermal leptogenesis imposes severe constraints on the seesaw parameters. If the temperatures where leptogenesis takes place are below about $\mathcal{O}(10^{12}\,\mathrm{GeV})$ in the SM and $(1+\tan^2\beta)\times\mathcal{O}(10^{12}\,\mathrm{GeV})$ in the MSSM, a flavour-dependent treatment of thermal leptogenesis is required [4, 5, 6] (for the MSSM case, see [7, 8]). An overview over the literature about flavour effects can be found, e.g., in [9]. To analyse the constraints which upper bounds on T_{RH} impose on thermal leptogenesis scenarios, the flavour-dependent treatment of leptogenesis has to be extended to include reheating. In this talk, we briefly summarise some recent results from Refs. [7, 8, 9].

ESTIMATION OF THE PRODUCED BARYON ASYMMETRY

In flavour-dependent thermal leptogenesis, the out-of-equilibrium decays of the heavy right-handed (s)neutrinos N_1 ($\widetilde{N}_1$) give rise to flavour-dependent asymmetries in the (s)lepton sector, which are then partly transformed via sphaleron conversion into a

CP957, *Particles, Strings, and Cosmology, 13*[th] *International Symposium*
edited by A. Rajantie, C. Contaldi, P. Dauncey, and H. Stoica

baryon asymmetry Y_B. As an example, we consider the case of the Minimal Supersymmetric extension of the Standard Model (MSSM). In general, the final baryon asymmetry via leptogenesis in the MSSM can be parameterised as

$$Y_B = \frac{10}{31}\sum_{\alpha}\hat{Y}_{\Delta_\alpha}\,, \quad \hat{Y}_{\Delta_\alpha} = \eta_\alpha\,\varepsilon_{1,\alpha}\left[Y^{\mathrm{eq}}_{N_1}(T \gg m_{N_1}) + Y^{\mathrm{eq}}_{\widetilde{N}_1}(T \gg m_{N_1})\right], \tag{1}$$

where $\hat{Y}_{\Delta_\alpha} \equiv Y_B/3 - Y_{L_\alpha}$ are the total (particle and sparticle) $B/3 - L_\alpha$ asymmetries, with Y_{L_α} the lepton number densities in the flavour $\alpha = e, \mu, \tau$. Y^{eq} are the equilibrium number densities, normalised to the entropy density, $\varepsilon_{1,\alpha}$ are the flavour-dependent decay asymmetries and η_α the flavour-dependent efficiency factors. The latter are calculated by solving the set of flavour-dependent coupled Boltzmann equations, describing the evolution of the number densities as a function of temperature. The flavour-dependent Boltzmann equations with reheating, which we will consider in the following, are given explicitly in the appendix of [8]. We restrict ourselves here to the simplest case of thermal leptogenesis, where only the lightest right-handed (s)neutrinos are produced in the thermal bath, and assume that the other particles participating in the (type I or type II) seesaw mechanism are much heavier.

FEATURES OF THE FLAVOUR-DEPENDENT TREATMENT

The most important properties of the flavour-dependent Boltzmann equations are:

- **Individual Boltzmann equations for the flavour-specific asymmetries** $\hat{Y}_{\Delta_\alpha}$ are included, instead of using the approximation of a one single equation for $\sum_\alpha \hat{Y}_{\Delta_\alpha}$.
- The term in each Boltzmann equations for $\hat{Y}_{\Delta_\alpha}$ which corresponds to the production of the asymmetries from the decay of the right-handed neutrinos is weighted by the **flavour-specific decay asymmetry** $\varepsilon_{1,\alpha}$.
- The term in each Boltzmann equations for $\hat{Y}_{\Delta_\alpha}$ which corresponds to the washout of a produced asymmetry is flavour-dependent, and typically governed by the **flavour-dependent washout parameters** $\widetilde{m}_{1,\alpha}$ (defined e.g. in [7]). It contains contributions proportional to $A_{\alpha\beta}Y_{\Delta_\beta}$, where β runs over the independent asymmetries for a given temperature regime. The matrix A is defined via the relation $\hat{Y}_\alpha = \sum_\beta A_{\alpha\beta}\hat{Y}_{\Delta_\alpha}$, with $\hat{Y}_\alpha = Y_\alpha + Y_{\widetilde{\alpha}}$ being the combined densities for lepton and slepton doublets and has been calculated for the MSSM case in [7]. The (small) off-diagonal entries of A induce a **coupling between the differential equations for the flavour asymmetries** $\hat{Y}_{\Delta_\alpha}$ which can be important in some cases.
- For reheating, we have assumed (the most simple and model-independent case) that the decay of the inflaton reheats only MSSM particles. This and other assumptions and limitations of the used approach are discussed, e.g., in detail in Sec. 3.1.3 of [8]. Our results for the **efficiency factor** η_α **as a function of** $|A_{\alpha\alpha}|\widetilde{m}_{1,\alpha}$ **and** m_{N_1}/T_{RH} are shown in Fig. 1 (from [8]; for publicly available numerical results, see [10]) where we have neglected the small off-diagonal elements of the matrix A which

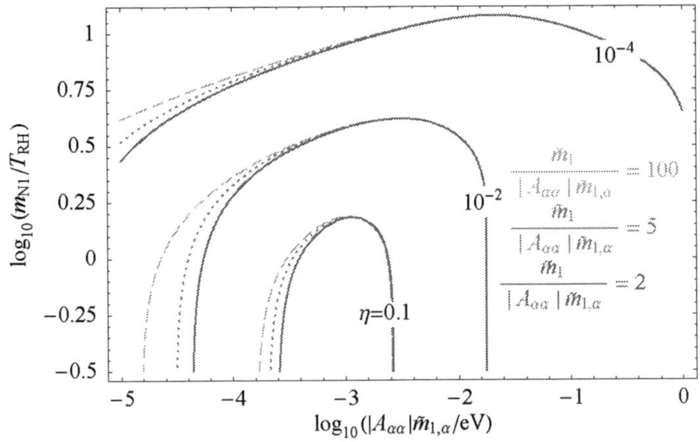

FIGURE 1. Contour lines for the flavour-dependent efficiency factor η in the MSSM as a function of $|A_{\alpha\alpha}|\widetilde{m}_{1,\alpha}$ (no sum over α) and m_{N_1}/T_{RH}, for fixed values of $\widetilde{m}_1/(|A_{\alpha\alpha}|\widetilde{m}_{1,\alpha})$ [8]. We take $\widetilde{m}_1/(|A_{\alpha\alpha}|\widetilde{m}_{1,\alpha}) = 2,5$ and 100, respectively corresponding to the solid, dotted and dashed contour lines.

couples the Boltzmann equations for the $\hat{Y}_{\Delta_\alpha}$. The plot illustrates how η_α is reduced if $m_{N_1} \gg T_{\mathrm{RH}}$, for instance by more than three orders of magnitude if $m_{N_1} \approx 10 \times T_{\mathrm{RH}}$. It is optimal for values of $|A_{\alpha\alpha}|\widetilde{m}_{1,\alpha} \approx m^*$, with $m^* = \sin^2\beta \times 1.58 \times 10^{-3}$ eV [7], and quickly drops for either smaller or larger $|A_{\alpha\alpha}|\widetilde{m}_{1,\alpha}$. The flavour-dependence becomes visible for $|A_{\alpha\alpha}|\widetilde{m}_{1,\alpha} < m^*$, where the efficiency exhibits a less pronounced drop if $\widetilde{m}_1/(|A_{\alpha\alpha}|\widetilde{m}_{1,\alpha}) \gg 1$ (with $\widetilde{m}_1 = \Sigma_\alpha \widetilde{m}_{1,\alpha}$).

BOUNDS ON T_{RH}: TYPE I VS. TYPE II LEPTOGENESIS

Comparing the type II and type I seesaw cases, in the latter the amount of produced baryon asymmetry becomes suppressed with increasing mass scale of the light neutrinos, either due to a factor $(m^*/m^\nu_{\mathrm{min}})^{1/2}$ in the decay asymmetries or due to a non-optimal washout parameter much larger than m^*, whereas in the former case leptogenesis becomes more efficient (for a detailed analysis, see [9]). One interesting consequence is that the lower bound on the mass m_{N_1} of the lightest right-handed neutrino from thermal leptogenesis in the type II seesaw case can be relaxed compared to the bound in the type I scenario. The relaxed bound on m_{N_1} furthermore leads to a relaxed lower bound on T_{RH} in the type II seesaw framework, as shown in Fig. 2 (taken from [9]). While the type II bound decreases by about an order of magnitude when the neutrino mass scale increases to 0.4 eV, the type I bound increases. Upper bounds on T_{RH} can thus induce constraints on the neutrino mass scale, i.e. on $m^\nu_{\mathrm{min}} = \min(m_{\nu_1}, m_{\nu_3}, m_{\nu_3})$. For instance, with $T_{\mathrm{RH}} \leq 5 \times 10^9$ GeV, values of m^ν_{min} in the approximate range $[0.01\,\mathrm{eV}, 0.32\,\mathrm{eV}]$ would be incompatible with thermal leptogenesis in the type I seesaw framework.

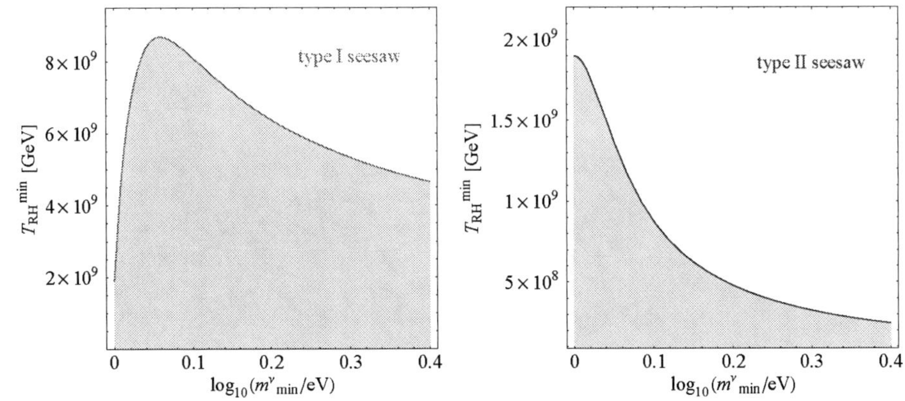

FIGURE 2. Lower bound on the reheat temperature T_{RH} in thermal type I (left panel) and in type II leptogenesis (right panel) as a function of the mass of the lightest neutrino $m^\nu_{\min} = \min(m_{\nu_1}, m_{\nu_3}, m_{\nu_3})$, in the MSSM with $\tan\beta = 30$ [9]. The used assumptions are specified and discussed in Sec. 3.1.3 of [8].

SUMMARY AND CONCLUSIONS

We have reviewed recent progress in flavour-dependent leptogenesis with reheating. Including reheating in leptogenesis studies is required in order to analyse the implications of upper bounds on the reheat temperature T_{RH} for models of thermal leptogenesis. We have solved the flavour-dependent Boltzmann equations with reheating to obtain the flavour-dependent efficiency factor η_α (see also [10]), which allows convenient estimation of the produced baryon asymmetry. We also presented lower bounds on T_{RH} from the requirement of successful thermal leptogenesis in type I and type II seesaw scenarios.

ACKNOWLEDGMENTS

I would like to thank Steve F. King, Antonio Riotto and Ana M. Teixeira for the collaboration in the work presented here.

REFERENCES

1. M. Fukugita and T. Yanagida, Phys. Lett. **174B** (1986), 45.
2. D. N. Spergel *et al.*, arXiv:astro-ph/0603449.
3. For a review: W. Buchmüller, R. D. Peccei and T. Yanagida, Ann. Rev. Nucl. Part. Sci. **55** (2005) 311.
4. R. Barbieri, P. Creminelli, A. Strumia and N. Tetradis, Nucl. Phys. B **575** (2000) 61.
5. A. Abada, S. Davidson, F. X. Josse-Michaux, M. Losada and A. Riotto, JCAP **0604** (2006) 004.
6. E. Nardi, Y. Nir, E. Roulet and J. Racker, JHEP **0601** (2006) 164.
7. S. Antusch, S. F. King and A. Riotto, JCAP **0611** (2006) 011.
8. S. Antusch and A. M. Teixeira, JCAP **0702** (2007) 024.
9. S. Antusch, Phys. Rev. D **76** (2007) 023512.
10. Our numerical results for $\eta_\alpha(\widetilde{m}_{1,\alpha}, m_{N_1}/T_{\mathrm{RH}})$ from [8] (as a *table* as well as in form of a *Mathematica function*) are available at **http://www.newphysics.eu/leptogenesis/**.

Suppression of Meszaros' Effect in coupled DE

Silvio A. Bonometto and Roberto Mainini

Dipart. di Fisica, Università di Milano–Bicocca, Piazza della Scienza 3, 120126 Milano (Italy) & INFN, Sezione di Milano–Bicocca

Abstract. A phaenomenological DM–DE coupling could indicate their common origin. Various constraint however exist to such coupling; here we outline that it can suppress Meszaros' effect, yielding transfered spectra with a softer bending above $k_{hor,eq}$. It could be therefore hard to reconcile these models with both CMB and deep sample data, using a constant spectral index.

Keywords: Cosmology: theory – dark energy
PACS: 98.80.-k, 98.65.-r

A tenable cosmological model must include at least two dark components, Dark Matter (DM) and Dark Energy (DE); yet only hypotheses on their nature exist; if DM and DE are physically unrelated, their presently similar densities are purely accidental.

Attempts to overcome this conceptual deadlock led to suggest, first of all, that DE has a dynamical nature [1]. Interactions between DM and dynamical DE [2], [3] might then partially cure the problem, keeping close values for their densities up to large redshift. This option could also be read as an approach to a deeper reality, whose physical features could emerge from phenomenological limits to coupling strength and shape.

A longer step forward was attempted by [4], suggesting that DM and DE are a single complex scalar field, being its quantized phase and modulus (*dual– axion* approach). Instead of introducing new parameters and limiting them through data fitting, this option, although cutting the available degrees of freedom, still allows to fit CMB constraints [5].

Here, however, we keep on the phenomenological side and discuss constraints to DM–DE interactions. This will have a fallout also on the *dual– axion* approach, which does face a problem, because of the feature of the DM–DE coupling it causes.

Any baryon–DE coupling is ruled out by consequences similar to modifying gravity. Limits are looser for DM–DE coupling, whose consequences appear only over cosmological distances. Here we outline a new constraint on this coupling, arising from the early behavior of fluctuations, over scales destined to evolve into cosmic structures.

Fluctuations over these scales enter the horizon before matter–radiation equality and their growth is initially inhibited by the radiative component, then still behaving as a single fluid together with baryons. While fluctuations in the fluid behave as sonic waves, DM self gravitation is just a minor dynamical effect is respect to cosmic expansion. This freezing of fluctuation amplitudes until equality is known as *Meszaros effect*.

Here we wish to outline that DM–DE coupling can damp Meszaros effect, so that fluctuation growth, between the entry in the horizon and equality, is significantly enhanced. As a matter of fact, fluctuation freezing is essential, in shaping the transfered spectrum.

The freezing or its damping only marginally affect baryons and radiation. Therefore, while the transfer function changes, CMB spectra keep almost unaffected. All details

CP957, *Particles, Strings, and Cosmology, 13th International Symposium*
edited by A. Rajantie, C. Contaldi, P. Dauncey, and H. Stoica

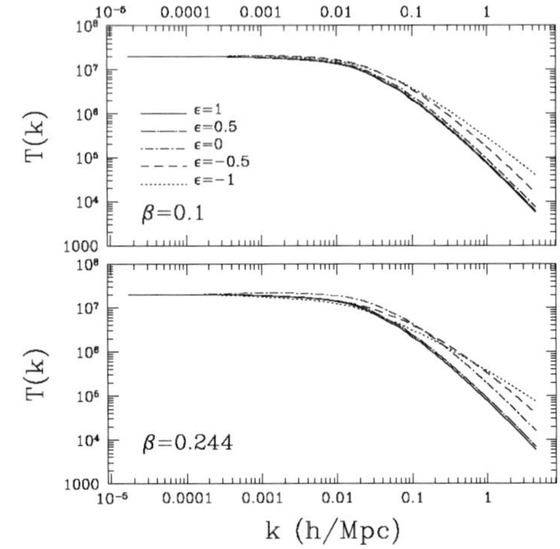

FIGURE 1. Transfer functions for different behaviors of DM–DE coupling with redshift and/or for different coupling normalization. The case $\varepsilon = 0$ corresponds to redshift independent coupling intensity. The case $\varepsilon = -1$ with $\beta = 0.244$ correspond to a coupling $C = 1/\phi$. Besides of the different slopes, notice the dependence on the model of the bending scale and, in particular, its dependence on the coupling strength, also in constant coupling models (dash–dotted lines).

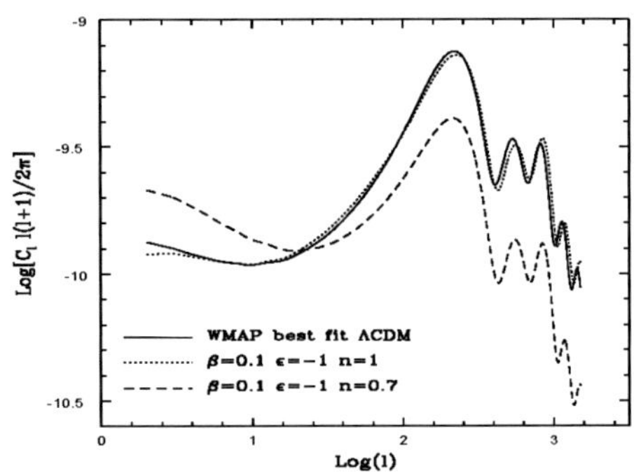

FIGURE 2. Anisotropy spectrum of the ΛCDM model yielding the best fit to WMAP3 data compared with the spectra for coupled models with $\beta = 0.1$ and $\varepsilon = 1$, for $n = 1$ and $n = 0.7$. In the former case one can expect to recover a reasonable fit to data by adjusting other model parameters. Taking $n = 0.7$, a value just acceptable to fit deep sample data, any fitting to CMB anisotropy data is apparently excluded.

on the way how these results are obtainable can be found in [6]. Let us however outline that what we wish to outline is a rather major effect, which allows to discard a class of models, *a priori*; no general data fitting, constraining parameters and/or showing specific model advantages, needs then to be performed here.

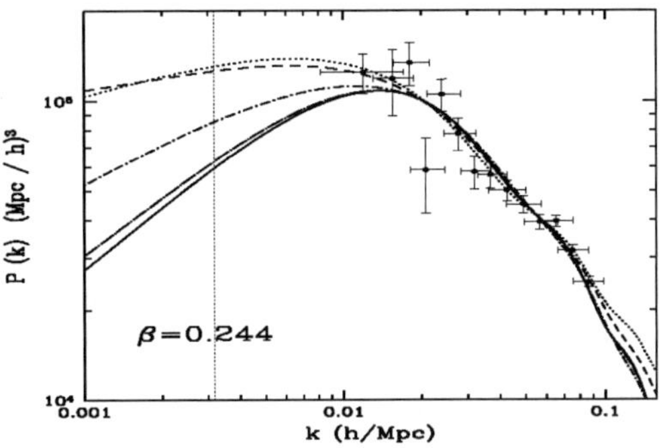

FIGURE 3. Comparison with SDSS digital survey data. Different curves refer to different ε's with the solid line ($\varepsilon = 1$) essentially coinciding with an uncoupled model. Constant coupling models ($\varepsilon = 0$) are described by dot–dashed curves. Negative ε's yield a further decrease of n. The vertical dotted line is the approximate scale where the Sachs & Wolfe C_l plateau begins.

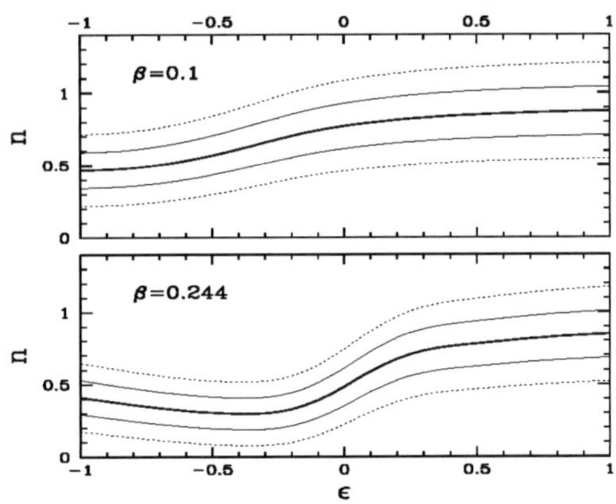

FIGURE 4. 1– and 2–σ intervals of n, when ε varies, for the β values in the frames

Accordingly, we keep to cosmological parameter values ensuing from WMAP3 best–fit [7], although deduced by assuming a ΛCDM model. In particular, we shall take $\Omega = 1$; the present value of the cold DM (baryon) density parameter will be $\Omega_{o,c} = 0.224$ ($\Omega_{o,b} = 0.044$); the Hubble parameter will be $h = 0.704$.

Our analysis here will however be restricted to SUGRA potentials [8]

$$V(\phi) = (\Lambda^{\alpha+4}/\phi^{\alpha})\exp(4\pi\phi^2/m_p^2) \tag{1}$$

with $\Lambda = 100$ GeV. Here $m_p = G^{-1/2}$ is the Planck mass.

The background equations for coupled DE and DM, using the conformal time τ, read

$$\ddot{\phi} + 2(\dot{a}/a)\dot{\phi} + a^2 V_{,\phi} = C(\phi)a^2\rho_c\ , \quad \dot{\rho}_c + 3(\dot{a}/a)\rho_c = -C(\phi)\dot{\phi}\rho_c \qquad (2)$$

where the coupling is set by

$$C(\phi) = 4\sqrt{\pi/3}(\beta/m_p)\,(\phi/m_p)^{\varepsilon} = 4\sqrt{\pi/3}(\tilde{\beta}/m_p) \qquad (3)$$

and therefore parametrized by β, ε or $\tilde{\beta}$. The *dual–axion* model naturally predicts a coupling $C = 1/\phi$, consistent with eq. (3) if $\varepsilon = -1$ and $\beta \simeq 0.244$.

Transfer functions are shown in Fig. 1 for $\beta = 0.1$ and 0.244 and various ε's.

The suppression of fluctuation freezing is obviously stronger for greater β (and increasingly negative ε values). For $\varepsilon = -1$, enclosing the case $C = 1/\phi$ when $\beta = 0.244$, the steepness of the transfer function, for $k > k_{hor,eq}$ is much reduced. The effect is still significant also for $\varepsilon = -0.5$, namely when $\beta = 0.244$.

While this occur, the CMB anisotropy spectum keeps a regular behavior, as is shown in Figure 2; here we compare WMAP3 data with C_l for $\beta = 0.1$, $\varepsilon = -1$ and $n = 1$ or 0.7 . Transfered spectra, compared with SDSS data (see Fig. 3), show in fact that $n \sim 0.5$–0.7 is needed to fit data.

In the case $\varepsilon \neq 0$, the discrepancy from unity of the spectral index n, assumed to be constant, is a measure of the distortion caused by the suppression of Meszaros effect.

If we take $n = 0.85$ at $1-\sigma$ as a threshold to discard a model, no $\varepsilon < 0$ model is allowed with $\beta = 0.244$, while $\varepsilon < -0.16$ are also inhibited with $\beta = 0.1$. At $2-\sigma$'s the situation is not much improved for $\beta = 0.244$, while lower values of ε are admitted for $\beta = 0.1$.

In particular, $C = 1/\phi$, as for the *dual–axion* model, is outside the range indicated. Modification to make this model consistent with data were however proposed [9].

The analysis was extended to models with positive ε, for which coupling rises while ϕ increases. A large deal of these models is apparently allowed.

REFERENCES

1. Wetterich C. 1988, Nucl.Phys.B 302, 668; Ratra B. & Peebles P.J.E., 1988, Phys.Rev.D 37, 3406; Peebles P.J.E. & Ratra B., 2003, Rev.Mod.Phys. 75, 559
2. Ellis J., Kalara S., Olive K.A. & Wetterich C., 1989, Phys. Lett. B228, 264; Wetterich C., 1995, A&A 301, 321 Amendola L., 2000, Phys.Rev. D62, 043511 Gasperini M., Piazza F.& Veneziano G., 2002, Phys.Rev. D65, 023508
3. Casas J.A., Garcia–Bellido J & Quiros M., 1992, Class.Quant.Grav. 9, 1371; Anderson G.W . & Carroll S.M., Procs. of "COSMO-97, First International Workshop on Particle Physics and the Early Universe", Ambleside, England, September 15-19, 1997, astro-ph/9711288; Bartolo N. & Pietroni M., 2000, Phys.Rev. D61, 023518; Pietroni M., 2003, Phys.Rev D67, 103523; Chimento L.P., Jakubi A.S., Pavon D. & Zimdahl W.,2003, Phys.Rev D67, 083513; Rhodes C.S., van de Bruck C, Brax P., & Davis A.C., 2003, Phys.Rev. D68, 083511; Farrar G.R. & Peebles P.J.E., 2004, ApJ 604, 1 Gromov A., Baryshev Y. & Teerikorpi P., 2004, A&A, 415, 813
4. Mainini R. & Bonometto S.A., 2004, Phys.Rev.Lett. 93, 121301
5. Mainini R., Colombo L. & Bonometto S.A., 2005, ApJ 632, 691
6. Mainini R. & Bonometto S., 2007, astro-ph/0703303, JCAP (in press).
7. Spergel D.N. et al. 2006, ApJ (in press), astro-ph/0603449
8. Brax P. & Martin J., 1999, Phys.Lett., B468, 40; Brax P. & Martin J., 2001, Phys.Rev. D61, 10350; Brax P., Martin J., Riazuelo A., 2000, Phys.Rev. D62, 103505
9. Mainini R. & Bonometto S., 2007, arxiv/0709.0174, JCAP (in press).

Cosmological implications of a supersymmetric extension of the Brans-Dicke theory

Riccardo Catena

Deutsches Elektronen-Syncrotron (DESY), 22603 Hamburg, Germany

Abstract. In the Brans-Dicke theory the Planck mass is replaced by a dynamical scalar field. We consider here the supersymmetric analogous of this mechanism replacing in the supergravity Lagrangian the Planck mass with a chiral superfield. This analysis is motivated by the research of possible connections between supersymmetric Dark Matter scenarios and Dark Energy models based on Brans-Dicke-like theories. We find that, contrary to the original Brans-Dicke theory, in its supersymmetric analogous the gravitational sector does not couple to the matter sector in a universal metric way. As a result, violations of the weak equivalence principle could be present in such a scenario.

Keywords: Scalar-Tensor theories, Cosmology, Supersymmetry
PACS: 04.50.+h, 04.65.+e

INTRODUCTION

In the Brans-Dicke approach to the gravitational interaction, the Planck mass is replaced in the Lagrangian by a dynamical scalar field φ [1]. Even though, as a theory of gravity, the original Brans-Dicke theory is at present very constrained by solar system measurements [2], its modern versions, namely Scalar-Tensor (ST) theories [3] – where a non trivial potential is associated to φ – can still pass all cosmological and astrophysical bounds leading at the same time to interesting and testable predictions [4].

Like in General Relativity, in ST theories matter couples to gravity in a universal metric way *i.e.* all matter fields feel the gravitational interaction only through *one and the same* metric. Because of this property, ST theories are by construction protected by any violation of the weak equivalence principle and even ultra-light degrees of freedom mediating long range forces are not in this framework phenomenologically dangerous [5]. Such a feature makes ST theories interesting arenas for dynamical Dark Energy model building[6, 7].

However, a realistic cosmological scenario can be only achieved when Dark Matter is consistently included to the picture [8]. Among different theories providing suitable Dark Matter candidates, Supersymmetry is certainly one of the most remarkable [9].

The interesting possibility to relate a ST interpretation of Dark Energy to a supersymmetric description of Dark Matter leads to study supersymmetric extensions of ST theories. This is the topic of the present work.

We consider here the supersymmetric analogous of the BD idea: we replace in the supergravity Lagrangian the Planck mass with a chiral superfield, the "Planck superfield". Such a replacement defines the "natural" supersymmetric extension of the BD

CP957, *Particles, Strings, and Cosmology, 13th International Symposium*
edited by A. Rajantie, C. Contaldi, P. Dauncey, and H. Stoica

theory. Let us refer to it as the Minimal Supersymmetric Brans-Dicke theory (MSBD) to distinguish it from other possible approaches. We find that, contrary to the original BD theory, in the MSBD the gravitational sector does not couple to the matter sector in a universal metric way. As a result, possible violations of the weak equivalence principle could make the minimal supersymmetric extension of the BD idea phenomenologically inconsistent.

The plan of this work is as follows. In the next section we review the BD model and the concept of universal metric coupling. The third section is devoted to the MSBD theory; we will specially underline the differences between its phenomenology and the one of the original BD theory. The results are finally discussed in the Conclusions. For notation we refer in the following to [10, 11].

THE BRANS-DICKE THEORY AND THE UNIVERSAL METRIC COUPLING

In General Relativity the coupling between gravity and matter is described by the following Lagrangian

$$\mathscr{L}_{EH} = -\frac{1}{2} e M_{Pl}^2 \mathscr{R} + \mathscr{L}_M[e_m^a, \Psi], \tag{1}$$

where $e \equiv det(e_m^a)$, $\mathscr{R}$ is the Ricci scalar and Ψ symbolically represents all matter fields involved in the theory. In the BD approach to the gravitational interaction the Planck mass appearing in eq. (1) becomes dynamical by means of the substitution

$$M_{Pl}^2 \Longrightarrow \varphi^2(y^m), \tag{2}$$

where $\varphi(y^m)$ is a real scalar field. As a consequence eq. (1) is replaced by

$$\begin{aligned} \mathscr{L}_{BD} &= \mathscr{L}_\varphi[e_m^a, \varphi] + \mathscr{L}_M[e_m^a, \Psi] \\ &= -\frac{1}{2} e \left(\varphi^2 \mathscr{R} + \omega\, \partial_m \varphi \partial^m \varphi\right) + \mathscr{L}_M[e_m^a, \Psi], \end{aligned} \tag{3}$$

where the factor ω that multiplies the kinetic term of φ has to be tuned to fit the post-newtonian bounds [2]. Eq. (3) gives the so called "Jordan frame" formulation of the theory. In this frame the BD scalar does not appear in the matter Lagrangian and particle physics is just the standard one. The theory can be formulated in other frames related to the Jordan one by a Weyl rescaling of the vielbein such as $e_m^a \to e_m^a e^{l(\varphi)}$, where $l(\varphi)$ is some φ-dependent function. In these alternative formulations the matter Lagrangian acquires an explicit functional dependence from φ, *i.e.* $\mathscr{L}_M = \mathscr{L}_M[e_m^a e^{l(\varphi)}, \Psi]$. However, the inverse Weyl rescaling $e_m^a \to e_m^a e^{-l(\varphi)}$ always brings back the theory to its original version in which particle physics is just the standard one.

Eq. (3) shows that in the BD theory all matter fields feel the gravitational interaction through the same vielbein, the Jordan frame vielbein. For this reason such a matter-gravity coupling is also called *universal and metric*. This is a non trivial property and has very important phenomenological implications. It can be shown, for instance, that in a theory where matter couples to gravity in a universal metric way the weak equivalence principle is satisfied by construction [5].

THE MINIMAL SUPERSYMMETRIC BRANS-DICKE THEORY

Eq. (2) gives a prescription to construct the BD Lagrangian starting from the Einstein-Hilbert one. In this section we apply an analogous prescription to the supergravity Lagrangian

$$\mathscr{L}_{sg} = -3M_{Pl}^2 \int d^2\theta\, 2\mathscr{E}R + \mathscr{L}_M[H,\Psi] + \text{h.c.}, \tag{4}$$

where H is the supergravity multiplet, $\mathscr{E}$ is the chiral density and R represents the curvature superfield, defined as the covariant derivative of the spin connection.
Let us start introducing a chiral superfield Φ with components given by the power series expansion $\Phi(y^m,\theta_\alpha) = A(y^m) + \sqrt{2}\,\theta^\alpha\chi_\alpha(y^m) + \theta^\alpha\theta_\alpha F(y^m)$, where $A(y^m)$ and $F(y^m)$ are complex scalars and $\chi_\alpha(y^m)$ a Weyl spinor. We will call Φ the Planck superfield. This dynamical object allows the natural supersymmetric extension of the substitution (2)

$$M_{Pl}^2 \Longrightarrow \Phi^2(y^m,\theta_\alpha). \tag{5}$$

Applying the substitution (5) to eq. (4) one finds

$$\begin{aligned}
\mathscr{L}_{MSBD} &= \mathscr{L}_\Phi[H,\Phi] + \mathscr{L}_M[H,\Psi] \\
&= -3\int d^2\theta\,\Phi^2\, 2\mathscr{E}R - \\
&- \frac{1}{8}\int d^2\theta\, 2\mathscr{E}\left(\bar{\mathscr{D}}_{\dot\alpha}\bar{\mathscr{D}}^{\dot\alpha} - 8R\right)\Phi^\dagger\Phi + \\
&+ \mathscr{L}_M[H,\Psi] + \text{h.c.},
\end{aligned} \tag{6}$$

where in the third line, in analogy with eq. (3), we introduced a kinetic term for Φ. To be as general as possible we do not assume any particular form for $\mathscr{L}_M$.

Eq. (6) defines the Minimal Supersymmetric Brans Dicke theory (MSBD). Its invariance under supergravity transformations follows from the properties of chiral densities. By definitions, chiral densities transform like total derivatives in the space (y^m,θ_α) and the product of a chiral density and a chiral superfield is again a chiral density [10]. Moreover, the superfields $(\mathscr{D}\mathscr{D} - 8R)\,\Phi^\dagger\Phi$ and Φ^2 are chiral if Φ is chiral. This proves the invariance of the Lagrangian (6) under supergravity transformations.

Let us focus now on its phenomenology. As it was shown in [11], the component fields expansion of eq. (6) gives rise to a Lagrangian with the following structure

$$\mathscr{L}_{MSBD} = \mathscr{L}_\Phi[e_m^a,\psi_\alpha^a,b^a,M,A,\chi_\alpha,F] + \mathscr{L}_M[e_m^a,\psi_\alpha^a,b^a,M,\Psi], \tag{7}$$

where we introduced the gravitino ψ_α^a and two auxiliary fields: a vector b^a and a scalar M. Eq. (7) is the supersymmetric version of eq. (3). The crucial difference between the two Lagrangians is that in the supersymmetric one $\mathscr{L}_M$ and $\mathscr{L}_\Phi$ communicate also through the auxiliary fields b^a and M. This has deep phenomenological consequences when the auxiliary fields are removed by means of their equations of motion. To show this point, let us write the general solution of the equations of motion for M and b^a as follows

$$\begin{aligned}
b^a &= h_1(\ldots,A,\chi_\alpha), \\
M &= h_2(\ldots,A,\chi_\alpha),
\end{aligned} \tag{8}$$

where h_1 and h_2 are two appropriate functions of the fields involved in the theory. In eq. (8) we underlined the crucial dependence of h_1 and h_2 from A and χ_α. Now, replacing the solutions (8) in the Lagrangian (7), the degrees of freedom of the Planck multiplet explicitly appear in the matter Lagrangian. Since no Weyl rescaling of the vielbein can remove the auxiliary fields from $\mathscr{L}_M$, it follows that the Planck multiplet couples intrinsically to matter. Therefore, there is no way to write the matter Lagrangian as $\mathscr{L}_M[e^a_m, \psi^a_\alpha, \Psi]$ by means of a suitable vielbein redefinition of the form $e^a_m \to e^a_m e^{l(A,\chi_\alpha,F)}$, where l is an appropriate function of the components of Φ. In other words, a Jordan frame does not exist for such a theory. The main consequence is that in the MSBD theory the weak equivalence principle is not satisfied by construction and time variations of masses and couplings are not under control. Explicit expressions for eqs. (7) and (8) can be found in [11].

CONCLUSIONS

In this work we have studied the minimal supersymmetric extension of the BD theory (MSBD) defined by eq. (6). The underlying motivation was the research of possible connections between a Scalar-Tensor interpretation of Dark Energy and a supersymmetric description of Dark Matter. Eq. (6) is obtained replacing the Planck mass with a chiral superfield in the supergravity Lagrangian (4). We called this extra superfield the Planck superfield. Although this approach looks very natural, the resulting phenomenology is radically different from the one of the original BD theory. In the MSBD theory the extra degrees of freedom of the Planck superfield intrinsically couple to matter and a Jordan frame formulation can not be achieved through a suitable vielbein redefinition. As a consequence, this theory does not satisfy the weak equivalence principle by construction. This conclusion could make the minimal supersymmetric extension of the BD idea phenomenologically inconsistent.

REFERENCES

1. C. Brans and R. H. Dicke, Phys. Rev. **124** (1961) 925.
2. B. Bertotti, L. Iess and P. Tortora, Nature **425** (2003) 374.
3. P. Jordan, *Schwerkaft und Weltall* (Vieweg, Braunschweig, 1955); M. Fierz, Helv. Phys. Acta **29**, 128 (1956); T. Damour and K. Nordtvedt, Phys. Rev. **D48**, 3436 (1993); T. Damour and A.M. Polyakov, Nucl. Phys. **B423**, 532 (1994).
4. R. Catena, M. Pietroni and L. Scarabello, Phys. Rev. D **70**, 103526 (2004). R. Catena, M. Pietroni and L. Scarabello, [arXiv:astro-ph/0604492].
5. T. Damour, "Gravitation, experiment and cosmology," [arXiv:gr-qc/9606079].
6. N. Bartolo and M. Pietroni, Phys. Rev. D **61** (2000) 023518; G. Esposito-Farese and D. Polarski, Phys. Rev. D **63**, 063504 (2001); R. Catena, N. Fornengo, A. Masiero, M. Pietroni and F. Rosati, Phys. Rev. D **70**, 063519 (2004).
7. E. J. Copeland, M. Sami and S. Tsujikawa, Int. J. Mod. Phys. D **15**, 1753 (2006).
8. D. N. Spergel *et al.* [WMAP Collaboration], Astrophys. J. Suppl. **170**, 377 (2007).
9. W. de Boer, "Is dark matter supersymmetric?," *Prepared for Cargese School of Particle Physics and Cosmology: the Interface, Cargese, Corsica, France, 4-16 Aug 2003*.
10. Julius Wess and Jonathan Bagger, *Supersymmetry and Supergravity*, Princeton University Press.
11. R. Catena, Phys. Rev. D **75**, 044019 (2007).

Accounting for the Unresolved X-ray Background with Sterile Neutrino Dark Matter

D.T. Cumberbatch and Joseph Silk

Department of Astrophysics, University of Oxford, Keble Road, Oxford, OX1 3RH

Abstract. We consider a scenario where keV sterile neutrinos constitute all of the currently inferred dark matter abundance, whose radiative decays could potentially account for the flux contributions to the X-ray background (XRB) by unresolved sources. Here we apply *integrated flux* methods to results from the observations of the North/South *Chandra deep fields* (CDF-N/S) in order to deduce constraints on the sterile neutrino mass-mixing parameters.

Keywords: Dark Matter, Sterile Neutrinos, X-rays
PACS: 95.35.+d, 14.60.Pq,14.60.St

INTRODUCTION

The proportion of the cosmological energy density manifested as dark matter is now well determined by observations to within 10% [1], yet its nature remains a subject of extensive debate. Recent astrophysical observations, such as indications of central cores in low-mass galaxies (see e.g. [2]), and the deficiency of satellites observed in Milky Way-sized galaxies (see e.g. [3]), indicate possible shortcomings of the [Λ]CDM paradigm and have boosted interest in a warm dark matter (WDM) scenario which may alleviate these "small scale problems" (see e.g. [4]). One of the most popular WDM candidates are right-handed ("sterile") neutrinos ν_s, since they naturally arise in many extensions of the Standard Model (see e.g. [5]). Other motivations for sterile neutrinos to possess mass-mixing parameters necessary for them to be produced via oscillations (with the familiar "active" neutrino species) include the enhanced production of neutral hydrogen before reionization [6] and a possible explanation for the baryon asymmetry of the Universe [5].

Sterile neutrinos can be produced through non-resonant oscillations, the simplest being the popular Dodelson-Widrow (DW) mechanism [7], or resonant oscillations for cosmologies which include a finite lepton number asymmetry. In both scenarios, sterile neutrinos are predominantly produced during the QCD epoch. Unfortunately, uncertainties relating to hadronic interactions during this time result in uncertainties in the relationship between the mass-mixing parameters and the relic abundance of sterile neutrino dark matter [5, 8].

Direct constraints on the mass-mixing parameters can be obtained by exploiting X-ray observations since keV-mass sterile neutrinos posses a radiative decay channel, yielding photons potentially detectable in astrophysical X-ray sources, including the diffuse cosmic X-ray background [9], the Milky Way [10, 11] as well as being inferred from the X-ray spectra of nearby galaxies [12] or clusters (see e.g. [13]). *Indirect* constraints can be achieved from Ly-α forest measurements [14], since they are sensitive tracers of the

CP957, *Particles, Strings, and Cosmology, 13th International Symposium*
edited by A. Rajantie, C. Contaldi, P. Dauncey, and H. Stoica

primordial density fluctuations on the smallest scales where WDM typically suppresses clustering.

Assuming that sterile neutrinos constitute all of the dark matter and are produced solely through the DW mechanism, recent X-ray analyses (see e.g. [9, 11, 10, 12, 13]) deduce upper limits on their mass, $m_s \simeq [3-8]$ keV. Conversely, the latest Ly-α analyses [15] of the high redshift power spectra from the SDSS [16] furnish constraints $m_s < (10-13)$ keV. Hence, these results seem to exclude scenarios where all dark matter consists of sterile neutrinos produced through the DW mechanism. (However, such evidence does not preclude the possibility of a sub-dominant sterile neutrino dark matter component [17].) However, hidden systematic effects within SDSS measurements of the Ly-α flux power spectrum could relax the severity of the corresponding constraints inferred from the 1-D matter power spectrum from the Ly-α forest [11].

In these proceedings we apply an *integrated flux* method to the CDF-N/S spectra of the XRB in order to constrain the radiative decay rate of DW sterile neutrino dark matter.

FLUX FROM STERILE NEUTRINO DECAY

Sterile neutrinos possess a radiative decay channel into an active neutrino and a photon with energy $E_\gamma = m_s/2$. For Majorana neutrinos, the radiative decay rate [18, 19] can be expressed as

$$\Gamma_\gamma \simeq 1.38 \times 10^{-22} \sin^2 2\theta_s \left(\frac{m_s}{1\,\mathrm{keV}}\right)^5 \mathrm{s}^{-1}, \tag{1}$$

where θ_s is the mixing angle in vacuum between the sterile and the active neutrino species. (For Dirac neutrinos the above rate is halved [19].)

The predicted signal from such decays originates from both extra-galactic (EG) neutrinos, as well as those residing in the Milky Way (MW) halo. The EG contribution can be evaluated assuming a uniform distribution of neutrinos in the visible Universe up to very small redshifts. The differential energy flux (energy flux per unit energy, integrated over the field of view (FOV), $\Omega_{\rm FOV}$, of the detector) can be expressed as [23]

$$\varphi_E^{\rm EG} \simeq \frac{\Gamma_\gamma}{4\pi m_s} \frac{\Omega_{\rm dm}\, \rho_c\, \Omega_{\rm FOV}}{H(m_s/2E_\gamma - 1)}, \tag{2}$$

where ρ_c is the present critical density, $\Omega_{\rm dm} \simeq 0.21$ is the dark matter density parameter, and $H(z)$ is the Hubble function, where in this study we adopt a flat Λ-matter dominated universe with $H(0) = 73$ km s^{-1} Mpc^{-1}, as well as matter and dark energy density parameters equal to $\Omega_{\rm m} \simeq 0.24$ and $\Omega_\Lambda \simeq$ 0.76 respectively [1].

The sterile neutrinos in the Galactic halo give rise to the differential energy flux

$$\varphi_E^{\rm MW} = \frac{\Gamma_\gamma}{4\pi m_s} \int_{\rm FOV} \int_{\rm l.o.s.} \rho_{\rm dm}(x,\Omega)\, dx\, d\Omega \equiv \frac{\Gamma_\gamma}{4\pi m_s} \int_{\rm FOV} S_{\rm dm}(\Omega)\, d\Omega \equiv \frac{\Gamma_\gamma}{4\pi m_s} \Omega_{\rm FOV} \bar{S}_{\rm dm}, \tag{3}$$

which depends on the integral over the dark matter density $\rho_{\rm dm}$ along the line of sight (l.o.s.), $S_{\rm dm}$.. Owing to the small variation of $S_{\rm dm}$ over the small FOV of *Chandra* ($\sim$ 5' circular area) in the directions of interest, we can approximately replace $S_{\rm dm}$ (i.e. the average value of $S_{\rm dm}$ over the relevant FOV), with the value of $S_{\rm dm}$ at the centre of each FOV.

We calculate the flux contributions from decays within the Galactic halo using the Navarro-Frenk-White (NFW) density profile $\rho_{\rm dm}(r) = \rho_s\ (r/r_s)^{-1}(1+(r/r_s))^{-2}$ [20], where ρ_s and r_s are the scale density and radius respectively. We adopt the recent evaluation of the virial mass of the MW halo, $0.6 \times 10^{12}\, M_\odot < M_{\rm vir.} < 2.0 \times 10^{12}\, M_\odot$, from [22], which is also consistent with $R_{\rm vir.} = 255$ kpc, $R_{\rm vir.}/r_s = 18$ and $R_\odot = 8$ kpc.

Adopting the above, we obtained values for the dark matter surface density of $(S_{\rm dm})_{\rm CDF-N} = [0.0115, 0.0384]\ {\rm g\,cm^{-2}}$ in the direction of CDF-N (with Galactic coordinates $(l,b)_{\rm CDF-N} = (125.89°, 54.83°)$), and $(S_{\rm dm})_{\rm CDF-S} = [0.0111, 0.0369]\ {\rm g\,cm^{-2}}$ in the direction of CDF-S (with Galactic coordinates $(l,b)_{\rm CDF-S} = (223.57°, -54.44°)$), where the lower (higher) extremes of each the ranges stated correspond to the lowest (highest) values of the MW virial mass mentioned above.

RESULTS

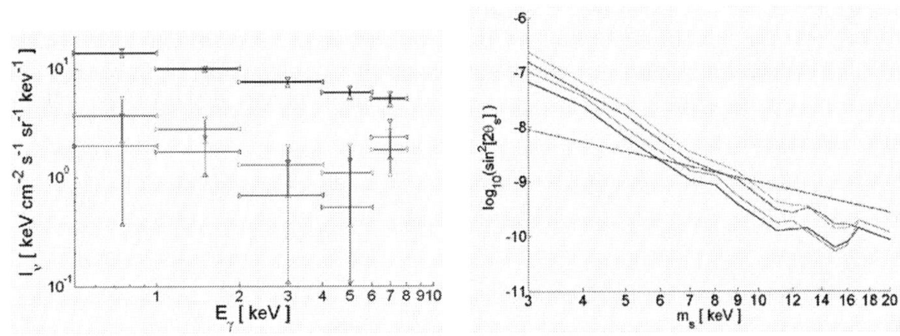

FIGURE 1. *Left* - XRB spectra in terms of the total specific intensity $I_\nu = d\varphi_E/d\Omega$ (*thick-black crosses*), and contributions from unresolved sources (*thin-black crosses (CDF-N), thick-grey crosses (CDF-S)*). The errors displayed are taken to be at the 1σ level. *Right* - 1 σ upper limits on $\sin^2(2\theta_s)$ using values of the Galactic halo virial mass of $M_{\rm vir.} = 0.6 \times 10^{12} M_\odot$ (*dashed lines*) and $M_{\rm vir.} = 0.6 \times 10^{12} M_\odot$ (*solid lines*), for CDF-N (*black lines*) and CDF-S (*grey lines*). The *dot-dashed line* indicates the parameter space permitted for DW sterile neutrinos constituting all of the currently inferred dark matter relic abundance.

Here, we utilise results from the analysis of the CDF observations by Worsley *et al.* [21]. Fig. 1 (*left*) displays the *total* XRB spectra aswell as the contributions attributed to unresolved sources. The errors displayed are taken to be at the 1σ level. We then derive upper limits on the radiative decay rate of sterile neutrinos by applying the following spectral analysis.

For each selected value of m_s, we add the corresponding EG and MW contributions (after accounting for signal broadening, owing to *Chandra's* finite energy resolution, (which we conservatively estimate as $\Delta E/E \geq 0.03$ [24]), and limit the decay rate (1) by invoking the following criterion

$$\int_{\Delta E_{\rm dp}} I_\nu^{CDF-N/S}\ dE \geq \int_{\Delta E_{\rm dp}} \left(I_\nu^{\rm EG} + I_\nu^{\rm MW}\right)\ dE \tag{4}$$

where the left-hand integral is evaluated over the *upper limit* of each of the data points displayed, each with energy range $\Delta E_{\rm dp}$. In fig. 1 (*right*), we display the resulting 1σ upper limits on $\sin^2(2\theta_s)$ as a function of m_s, obtained using values of the Galactic halo virial mass $M_{\rm vir.} = 0.6\times10^{12}M_\odot$ (*dashed lines*) and $M_{\rm vir.} = 0.6\times10^{12}M_\odot$ (*solid lines*). In addition, we plot a contour indicating the parameter space permitted for sterile neutrinos constituting all of the currently inferred dark matter relic abundance when generated solely via the DW mechanism (*dot-dashed line*).

Hence, using fig. 1 (*right*), the 1σ upper mass limits for DW sterile neutrino dark matter are

$$m_s < \begin{cases} [5.95, 8.30] & \text{keV for CDF}-\text{N observations} \\ [6.65, 9.49] & \text{keV for CDF}-\text{S observations} \end{cases}$$

where once again, the lowest (highest) extremes of each mass range correspond to the lowest (highest) values mentioned above for the virial mass of the Galactic halo.

In conclusion, the above constraints are slightly less stringent than those obtained from other related works which utilise a line signal non-detection spectral analysis of the XRB data (see e.g. [11]). However, we consider that such techniques yield overly stringent results when the energy resolution of the data is not sufficient enough to resolve line signals from local contributions. If so, then such results should be viewed as absolute lower/upper bounds on the sterile neutrino mass/mixing-angle, whereas the results obtained using the above *integrated flux* method should be considered to be the converse bounds on these parameters.

REFERENCES

1. D.N. Spergel *et al.* [WMAP Collaboration], Astrophys. J. Suppl. **170**, 377 (2007).
2. J.J. Dalcanton and C.J. Hogan, Astrophys. J. **561**, 35 (2001).
3. A.A. Klypin, A.V. Kravtsov, O. Valenzuela and F. Prada, Astrophys. J. **522**, 82 (1999).
4. P. Bode, J.P. Ostriker and N. Turok, Astrophys. J. **556**, 93 (2001).
5. T. Asaka, S. Blanchet and M. Shaposhnikov, Phys. Lett. B **631**, 151 (2005).
6. P.L. Biermann and A. Kusenko, Phys. Rev. Lett. **96**, 091301 (2006).
7. S. Dodelson and L.M. Widrow, Phys. Rev. Lett. **72**, 17 (1994).
8. T. Asaka, M. Laine and M. Shaposhnikov, JHEP **0606**, 053 (2006).
9. A. Boyarsky, A. Neronov, O. Ruchayskiy and M. Shaposhnikov, MNRAS **370**, 213 (2006).
10. A. Boyarsky, J. Nevalainen and O. Ruchayskiy, arXiv:astro-ph/0610961.
11. K.N. Abazajian *et al.* Phys. Rev. D **75**, 063511 (2007).
12. C.R. Watson, J.F. Beacom, H. Yuksel and T.P. Walker, Phys. Rev. D **74**, 033009 (2006).
13. S. Riemer-Sorensen, K. Pedersen, S.H. Hansen and H. Dahle, arXiv:astro-ph/0610034.
14. V.K. Narayanan, D.N. Spergel, R. Dave and C.P. Ma, Astrophys. J. Lett. **543**, L103 (2000).
15. U. Seljak, A. Makarov, P. McDonald and H. Trac, Phys. Rev. Lett. **97**, 191303 (2006).
16. P. McDonald *et al.*, Astrophys. J. Suppl. **163**, 80 (2006).
17. A. Palazzo, D. T. Cumberbatch, A. Slosar and J. Silk, arXiv:0707:1495 [astro-ph]
18. P.B. Pal and L. Wolfenstein, Phys. Rev. D **25**, 766 (1982).
19. V.D. Barger, R.J.N. Phillips and S. Sarkar, Phys. Lett. B **352**, 365 (1995)
20. j. F. Navarro, C. S. Frenk and S. D. M. White, Astrophys. J. **462**, 563 (1996).
21. M. A. Worsley *et al.* MNRAS **357**, 1281 (2005).
22. G. Battaglia *et al.*, MNRAS **364**, 433 (2005) [Erratum-ibid. **370**, 1055 (2006)].
23. E. Masso and R. Toldra, Phys. Rev. D **60**, 083503 (1999).
24. NASA X-ray & Gamma-ray Astronomy Data Archive (HEASARC), (http://heasarc.gsfc.nasa.gov/).

Solar System Constraints on Gauss-Bonnet Dark Energy

Stephen C. Davis

Lorentz Institute, Postbus 9506, 2300 RA Leiden, The Netherlands

Abstract. Quadratic curvature Gauss-Bonnet gravity may be the solution to the dark energy problem, but a large coupling strength is required. This can lead to conflict with laboratory and planetary tests of Newton's law, as well as light bending. The corresponding constraints are derived. If applied directly to cosmological scales, the resulting bound on the density fraction is $|\Omega_{GB}| \lesssim 3.6 \times 10^{-32}$.

Keywords: dark energy, experimental gravity tests, string cosmology
PACS: 04.25.Nx, 04.50.+h, 04.80.Cc, 95.36.+x

GAUSS-BONNET GRAVITY AND THE SOLAR SYSTEM

Corrections to Einstein gravity, such as the string-motivated Gauss-Bonnet term $\mathscr{L}_{GB} = R^2 - 4R_{\mu\nu}R^{\mu\nu} + R_{\mu\nu\rho\sigma}R^{\mu\nu\rho\sigma}$ could explain the current accelerated expansion of our universe. On its own, in four dimensions, the Gauss-Bonnet term does not contribute to the gravitational field equations. Coupling it to a scalar field will produce a non-trivial effect, which could act as effective dark energy. Including the corresponding higher order scalar kinetic terms, we obtain the ghost-free, quadratic curvature, gravitational Lagrangian

$$\mathscr{L} = \sqrt{-g}\left\{R - (\nabla\phi)^2 + \xi_1\mathscr{L}_{GB} + \xi_2 G^{\mu\nu}\nabla_\mu\phi\nabla_\nu\phi + \xi_3(\nabla\phi)^2\nabla^2\phi + \xi_4(\nabla\phi)^4\right\}. \quad (1)$$

The gravity modifications will not only be felt at cosmological scales, but also within the solar system where high precision gravitational measurements have been performed. The fields there are relatively weak and slowly varying, allowing us to use the approximate post-Newtonian metric [1]

$$ds^2 = -(1+2\Phi/c^2)(c\,dt)^2 + (1-2\Psi/c^2)dx^i dx_i + \mathscr{O}(\varepsilon^{3/2}), \quad (2)$$

with $\Phi, \Psi \sim c^2\varepsilon$, and $\partial_t \sim \varepsilon^{1/2}$. We take $\phi = \phi_0 + \mathscr{O}(\varepsilon)$, with ϕ_0 a constant. For standard Einstein gravity $\Phi = \Psi = -Gm/r$. We find the expansion parameter satisfies $\varepsilon \lesssim 10^{-5}$.

A perturbative analysis of the gravitational field equations can now be performed. However, it should be noted that if the field-dependent couplings $\xi_i(\phi)$ are to produce cosmological acceleration, they must be large. With this in mind we make no assumptions on the relative magnitude of $\xi_i(\phi_0)$ and ε. For simplicity we will assume the $\xi_i(\phi)$, and all of their derivatives, are of the same order. This is the case, for example, when ϕ arises from a toroidal compactification of a higher dimensional space [2].

To leading order in ε, the scalar field equation is

$$c^4\Delta\phi = -4\xi_1'\mathscr{D}(\Phi,\Psi) + \mathscr{O}(\varepsilon^2, \xi_i\varepsilon^3/r^2), \quad (3)$$

CP957, *Particles, Strings, and Cosmology, 13th International Symposium*
edited by A. Rajantie, C. Contaldi, P. Dauncey, and H. Stoica

where $\xi_1' = \partial\xi_1/\partial\phi$, evaluated at $\phi = \phi_0$. The Einstein equations take the form

$$\Delta\Phi = 4\pi G\rho_m - 2\xi_1'\mathscr{D}(\Phi+\Psi,\phi) + \mathscr{O}(\varepsilon^2, \xi_i\varepsilon^3/r^2)\,, \quad (4)$$

$$\Delta\Psi = 4\pi G\rho_m - 2\xi_1'\mathscr{D}(\Psi,\phi) + \mathscr{O}(\varepsilon^2, \xi_i\varepsilon^3/r^2)\,, \quad (5)$$

with ρ_m the matter energy density in the solar system. We have defined the operators

$$\Delta X = \sum_i X_{,ii}\,, \qquad \mathscr{D}(X,Y) = \sum_{i,j} X_{,ij}Y_{,ij} - \Delta X \Delta Y\,. \quad (6)$$

with $i, j = 1,2,3$. To leading order, the Gauss-Bonnet term is $\mathscr{L}_{GB} = 8\mathscr{D}(\Phi,\Psi)$.

To agree with observation, the solution to the above equations must be close to the usual Einstein gravity results. We can therefore assume $\Phi = -Gm/r + \mathscr{O}(\xi_i)$, etc., from which we obtain, to leading order [3]

$$\phi = \phi_0 - 2\xi_1' \frac{(Gm)^2}{c^4 r^4}\,, \quad \Phi = -\frac{Gm}{r} - \frac{64}{7}\frac{\xi_1'^2 (Gm)^3}{c^4 r^7}\,, \quad \Psi = -\frac{Gm}{r} - \frac{32}{7}\frac{\xi_1'^2 (Gm)^3}{c^4 r^7}\,. \quad (7)$$

We see there are mass-dependent, $1/r^7$ corrections, which are not covered by the usual parametrised post-Newtonian formalism [1]. This is in agreement with [4], but not [5] (which does not allow for the possibility that the couplings ξ_i, could be large).

Planetary motion

Planets in our solar system experience a gravitational acceleration $g_{\text{acc}} = -Gm/r^2$, resulting in elliptical orbits with period $2\pi\sqrt{a^3/(Gm)}$, where a is the semi-major axis of the planet and m is the sun's mass. Corrections to the Newtonian potential alter the effective mass felt by the planets [6, 7]. From (7) we obtain [3]

$$g_{\text{acc}}(r) = -\frac{d\Phi}{dr} = -\frac{Gm}{r^2}\left[1 - \frac{64\xi_1'^2 r_g^2}{r^6}\right] \equiv -\frac{G(m+\delta m)}{r^2} \quad (8)$$

where $r_g \equiv Gm/c^2 \approx 1.5\,\text{km}$ is gravitational radius of the sun. To agree with observation, the correction must be smaller than the uncertainty in a, so $\delta m/m < 3\delta a/a$.

The strongest bound comes from Mercury (with $a \approx 5.8 \times 10^7\,\text{km}$ and $\delta a \approx 0.11\,\text{m}$ [8])

$$|\xi_1'| \lesssim \left.\frac{\sqrt{3a^5\delta a}}{8r_g}\right|_{☿} \approx 3.8 \times 10^{16}\,\text{km}^2\,. \quad (9)$$

Applying this directly to Gauss-Bonnet density fraction [2] in cosmology, we find

$$|\Omega_{GB}| = \left|4\xi_1' H \frac{d\phi}{dt}\right| \lesssim 8.8 \times 10^{-30} \quad (10)$$

if $d\phi/dt \sim H$, and if $\xi_1'(\phi)$ has comparable values on local and cosmological scales. This value is far short the 0.7 required to solve the dark energy problem.

For a cosmological constant $\Phi = -Gm/r - r^2c^2\Lambda/6 + \cdots$. The corresponding bound comes from Mars [7] ($a \approx 2.3 \times 10^8$ km, $\delta a \approx 0.66$m [8]) and is

$$|\Lambda| \lesssim \left.\frac{9r_g\delta a}{a^4}\right|_{♂} \approx 1.2 \times 10^{-34}\,\text{km}^{-2}. \tag{11}$$

This implies $\Omega_\Lambda = \Lambda/(3H^2) \lesssim 7.3 \times 10^{11}$, which is vastly weaker than the corresponding cosmological constraint ($\Omega_\Lambda \lesssim 1$).

Cassini spacecraft

An even stronger constraint is obtained from signals between the earth (at $r_\oplus \approx 1.5 \times 10^8$ km) and Cassini spacecraft (at $r_e \approx 1.3 \times 10^9$ km) as it travelled to Saturn. For a round trip, the sun's gravitational field produces a time delay in the signals of [3]

$$c\Delta t = 2\int_{\text{ray}} \left(\sqrt{\frac{g_{xx}}{-g_{tt}}} - 1\right) dx \approx -2\int_{\text{ray}} (\Phi + \Psi)\, dx \approx 4r_g \ln\frac{r_\oplus r_e}{4b^2} + \frac{1024\xi_1'^2 r_g^3}{b^6}, \tag{12}$$

where the impact parameter b, is the smallest value of r on the signal's path. In 2002 it fell to its lowest value, $b \approx 1.1 \times 10^6$ km.

Rather than directly measure Δt, the Cassini experiment actually found the frequency shift in the signal [9]

$$y_{\text{gr}} = \frac{d\Delta t}{dt} \approx \frac{d\Delta t}{db}\frac{db}{dt} = -\frac{10^{-5}\,\text{s}}{b}\frac{db}{dt}\left[2 + (2.1 \pm 2.3) \times 10^{-5}\right]. \tag{13}$$

Requiring the Gauss-Bonnet correction (12) to be within the measured range (13) implies the bounds

$$|\xi_1'| \lesssim 1.6 \times 10^{14}\,\text{km}^2, \qquad |\Omega_{GB}| \lesssim 3.6 \times 10^{-32}. \tag{14}$$

A table-top laboratory test of Newton's law

Laboratory tests will also constrain modified gravity, as we will illustrate with the experiment described in [10]. It consists of a 60 cm copper bar, suspended at its midpoint by a tungsten wire. Two 7.3 kg masses 105 cm from the bar produce a torque N_{105} on the bar, and an $m \approx 43$ g mass 5 cm to the side of bar produces a comparable torque $-N_5$. By changing the positions of the masses, the ratio $R = N_{105}/N_5$ was determined and compared to theory

$$\delta_R = \frac{R_{\text{expt}}}{R_{\text{Newton}}} - 1 = (1.2 \pm 7) \times 10^{-4}. \tag{15}$$

The Gauss-Bonnet term affects all the masses, and gives cross-terms due its non-linearity. However we can ignore these complications, and just use (7) for the mass m, since it gives the dominant correction. A mass at $\vec{X} = (X, Y, Z)$ produces a torque

$$N = \int_{\rm bar} d^3x\,(\vec{x}\wedge\vec{F})_z \propto \int_{\rm bar} d^3x \frac{yX - xY}{r}\frac{d\Phi}{dr}\bigg|_{r=|\vec{X}-\vec{x}|}. \tag{16}$$

We find $\delta N_5/N_5 \approx -0.003(Gm\xi_1')^2 c^{-4}\,\mathrm{cm}^{-6}$. Requiring $\delta N_5/N_5 < \delta_R$, gives the bound $|\xi_1'| \lesssim 1.3\times10^{16}\,\mathrm{km}^2$, which is comparable to the planetary constraint (9).

DISCUSSION

Extrapolating solar system constraints to cosmological scales suggests that the density fraction Ω_{GB} is far too small to solve the dark energy problem. However our analysis features many assumptions, which while credible, could be violated and thus offer a way round the constraints. Clearly at least one of them must be broken if Einstein-Gauss-Bonnet gravity is to explain the acceleration of our universe.

In particular, we applied solar system results directly to cosmological scales. This assumes no significant spatial or temporal evolution of the field ϕ. Significant variation in the couplings ξ_i seems to offer the best way to save Gauss-Bonnet dark energy. Another possibility is that ϕ couples differently to dark matter and baryons, which will also break the relation between the two scales.

Instead, it may be that our assumptions on the form of the theory should be changed. The scalar field could be coupled directly to the Einstein-Hilbert term, as in Brans-Dicke gravity. Additionally, the couplings ξ_i and their derivatives could be of different orders. Both these changes open up the possibility of the corrections to Einstein gravity cancelling within the solar system. Alternatively ϕ could be given a large mass, which would suppress the quadratic curvature effects, as they operate via the scalar field. However this is also likely to inhibit acceleration.

ACKNOWLEDGMENTS

I am grateful to the Netherlands Organisation for Scientific Research (NWO) for financial support, to my collaborators C. Charmousis and L. Amendola, and of course to the conference organisers themselves.

REFERENCES

1. C. M. Will, arxiv:gr-qc/0510072.
2. L. Amendola, C. Charmousis and S. C. Davis, JCAP **0612**, 020 (2006) [hep-th/0506137].
3. L. Amendola, C. Charmousis and S. C. Davis, arxiv:0704.0175 [astro-ph].
4. G. Esposito-Farese, arxiv:gr-qc/0306018; AIP Conf. Proc. **736**, 35 (2004) [arxiv:gr-qc/0409081].
5. T. P. Sotiriou and E. Barausse, Phys. Rev. D **75**, 084007 (2007) [arxiv:gr-qc/0612065].
6. J. D. Anderson, *et al.* Astrophys. J. **342**, (1989) 539.
7. M. Sereno and P. Jetzer, Phys. Rev. D **73**, 063004 (2006) [arxiv:astro-ph/0602438].
8. E. V. Pitjeva, Solar System Research **39**, 176 (2005).
9. B. Bertotti, L. Iess and P. Tortora, Nature **425**, 374 (2003).
10. J. K. Hoskins, R. D. Newman, R. Spero and J. Schultz, Phys. Rev. D **32**, 3084 (1985).

Big Bang nucleosynthesis as a probe of varying fundamental "constants"

Thomas Dent, Steffen Stern and Christof Wetterich

Institut für Theoretische Physik, University of Heidelberg, Germany

Abstract. We analyze the effect of variation of fundamental couplings and mass scales on primordial nucleosynthesis in a systematic way. The first step establishes the response of primordial element abundances to the variation of a large number of nuclear physics parameters, including nuclear binding energies. We find a strong influence of the n-p mass difference, of the nucleon mass and of A=3,4,7 binding energies. A second step relates the nuclear parameters to the parameters of the Standard Model of particle physics. The deuterium, and, above all, 7Li abundances depend strongly on the average light quark mass. We calculate the behaviour of abundances when variations of fundamental parameters obey relations arising from grand unification. We also discuss the possibility of a substantial shift in the lithium abundance while the deuterium and 4He abundances are only weakly affected.

Keywords: Big Bang Nucleosynthesis, varying constants
PACS: 98.80.Ft

INTRODUCTION

In the Standard Model of particle physics the constancy of its more than 20 "free" parameters (coupling strengths, particle masses) over space and time is assumed. However, in high-energy theories like string theory the Standard Model Lagrangian appears as an effective low-energy theory whose parameters are fixed dynamically. More generally, variations may arise due to the coupling of Standard Model particles to a scalar field whose cosmological value depends on time [1] and/or the environment. The possibility of varying couplings has been discussed since the first suggestions of dynamical dark energy [2] due to a scalar field evolving over recent cosmological epochs.

The observational situation is contradictory, with null results from laboratory experiments [3] and the Oklo natural nuclear reactor [4], and both null and nonzero results from absorption spectra of interstellar clouds at redshifts $z = 0.5$–4 ([5, 6, 7]), which probe the fine structure constant α and the proton-electron mass ratio $\mu \equiv m_p/m_e$.

It is desirable to have bounds at widely different redshifts and in different environments to probe possible variations in a model-independent way. We consider variations at the epoch of Big Bang nucleosynthesis (BBN) ($z \simeq 10^{10}$), which is currently the earliest time at which theories of nuclear and particle physics can be compared to cosmological observation in a controllable way.

CP957, *Particles, Strings, and Cosmology, 13th International Symposium*
edited by A. Rajantie, C. Contaldi, P. Dauncey, and H. Stoica

BIG BANG NUCLEOSYNTHESIS

As the temperature of the cooling Universe drops below $T \approx 0.1$ MeV, neutrons and protons combine to form light elements up to mass number $A = 7$. The physics of BBN has already been described 40 years ago [8] and its numerical simulation has been improved since then. It turns out that, comparing observational values of primordial abundances with those obtained from theory simulations, the predictions of standard BBN theory match the observations quite well for deuterium and ^{4}He, but deviate from those of ^{7}Li by about a factor of 3. Amongst others, variations of constants might provide a solution to this "lithium problem". Hence, the intent of our work is to reveal the dependence of BBN results on variations of fundamental physics parameters.

Variation of parameters

As BBN is a low-energy process, the relevant parameters are rather effective low-energy constants than the parameters of the Standard Model. To cope with this circumstance, we use a 2-step approach. First, we study the dependence of primordial abundances Y_a with $a = (\mathrm{D}, {}^3\mathrm{He}, {}^4\mathrm{He}, {}^6\mathrm{Li}, {}^7\mathrm{Li})$ on the variation of a set of low-energy nuclear physics parameters X_i. We define a response matrix C with matrix elements

$$c_{ai} = \frac{\partial \ln Y_a}{\partial \ln X_i}, \tag{1}$$

extracted by varying X_i independently in the BBN code. This includes variation of the reaction cross-sections and rates which have a physical dependence on X_i. We vary with respect to 13 "nuclear" parameters X_i, given in Table 1, where G_N is the gravitational constant, τ_n the neutron lifetime, α the fine structure constant, m_e the electron mass, m_N the average nucleon mass, Q_N the neutron-proton mass difference and B_i the binding energies of elements i. It is remarkable that large numbers, which indicate a strong dependence on variation of parameters, dominantly appear in the lithium columns. Thus a solution of the lithium problem with varying constants seems feasible.

The second step involves the relation between the nuclear parameters X_i and a set of Standard Model parameters G_k, encoded in a second response matrix F with entries

$$f_{ik} = \frac{\partial \ln X_i}{\partial \ln G_k}. \tag{2}$$

For most of nuclear physics parameters, this connection is roughly known to first order. Details can be found in [9]. We consider six fundamental parameters G_k, given in Table 2, where $\delta_q \equiv m_d - m_u$ is the light quark mass difference, $\hat{m} \equiv (m_d + m_u)/2$ the averaged light quark mass and $\langle \phi \rangle$ denotes the Higgs v.e.v. The variation of abundances with respect to fundamental parameters G_k is then given by the matrix R (obtained by simple matrix multiplication: $R = CF$) with elements r_{ak},

$$\frac{\Delta Y_a}{Y_a} = r_{ak} \frac{\Delta G_k}{G_k}. \tag{3}$$

Table 2 shows the resulting dependences of abundances on fundamental parameters.

TABLE 1. Response matrix C, dependence of abundances Y_a on nuclear parameters X_i.

$\partial \ln Y_a / \partial \ln X_i$	D	^{3}He	^{4}He	^{6}Li	^{7}Li
G_N	0.94	0.33	0.36	1.4	-0.72
α	2.3	0.79	0.00	4.6	-8.1
τ_n	0.41	0.15	0.73	1.4	0.43
m_e	-0.16	-0.02	-0.71	-1.1	-0.82
Q_N	0.83	0.31	1.55	2.9	1.00
m_N	3.5	0.11	-0.07	2.0	-12
$B_{\rm D}$	-2.8	-2.1	0.68	-6.8	8.8
$B_{\rm T}$	-0.22	-1.4	0	-0.20	-2.5
$B_{\rm 3He}$	-2.1	3.0	0	-3.1	-9.5
$B_{\rm 4He}$	-0.01	-0.57	0	-59	-57
$B_{\rm 6Li}$	0	0	0	69	0
$B_{\rm 7Li}$	0	0	0	0	-6.9
$B_{\rm 7Be}$	0	0	0	0	81

TABLE 2. Response matrix R, dependence of abundances Y_a on fundamental parameters G_k.

$\partial \ln Y_a / \partial \ln G_k$	D	^{3}He	^{4}He	^{6}Li	^{7}Li
$G_{\rm N}$	0.94	0.33	0.36	1.4	-0.72
α	3.6	0.95	1.9	6.6	-11
$\langle\phi\rangle$	1.6	0.60	2.9	5.5	1.7
m_e	0.46	0.21	0.40	0.97	-0.17
δ_q	-2.9	-1.1	-5.1	-9.7	-2.9
$\hat{m}$	17	5.0	-2.7	-6	-61

Grand unified theories

To reduce the number of free parameters, we consider unified scenarios where the variations of fundamental couplings satisfy certain relations. Our examples involve a grand unified theory with unified coupling α_X, broken at the scale M_X to the Standard Model symmetry: the observable couplings of QCD and electromagnetism are then related to α_X via renormalization group flow and electroweak symmetry breaking. For simplicity we will take the Planck mass fixed relative to the unification scale, thus $\Delta(M_{\rm P}/M_X) = 0$, and the fermion Yukawa couplings to be constant, thus the electron and quark masses are proportional to $\langle\phi\rangle$. The variation of $\langle\phi\rangle$ is parameterised as

$$\frac{\langle\phi\rangle}{M_X} = \text{const.} \left(\frac{\Lambda_c}{M_X}\right)^{\gamma}. \tag{4}$$

We have studied the following three cases:

- $\gamma = 0$ $\quad \Delta\ln(G_N, \alpha, \langle\phi\rangle, m_e, \delta_q, \hat{m}) = (65, 1, -32, -32, -32, -32)\Delta\ln\alpha$
- $\gamma = 1$ $\quad \Delta\ln(G_N, \alpha, \langle\phi\rangle, m_e, \delta_q, \hat{m}) = (78, 1, 0, 0, 0, 0)\Delta\ln\alpha$
- $\gamma = 1.5$ $\quad \Delta\ln(G_N, \alpha, \langle\phi\rangle, m_e, \delta_q, \hat{m}) = (87, 1, 22, 22, 22, 22)\Delta\ln\alpha$

The resulting variations are shown in Fig.1 (dashed, dotted and solid line, respectively).

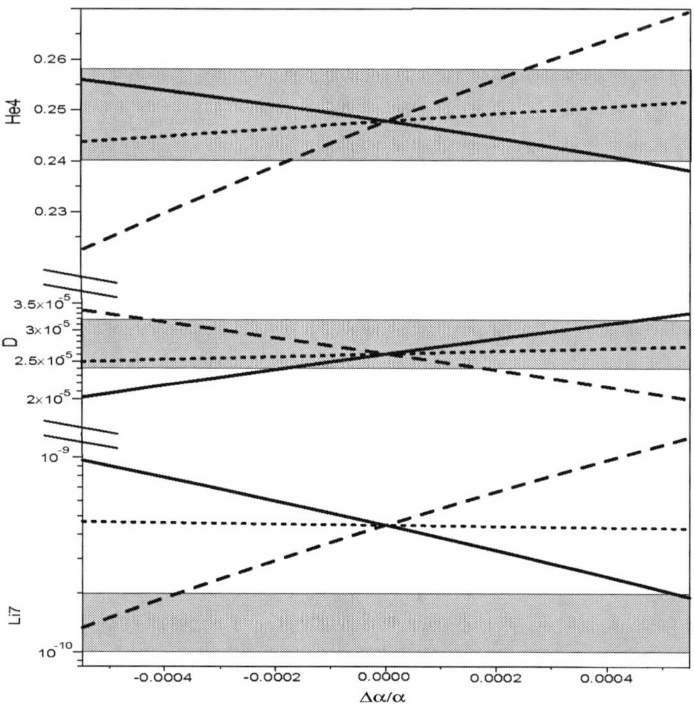

FIGURE 1. Variation of abundances with α in three GUT scenarios. Highlighted regions give the observational 1σ limits.

In the first scenario, a fractional variation of α by about -2.5×10^{-4} would bring abundances within 2σ observational bounds. In the third scenario, a variation of α by about $+5 \times 10^{-4}$ brings theory and observation into agreement within 1σ bounds. Hence the lithium problem may be considerably ameliorated by allowing variable "constants".

REFERENCES

1. J. D. Bekenstein, Phys. Rev. D **25** (1982) 1527.
2. C. Wetterich, *Nucl. Phys.* B **302** (1988) 645, *Nucl. Phys.* B **302** (1988) 668; B. Ratra and P. J. E. Peebles, *Phys. Rev.* D **37** (1988) 3406.
3. E. Peik, B. Lipphardt, H. Schnatz, C. Tamm, S. Weyers and R. Wynands, arXiv:physics/0611088.
4. Yu. V. Petrov, A. I. Nazarov, M. S. Onegin, V. Y. Petrov and E. G. Sakhnovsky, Phys. Rev. C **74**, 064610 (2006) [arXiv:hep-ph/0506186].
5. M. T. Murphy *et al.*, Lect. Notes Phys. **648** (2004) 131; M. T. Murphy, J. K. Webb and V. V. Flambaum, *Mon. Not. Roy. Astron. Soc.* **345** (2003) 609.
6. E. Reinhold *et al.*, *Phys. Rev. Lett.* **96** (2006) 151101.
7. S. A. Levshakov, M. Centurion, P. Molaro and S. D'Odorico, preprint astro-ph/0408188; S A. Levshakov *et al.*, preprint astro-ph/0703042.
8. R. V. Wagoner, W. A. Fowler and F. Hoyle, Astrophys. J. **148**, 3 (1967).
9. T. Dent, S. Stern and C. Wetterich, preprint arXiv:0705.0696 [astro-ph].

Density Perturbations in the Universe from Massive Vector Fields

K. Dimopoulos

Physics Department, Lancaster University,Lancaster LA1 4YB, U.K.

Abstract. I discuss the possibility of using a massive vector field to generate the density perturbation in the Universe. I find that a scale-invariant superhorizon spectrum of vector field perturbations is possible to generate during inflation. The associated curvature perturbation is imprinted onto the Universe following the curvaton scenario. The mechanism does not generate a long-range anisotropy because an oscillating massive vector field behaves as a pressureless isotropic fluid.

Keywords: Cosmology, Inflation, Curvature Perturbation, Particle Production, Vector Fields
PACS: 98.80.Cq

INTRODUCTION

The use of scalar fields in theoretical cosmology is extensive. For example, the dynamics of inflation is taken to be controlled by a scalar field (the inflaton) [1]. In fact, in many inflation models several scalar fields are employed (e.g. hybrid inflation, assisted inflation). Alternatives to minimal inflation, such as the curvaton paradigm, also use scalar fields [2]. Scalar fields have been used in other aspects of cosmology such as Dark Energy (quintessence) or baryogenesis (Affleck-Dine mechanism). In fact one might wonder whether we could do any cosmology without scalar fields.

Considering that scalar fields play an important role in the Early Universe is well motivated by the theory. Indeed, scalar fields are ubiquitous in theories beyond the standard model such as Supersymmetry (scalar partners, flat directions) and String Theory (moduli fields). However, *no scalar field has ever been observed.* This means that, designing models using unobserved scalar fields undermines their predictability and falsifiability, despite the recent precision of the observational data.

Given the above addiction of cosmologists to scalar fields and the approaching collider experiments which might disprove some of our favourite theories, it is necessary to explore other means of tackling the cosmological problems. Here we discuss the possibility of generating the curvature perturbation in the Universe using a massive vector field.

Despite the fact that we do have evidence of the existence of both massive and massless vector fields their use in cosmology is limited. This is so because of the following fundamental prejudice against them: *Were a vector field to dominate the Universe it would generate a large-scale anisotropy, in conflict with the CMB observations.* A further prejudice applies to particle production of vector fields during inflation: *Particle production of a light vector field is negligible because a massless vector field is conformally invariant and is not gravitationally generated during inflation.* We will show that both these prejudices can be evaded. Moreover, there is evidence that some weak large-scale anisotropy might be present in the CMB ("Axis of Evil" [3]).

CP957, *Particles, Strings, and Cosmology, 13th International Symposium*
edited by A. Rajantie, C. Contaldi, P. Dauncey, and H. Stoica

MASSIVE ABELIAN VECTOR FIELDS

The equations of motion

Consider a massive Abelian vector field with Lagrangian density

$$\mathscr{L} = -\frac{1}{4}F_{\mu\nu}F^{\mu\nu} + \frac{1}{2}m^2 A_\mu A^\mu \,, \tag{1}$$

where, the field strength tensor is $F_{\mu\nu} = \partial_\mu A_\nu - \partial_\nu A_\mu$. The equations of motion are: $[\partial_\mu + (\partial_\mu \ln\sqrt{-g})](\partial^\mu A^\nu - \partial^\nu A^\mu) + m^2 A^\nu = 0$, where g is the determinant of the metric tensor $g_{\mu\nu}$. We assume that inflation lasts long enough to homogenise the Universe and inflate away its spatial curvature before the cosmological scales exit the horizon. Hence, we can use the flat-FRW metric: $ds^2 = dt^2 - a^2(t)dx^i dx^i$. Similarly, we expect inflation to homogenise the vector field too, so that $A_\mu = A_\mu(t)$. Using the above, the equations of motion of the spatial components of the homogeneous vector field are [4]

$$\ddot{A}_i + H\dot{A}_i + m^2 A_i = 0\,, \tag{2}$$

while for the temporal component we have $A_t(t) = 0$, where the dot denotes time derivative. Perturbing the field equations and switching to momentum space we obtain the equations of motion of the Fourier components of the vector field perturbations [4]

$$\left[\partial_t^2 + H\partial_t + m^2 + \left(\tfrac{k}{a}\right)^2\right]\delta\mathscr{A}_i^{\perp} = 0 \tag{3}$$

$$\left[\partial_t^2 + \left(1 + \tfrac{2k^2}{k^2+(am)^2}\right)H\partial_t + m^2 + \left(\tfrac{k}{a}\right)^2\right]\delta\mathscr{A}_i^{\parallel} = 0\,, \tag{4}$$

where $\delta\mathscr{A}_i^{\parallel} \equiv k_i(k_j\delta\mathscr{A}_j)/k^2$ is the component parallel to k_i and $\delta\mathscr{A}_i^{\perp} \equiv \delta\mathscr{A}_i - \delta\mathscr{A}_i^{\parallel}$, with $k^2 \equiv k_i k_i$. Using the above equations we can investigate particle production of A_μ during inflation. Our aim is to generate a scale invariant spectrum of vector field perturbations. We will ignore hereafter the longitudinal component, since $\delta\mathscr{A}_i^{\parallel}$ can generate such a spectrum only under the condition $m \ll e^{-N_*}H_*$ [4], where $N_* \simeq 40-60$ is the number of e-folds before the end of inflation corresponding to the cosmological scales and H_* is the inflationary Hubble scale. It turns out that this condition is impossible to meet if the field is to generate the curvature perturbation in the Universe [4]. Hence, in the following we focus on the transverse component $\delta\mathscr{A}_i^{\perp}$ and we drop the $\perp$ symbol.

Particle production

Solving Eq. (3) with boundary condition $\lim_{k/aH\to\infty}\delta\mathscr{A}_k = e^{ik/aH}/\sqrt{2k}$, (i.e. matching to the Bunch-Davies vacuum at early times so that the perturbations start out as vacuum fluctuations) we get [4]

$$\delta\mathscr{A}_k = \sqrt{\frac{\pi}{aH}}\,\frac{e^{i\pi\left(\nu-\frac{1}{2}\right)/2}}{1-e^{i2\pi\nu}}\left[J_\nu(k/aH) - e^{i\pi\nu}J_{-\nu}(k/aH)\right], \tag{5}$$

where with J_ν we denote Bessel functions of the first kind and $\nu \equiv \sqrt{\frac{1}{4} - (m/H)^2}$. The power spectrum of the generated perturbations is obtained in the limit of late times, when the perturbations have grown to superhorizon scales. Using the above we find [4]

$$\mathscr{P}_{\mathscr{A}} \equiv \frac{k^3}{2\pi^2} \lim_{k/aH \to 0} |\delta \mathscr{A}_k|^2 \approx \frac{8\pi |\Gamma(1-\nu)|^{-2}}{(1-\cos 2\pi\nu)} \left(\frac{aH}{2\pi}\right)^2 \left(\frac{k}{2aH}\right)^{3-2\nu} \tag{6}$$

For a light field $\nu \simeq \frac{1}{2}$ and so $\mathscr{P}_{\mathscr{A}}{}^{\text{vac}} \simeq (k/2\pi)^2$. This is the vacuum spectrum, that can be readily obtained using the Bunch-Davies vacuum expression for $\delta\mathscr{A}_k$. Thus, in the massless limit, we recover the conformal invariant result where there is no particle production. However, what we are after is a scale invariant spectrum. From Eq. (6) it is evident that scale invariance requires $\nu \simeq \frac{3}{2}$, which is equivalent with: $m^2 \approx -2H^2$.

This condition can be understood by noting that A_μ is the *comoving* vector field. Indeed, the mass term in Eq. (1), when using the flat-FRW metric, becomes $\delta\mathscr{L}_m = \frac{1}{2}m^2 A_\mu A^\mu = \frac{1}{2}m^2(A_t^2 - a^{-2}A_iA_i)$. Since the Lagrangian is a physical quantity we find that the components of the physical vector field are $W_i \equiv A_i/a$. Introducing these into Eq. (2) during inflation we obtain: $\ddot{W}_i + 3\dot{W}_i + (2H^2 + m^2)W_i = 0$ [4]. Hence, the equation of motion is identical to the one of a scalar field with mass $\tilde{m}^2 = 2H^2 + m^2$. Thus, a scale invariant spectrum is generated when the *physical* vector field becomes light $\tilde{m} \to 0$, i.e. its Compton wavelength exceeds the horizon. The power spectrum, in this case, is given by the Hawking temperature $\mathscr{P}_{\mathscr{W}} = \mathscr{P}_{\mathscr{A}}/a^2 \approx (H/2\pi)^2$.

VECTOR CURVATON

We now employ the scale invariant superhorizon spectrum of vector field perturbations obtained above to generate the curvature perturbation. To retain isotropy, the vector field cannot be the inflaton. The vector field can act as a *curvaton* provided it safely dominates the Universe after inflation. Hence, studying its post-inflationary evolution is necessary.

The stress-energy tensor is: $T_{\mu\nu} = \frac{1}{4}g_{\mu\nu}F_{\rho\sigma}F^{\rho\sigma} - F_{\mu\rho}F_\nu^\rho + m^2\left(A_\mu A_\nu - \frac{1}{2}g_{\mu\nu}A_\rho A^\rho\right)$ [4]. This can be written as $T_\mu^\nu = \text{diag}(\rho_A, -p_\perp, -p_\perp, +p_\perp)$, which is similar to a perfect fluid with the crucial difference that the pressure along the longitudinal direction is of opposite sign to the transverse pressure $p_\perp$, where $\rho_A \equiv \rho_{\text{kin}} + V$ and $p_\perp \equiv \rho_{\text{kin}} - V$, with $\rho_{\text{kin}} \equiv -\frac{1}{4}F_{\mu\nu}F^{\mu\nu}$ and $V \equiv -\frac{1}{2}m^2A_\mu A^\mu$. If $p_\perp \neq 0$ then the fluid is anisotropic and cannot be allowed to dominate the Universe. This is why A_μ cannot be the inflaton.

However, suppose that, after the end of inflation a phase transition renders the mass ^{2} positive for the vector field. This not only restores Lorentz invariance in the vacuum, but allows the field to begin coherent oscillations after inflation, when $m > H(t)$, as evident from Eq. (2). During these quasi-harmonic oscillations we have on average $\overline{\rho_{\text{kin}}} \approx \overline{V}$ [4], i.e. $\overline{p_\perp} \approx 0$. Hence, the coherently oscillating homogeneous massive Abelian vector field behaves as pressureless *isotropic* matter. Consequently, its density decreases as $\rho_A \propto a^{-3}$ [4], so that it can dominate the radiation background without introducing significant anisotropy. Therefore, the vector field can act as a curvaton and impose, upon domination, its own curvature perturbation onto the Universe. Hence, the observed curvature perturbation $\zeta = 5 \times 10^{-5}$ can be attributed to the vector curvaton field, i.e.

$$\zeta = \zeta_A \equiv -H\frac{\delta\rho_A}{\dot{\rho}_A} = \frac{1}{3}\frac{\delta\rho_A}{\rho_A}\bigg|_{\rm dec} \simeq \frac{2}{3}\frac{\delta\bar{A}}{A}\bigg|_{\rm dec} = \frac{2}{3}\frac{\delta\bar{A}}{A}\bigg|_{\rm osc} \approx \frac{2}{3}\frac{a_{\rm osc}}{a_*}\frac{\delta A}{A}\bigg|_* \simeq \frac{H_*}{3\pi W_{\rm osc}}, \quad (7)$$

where we defined ζ_A on a foliage of spacetime along spatially flat hypersurfaces and we used that: $\rho_A \propto a^{-3}$ at the decay time of the field (denoted by 'dec'); $\rho_A^{\rm dec} \approx 2\overline{V} \propto \overline{A^2}$; $\delta A/A \simeq$ const. during oscillations but, before their onset (denoted by 'osc'), A is frozen ($A_{\rm osc} \approx A_*$) and $\delta A/A \propto a$ [4]; $\delta A_* \simeq a_* H_*/2\pi$ and $W_{\rm osc} = (A/a)_{\rm osc} \approx A_*/a_{\rm osc}$ with '*' denoting the epoch when the cosmological scales exit the horizon.

As shown in Eq. (7), $\zeta_A \propto \delta\rho_A/\rho_A$. Since ρ_A is a scalar quantity, the generated perturbations are *scalar and not vector* in nature, despite originating from a vector field.

CONCLUSIONS

A vector field can generate the curvature perturbation in the Universe. To obtain a superhorizon spectrum of perturbations, its mass should satisfy the condition: $m^2 \approx -2H^2$. After inflation, the vector field can act as a curvaton provided $m > H$ (i.e. the vacuum is Lorentz invariant). In this case the vector field undergoes quasi-harmonic oscillations, during which it acts as a pressureless *isotropic* fluid. Hence, it can dominate the Universe and impose its own, *scalar* curvature perturbation without introducing any anisotropy.

The challenge is to realise the above conditions in theories beyond the standard model. In supergravity, one may dispense with the requirement that $m^2 < 0$ during inflation. Indeed, the gauge kinetic function f of vector fields is expected to vary significantly during inflation. This is because supergravity corrections to the potential give the scalar fields masses of order H_* [5], which means that they are fast-rolling down the potential slopes resulting in $\dot{f}/f \sim H$. It can be shown that, if $f \propto a^2$, then a light vector field can obtain a scale invariant spectrum of perturbations without the need of $m^2 < 0$ [6].

ACKNOWLEDGMENTS

This work was supported (in part) by the European Union through the Marie Curie Research and Training Network "UniverseNet" (MRTN-CT-2006-035863) and by STFC (PPARC) Grant PP/D000394/1.

REFERENCES

1. A. R. Liddle and D. H. Lyth, *Cosmological Inflation and Large Scale Structure*, (Cambridge University Press, Cambridge U.K., 2000).
2. D. H. Lyth and D. Wands, Phys. Lett. B **524** (2002) 5; K. Enqvist and M. S. Sloth, Nucl. Phys. B **626** (2002) 395; K. Dimopoulos and D. H. Lyth, Phys. Rev. D **69** (2004) 123509.
3. K. Land and J. Magueijo, Phys. Rev. Lett. **95** (2005) 071301.
4. K. Dimopoulos, Phys. Rev. D **74** (2006) 083502.
5. M. Dine, L. Randall and S. Thomas, Nucl. Phys. B **458** (1996) 291; Phys. Rev. Lett. **75** (1995) 398.
6. K. Dimopoulos, Phys. Rev. D 77 (2007) (in press); arXiv:0705.3334 [hep-ph].

Cosmological tests of generalized RS brane-worlds with Weyl fluid

László Á. Gergely, Zoltán Keresztes and Gyula M. Szabó

Departments of Theoretical and Experimental Physics, University of Szeged, Dóm tér 9, Szeged 6720, Hungary

Abstract. A class of generalized Randall-Sundrum type II (RS) brane-world models with Weyl fluid are confronted with the Gold supernovae data set and BBN constraints. We consider three models with different evolutionary history of the Weyl fluid, characterized by the parameter α. For $\alpha = 0$ the Weyl curvature of the bulk appears as dark radiation on the brane, while for $\alpha = 2$ and 3 the brane radiates, leaving a Weyl fluid on the brane with energy density decreasing slower than that of (dark) matter. In each case the contribution Ω_d of the Weyl fluid represents but a few percent of the energy content of the Universe. All models fit reasonably well the Gold2006 data. The best fit model for $\alpha = 0$ is for $\Omega_d = 0.04$. In order to obey BBN constraints in this model however, the brane had to radiate at earlier times.

Keywords: brane-worlds, Weyl-fluid, supernovae
PACS: 43.35.Ei, 78.60.Mq

INTRODUCTION

The ΛCDM model according to which our Universe is a Friedmann-Lemaitre-Robertson-Walker (FLRW) space-time with flat spatial sections containing approximately 3% baryonic matter, 24% cold dark matter, the rest being given by the contribution of a cosmological constant Λ seems to be in excellent agreement with current observational data. As the dark sector (dark matter and dark energy - a generalization of the vacuum expectation energy represented by the cosmological constant) remains unknown, alternate gravitational theories have been advanced.

The string-theory motivated brane-world models contain our observable Universe as a time-evolving 3-dimensional brane embedded in a 5-dimensional bulk. Standard model fields act on the brane, but gravity is allowed to leak into the fifth dimension [1], where non-standard model fields could also exist (for a review see [2]). The projection of the 5-dimensional Einstein equation onto the brane generates an effective Einstein equation with new source terms as compared to general relativity [3], [4].

Among them, the energy-momentum squared source term modifies early cosmology [5] and becomes important during the final stages of gravitational collapse [6]. It behaves as the dominant source term before the Bing Bang Nucleosynthesis (BBN). This quadratic source term is proportional to $1/\lambda$. The value of the brane tension λ is constrained by the deviation from the gravitational Newton-law still compatible with nowadays rigurous experiments [7]-[8]. The emerging high value of λ and the fast decrease

CP957, *Particles, Strings, and Cosmology, 13th International Symposium*
edited by A. Rajantie, C. Contaldi, P. Dauncey, and H. Stoica

of the square of the energy density of matter implies that in a cosmological context this source term can be safely ignored at present time.

The Weyl curvature of the bulk gives rise to a non-local bulk effect on the brane, appearing as fluid on the brane (the Weyl fluid). In the simplest case, when the bulk is static, the Weyl fluid is a radiation field (dark radiation). This situation represents an equilibrium configuration, without any energy exchange between the brane and the bulk. BBN constraints the amounts of the energy density of the dark radiation as $-1.02 \times 10^{-4} \leq \Omega_d \leq 2.62 \times 10^{-5}$ [9]. (Here Ω_d is the dimensionless dark radiation energy density parameter.) More generic Weyl fluids were also considered [8], [10]-[15]. Depending on how this Weyl fluid evolves, its present day amount can be either negligible or not. It is this aspect we wish to consider here, based on our previous analysis [16]-[17].

Various brane-world models were confronted with supernova data [18]-[21], however in all these models the contribution of Weyl fluid was dropped, assuming it was pure dark radiation during all stages all the cosmological evolution. In our analysis we keep the contribution of the (non-radiation like) Weyl fluid. We already gave the analytical expression in terms of elliptical integrals for the luminosity distance when the Weyl contribution is small [16] and can be considered a perturbation. [1] Then we have tested the models with Weyl fluid characterizing a bulk-brane energy exchange, by comparing their predictions with the best available supernova data [17]. Here we present additional analysis and strenghten our conclusions.

Our model consists of a spatially flat FLRW brane embedded symmetrically into a 5-dimensional Vaidya-anti de Sitter bulk. The latter has a cosmological constant $\widetilde{\Lambda}$, black holes with masses m on either sides of the brane and radiation. If the radiation is swithced off, m is constant, and the bulk becomes Schwarzschild-anti de Sitter. In this configuration the Weyl fluid appears as dark radiation on the brane (with energy density m/a^4). Any radiation escaping from the brane causes m to vary. The ansatz $m = m_0 a^\alpha$, with $\alpha = 2$, 3, comparable with structure formation has been recently advanced [15].

The brane coupling constant κ^2 is related to the coupling constant in the bulk $\widetilde{\kappa}^2$ and brane tension λ as $6\kappa^2 = \widetilde{\kappa}^4 \lambda$. The relation between the brane tension and the bulk and the brane cosmological constant Λ is $2\Lambda = \kappa^2 \lambda + \widetilde{\kappa}^2 \widetilde{\Lambda}$.

We introduce the following dimensionless quantities:

$$\Omega_\rho = \frac{\kappa^2 \rho_0}{3H_0^2}, \quad \Omega_\lambda = \frac{\kappa^2 \rho_0^2}{6\lambda H_0^2}, \quad \Omega_d = \frac{2m_0}{a_0^{4-\alpha} H_0^2}, \quad \Omega_\Lambda = \frac{\Lambda}{3H_0^2}, \tag{1}$$

where $\Omega_{tot} = \Omega_\Lambda + \Omega_\rho + \Omega_\lambda + \Omega_d$. Here H is the Huble parameter, ρ is the matter energy density on the brane and a is the scale factor. The subscript 0 denotes the present value of the respective quantities. The Friedmann equation written in these parameters becomes $H^2/H_0^2 = \Omega_\Lambda + \Omega_\rho a_0^3/a^3 + \Omega_d a_0^{4-\alpha}/a^{4-\alpha} + \Omega_\lambda a_0^6/a^6$. At present time this gives $\Omega_{tot} = 1$.

---

[1] We note that for a wide class of phantom Friedmann cosmologies similar analytical results in terms of elementary and Weierstrass elliptic functions for the luminosity distance are available [22].

The luminosity distance for the spatially flat FLRW brane becomes

$$d_L(z) = \frac{(1+z)\,a_0}{H_0} \int_{a_{em}}^{a_0} \frac{a\,da}{\left[\Omega_\Lambda a^6 + \Omega_\rho a_0^3 a^3 + \Omega_d a_0^{4-\alpha} a^{\alpha+2} + \Omega_\lambda a_0^6\right]^{1/2}} . \tag{2}$$

The above complicated integral has no analytical form in the majority of cases. However for small Ω_d and (as noted earlier)) vanishing Ω_λ this integral could be given analytically as $d_L^{\Lambda\lambda d} = d_L^{\Lambda\text{CDM}} + \Omega_d I_d$, where the coefficient I_d is an analytic expression of elementary functions and elliptic integrals of the first and second kind [16], having different forms depending on the actual value of the parameter α.

We have then compared the predictions of the models characterized by various values of α with the Gold 2006 supernovae data set [23], for the range $-0.1 < \Omega_d < 0.1$ up to

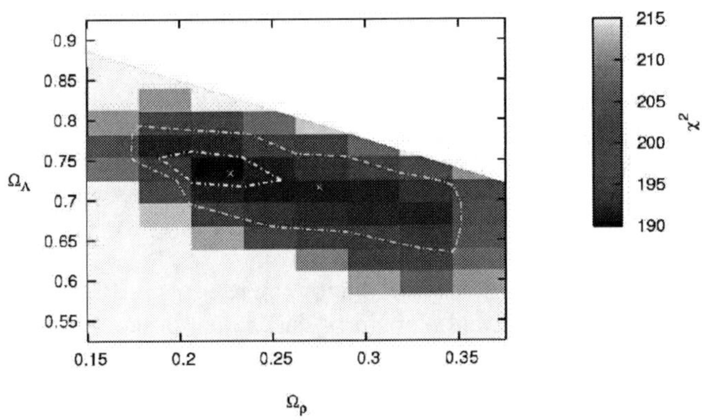

FIGURE 1. For $\alpha = 0$ the global minimum is at $\Omega_\Lambda = 0.735$, $\Omega_\rho = 0.225$ implying $\Omega_d = 0.04$. The ΛCDM model is at $\Omega_\Lambda = 0.725$ and $\Omega_m = 0.275$. The white area represents a forbidden parameter range.

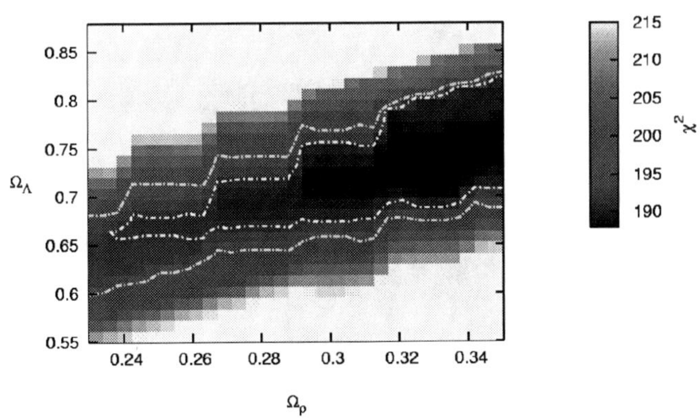

FIGURE 2. For $\alpha = 2$ a trough takes the place of the global minimum, showing that the compatibility of the Weyl fluid models with $\alpha = 2$ to supernova data does not depend crucially on the particular (small) value of Ω_d.

$z = 3$.

The data selects among the brane-world models with $\alpha = 0$ a global minimum at $\Omega_\Lambda = 0.735$, $\Omega_\rho = 0.225$ and $\Omega_d = 0.04$. We note that the value of Ω_ρ is in perfect agreement with the WMAP 3-year data [24]. The 1-σ and 2-σ confidence levels are shown of Fig1 in the $\Omega_\Lambda - \Omega_\rho$ plane. The ΛCDM model is contained in the 2-σ cofidence level at $\Omega_\Lambda = 0.725$, $\Omega_\rho = 0.275$. The forbidden (white) region appears because the Friedmann equation combined with $\Omega_{tot} = 1$ gives constraints on the allowed range of $\Omega_d - \Omega_\rho$ [17].

The cases $\alpha = 2$ and $\alpha = 3$ are represented on Fig2 and Fig3, respectively. In these cases the determination of the global minimum becomes unreliable, as the 1-σ contours become much elongated. Instead of a peak we have a "trough", which lies aslope in the Ω_Λ-Ω_ρ plane. Together with the increase in α the 1-σ contour becomes more elongated and turning counter-clock-wise.

This feature of the 1-σ contours indicates that with increasing α the models become more likely to be compatible with the available supernova data, irrelevant of the exact (but small) value of Ω_d. This practically means that any small amount of Weyl fluid with $\alpha = 2, 3$ is perfectly compatible with supernova data.

In the investigated models the Weyl fluid is either dark radiation ($\alpha = 0$) or describes a situation when the brane radiates into the bulk, feeding the bulk black holes ($\alpha = 2, 3$). In the latter cases the energy density of the Weyl fluid decreases slower, than that of matter, therefore a sizeable Weyl fluid contribution nowasays is perfectly compatible with BBN constraints.

For $\alpha = 0$ the Weyl fluid evolves as radiation, therefore the BBN constraints can be satisfied only with an infinitesimal amount of dark radiation nowadays. The preferred value of $\Omega_d = 0.04$ would give too much dark radiation in the past. However if we assume that $\alpha = 0$ is only a recent characteristic of the brane, this equilibrium situation being preceded by an epoch in which the brane is allowed to radiate, the BBN constraints can be obeyed with a small, but non-negligible amount of dark radiation nowadays. The BBN contraint [9] can be satisfied [17], if the brane radiates in the interval (z_1, z_*). Assuming $z_1 = 3$ the constraint for z_* with different value of α gives $z_* \geq 6114.20$ for

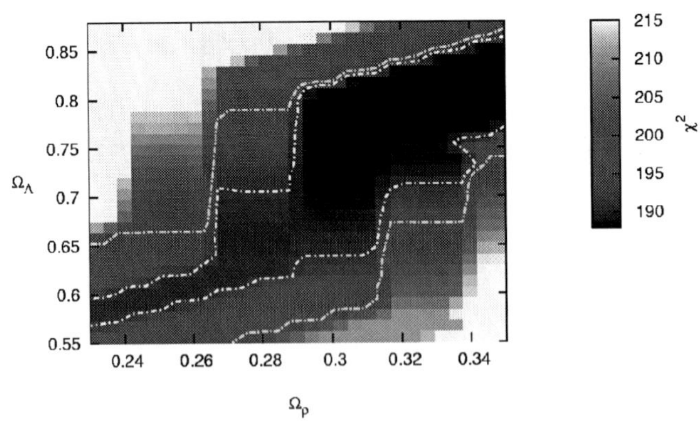

FIGURE 3. Same as on Fig2, but for the $\alpha = 3$ models.

$\alpha = 1$, $z_* \geq 155.40$ for $\alpha = 2$, $z_* \geq 45.08$ for $\alpha = 3$, $z_* \geq 24.01$ for $\alpha = 4$. Thus the known history of the Universe can be explained if the brane radiates during at least some period of the cosmological evolution. At early times the radiation of the brane leads to a black hole, which can further grow during structure formation.

None of the investigated models, compatible with the available supernova data and structure formation, can be excluded by present observational accuracy. The differences among the predictions of the models are however increasing with redshift, therefore future measurements of very far supernovae will be able to either support or falsify these cosmological models.

Acknowledgments: this work was supported by OTKA grants no. 46939 and 69036. LÁG and GyMSz were further supported by the János Bolyai Fellowship of the Hungarian Academy of Sciences.

REFERENCES

1. L. Randall and R. Sundrum, *An Alternative to Compactification*, 1999 *Phys. Rev. Lett.* **83**, 4690
2. R. Maartens, *Brane-world Gravity*, 2004 *Living Rev. Rel.* **7** 1
3. T. Shiromizu, K. Maeda and M. Sasaki, *The Einstein Equations on the 3-Brane World*, 2000 *Phys. Rev.* D **62** 024012
4. L. Á. Gergely, *Generalized Friedmann branes*, 2003 *Phys. Rev.* D **68** 124011
5. P. Binétruy, C. Deffayet, U. Ellwanger and D. Langlois, *Brane cosmological evolution in a bulk with cosmological constant*, 2000 *Phys. Lett.* B **477**, 285
6. L. Á. Gergely, *Black holes and dark energy from gravitational collapse on the brane*, 2007 *JCAP* **07**(02) 027
7. J. C. Long et al., *New experimental limits on macroscopic forces below 100 microns*, 2003 *Nature* **421** 922; J. H. Gundlach, S. Schlamminger, C. D. Spitzer at al., *Laboratory Test of Newton´s Second Law for Small Acceleration*, 2007 *Phys. Rev. Lett.* **98** 150801; D. J. Kapner, T. S. Cook, E. G. Adelberger, *Tests of the Gravitational Inverse-Square Law below the Dark-Energy Lenght Scale*, 2007 *Phys. Rev. Lett.* **98** 021101
8. L. Á. Gergely and Z. Keresztes, *Irradiated asymmetricFriedmann branes*, 2006 *JCAP* **06**(01) 022
9. K. Ichiki, M. Yahiro, T. Kajino, M. Orito and G. J. Mathews, *Observational Constraints on Dark Radiation in Brane Cosmology*, 2002 *Phys. Rev.* **D 66**, 043521
10. D. Langlois, L. Sorbo, and M. Rodríguez-Martínez, *Cosmology of a brane radiating gravitons into the extra dimension*, 2002 Phys. Rev. Lett. **89** 171301
11. L. Á. Gergely, E. Leeper and R. Maartens, *Asymmetric radiating brane-world*, 2004 *Phys. Rev.* D **70** 104025
12. D. Jennings and I. R. Vernon, *Graviton emmission into non-Z2 symmetric brane world spacetimes*, 2005 *JCAP* **05**(07) 011
13. D. Jennings, I. R. Vernon, A-C. Davis and C. van de Bruck, *Bulk black holes radiating in non-Z2 brane-world spacetimes*, 2005 *JCAP* **05**(04) 013
14. Z. Keresztes, I. Képíró and L. Á. Gergely, *Semi-transparent brane-worlds*, 2006 *JCAP* **06**(05) 020
15. S. Pal, *Structure formation on the brane: A mimicry*, 2006 *Phys. Rev.* D **74** 024005
16. Z. Keresztes, L. Á. Gergely, B. Nagy and Gy. M. Szabó, *The luminosity-redshift relation in brane-worlds: I. Analytical results*, 2006 *astro-ph/0606698*
17. Gy. M. Szabó, L. Á. Gergely and Z. Keresztes, *The luminosity-redshift relation in brane-worlds: II. Confrontation with expreimental data*, 2007 *astro-ph/0702610*
18. U. Alam and V. Sahni, *Confronting Brane world Cosmology with Supernova data and Baryon Oscillations*, 2006 *Phys. Rev.* D **73** 084024
19. R. Lazkoz, R. Maartens and E. Majerotto, *Observational constraints on phantom-like braneworld cosmologies*, 2006 *Phys. Rev.*D **74** 083510
20. M. P. Dąbrowski, W. Godłowski and M. Szydłowski, *Brane universes tested against astronomical data*, 2004 *Int. J. Mod. Phys.* D **13** 1669

21. S. Fay, *Branes: cosmological surprise and observational deception*, 2006 *Astron. Astrophys.* **452** 781
22. M. P. Dąbrowski and T. Stachowiak, *Phantom Friedmann cosmologies and higher-order characteristics of expansion*, 2006 *Annals of Physics* **321** 771
23. A. G. Riess, L-G. Strolger, S. Casertano et al., *New Hubble Space Telescope Discoveries of Type Ia Supernovae at $z > 1$: Narrowing Constraints on the Early Behavior of Dark Energy* 2007 to appear in Astrophys. J. **656** 98
24. D. N. Spergel, R. Bean, Doré O et al., *Wilkinson Microwave Anisotropy Probe (WMAP) Three Year Results: Implications for Cosmology*, 2006 *astro-ph/0603449*

Multiple scalar particle decay and perturbation generation

Jinn-Ouk Gong

Harish-Chandra Research Institute, Chhatnag Road, Jhunsi, Allahabad 211 019, India

Abstract.
We study the evolution of the universe which contains a multiple number of non-relativistic scalar fields decaying into both radiation and pressureless matter. We present a powerful analytic formalism to calculate the matter and radiation curvature perturbations, and find that our analytic estimates agree with full numerical results within an error of less than one percent. Also we discuss the isocurvature perturbation between matter and radiation components, which may be detected by near future cosmological observations, and point out that it crucially depends on the branching ratio of the decay rate of the scalar fields and that it is hard to make any model independent predictions.

Keywords: Curvaton, cosmological perturbation theory
PACS: 98.80.Cq

Nowadays it is widely accepted that the primordial curvature perturbation ζ is the origin of the temperature anisotropy in the cosmic microwave background (CMB) and large scale structure in the present observable universe. For single and multi field inflation cases it is well known how to calculate the power spectrum $\mathscr{P}_\zeta$ of ζ produced during inflation with good enough accuracy [1]. However, during inflation, *any* scalar fields with their masses being smaller than the Hubble scale acquire almost scale invariant fluctuations and they may be responsible for the observed structure. This "curvaton" scenario [2] may find its natural accommodation in the context of multi-field inflation: e.g. recently it is shown that at the end of multi-field inflation some light inflaton fields retain non-zero amplitudes depending on the mass distribution [3], and they may give the right observed spectral index [4]. Our purpose is to consider the case where multiple "curvaton" fields are responsible for the generation of the curvature perturbation after inflation, and study the evolution of ζ.

We assume that the curvatons (σ_i) decay into both radiation (γ) and non-relativistic matter (m) with constant decay rates $\Gamma_\gamma^{(i)}$ and $\Gamma_m^{(i)}$ respectively, which are fixed by underlying physics. With the energy transfer equation of σ_i given by

$$Q_i = -(\Gamma_\gamma^{(i)} + \Gamma_m^{(i)})\rho_i \equiv -\Gamma^{(i)}\rho_i\,, \tag{1}$$

we can find the evolution of the component α by solving the continuity equation of the corresponding component including energy transfer

$$\dot{\rho}_\alpha = -3H(\rho_\alpha + p_\alpha) + Q_\alpha\,, \tag{2}$$

CP957, *Particles, Strings, and Cosmology, 13th International Symposium*
edited by A. Rajantie, C. Contaldi, P. Dauncey, and H. Stoica

as well as that of ζ_α, the curvature perturbation of α, which can be written using the total curvature perturbation

$$\zeta = -H\frac{\delta\rho}{\dot{\rho}} = \sum_\alpha \frac{\dot{\rho}_\alpha}{\dot{\rho}}\zeta_\alpha\,, \tag{3}$$

and the evolution equation of ζ on large scales [5]

$$\dot{\zeta} = -\frac{H}{\rho+p}\delta p_{\rm nad}\,, \tag{4}$$

where $\delta p_{\rm nad} \equiv \delta p - \delta\rho\,\dot{p}/\dot{\rho}$ is the non-adiabatic pressure perturbation. Note that the difference between any two components gives an isocurvature perturbation [6]

$$\mathscr{S}_{\alpha\beta} = 3(\zeta_\alpha - \zeta_\beta)\,. \tag{5}$$

We can analytically study the evolution of the curvature perturbation under the assumption that there is no interaction between components until the curvaton fields decay and that the decay of each curvaton is instantaneous, i.e. "sudden decay approximation". For our purpose in this section, we write the total energy density according to the origin,

$$\begin{aligned}\rho &= \rho_{\gamma 0} + \rho_{m0} + \sum_i\left(\rho_{\gamma i} + \frac{\Gamma_\gamma^{(i)}}{\Gamma^{(i)}}\rho_i\right) + \sum_i\left(\rho_{mi} + \frac{\Gamma_m^{(i)}}{\Gamma^{(i)}}\rho_i\right)\\ &\equiv \rho_{\gamma 0} + \rho_{m0} + \sum_i\widetilde{\rho}_{\gamma i} + \sum_i\widetilde{\rho}_{mi}\,.\end{aligned} \tag{6}$$

First, we can see that for the composite density $\widetilde{\rho}_{mi}$, the energy transfer is zero ($\widetilde{Q}_{mi} = 0$) and the equation of state is unique. Thus $\widetilde{\zeta}_{mi} = \delta\widetilde{\rho}_{mi}/(3\widetilde{\rho}_{mi})$ is conserved on large scales [7]. Therefore, connecting the initial and final $\widetilde{\zeta}_{mi}$, we can simply find

$$\zeta_m^{(\rm out)} = \left(\rho_{m0}^{(\rm in)}\zeta_{m0}^{(\rm in)} + \sum_i\frac{\Gamma_m^{(i)}}{\Gamma^{(i)}}\rho_i^{(\rm in)}\zeta_i^{(\rm in)}\right)\left[\rho_{m0}^{(\rm in)} + \sum_j\frac{\Gamma_m^{(j)}}{\Gamma^{(j)}}\rho_j^{(\rm in)}\right]^{-1} \equiv \sum_{i=0}^n s_i\zeta_i^{(\rm in)}\,. \tag{7}$$

So we can see that the final matter curvature perturbation is completely determined by the decay rate and the initial energy density $\rho_i^{(\rm in)}$, or equivalently, initial density parameter $\Omega_i^{(\rm in)}$ of each curvaton field and that of pre-existing matter as shown above.

Now we turn to the curvature perturbation in the radiation component. Note that although the energy transfer of $\widetilde{\rho}_{\gamma i}$ is zero, its equation of state is not unique and thus the corresponding curvature perturbation $\widetilde{\zeta}_{\gamma i}$ evolves on large scales. Therefore, unlike $\widetilde{\rho}_{mi}$ we cannot simply connect the initial curvature perturbations in the curvaton fields to the final one in radiation, but rather we have to get through the moments of decay. Using the fact that due to the sudden decay approximation the energy density is the same at the moment of decay, we can find that after i-th curvaton σ_i decays, the curvature perturbation in the radiation component is constant until the decay of $(i+1)$-th curvaton, and is written as

$$\zeta_\gamma^{>i} \approx \left(1 - f_i^{(\rm dec)}\right)\zeta_\gamma^{>i-1} + f_i^{(\rm dec)}\zeta_i^{(\rm in)}\,, \tag{8}$$

where

$$f_i^{(\mathrm{dec})} = \frac{3\rho_{\gamma i}/\rho_{\gamma 0}}{4\sum_{k=0}^{i-1}\rho_{\gamma k}/\rho_{\gamma 0} + 3\rho_{\gamma i}/\rho_{\gamma 0}}. \tag{9}$$

Therefore, after all the n curvatons decay, we find the final curvature perturbations in radiation as

$$\begin{aligned}\zeta_\gamma^{(\mathrm{out})} &\approx \left(1 - f_n^{(\mathrm{dec})}\right)\zeta_\gamma^{>n-1} + f_n^{(\mathrm{dec})}\zeta_n^{(\mathrm{in})} \\ &= \left(1 - f_n^{(\mathrm{dec})}\right)\left(1 - f_{n-1}^{(\mathrm{dec})}\right)\zeta_\gamma^{>n-2} + \left(1 - f_n^{(\mathrm{dec})}\right) f_{n-1}^{(\mathrm{dec})}\zeta_{n-1}^{(\mathrm{in})} + f_n^{(\mathrm{dec})}\zeta_n^{(\mathrm{in})} \\ &= \cdots \equiv \sum_{i=0}^{n} r_i \zeta_i^{(\mathrm{in})}. \end{aligned} \tag{10}$$

Here, the ratio $\rho_{\gamma i}/\rho_{\gamma 0}$ which determines $f_i^{(\mathrm{dec})}$ and hence $\zeta_\gamma^{(\mathrm{out})}$ is given by

$$\frac{\rho_{\gamma i}}{\rho_{\gamma 0}} = \alpha_i \int_0^\infty y e^{-u_i} du_i, \tag{11}$$

where

$$\alpha_i \equiv \frac{\Gamma_\gamma^{(i)}}{\Gamma^{(i)}} \frac{\Omega_i^{(\mathrm{in})}}{\Omega_{\gamma 0}^{(\mathrm{in})}} x_H^{-1/2}, \tag{12}$$

with $x_H \equiv \Omega_{\gamma 0}^{(\mathrm{in})\,-1/2}\Gamma^{(1)}/H^{(\mathrm{in})}$, is some constant. The detail of the derivation is given in Ref. [8].

Now, we consider a simple case where there are two curvaton fields decaying into both radiation and matter. The results are given in Fig. 1 and Table 1, and the accuracy is good with $\mathcal{O}(1\%)$ error. Also we confirmed that our analytic formalism works very well with arbitrary number of the curvaton fields, e.g. for five fields the error is of $\mathcal{O}(0.1\%)$.

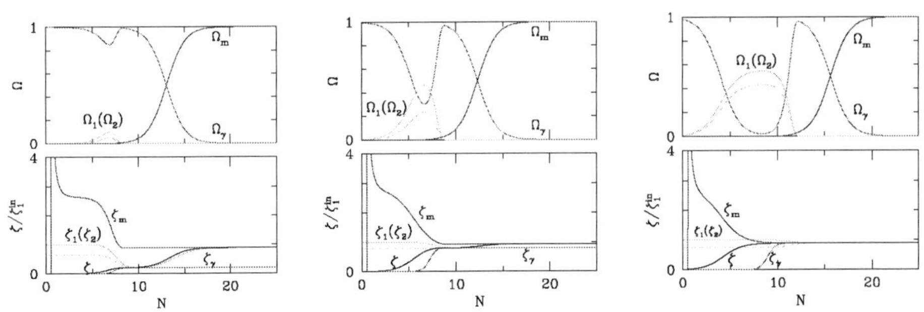

FIGURE 1. The evolution of density parameters (upper row) and curvature perturbations (lower row) for three cases of two curvaton decays: as shown, the energy densities of the two curvaton fields are sub-dominant (left panel), dominant (right panel) and comparable (middle panel) to the radiation energy density. The details are given in Table 1.

From this, one may tempted to conclude that there will be negligible isocurvature perturbation between matter and radiation if the curvatons dominate before they decay,

TABLE 1. The analytic and numerical results for the cases shown in Fig. 1. In the upper half of the table, we give the initial values used in the calculation and in the lower half we compare the analytic estimates with the numerical results.

		left panel	middle panel	right panel
$\zeta_2^{(\text{in})}/\zeta_1^{(\text{in})}$		0.65	0.8	0.75
$\Gamma_\gamma^{(i)}/H^{(\text{in})}, \Gamma_m^{(i)}/H^{(\text{in})}$		$10^{-6}, 10^{-8}$	$10^{-6}, 10^{-8}$	$10^{-8}, 10^{-10}$
$\Omega_i^{(\text{in})}$		$10^{-3.7}, 10^{-4.0}$	$10^{-2.5}, 10^{-2.8}$	$10^{-2.0}, 10^{-2.1}$
$\zeta_\gamma^{(\text{out})}/\zeta_1^{(\text{in})}$	analytic	0.200202	0.816117	0.888255
	numerical	0.195615	0.792049	0.887648
$\zeta_m^{(\text{out})}/\zeta_1^{(\text{in})}$	analytic	0.883149	0.933227	0.889327
	numerical	0.883149	0.933228	0.889328

because they are both generated due to the decay of the curvaton fields. This is not true when there are a number of curvaton fields: the final isocurvature perturbation is dependent on the background parameters such as curvaton densities and decay rates. If $\Gamma_m^{(i)}$ of the curvaton with large energy density is too small, although ζ_m receives contribution from the decay product of that field and this gives a rise of ζ_m, this rise is never enough to catch up ζ_γ to make $\mathscr{S}_{m\gamma}^{(\text{out})}$ vanishing if the branching ratio is very small as in this case. This is reminiscent of multi-field inflation: in multi-field inflation, there is no unique prediction on the isocurvature perturbation produced during inflation. The detail depends on the inflaton trajectory in the field space. Likewise, generally we can hardly make any definite prediction on the isocurvature perturbation without the detail. We can determine $\mathscr{S}_{m\gamma}^{(\text{out})}$ which may be detected in the CMB observations only when we have detailed information on the curvaton fields.

ACKNOWLEDGMENTS

It is my pleasure to thank Ki-Young Choi for a pleasant collaboration on this topic. Also I am grateful to Arttu Rajantie and the organisers of PASCOS-07 at Imperial College.

REFERENCES

1. See, e.g. J. O. Gong and E. D. Stewart, Phys. Lett. B **510**, 1 (2001) ; J. O. Gong and E. D. Stewart, Phys. Lett. B **538**, 213 (2002) ; J. Choe, J. O. Gong and E. D. Stewart, JCAP **0407**, 012 (2004).
2. D. H. Lyth and D. Wands, Phys. Lett. B **524**, 5 (2002) ; T. Moroi and T. Takahashi, Phys. Lett. B **522**, 215 (2001).
3. J. O. Gong, Phys. Rev. D **75**, 043502 (2007).
4. J. O. Gong, arXiv:0706.3599 [astro-ph].
5. D. Wands, K. A. Malik, D. H. Lyth and A. R. Liddle, Phys. Rev. D **62**, 043527 (2000).
6. K. A. Malik, D. Wands and C. Ungarelli, Phys. Rev. D **67** (2003) 063516.
7. S. Gupta, K. A. Malik and D. Wands, Phys. Rev. D **69** (2004) 063513.
8. K. Y. Choi and J. O. Gong, JCAP **0706**, 007 (2007).

Non-perturbative Effect on Thermal Relic Abundance of Dark Matter

J. Hisano*, S. Matsumoto†, M. Nagai*, O. Saito** and M. Senami*

*ICRR, University of Tokyo, Kashiwa 277-8582, Japan
†Institute for International Advanced Interdisciplinary Research, Tohoku University, Sendai 980-8578, Japan
**Theory Group, KEK, Tsukuba 305-0801, Japan

Abstract. We point out that thermal relic abundance of dark matter is strongly altered by a non-perturbative effect called the Sommerfeld enhancement, when constituent particles of the dark matter are $SU(2)_L$ non-singlet and much heavier than the weak gauge bosons. Typical candidates are the heavy wino and higgsino. We investigate the non-perturbative effect on the relic abundance of dark matter for the wino as an example. We show that its thermal abundance is reduced by 50% compared to the perturbative result. The wino mass consistent with the observed dark matter abundance turns out to be 2.7 TeV $< m <$ 3.0 TeV. This article is based on Ref. [1].

Keywords: Dark Matter, Relic abundance, Wino, Sommerfeld effect
PACS: 95.35.+d, 98.80.-k

INTRODUCTION

Some astrophysical observations have confirmed that non-baryonic cold dark matter (DM) amounts to 20% of the energy of our universe. The existence of DM forces us to consider physics beyond the standard model (SM), because the SM has no candidate for DM. Many models have been proposed for providing DM candidates. Among them, models with weakly interacting massive particles (WIMPs), have an advantage, since the WIMPs can explain the observed abundance naturally in the thermal relic scenario.

In the scenario, the annihilation cross sections of DM play a central role in evaluation of the abundance. Perturbative calculation of the cross sections is sufficient in usual cases. However, a non-perturbative effect on the cross sections has to be included, when DM particles are $SU(2)_L$ non-singlet and much heavier than the weak gauge bosons. Even the weak interaction is a long-distance force for non-relativistic two-bodies states of such heavy particles. The wave functions for these states are modified from the plane waves by the interaction, and the annihilation cross sections are affected. The DM particles are non-relativistic at the freeze-out temperature. The attractive channels due to the weak force enhance the annihilation cross sections, compared with the perturbative ones, which is called the Sommerfeld enhancement.

In this article, we point out that the non-perturbative effect strongly alters the DM relic abundance from that in perturbative calculation. We investigate the relic abundance of the wino including this effect and show that the abundance is reduced by 50% due to the effect. We also mention non-perturbative effects on other DM candidates.

CP957, *Particles, Strings, and Cosmology, 13th International Symposium*
edited by A. Rajantie, C. Contaldi, P. Dauncey, and H. Stoica

SOMMERFELD ENHANCEMENT IN WINO ANNIHILATION

Winos are the $SU(2)_L$ triplet gauginos in the SUSY SM. The mixing with higgsinos is small for heavy winos, since it is suppressed by the electroweak scale. So, in this article the pure wino-like neutralino is considered. Thus, the neutral wino ($\tilde{\chi}^0$) is highly degenerate with charged wino ($\tilde{\chi}^-$) in mass when their masses are heavy enough. For the wino mass (m) of the order of 1 TeV, the mass difference between $\tilde{\chi}^0$ and $\tilde{\chi}^-$ is dominated by a radiative correction [2] and calculated as $\delta m \simeq 0.17$ GeV.

There are four processes related to the calculation of the wino relic abundance under CP conservation. The non-perturbative effect can significantly alter cross sections [2, 3, 4], so the effect of this enhancement on the relic abundance of DM is investigated in this article.

Two-bodies effective Lagrangian, describing relative motion between two particles in the two-bodies system, is useful to evaluate the non-perturbative effect in the annihilation [1]. In Ref. [1], we show the Lagrangian in detail. For derivation of the effective Lagrangian and the evaluation of the annihilation cross section using it, also see Ref. [2]. All processes have attractive forces except for that of $\tilde{\chi}^-\tilde{\chi}^-$. The overlap between the wave functions of the incident particles are increased compared to the case without attractive force (perturbative calculation), and it leads to enhancement of the annihilation cross sections. Once the two-bodies effective Lagrangian is obtained, annihilation cross sections including the non-perturbative effect can be calculated as the Sommerfeld enhancement factor.

SOMMERFELD EFFECT ON DARK MATTER ABUNDANCE

Now we evaluate the thermal relic abundance of the wino DM, including the non-perturbative effect. In the evaluation we have to include coannihilation processes in addition to $\tilde{\chi}^0\tilde{\chi}^0$ annihilation. Under reasonable assumptions, the relic density obeys the following Boltzmann equation,

$$\frac{dY}{dx} = -\frac{\langle\sigma_{\text{eff}}v\rangle}{Hx}\left(1-\frac{x}{3g_{*s}}\frac{dg_{*s}}{dx}\right)s\left(Y^2-Y_{\text{eq}}^2\right) , \tag{1}$$

where $x = m/T$, $Y = n/s$ and g_{*s} is the relativistic degrees of freedom of the thermal bath. Here, n is the sum of the number densities of $\tilde{\chi}^0$, $\tilde{\chi}^-$, and $\tilde{\chi}^+$. The equilibrium abundance Y_{eq}, the entropy density s and the Hubble parameter H are given in Ref. [1].

The most important quantity to determine the abundance is the thermally averaged effective annihilation cross section, $\langle\sigma_{\text{eff}}v\rangle$, which is derived by averaging the degrees of freedom of $\tilde{\chi}^0$ and $\tilde{\chi}^-$. In Fig. 1, the enhancement of $\langle\sigma_{\text{eff}}v\rangle$ to that in the perturbative calculation is shown as a function of T. For comparison, the cross section in a perturbative calculation is also shown as a dotted line. The decoupling of $\tilde{\chi}^\pm$ occurs at $x \sim 10^5$. The enhancement by the non-perturbative effect is more significant for lower temperature. Since $\tilde{\chi}^0$ and $\tilde{\chi}^-$ are more non-relativistic for lower temperature, the long-range force strongly modifies their wave functions and alters the cross section significantly.

After solving the Boltzmann equation numerically, we obtain the DM mass fraction in the present universe. The result is shown in Fig. 2. In the left figure, the ratio of the yield

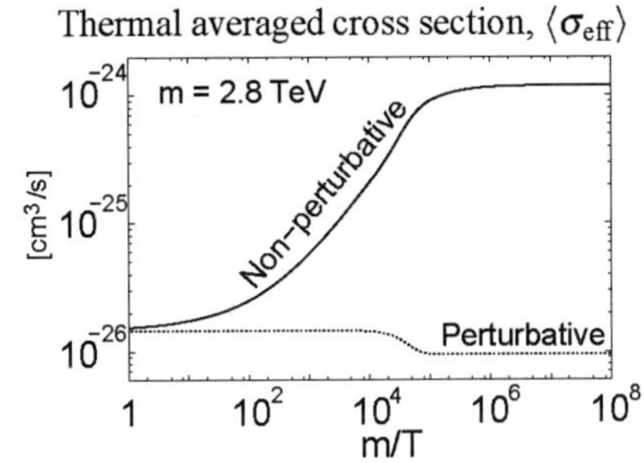

FIGURE 1. Temperature dependence with $m = 2.8$ TeV. The perturbative result is also shown as a dotted line for comparison.

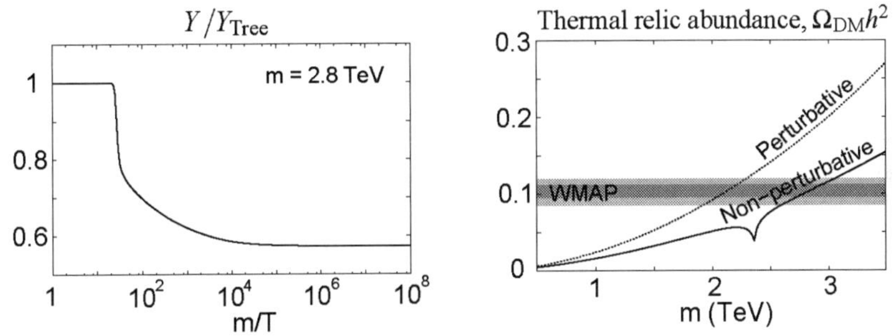

FIGURE 2. Ratio of yield with the non-perturbative effect to that in the perturbative calculation (left figure). Thermal relic abundance of $\tilde{\chi}^0$ as a function of the wino mass (right figure). Allowed regions at 1(2) σ levels are also shown as the dark (light) shaded area.

with the enhancement to perturbative result is shown as a function of T. The enhancement of the cross section at the departure from the equilibrium decreases the abundance by 20-30%, and it leads to quick deviation of the yield from the perturbative one after decoupling. Since the annihilation cross section increases for lower temperature by the Sommerfeld enhancement, the sudden freeze-out phenomenon does not occur and the yield continues to be reduced by the non-perturbative effect even for $x > 100$ compared to the perturbative one. The resultant abundance of $\tilde{\chi}^0$ is reduced by 50% compared to the perturbative one. In the right figure, the abundance of $\tilde{\chi}^0$ in the present universe is shown in terms of Ωh^2. The allowed region of m by the WMAP [5] is shifted by 600 GeV and turns out to be 2.7 TeV $< m <$ 3.0 TeV.

SUMMARY AND DISCUSSION

In this article, we pointed out that the thermal relic abundance of DM, which is $SU(2)_L$ non-singlet and much heavier than the weak gauge bosons, can be strongly reduced by the non-perturbative effect. We investigated the non-perturbative effect on the relic abundance of wino as an example. Compared with the perturbative result, this effect reduces the abundance by about 50%. As a result, the $\tilde{\chi}^0$ abundance is consistent with observed abundances when 2.7 TeV $< m <$ 3.0 TeV.

The non-perturbative effect can change relic abundances of other DM candidates with $SU(2)_L$ charge and heavy mass, such as higgsino. The non-perturbative effect on the abundance of higgsino is expected to be roughly 10%, since winos are $SU(2)_L$ triplet, while higgsinos are doublet. Therefore, non-perturbative effect may change the abundance by O(10%) for other $SU(2)_L$ doublet candidates.

The Sommerfeld enhancement occurs reasonably for particles with electric charge. In fact the non-perturbative effect through photon exchanges for charged particle annihilation can change the abundance by about 10%. Therefore, one may think that the relic abundance of DM will be changed e.g. in the stau coannihilation region. However, the non-perturbative effect for stau does not change the DM abundance. In stau annihilation, $\sigma(\tilde{\tau}^+\tilde{\tau}^-) \simeq \sigma(\tilde{\tau}^+\tilde{\tau}^+)$ and the Sommerfeld enhancement is positive (negative) for the former (latter) process. Hence, the enhancement is almost canceled and the change of the abundance is 1% at most.

Kaluza-Klein (KK) right-handed leptons ($E^{(1)}$) in universal extra dimension models [6] are highly degenerate with the lightest KK particle in mass and $\sigma(E^{(1)}\bar{E}^{(1)}) > \sigma(E^{(1)}E^{(1)})$ [7]. Hence, the enhancement could be expected to change the abundance. The change of the abundance is within 4% since $\sigma(E^{(1)}\bar{E}^{(1)})$ contributes to the effective annihilation cross section by 40% at most.

Finally, we comment on the non-perturbative effect of colored particles. Although the enhancement for colored particles are very effective [8], it may be very complicated due to the QCD phase transition, which is discussed in Ref. [9]. Furthermore, colored particles can not be DM candidates and are not expected to be degenerate with a DM particle in mass due to large radiative corrections by the strong interaction. Hence, this subject is beyond the scope of this article.

ACKNOWLEDGMENTS

This work is supported in part by the Grant-in-Aid for Science Research, Ministry of Education, Science and Culture, Japan (No.1803422 and 15540255 for JH, 16081211 for SM and 18840011 for MS). Also, the work of MN is supported in part by JSPS.

REFERENCES

1. J. Hisano, S. Matsumoto, M. Nagai, O. Saito and M. Senami, Phys. Lett. B **646** (2007) 34.
2. J. Hisano, S. Matsumoto, M. M. Nojiri and O. Saito, Phys. Rev. D **71**, 063528 (2005).
3. J. Hisano, S. Matsumoto and M. M. Nojiri, Phys. Rev. Lett. **92**, 031303 (2004).
4. J. Hisano, S. Matsumoto, O. Saito and M. Senami, Phys. Rev. D **73**, 055004 (2006).
5. D. N. Spergel *et al.*, arXiv:astro-ph/0603449.
6. T. Appelquist, H. C. Cheng and B. A. Dobrescu, Phys. Rev. D **64**, 035002 (2001).
7. For an example, see, M. Kakizaki, S. Matsumoto and M. Senami, Phys. Rev. D **74**, 023504 (2006). M. Kakizaki, S. Matsumoto, Y. Sato and M. Senami, Nucl. Phys. B **735**, 84 (2006).
8. H. Baer, K. m. Cheung and J. F. Gunion, Phys. Rev. D **59**, 075002 (1999).
9. A. Arvanitaki, C. Davis, P. W. Graham, A. Pierce and J. G. Wacker, Phys. Rev. D **72**, 075011 (2005).

Cosmological constant from gauge fields on extra dimensions

Choon-Lin Ho[1], Hing-Tong Cho[1], Kin-Wang Ng[2]

[1]*Department of Physics, Tamkang University, Tamsui 25137, Taiwan, R.O.C.*
[2]*Institute of Physics, Academia Sinica, Nankang, Taipei, Taiwan 115, R.O.C.*

Abstract. We present a new model of dark energy which could explain the observed accelerated expansion of our Universe. We show that a five-dimensional Einstein-Yang-Mills theory defined in a flat Friedmann-Robertson-Walker universe compactified on a circle possesses degenerate vacua in four dimensions. The present Universe could be trapped in one of these degenerate vacua. With the natural requirement that the size of the extra dimension could be of the GUT scale or smaller, the energy density difference between the degenerate vacua and the true ground state can provide us with just the right amount of dark energy to account for the observed expansion rate of our Universe.

Keywords: Cosmological constant, dark energy, Hosotani mechanism
PACS: 98.80.Cq, 98.80.Es, 11.25.Mj, 11.15.Ex

Introduction.– It is now generally agreed that the biggest unsolved problem in astronomy and cosmology is the newly discovered fact that our Universe is undergoing a stage of accelerated expansion. Recent astrophysical and cosmological observations such as structure formation, type Ia supernovae, gravitational lensing, and cosmic microwave background anisotropies have concordantly predicted a spatially flat universe containing a mixture of matter and a dominant smooth component, dubbed the "dark energy", which provides an anti-gravity force to accelerate the cosmic expansion [1]. The simplest candidate for this invisible component carrying a sufficiently large negative pressure for the anti-gravity is a true cosmological constant. Here we will address the dark energy problem by considering non-Abelian gauge theories on compact extra dimensions [2]. We shall show how a five-dimensional (5d) gauge theory based on the Hosotani symmetry breaking mechanism [3] could naturally give rise to a small but finite cosmological constant that plays the role of dark energy.

Hosotani mechanism in 5d cosmology.– The Hosotani mechanism is a non-Higgs-type symmetry-breaking mechanism which has been widely discussed in the literature. Its main idea is that in a multiply connected spacetime manifold, the vanishing of the field strength $F_{MN} = 0$ of gauge fields A_M in a vacuum does not necessarily imply the vanishing of the gauge fields, and $A_M \neq 0$ will imply gauge symmetry breaking in general. This mechanism has been extensively employed in superstring phenomenology and in Kaluza-Klein cosmology.

In Ref. [4] we have applied the Hosotani mechanism to Kaluza-Klein cosmology in connection with the problem of vacuum stability [4]. There we considered a 5d Einstein-Yang-Mills theory, with massless fermions, defined in a flat Friedmann-Robertson-Walker universe compactified on a circle. The spacetime metric is given by

$$ds^2 = g_{MN}dx^M dx^N \qquad (M = 0,1,2,3,5) = dt^2 - a^2(t)d\vec{x}^2 - b^2(t)dx_5^2, \tag{1}$$

CP957, *Particles, Strings, and Cosmology, 13th International Symposium*
edited by A. Rajantie, C. Contaldi, P. Dauncey, and H. Stoica

and the action consists of gravity, $SU(2)$ gauge fields, and a fermionic sector L_f which contains N_f massless gauged fermions and n_f massless free fermions:

$$S = -\int d^5x\sqrt{|g_{MN}|}\left(\frac{R}{16\pi G}+\frac{1}{2}\mathrm{Tr}F_{MN}F^{MN}+L_f\right). \tag{2}$$

The Casimir energy of the system was computed by evaluating the one-loop effective potentials of the gauge fields and the fermions in the backgrounds defined by the metric (1) and the classical gauge fields of the form $A_\mu = 0, \quad A_5 = \phi(t)\sigma^3, \ \mu = 0,1,2,3$, where A_5 is the fifth component along the circle and σ is the Pauli matrix. This amounts to a total effective potential of the system denoted by $V(bR_0, \phi)$, where R_0 is the final radius of the circle. It was shown that the Einstein's equations for $a(t)$, $b(t)$ and $\phi(t)$, derived from the action (2), admit static vacuum solutions with stable compactification. These solutions with $\dot{a}_0 = \dot{b}_0 = \dot{\phi}_0 = 0$ are given by the global minima of $V(bR_0, \phi)$ determined by

$$2g_5 b_0 R_0 \phi_0 = \frac{r}{2} \pmod{r}, \tag{3}$$

where g_5 is the 5d $SU(2)$ gauge coupling and $r=1$ and $r=2$ correspond respectively to periodic fermions in the adjoint and fundamental representations. In the case of adjoint fermions, the vacuum states correspond to $U(1)$ symmetry, hence $SU(2)$ gauge symmetry is dynamically broken.

Dark energy.– Now we turn to the dark energy problem. We note that the observed value for the vacuum-like energy density is $\rho_\Lambda \simeq 0.7\rho_c \simeq 1.6h^2 \times 10^{-120}M_P^4$, where the Planck mass $M_P = (8\pi G)^{-1/2} = 2.44 \times 10^{18}$GeV, and $h \simeq 0.7$ is the present Hubble parameter defined by $H_0 = 8.76h \times 10^{-61}M_P$ [1]. We assume that the cosmological constant absolutely vanishes in a true ground state with lowest possible energy density. Then, we will show that the vacuum states in the model considered above are indeed metastable due to quantum tunneling effects and will eventually settle down to this true ground state. If the present Universe is still in one of these quasiground states, then the energy density difference above the true ground state can provide us with a small but finite cosmological constant. This idea has also been put forth by considering the topological vacua in a 4d $SU(2)$ Yang-Mills-Higgs theory which is spontaneously broken to $U(1)$ via the Higgs mechanism with a Higgs potential [5]. Unfortunately, the existence of massless or light gauged fermions would suppress the tunneling [6] and thus spoil the idea.

Without loss of generality, let us consider a universe corresponding to the vacuum state in the case of periodic fundamental fermions with $2g_5R_0\phi_0 = 1$, where we have set $a_0 = b_0 = 1$ and $r=2$ in Eq. (3). Presumably, the Universe has undergone the compactification of the circle with initial conditions at some early time t_i (for instance, given by $a(t_i)$, $b(t_i)$, and $\phi(t_i) \simeq 0$), and rolls down the effective potential V to this vacuum state. At energies below $1/R_0$, the 4d effective action for the zero Fourier modes of gauge fields in Eq. (2) is given by

$$S_{\mathrm{eff}}^{\mathrm{gauge}} = -\int d^4x\left(\frac{1}{2}\mathrm{Tr}\tilde{F}_{\mu\nu}\tilde{F}^{\mu\nu}+\mathrm{Tr}\tilde{F}_{\mu5}\tilde{F}^{\mu5}\right). \tag{4}$$

Note that we have rescaled $A_M = \tilde{A}_M/\sqrt{2\pi R_0}$ and $g_5 = g_4\sqrt{2\pi R_0}$, where g_4 is the dimensionless 4d $SU(2)$ gauge coupling. Since $\tilde{A}_\mu$ is independent of x_5 in the effective action, so $\partial_5\tilde{A}_\mu = 0$ and $\tilde{F}_{\mu 5}$ reduces to the covariant derivative of $\tilde{A}_5$. Hence, by rewriting $\tilde{A}_5 = \Phi$, we have

$$S_{\rm eff}^{\rm gauge} = \int d^4x \left(-\frac{1}{2}{\rm Tr}\tilde{F}_{\mu\nu}\tilde{F}^{\mu\nu} + {\rm Tr}(D_\mu\Phi)(D^\mu\Phi) \right). \tag{5}$$

This resembles the $SU(2)$ Yang-Mills-Higgs system without a Higgs potential. Although the gauge symmetry is unbroken for the vacuum states given by Eq. (3), the fifth component obtains a non-zero vacuum expectation value $\langle\Phi\rangle = \tilde{\phi}_0\sigma^3$, where $\tilde{\phi}_0 = 1/(2g_4R_0)$, and the gauged fermions acquire huge mass equal to $1/(2R_0)$ [4].

It is well known [6] that a non-Abelian gauge, bears degenerate perturbative vacua classified in terms of the winding number n and denoted by $|n\rangle$. Let us consider the 4d $SU(2)$ Yang-Mills theory with gauge coupling g. While the degenerate vacua are each separated by an energy barrier, the change of the winding number can take place through quantum tunneling from one vacuum to another. As such, the true ground state, so-called the θ vacuum, can be constructed as $|\theta\rangle = \sum_{n=-\infty}^{\infty} e^{in\theta}|n\rangle$, where θ is a real parameter. Using the dilute instanton gas approximation [6], it was found that the energy density difference between each θ vacuum and the perturbative vacuum is given by $\langle\theta|\rho_\Lambda|\theta\rangle - \langle n|\rho_\Lambda|n\rangle = -2Ke^{-S_0}\cos\theta$. Here $S_0 = 8\pi^2/g^2$ is the Euclidean action for an instanton solution which corresponds to the quantum tunneling from $|n\rangle$ to $|n\pm 1\rangle$, and K is a positive determinantal prefactor. This equation shows that the θ vacuum with the lowest energy is given by $\theta = 0$. Let us assume that the $\theta = 0$ vacuum is the true ground state where the vacuum energy vanishes. Then, after normalizing $\langle\theta = 0|\rho_\Lambda|\theta = 0\rangle = 0$, we find that the vacuum energy density in each perturbative vacuum is $\langle n|\rho_\Lambda|n\rangle = 2Ke^{-S_0}$. This vacuum energy will manifest as a cosmological constant provided that we still live in one of these perturbative vacua. To guarantee this condition, the tunneling probability from any one of these perturbative vacua to the true ground state in the current horizon volume in the cosmic age must satisfy $\Gamma H_0^{-4} \le 1$, where Γ is the tunneling rate per unit volume per unit time given by $\Gamma \simeq \rho_\Lambda e^{-S_0}$.

Unfortunately, for a pure $SU(2)$ gauge theory which is scale-invariant, the size of the instanton ρ is arbitrary, and so the prefactor K which involves an integral over instanton sizes diverges as ρ goes to infinity. However, if the $SU(2)$ gauge field is coupled to a Higgs scalar with isospin q and vacuum expectation value v, the quantum tunneling will proceed via a constrained instanton whose size is cut off at a scale v^{-1}. The Higgs contribution to the Euclidean action of the constrained instanton is approximately given by $S_H = 4\pi^2 q\rho^2 v^2$, rendering K a finite quantity. Including N_f gauged fermions with mass $m_f = gv$,, we find that

$$\rho_\Lambda \simeq c^2\left(\frac{v}{\alpha}\right)^4 \alpha^{\frac{N_f}{2}} e^{-2\pi/\alpha}, \quad \Gamma \simeq c^2\left(\frac{v}{\alpha}\right)^4 \alpha^{\frac{N_f}{2}} e^{-4\pi/\alpha}, \tag{6}$$

where $\alpha = g^2/(4\pi)$ is the gauge coupling strength at the energy scale v, and

$$c^2 = 4\pi^2 e^{c_2}(4\pi)^{\frac{N_f}{2}} \frac{\Gamma[(N_f+c_1-4)/2]}{(4\pi^2 q)^{(N_f+c_1-4)/2}}, \tag{7}$$

with $c_1 = 20/3 - 2N_f/3$ and $c_2 = 5.96 - 0.36N_f$. The requirements that $\rho_\Lambda = 1.6h^2 \times 10^{-120}M_P^4$ and $\Gamma H_0^{-4} \leq 1$ give

$$\frac{\pi}{2\alpha} + \left(1 - \frac{N_f}{8}\right)\ln\alpha = \ln\left(\frac{v}{M_P}\right) + 30\ln 10 - \frac{1}{2}\ln\left(\frac{1.27h}{c}\right), \tag{8}$$

$$\frac{v}{M_P} \geq 1.45\alpha^{1-\frac{N_f}{8}}/\sqrt{c}. \tag{9}$$

For the minimal value of v in Eq. (9) and $h = 0.7$, we obtain $\alpha \simeq 1/44.25$. Let us assume that $q = 1$. Then, we have $v \simeq 3.8 \times 10^{16}$GeV for $N_f = 1$ and $v \simeq 8.3 \times 10^{16}$GeV for $N_f = 3$. If we take $v = M_P$ in Eq. (8), we will obtain $\alpha = 1/46.9$ for $N_f = 1$ and $\alpha = 1/46.4$ for $N_f = 3$

To apply the above results to our model (5), we make the replacements $g = g_4$ and $v = \tilde{\phi}_0 = 1/(2g_4R_0)$. Thus, we find that $R_0^{-1} \geq 4.1 \times 10^{16}$GeV, which indicates that the maximal size of the compact extra dimension is of the order of the GUT scale. This result is obtained without fine-tuning. The smallness of the cosmological constant is related to the tunneling rate between the various vacua of the theory which is also very tiny.

Summary.– We have constructed a model of dark energy based on a 5d Einstein-Yang-Mills theory. The presence of dark energy comes out rather naturally in our model just from the assumption that there exist extra dimensions. Dynamical fermion mass generation by the Hosotani mechanism and the constrained instantons related to the vacuum structure of the gauge field are both immediate consequences. As long as the size of the extra dimension is between the GUT and the Planck scales, the resulting cosmological constant will then be just of the right amount to account for the dark energy content in our Universe. No other ingredients are needed to achieve this goal.

Work supported in part by the National Science Council, Taiwan, ROC under the Grants NSC 94-2112-M-032-009 (H.T.C.), NSC 94-2112-M-032-007 (C.L.H.), and NSC 94-2112-M-001-024 (K.W.N.).

REFERENCES

1. See, e.g., J. Gunn, C. Frenk, A. Riess, A. Refregier, and L. Page, in *Proc. of the 22nd Texas Symposium on Relativistic Astrophysics*, published on the SLAC Electronic Conference Proceedings Archive (http://www.slac.stanford.edu/econf/).
2. H.T. Cho, C.L. Ho and K.W. Ng, Phys. Lett. B **643**, 71 (2006).
3. Y. Hosotani, Phys. Lett. B **126**, 309 (1983); Ann. Phys. **190**, 233 (1989).
4. C.-L. Ho and K.-W. Ng, Phys. Rev. D **43**, R3107 (1991).
5. J. Yokoyama, Phys. Rev. Lett. **88**, 151302 (2002).
6. C. G. Callan, R. F. Dashen, and D. J. Gross, Phys. Lett. B **63**, 334 (1976); R. Jackiw and C. Rebbi, Phys. Rev. Lett. **37**, 172 (1976); G. 't Hooft, Phys. Rev. D **14**, 3432 (1976); **18**, 2199 (1978)(E); and Erratum II in C. Rebbi and G. Soliani (Eds.), *Solitons and Particles* (World Scientific, Singapore, 1984), pg. 765; I. Affleck, Nucl. Phys. B **191**, 429 (1981).

Quasi-exactly solvable quasinormal modes

Choon-Lin Ho and Hing-Tong Cho

Department of Physics, Tamkang University, Tamsui 25137, Taiwan, R.O.C.

Abstract. We consider quasinormal modes with complex energies from the point of view of the theory of quasi-exactly solvable (QES) models. We demonstrate that it is possible to find new potentials which admit exactly solvable or QES quasinormal modes by suitable complexification of parameters defining the QES potentials. Particularly, we obtain one QES and four exactly solvable potentials out of the five one-dimensional QES systems based on the $sl(2)$ algebra.

Keywords: Quasi-exact solvability, quasinormal modes
PACS: 03.65.-w, 03.65.Nk, 03.65.Fd, 04.70.-s

Introduction.– Quasinormal modes (QNM) arise as perturbations of stellar or black hole spacetimes [1]. They are solutions of the perturbation equations that are outgoing to spatial infinity and the event horizon. Generally, these conditions lead to a set of discrete complex eigenfrequencies, with the real part representing the actual frequency of oscillation and the imaginary part representing the damping. QNM carry information of black holes and neutron stars, and thus are of importance to gravitational-wave astronomy. In fact, these oscillations, produced mainly during the formation phase of the compact stellar objects, can be strong enough to be detected by several large gravitational wave detectors under construction. Recently, QNM of particles with different spins in black hole spacetimes have also received much attention.

Owing to the intrinsic complexity in solving the perturbation equations in general relativity with the appropriate boundary conditions, one has to resort to various approximation methods, eg., the WKB method, the phase-integral method etc., in obtaining QNM solutions. It is therefore helpful that one can get some insights from exact solutions in simple models, such as the inverted harmonic oscillator and the Pöschl-Teller potential. Unfortunately, the number of exactly solvable models is rather limited.

Recently, in non-relativistic quantum mechanics a new class of potentials which are intermediate to exactly solvable ones and non-solvable ones has been found. These are called quasi-exactly solvable (QES) problems for which it is possible to determine analytically a part of the spectrum but not the whole spectrum [2, 3, 4, 5]. The discovery of this class of spectral problems has greatly enlarged the number of physical systems which we can study analytically. In the last few year, QES theory has also been extended to the Pauli and Dirac equations.

In [6] we have considered solutions of QNM based on the Lie-algebraic approach of QES theory. There we demonstrate that, by suitable complexification of some parameters of the generators of the $sl(2)$ algebra while keeping the Hamiltonian Hermitian, we can indeed obtain potentials admitting exact or quasi-exact QNMs. Such consideration has not been attempted before in studies of QES theory. Our work represents a direct opposite of the work in [7], where QES *real energies* were obtained from a *non-Hermitian*

CP957, *Particles, Strings, and Cosmology, 13th International Symposium*
edited by A. Rajantie, C. Contaldi, P. Dauncey, and H. Stoica

PT-symmetric Hamiltonian.

QES Theory.– Let us briefly review the essence of the Lie-algebraic approach [2, 3, 5] to QES models. Consider a Schrödinger equation $H\psi = E\psi$ with Hamiltonian $H = -d_x^2 + V(x)$ ($d_x \equiv d/dx$) and wave function $\psi(x)$. Here x belongs either to the interval $(-\infty,\infty)$ or $[0,\infty)$. Now suppose we make an "imaginary gauge transformation" on the function ψ: $\psi(x) = \chi(x)e^{-g(x)}$, where $g(x)$ is called the gauge function. For physical systems which we are interested in, the phase factor $\exp(-g(x))$ is responsible for the asymptotic behaviors of the wave function so as to ensure normalizability. The function $\chi(x)$ satisfies a Schrödinger equation with a gauge transformed Hamiltonian $H_g = e^g H e^{-g}$. Suppose H_g can be written as a quadratic combination of the generators J^a of some Lie algebra with a finite dimensional representation. Within this finite dimensional Hilbert space the Hamiltonian H_g can be diagonalized, and therefore a finite number of eigenstates are solvable. Then the system described by H is QES. For one-dimensional QES systems the most general Lie algebra is $sl(2)$, and H_g can be expressed as

$$H_g = \sum C_{ab} J^a J^b + \sum C_a J^a + \text{real constant}\ , \tag{1}$$

where C_{ab}, C_a are taken to be *real constants* in [3, 5]. The generators J^a of the $sl(2)$ Lie algebra take the differential forms: $J^+ = z^2 d_z - nz$, $J^0 = z d_z - n/2$, $J^- = d_z$ ($n = 0, 1, 2, \ldots$). The variables x and z are related by some function to be described later. n is the degree of the eigenfunctions χ, which are polynomials in a $(n+1)$-dimensional Hilbert space with the basis $\langle 1, z, z^2, \ldots, z^n \rangle$.

Substituting the differential forms of J^a into Eq. (1), one sees that every QES operator H_g can be written in the canonical form : $H_g = -P_4(z)d_z^2 + P_3(z)d_z + P_2(z)$, where $P_k(z)$ are k-th degree polynomial in z with real coefficients related to the constants C_{ab} and C_a. The relation between H_g and the standard Schrödinger operator H fixes the required form of the gauge function g and the transformation between the variable x and z. Particularly, $x = \int^z dy/\sqrt{P_4(y)}$. Analysis on the inequivalent forms of real quartic polynomials P_4 thus give a classification of all $sl(2)$-based QES Hamiltonians [3, 5]. If one imposes the requirement of non-periodic potentials, then there are only five inequivalent classes, which are called case 1 to 5 in [5].

Our main observation is this. If some of the coefficients in $P_k(z)$ are allowed to be complex while keeping $V(x)$ real, then all the five cases classified in [5] can indeed support QES/exact quasinormal modes. We shall illustrate this using one of the cases below.

QES QNM.– We consider Case 3 in [5], which corresponds to Class I potential in Turbiner's scheme [3]. There are two subclasses in this case, namely, Case (3a) and (3b). We shall present the analysis of QNM potential for case (3a). The other case turns out to give the same potential with a suitable choice of the parameters. The potential in case (3a) has the form (in this paper we adopt the notation of [5]):

$$V(x) = Ae^{2\sqrt{\nu}x} + Be^{\sqrt{\nu}x} + Ce^{-\sqrt{\nu}x} + De^{-2\sqrt{\nu}x}\ , \tag{2}$$

where $x \in (-\infty,\infty)$ and ν is a positive scale factor. Note that $V(x)$ is defined up to a real constant, which we omit for simplicity, as it merely shifts the real part of the energy. This remark also applies to the other cases. $V(x)$ in Eq. (2) reduces to the exactly solvable Morse potentials when $A=B=0$, or $C=D=0$. This potential is QES when the coefficients are related by

$$\begin{aligned} A &= \frac{\hat{b}^2}{4\nu}, \quad B=\frac{\hat{c}+(n+1)\nu}{2\nu}\,\hat{b}, \\ C &= \frac{\hat{c}-(n+1)\nu}{2\nu}\,\hat{d}, \quad D=\frac{\hat{d}^2}{4\nu}, \\ & n=0,1,2\ldots \end{aligned} \tag{3}$$

Here $\hat{b}$, $\hat{c}$, $\hat{d}$ are arbitrary real constants. For each integer $n\geq 0$, there are $n+1$ exactly solvable eigenfunctions in the $(n+1)$-dimensional QES subspace:

$$\psi_n(x)=\exp\left[\frac{\hat{b}}{2\nu}e^{\sqrt{\nu}x}+\frac{\hat{c}-n\nu}{2\sqrt{\nu}}x-\frac{\hat{d}}{2\nu}e^{-\sqrt{\nu}x}\right]\chi_n(e^{\sqrt{\nu}x})\,. \tag{4}$$

Here $\chi_n(z)$ is a polynomial of degree n in $z=\exp(\sqrt{\nu}x)$. To guarantee normalizability of the eigenfunctions, the real constants $\hat{b}$, $\hat{c}$, $\hat{d}$, ν and n must satisfy certain relations [5].

We want to see if we can get QNM solutions if we allow some parameters to be complex, while still keeping the potential $V(x)$ real. This latter requirement restricts the possible values of the parameters, and hence the forms of QES potential admitting quasinormal modes. For the case at hand, we find that one possible choice of values of $\hat{b}$, $\hat{c}$ and $\hat{d}$ is:

$$\hat{b}=ib,\ \hat{c}=-(n+1)\nu,\ \hat{d}=d,\ \ b,d:\text{real constants}\,. \tag{5}$$

The potential Eq. (2) becomes

$$V_n(x)=-\frac{b^2}{4\nu}e^{2\sqrt{\nu}x}-(n+1)\,de^{-\sqrt{\nu}x}+\frac{d^2}{4\nu}e^{-2\sqrt{\nu}x}\,, \tag{6}$$

and the wave function Eq. (4) becomes

$$\begin{aligned} \psi_n(x) &= \exp\left[\frac{ib}{2\nu}e^{\sqrt{\nu}x}-\left(n+\frac{1}{2}\right)\sqrt{\nu}x-\frac{d}{2\nu}e^{-\sqrt{\nu}x}\right] \\ &\quad\times\chi_n(e^{\sqrt{\nu}x})\,. \end{aligned} \tag{7}$$

$V(x)$ approaches $\mp\infty$ as $x\to\pm\infty$ respectively: it is unbounded from below on the right. For small positive d and sufficiently large n, $V(x)$ can have a local minimum and a local maximum. In this case, the well gets shallower as d increases at fixed value of n, or as n decreases at fixed d. We emphasize here that for different value of n, each $V(x)$ represents a different QES potential admitting $n+1$ QES solutions. Since $V(x)\to\infty$ as $x\to-\infty$, the wave function must vanish in this limit. This means $d>0$ from Eq. (7). For the outgoing boundary condition, we must take $b>0$. Before we go on, we note here that there is another possible choice of the parameters, namely,

$$\hat{b}=-b,\ \hat{c}=(n+1)\nu,\ \hat{d}=id,\ \ b,d:\text{real constants}\,. \tag{8}$$

However, this choice leads to a potential related to Eq. (6) by the reflection $x \mapsto -x$. Hence, we will only discuss the potential in Eq. (6) here.

To see that the wave functions $\psi_n(x)$ do represent quasinormal modes, we determine the corresponding energy E_n. This is easily done by solving the eigenvalue problem of the polynomial part $\chi_n(z)$ of the wave function. From the Schrödinger equation we find that for $n = 0$, the energy is $E_0 = -\nu/4 - ibd/2\nu$. This clearly shows that the only QES solution when $n = 0$ is a QNM with an energy having a negative imaginary part (recall that $b, d, \nu > 0$). For $n = 1$, we have two QES solutions. Their energies are $E_1 = -5\nu/4 - ibd/2\nu \pm \sqrt{\nu^2 - ibd}$. Again we have two QNM modes. One can proceed accordingly to obtain $n+1$ QNM modes with higher values of n. However, for large n, computation becomes tedious, and one has to resort to numerical means.

One can do the same for the other four cases listed in [5]. It is found that these cases admits exact QNM solutions.

Summary.– To summarize, we have demonstrated that it is possible to extend the usual QES theory to accommodate QNM solutions, by complexifying certain parameters defining the QES potentials. We found that the five $sl(2)$-based QES systems listed in [5] can be so extended. While one of these cases admits QES QNM, the other four cases give exact QNM solutions. It is hoped that our work would motivate the search of many more exact/quasi-exact systems of QNM in QES theories based on higher Lie algebras, and in higher dimensions.

This work was supported in part by the National Science Council of the Republic of China under the Grants NSC 94-2112-M-032-009 (H.T.C.) and NSC 94-2112-M-032-007 (C.L.H.).

REFERENCES

1. For a good review, see eg.: K.D. Kokkotas and B.G. Schmidt, Living Rev. Rel. **2**, 2 (1999); S. Chandrasekhar, *The mathematical theory of black holes* (Clarendon, Oxford, 1983).
2. A. Turbiner and A.G. Ushveridze, Phys. Lett. A **126**, 181 (1987).
3. A.V.Turbiner, Comm. Math. Phys. **118**, 467 (1988).
4. A.G. Ushveridze, Sov. Phys.-Lebedev Inst. Rep. **2**, 50, 54 (1988); *Quasi-exactly solvable models in quantum mechanics* (IOP, Bristol, 1994).
5. A. González, N. Kamran and P.J. Olver, Comm. Math. Phys. **153**, 117 (1993).
6. H.-T. Cho and C.-L. Ho, J.Phys. A **40**, 1325 (2007).
7. C.M. Bender and and S. Boettcher, J. Phys. A **31**, L273 (1998).

Increasing the Effective Number of Neutrinos with Decaying Particles

Kazuhide Ichikawa*, Masahiro Kawasaki*, Kazunori Nakayama*, Masato Senami* and Fuminobu Takahashi†

*Institute for Cosmic Ray Reseach, University of Tokyo, Kashiwa 277-8582, Japan
†Deutsches Elektronen Synchrotron DESY, Notkestrasse 85, 22607, Hamburg, Germany

Abstract. We present a model of decaying particles to increase the effective number of neutrinos N_ν after big bang nucleosynthesis but before the structure formation begins. We point out that our scenario solves the discrepancy between the constraints on N_ν from these two epochs, As an example, we consider saxion decay into two axions.

Keywords: cosmological neutrinos, axions, cosmologies of theories beyond the standard model
PACS: 98.80.Cq,14.80.Mz

INTRODUCTION

In the standard cosmology, the total radiation energy density of the universe consists of relic photons and three generations of neutrinos. In terms of the effective number of neutrinos N_ν, which parameterizes the radiation energy density of the universe, the standard prediction is $N_\nu \simeq 3.046$. But if there exist unknown light particle species, N_ν can deviate from the standard prediction. If we regard N_ν as a free cosmological parameter, various observations can constrain it. One of such constrains comes from Big-Bang nucleosynthesis (BBN). The modification on N_ν changes the Hubble expansion rate at the epoch of the weak interaction freeze-out, which leads to modified primordial ^{4}He abundance. Most of the recent analysis of the primordial ^{4}He abundance [1, 2, 3, 4] are consistent with $N_\nu \sim 3$ [1]. On the other hand, the nonstandard value of N_ν also affects the structure formation and can be constrained from combined cosmological measurements such as CMB anisotropy, galaxy clustering, weak lensing and Lyman-α forests. There are now many works which treat such subjects. Recent analysis are found in Refs. [5, 6, 7, 8, 9] (see also [10]), and the result is that somewhat large value of N_ν is favored in particular when the Lyman-α data is included. In fact, the central value distributes around $N_\nu \sim 4-5$.

These observational results imply the possibility that the effective number of neutrinos increased after BBN but before the structure formation begins. We show that decaying particles motivated by some particle physics models can take the role [11]. This solves not only the observational discrepancy on N_ν between these two epochs, but also the inconsistency of matter power spectrum amplitude between WMAP and small scale measurements such as Lyman-α forests [11].

---

[1] Although the analysis of Ref. [4] shows small deviation from $N_\nu \simeq 3$, but $\Delta N_\nu \sim 1$ is still disfavored.

CP957, *Particles, Strings, and Cosmology, 13th International Symposium*
edited by A. Rajantie, C. Contaldi, P. Dauncey, and H. Stoica

THE MODEL

One of the most promising solutions to the strong CP problem in the quantum chromo dynamics (QCD) is the Peccei-Quinn (PQ) mechanism [12], which involves a pseudo Nambu-Goldstone boson a, the axion, associated with the spontaneous PQ symmetry breaking. In a supersymmetric theory, the axion forms a supermultiplet, including a fermionic superpartner $\tilde{a}$, the axino, and a scalar partner s, the saxion. In general, the saxion acquires a mass of order $m_{3/2}$ in the presence of the supersymmetry (SUSY) breaking [13]. (Here $m_{3/2}$ is the gravitino mass.) In a class of models, the saxion mainly decays into a pair of the axions [14], and these axions contribute to the extra effective number of neutrinos without disturbing the BBN results. However, since the saxion may also decay into two photons, we need to examine whether the produced photons do not spoil the standard cosmology.

The lifetime of the saxion is given as

$$\tau_s(s \to 2a) \simeq 1.3 \times 10^5 \sec \left(\frac{m_s}{100\,\mathrm{MeV}}\right)^{-3} \left(\frac{F_a}{10^{12}\,\mathrm{GeV}}\right)^2, \qquad (1)$$

where F_a denotes the PQ breaking scale. The saxion begins to oscillate coherently when the Hubble parameter H becomes equal to $m_s \sim m_{3/2}$ with initial amplitude δs_i. If the reheating is completed after the oscillation of the saxion commences, the increase in the effective number of neutrinos by the saxion decay is expressed as

$$\Delta N_\nu \simeq 3.0 \left(\frac{m_s}{10\,\mathrm{MeV}}\right)^{-3/2} \left(\frac{F_a}{10^{12}\,\mathrm{GeV}}\right)^3 \left(\frac{\delta s_i}{F_a}\right)^2 \left(\frac{T_R}{10^6\,\mathrm{GeV}}\right) \qquad (2)$$

where T_R denotes the reheating temperature. Thus it is possible to increase N_ν by order unity in this scenario.

Now let us consider the saxion decay into two photons. The branching ratio of two photon decay is $B_\gamma \simeq 1.7 \times 10^{-7}$. The injected photons may destroy the light elements and change the result of BBN for $m_s \gtrsim 40\,\mathrm{MeV}$ [15], while for $m_s \lesssim 40\,\mathrm{MeV}$, the injected photons can never dissociate ^{4}He nuclei. On the other hand, the saxion can also decay into two gluons which produce hadron jets for $m_s \gtrsim 1$ GeV. The hadronic branching ratio is estimated as $B_h = \alpha_s^2/\pi^2 \simeq 1.4 \times 10^{-3}$. Even in the case $m_s \lesssim 40$ MeV, the energy injection is constrained from the CMB. If the injected photons cannot reach chemical or kinetic equilibrium due to the small rate of interactions with background plasma, it leads to the distortion of the CMB blackbody spectrum which is constrained from observations [16].

Let us here comment on the thermal production of gravitinos. If the reheating temperature is too high, the thermally produced gravitinos may overclose the universe [17]. For $m_{3/2} \ll m_{\tilde{g}}$ ($m_{\tilde{g}}$ is the gluino mass), the gravitino abundance is inversely proportional to the gravitino mass, $\Omega_{3/2}^{\mathrm{TP}} h^2 \propto T_R/m_{3/2}$. Therefore, the thermal gravitino production sets the lower bound on the gravitino mass for a fixed reheating temperature. The axinos, in addition to the gravitinos, are also produced by thermal scattering. Their abundance is proportional to $\Omega_{\tilde{a}}^{\mathrm{TP}} h^2 \propto m_{\tilde{a}} T_R/F_a^2$ [18, 19]. If the axino mass ($m_{\tilde{a}}$) is too large and/or the reheating temperature (T_R) is too high, the thermally produced axino may overclose the universe.

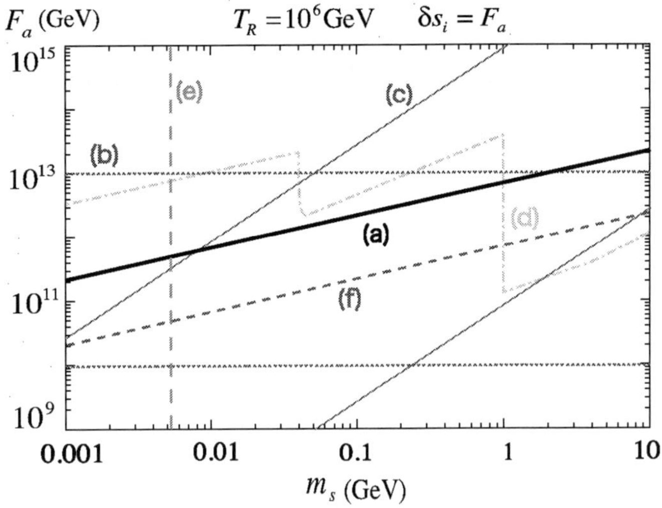

FIGURE 1.

In Fig. 1, we summarize all the constraints discussed above. Here we have taken $\delta s_i = F_a$ and $T_R = 10^6$ GeV as reference values. The thick solid black line labeled (a) shows $\Delta N_\nu = 1$. Note that the region above this line corresponds to $\Delta N_\nu \gtrsim 1$. The dotted blue lines (b) denote the astrophysical and cosmological constraints on the PQ scale, and we have set the initial misalignment angle $\theta = 0.1$. In order to satisfy 1 sec $\lesssim \tau_s \lesssim 10^8$ sec, the combination of parameters (m_s, F_a) must lie in the region between two thin solid red lines (c). The constraints from BBN and CMB provide an upper bound on F_a for a fixed m_s, as represented by the dot-dashed green line (d). The thermally produced gravitinos exceed the current observed dark matter abundance if $m_s \sim m_{3/2}$ is smaller than the value indicated by the vertical long-dashed yellow line (e). For $m_{\tilde{a}} = 0.01\, m_s$, the abundance of the thermally produced axinos exceeds the current observed dark matter abundance below the long-dashed purple line (f). For $m_{\tilde{a}} = m_s$, the constraint from the thermally produced axinos coincides with the line (a) accidentally, so it is not explicitly drawn. The region below (a) is excluded as long as $m_{\tilde{a}} \simeq m_s$. We have found regions for $\Delta N_\nu \sim 1$ consistent with all the constraints. For $\delta s_i = F_a$, they are $1\,\text{MeV} \lesssim m_s \lesssim 1\,\text{GeV}$, $F_a \sim 10^{12}$ GeV and 10^5 GeV $\lesssim T_R \lesssim 10^6$ GeV. Since the PQ scale F_a is close to the upper bound coming from the axion-oveclosure limit, the axion can also play a role of dark matter of the universe. Moreover, if $m_{\tilde{a}} \simeq m_s$, dark matter can be also explained by the thermally produced axinos. On the other hand, δs_i may be as large as the Planck scale. We also found consistent parameter region around $\delta s_i \sim 10^{16} - 10^{18}$ GeV, 1 MeV $\lesssim T_R \lesssim$ 100 MeV and 1 MeV $\lesssim m_s \lesssim$ 1 GeV. (See Ref. [11].)

CONCLUSIONS

We have presented a model of decaying particles to increase the effective number of neutrinos N_ν after BBN but before the structure formation begins. Other models are gravitino decay into axino + axion and Dirac right-handed sneutrino decay into gravitino + Dirac right-handed neutrino [11]. Such a scenario is motivated because nonstandard values of $N_\nu > 3$ is preferred by the combined data of the CMB, galaxy clustering and the Lyman-α forest [5, 7, 8, 9] whereas most of the recent analyses of primordial ^{4}He abundance favor standard $N_\nu = 3$ [1, 2, 3].

Since the discrepancy which we have addressed is about 2σ level, further data and studies are necessary in order to see whether inconsistency exists in the standard cosmological model or in the interpretation of one or more observations. We believe the ongoing works in the communities to understand sources of systematic errors will accomplish this task together with the future CMB experiments (the PLANCK sensitivity for N_ν is forecasted to be ~ 0.2, see e.g. [20]).

ACKNOWLEDGMENTS

The work of KI, MK and MS were supported in part by a Grant-in-Aid of the Ministry of Education, Culture, Sports, Science, and Technology, Government of Japan (No. 18840010 for KI, No. 18540254 and 14102004 for MK, and No. 18840011 for MS). KN would like to thank the Japan Society for the Promotion of Science for financial support. This work was also supported in part by JSPS-AF Japan-Finland Bilateral Core Program.

REFERENCES

1. K. A. Olive and E. D. Skillman, Astrophys. J. **617**, 29 (2004).
2. M. Fukugita and M. Kawasaki, Astrophys. J. **646**, 691 (2006).
3. M. Peimbert, V. Luridiana and A. Peimbert, arXiv:astro-ph/0701580.
4. Y. I. Izotov, T. X. Thuan and G. Stasinska, arXiv:astro-ph/0702072.
5. U. Seljak, A. Slosar and P. McDonald, JCAP **0610**, 014 (2006).
6. K. Ichikawa, M. Kawasaki and F. Takahashi, JCAP **0705**, 007 (2007).
7. G. Mangano, A. Melchiorri, O. Mena, G. Miele and A. Slosar, JCAP **0703**, 006 (2007).
8. M. Cirelli and A. Strumia, JCAP **0612**, 013 (2006).
9. J. Hamann, S. Hannestad, G. G. Raffelt and Y. Y. Y. Wong, arXiv:0705.0440 [astro-ph].
10. K. Ichikawa, arXiv:0706.3465 [astro-ph].
11. K. Ichikawa, M. Kawasaki, K. Nakayama, M. Senami and F. Takahashi, JCAP **0705**, 008 (2007).
12. R. D. Peccei and H. R. Quinn, Phys. Rev. Lett. **38**, 1440 (1977).
13. K. Rajagopal, M. S. Turner and F. Wilczek, Nucl. Phys. B **358**, 447 (1991).
14. E. J. Chun and A. Lukas, Phys. Lett. B **357**, 43 (1995).
15. M. Kawasaki, K. Kohri and T. Moroi, Phys. Lett. B **625**, 7 (2005); Phys. Rev. D **71**, 083502 (2005).
16. D. J. Fixsen, E. S. Cheng, J. M. Gales, J. C. Mather, R. A. Shafer and E. L. Wright, Astrophys. J. **473**, 576 (1996).
17. T. Moroi, H. Murayama and M. Yamaguchi, Phys. Lett. B **303**, 289 (1993).
18. L. Covi, H. B. Kim, J. E. Kim and L. Roszkowski, JHEP **0105**, 033 (2001).
19. A. Brandenburg and F. D. Steffen, JCAP **0408**, 008 (2004).
20. K. Ichikawa and T. Takahashi, Phys. Rev. D **73**, 063528 (2006).

Neutrino mass constraint from CMB, BAO and SN

Kazuhide Ichikawa

Institute for Cosmic Ray Research, University of Tokyo, Kashiwa 277 8582, Japan

Abstract. We show that the cosmic microwave background (CMB) data of WMAP can give subelectronvolt limit on the neutrino mass. We investigate how much we can make it more stringent by using "standard ruler" measurements such as baryon acoustic oscillation (BAO) and type Ia supernovae (SN).

Keywords: neutrino mass, cosmic microwave background radiation
PACS: 98.70.Vc, 14.60.Pq

INTRODUCTION

The neutrino mass is directly searched by tritium β-decay experiments and the current upper bound for the electron neutrino mass is 2 eV (95% C.L.) [1]. At present, it is difficult to push down the limit in this way, but, as is well known, cosmological considerations give more stringent constraints.

For example, Ref. [2] obtained $m_\nu < 0.21$ eV and Ref. [3] obtained $m_\nu < 0.58$ eV, using the WMAP 1st year data combined with the galaxy power spectrum (the former used the data from the 2dFGRS and the latter from the SDSS main sample. The difference can be ascribed to the use of the bias information by Ref. [2]). By contrast, we found $m_\nu < 0.66$ eV from the WMAP 1st year data alone as reported in Ref. [4]. We note that this is the first to point out CMB data (the WMAP 1st year data) alone can give a sub-eV upper bound on the neutrino mass, which is comparable to the limits obtained from the CMB and galaxy clustering data combined.

Before Ref. [4] (and some time after that too), there seems to be a lack of consensus about whether the CMB experiment with the WMAP-level precision can derive a sub-eV neutrino mass limit and, in fact, the WMAP alone limit reported in Ref. [3], $m_\nu < 3.8$ eV, which allows 100% HDM was somewhat accepted (incidentally, the WMAP group did not report the WMAP 1st year data alone limit on the neutrino mass). We, on the contrary, have derived the upper limit 0.66 eV as quoted above from the same data by the χ^2 minimization method which is independent from the MCMC method adopted by Ref. [3]. Our conclusion is later confirmed by Refs. [5, 6, 7] (Ref. [5] does not report the WMAP alone limit in a number but judging from their likelihood figure, it looks less than 1 eV. Refs. [6] and [7] gives $m_\nu < 0.70$ eV and $m_\nu < 0.63$ eV respectively).

Below, we first discuss the WMAP alone limit comparing results from 1st year data and 3rd year data. Then, we will see how the limit is tightened by baryon acoustic oscillation (BAO) and/or type Ia supernovae (SN) data.

We here summarize our notations. We derive neutrino mass constraint in the flat

CP957, *Particles, Strings, and Cosmology, 13th International Symposium*
edited by A. Rajantie, C. Contaldi, P. Dauncey, and H. Stoica

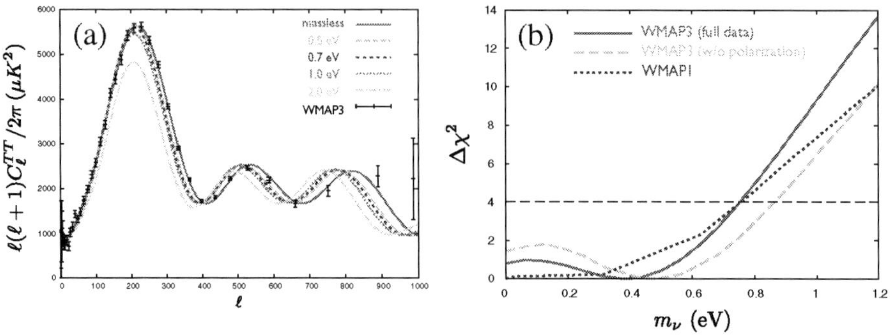

FIGURE 1. (a) Effects of massive neutrinos on the CMB TT power spectrum. The labels denote the neutrino mass for single species m_ν and we assume three degenerate species. When we vary the neutrino mass, we fix ω_b, ω_c, h and $\Omega_{\text{tot}} = 1$ so that Ω_Λ varies. We also fix the amplitude (normalization) for the primordial power spectrum A. Errorbars are from the WMAP 3-year observation. (b) $\Delta\chi^2$ of WMAP data as functions of the neutrino mass m_ν [4, 8]. $\Delta\chi^2 = 4$ roughly corresponds to 95% C.L. limit. Each curve represents different data sets. The blue dotted line use the WMAP 1st year data (TT+TE), the red solid line uses the full WMAP 3-year data including temperature and polarization and the green dashed line uses only temperature power spectrum of the WMAP 3-year data.

ΛCDM model with the power-law adiabatic perturbations. Namely, cosmological parameters we consider are: baryon density ω_b, matter density ω_m, hubble parameter h, reionization optical depth τ, spectral index of primordial spectrum n_s, its amplitude A and massive neutrino density ω_ν. Here, $\omega \equiv \Omega h^2$ where Ω is the energy density normalized to the critical density and $\omega_m \equiv \omega_b + \omega_c$ where ω_c is cold dark matter density (Caution that many literatures define $\omega_m \equiv \omega_b + \omega_c + \omega_\nu$ to include the massive neutrino in the matter density). ω_ν is related to neutrino masses by $\omega_\nu = \sum m_\nu/(94.1\text{ eV})$ and we assume three generations of massive neutrinos with degenerate masses so that $\omega_\nu = m_\nu/(31.4\text{ eV})$. The flatness condition is expressed as $\omega_m + \omega_\nu + \omega_\Lambda = h^2$ where ω_Λ is the cosmological constant energy density.

WMAP ALONE LIMIT

We begin with showing how CMB power spectrum is modified by increasing neutrino mass in Fig. 1 (a). The other cosmological parameters are fixed here. Also, we assume equal mass m_ν for three neutrino species, which is justified because, from the oscillation experiments, the mass differences are known to be much smaller than one eV. In this figure, we see horizontal shift, and suppression around the first peak. We will briefly discuss whether these variations are degenerate with other cosmological parameters.

As for the horizontal shift, this comes from the fact that the larger m_ν implies smaller amount of cosmological constant, since we assume that the universe is flat. Then the distance to the last scattering surface is shorter and the peaks move to smaller ℓ (to larger angular scales). However, this shift is easily cancelled by the shift in the hubble parameter. Therefore this does not produce a neutrino mass signal.

Next, the suppression of the 1st peak takes place only when m_ν is larger than about

0.6 eV. This corresponds to 0.3 eV in terms of photon temperature T. Meanwhile, the recombination takes place at $z \approx 1088$ or $T \approx 0.3$ eV. In other words, massive neutrinos become non-relativistic before the epoch of recombination if they are heavier than 0.6 eV. Therefore, only in this case, the neutrino mass can imprint a characteristic signal in acoustic peaks (to be specific, the matter-radiation equality occurs earlier due to less relativistic degrees of freedom and the enhancement of the 1st peak by the early-integrated Sachs-Wolfe effect is smaller).

In passing, notice that such separation of the neutrino mass effect into the horizontal shift and the suppression of the 1st peak for $m_\nu > 0.6$ eV can be accomplished by our parametrization that $\omega_m \equiv \omega_b + \omega_c$ is fixed (Ω_Λ is varied) when m_ν is varied.

This signal, however, could be accidentally mimicked by some combination of other cosmological parameters. So we searched a large cosmological parameter space in order to find the degree of degeneracy between m_ν and the other cosmological parameters. The results are shown in Fig. 1 (b) . We calculated WMAP χ^2 (log likelihood) as functions of m_ν. For each value of m_ν, we varied 6 other ΛCDM cosmological parameters to find minimum χ^2. In Fig. 1 (b), the different lines represent different data sets. The blue dotted line is the WMAP 1st year result which I mentioned in the introduction. The others use WMAP 3-year data. The red solid line uses full data sets including temperature and polarization and the green dashed line uses only temperature power spectrum. We find that three lines are quite similar.

In summary, we obtained the upper bound of 0.7 eV from WMAP alone and two points are worth mentioning. First is that the WMAP 3-year constraint is not improved from the 1st year limit. Next is that the polarization data does not contribute to the limit much. These results are reasonable because the neutrino mass (larger than 0.6 eV) characteristically modifies the acoustic peaks around 1st and 2nd peaks in the temperature power spectrum and these regions are already well measured by the WMAP 1st year.

We stress that this bound is robust in a sense that it is obtained from CMB data of the WMAP which is considered to be the cleanest cosmological data and that it is obtained from a single experiment. Also, CMB can be dealt with the linear perturbation theory so it does not suffer from non-linearity or biasing which appear in galaxy clustering data.

However, we have to combine other data sets to improve this limit. This is because, as we mentioned earlier, CMB is insensitive to the neutrino mass lighter than 0.6 eV (The lensed CMB is known to be sensitive to smaller neutrino masses if we obtained polarization data of next (or next-to-next) generation experiments, but we do not discuss this possibility here) , whose effects being absorbed in the shift of the hubble parameter.

WMAP + BAO/SN LIMIT

Then, we would like to know whether we can push down the CMB limit by constraining the Hubble constant indirectly by baryon acoustic oscillation (BAO) and/or type Ia supernovae (SN). As shown in Fig. 2, when we combine WMAP data with the correlation function data of the luminous red galaxy (LRG) sample [9], where the detection of the baryon acoustic peak was reported, the limit is improved to be about 0.2 eV. This is consistent with the limit obtained when combined with LRG power spectrum data of

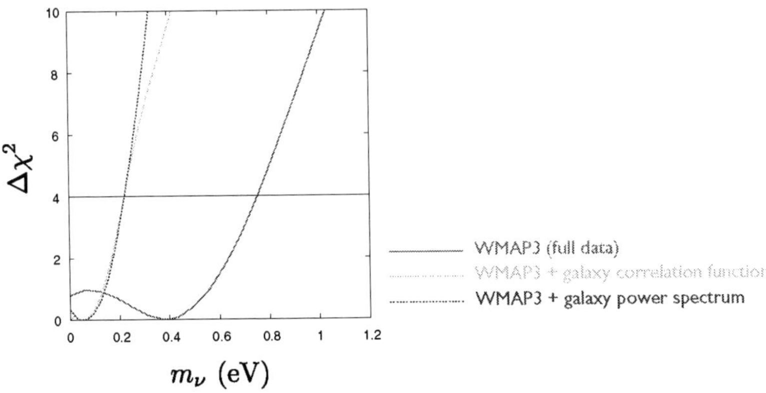

FIGURE 2. $\Delta\chi^2$ of WMAP data (solid line), WMAP + SDSS LRG correlation function data of [9] (dashed line) and WMAP + SDSS LRG power spectrum data of [10] (dotted line) as functions of the neutrino mass m_ν

Ref. [10]. However, the constraint comes from the distortion of overall shape of the correlation function (or power spectrum) rather than the peak position [11]. Since this limit is not coming from the peak position in the correlation function or the oscillatory feature in the power spectrum, it is not considered to be very robust (may suffer much from non-linear correction or biasing). We also study the constraint using WMAP combined with the SN data sets of Riess et al. or Davis et al. [12]. We find that SN does not help much to improve the limit and moreover it depends on the data set we adopt (as is the case for constraining dark energy parameters and the curvature of the universe. See e.g. [13]). We should wait for SN data to converge before using them to constrain neutrino masses.

REFERENCES

1. W. M. Yao *et al.* [Particle Data Group], J. Phys. G **33**, 1 (2006).
2. D. N. Spergel *et al.* [WMAP Collaboration], Astrophys. J. Suppl. **148**, 175 (2003).
3. M. Tegmark *et al.* [SDSS Collaboration], Phys. Rev. D **69**, 103501 (2004).
4. K. Ichikawa, M. Fukugita and M. Kawasaki, Phys. Rev. D **71**, 043001 (2005).
5. C. J. MacTavish *et al.*, Astrophys. J. **647**, 799 (2006).
6. S. Hannestad, arXiv:hep-ph/0602058.
7. J. Lesgourgues and S. Pastor, Phys. Rept. **429**, 307 (2006).
8. M. Fukugita, K. Ichikawa, M. Kawasaki and O. Lahav, Phys. Rev. D **74**, 027302 (2006).
9. D. J. Eisenstein *et al.* [SDSS Collaboration], Astrophys. J. **633**, 560 (2005).
10. M. Tegmark *et al.*, Phys. Rev. D **74**, 123507 (2006).
11. G. Huetsi, Astron. Astrophys. **459**, 375 (2006).
12. A. G. Riess *et al.*, arXiv:astro-ph/0611572; W. M. Wood-Vasey *et al.*, arXiv:astro-ph/0701041; T. M. Davis *et al.*, arXiv:astro-ph/0701510.
13. K. Ichikawa and T. Takahashi, Phys. Rev. D **73**, 083526 (2006); JCAP **0702**, 001 (2007); K. Ichikawa, M. Kawasaki, T. Sekiguchi and T. Takahashi, JCAP **0612**, 005 (2006).

Dark Energy-Dark Matter Interaction from the Abell Cluster A586 and violation of the Equivalence Principle

Morgan Le Delliou*, Orfeu Bertolami† and Francisco Gil Pedro**

* **Speaker**; *CFTC, U. Lisboa Av. Gama Pinto 2, 1649-003 Lisboa, Portugal*
†*Centro de Física dos Plasmas, IST, Lisbon, Portugal*
***Departamento de Física, Instituto Superior Técnico, Lisbon, Portugal*

Abstract. We find that the Abell Cluster A586 exhibits evidence of the interaction between dark matter and dark energy and argue that this interaction suggests evidence of violation of the Equivalence Principle. This violation is found in the context of two different models of dark energy-dark matter interaction.

Keywords: Cosmology – dark matter – dark energy – Abell Cluster A586 – Equivalence principle
PACS: 98.80.-k,98.80.Cq,95.35.+d,95.36.+x,98.65.Dx,04.20.Cv

INTRODUCTION

Contemporary cosmology regards consensually the nature of dark energy and dark matter (hereafter DE and DM) to be crucial for a suitable description of universe's evolution and dynamics. Even though observations agree with the ΛCDM model, a deeper insight into DE and DM leads to more complex models – in particular, interaction between them. However, so far, no evidence of this interaction has been presented. In this work, we argue that Cluster A586 exhibits such evidence. Furthermore, we argue that this suggests evidence of violation of the Equivalence Principle (EP).

In what follows, we first set the formalism for DE-DM interaction and consider two very different, phenomenologically viable models: one based on *ad hoc* DE-DM interaction [1], the other on the generalized Chaplygin gas (GCG) model with explicit identification of DE and DM [5]. Our observational inferences are based on Cluster A586, given its stationarity, spherical symmetry and wealth of available observations [3]. We also compare our results with other cosmological observations [4].

INTERACTING MODELS

Our results are obtained in the context of two distinct phenomenologically viable models for the DE-DM interaction: a naturally unified model, the GCG [2], but also a less constrained interacting model with constant DE equation of state (hereafter EOS) parameter $\omega_{DE} = p_{DE}/\rho_{DE}$ (where p_X is pressure and ρ_X is energy density of X; see e.g. [1]).

We consider first a quintessence model with constant EOS. The Bianchi conservation equations for both DE and DM read (H denoting the Hubble parameter and ζ, the interaction strength)

$$\dot{\rho}_{DM} + 3H\rho_{DM} = \zeta H \rho_{DM}, \quad (1)$$

$$\dot{\rho}_{DE} + 3H\rho_{DE}(1+\omega_{DE}) = -\zeta H \rho_{DM}, \quad (2)$$

CP957, *Particles, Strings, and Cosmology, 13th International Symposium*
edited by A. Rajantie, C. Contaldi, P. Dauncey, and H. Stoica

where constant EOS parameter (ω_{DE}), and scaling (η) given by $\frac{\rho_{DE}}{\rho_{DM}} = \frac{\Omega_{DE_0}}{\Omega_{DM_0}} a^{\eta}$ are assumed, as in [1] (with density parameter $\Omega_X - \rho_X$ scaled with critical density – and a, the scale of the universe). Then, the coupling varies as $\zeta = -\frac{(\eta+3\omega_{DE})\Omega_{DE_0}}{\Omega_{DE_0}+\Omega_{DM_0}a^{-\eta}}$ and energy densities evolve as

$$\rho_{DM} = a^{-3}\rho_{DM_0}\left[\Omega_{DE_0}a^{\eta} + \Omega_{DM_0}\right]^{-\frac{(\eta+3\omega_{DE})}{\eta}}, \qquad \rho_{DE} = a^{\eta}\rho_{DE_0}\frac{\rho_{DM}}{\rho_{DM_0}}. \tag{3}$$

We turn now to the GCG model. It is defined by its unified EOS $p_{DE} = -A/(\rho_{DM} + \rho_{DE})^{\alpha}$, with A and α constant, and assuming DE constant EOS parameter $\omega_{DE} = -1$ [2]. This last splitting into DE and DM, together with Bianchi conservation imply a scaling behaviour with $\eta = 3(1+\alpha)$ and energy densities to be

$$\rho_{DM} = \rho_{DM_0}\frac{\rho_{DE}}{\rho_{DE_0}}a^{-3(1+\alpha)}, \qquad \rho_{DE} = \rho_{DE_0}\left(\frac{\Omega_{DE_0}+\Omega_{DM_0}}{\Omega_{DE_0}+\Omega_{DM_0}a^{-3(1+\alpha)}}\right)^{\frac{\alpha}{1+\alpha}}. \tag{4}$$

GENERALIZED LAYZER-IRVINE EQUATIONS

In this work we shall focus on the effect of interaction on clustering as revealed by the Layzer-Irvine equation. We write the kinetic and potential energy densities ρ_K and ρ_W of clustering DM in interacting models in terms of scale factor dependence

$$\rho_K \propto a^{-2}, \qquad \rho_W \propto a^{\zeta-1}, \tag{5}$$

and we use DM virialization dynamics in an expanding universe described by the generalised Layzer-Irvine equation [6]

$$\dot{\rho}_{DM} + (2\rho_K + \rho_W)H = \zeta H \rho_W, \tag{6}$$

with ζ, the coupling for the quintessence model (for the GCG, the scaling allows to just replace $\eta = 3(1+\alpha)$ and $\omega_{DE} = -1$). At equilibrium, $2\rho_K + \rho_W = \zeta\rho_W \neq 0$.

DETECTION OF INTERACTION FROM OBSERVATION OF A586

Using those models, we have interpreted observations from an Abell cluster [3] and compared it with other observations [4].

From [3], we extract: the total mass, $M_{Cluster} = (4.3 \pm 0.7) \times 10^{14} M_{\odot}$ (galaxies, DM and intra-cluster gas), radius, $R_{Cluster} = 422$ kpc at $z = 0.1708$ (angular radius $\Delta_{max} = 145''$) and cluster dispersion $\sigma_v = (1243 \pm 58)\, kms^{-1}$, from weak lensing. From photometry, we obtain $<R> = \frac{2}{N_{gal}(N_{gal}-1)}\sum_{i=2}^{N_{gal}}\sum_{j=1}^{i-1} r_{ij}$, the mean intergalactic distance[1], from declinations and right assentions of a galaxy sub-sample within the projected Δ_{max} (i.e. for galaxy i, $\sqrt{\alpha_{ci}^2 + \delta_{ci}^2} \leq \Delta_{max}$). Further assuming $\omega_{DE} = -1$ and $\Omega_{DE_0} = 0.72$, $\Omega_{DM_0} = 0.24$ [7], we get

$$\rho_K \simeq \frac{9}{8\pi}\frac{M_{Cluster}}{R_{Cluster}^3}\sigma_v^2 = (2.14 \pm 0.55) \times 10^{-10} Jm^{-3}, \tag{7}$$

$$\rho_W \simeq -\frac{3}{8\pi}\frac{G}{<R>}\frac{M_{Cluster}^2}{R_{Cluster}^3} = (-2.83 \pm 0.92) \times 10^{-10} Jm^{-3}, \tag{8}$$

---

[1] where $r_{ij}^2 = 2d^2\left[1 - \cos(\alpha_{ci} - \alpha_{cj})\cos\delta_{ci}\cos\delta_{cj} - \sin\delta_{ci}\sin\delta_{cj}\right]$, $\alpha_{ci} = \alpha_i - \alpha_{center}$ and $\delta_{ci} = \delta_i - \delta_{center}$

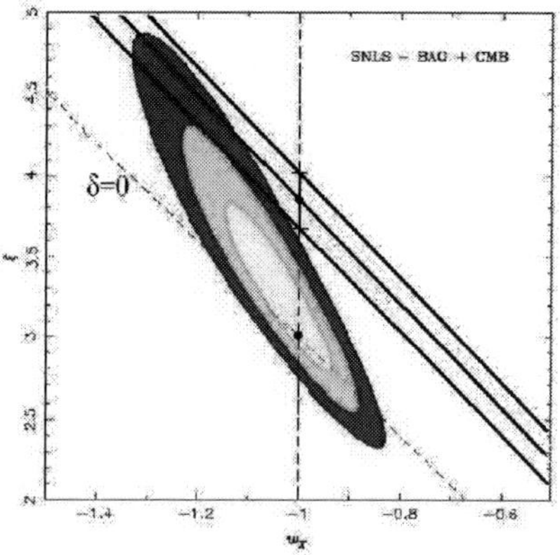

FIGURE 1. Superimposition of the probability contours for the interacting DE-DM model in the (ω_X, ξ) plane (denoted as (ω_{DE_0}, η) in [8]), marginalized over Ω_{DE_0} in [4] study of CMB, SN-Ia and BAO ($2.66 < \xi < 4.05$ at 95% C.L.) with the results extended from [8], based on the study of A586 cluster. The $\xi = -3\omega_X$ line corresponds to uncoupled models.

which allow us to obtain [8]

$$\eta = 3.82^{+0.18}_{-0.17} \neq -3\omega_{DE}, \qquad \alpha = 0.27^{+0.06}_{-0.06} \neq 0. \tag{9}$$

Notice that $\eta \neq -3\omega_{DE}$ signals the energy exchange between DM and DE. It is also remarkable that $\alpha \neq 0$ which implies the GCG description is not degenerate with ΛCDM ($\alpha = 0$). We mention that our results (see Fig. 1 and [11]) are consistent with the study of [4] where DE-DM interacting quintessence are analysed for compatibility with WMAP CMB [7], SNLS SN-Ia [9] and Baryon Acoustic Oscillations in SDSS [10].

VIOLATION OF EQUIVALENCE PRINCIPLE?

Given that the EP concerns the way matter falls in the gravitational field, considering the clustering of matter against the cosmic expansion and the interaction with DE seems to be a logical way to test its validity. Both models *predict* departure of homogeneous DM from dust behaviour and have effects that can be interpreted as violation of EP.

The non-dust evolution of DM density leads to evolution of the bias parameter, $b = \rho_B/\rho_{DM}$, at the homogeneous level on cosmological timescales [8] (Fig. 2a). Other astrophysical effects also affect the bias so the detection of this drift would require statistics over different z ranges.

If we attribute the non-dust ρ_{DM} to dust-like DM particles mass,

$$m_{DM}(a) = m_{DM,0}\left[\Omega_{DE_0} a^{\eta} + \Omega_{DM_0}\right]^{-\frac{\eta + 3\omega_{DE_0}}{\eta}}, \tag{10}$$

gravity is then Baryon/DM composition dependent as seen in the two particles potential $U_{(1-2)} = -\frac{G m_1 m_2}{r_{12}}$. We can now assign the time evolution to a varying G, as seen in Fig. (2b)[11], to compare with simulations of the type done by [12].

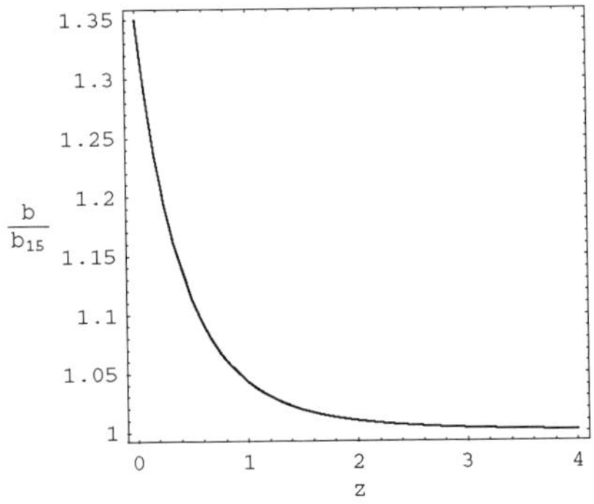

(a)

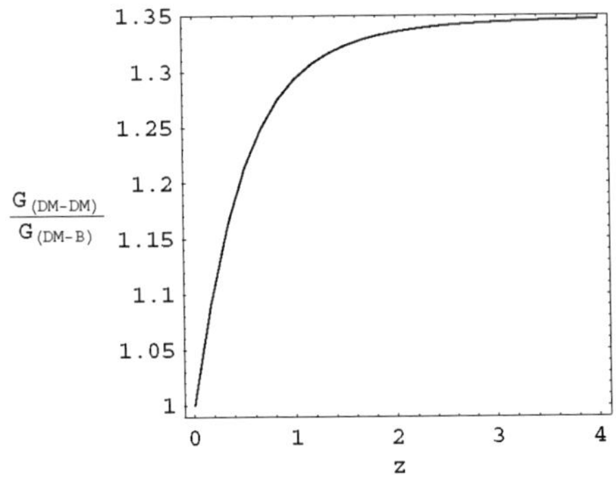

(b)

FIGURE 2. (a) Normalized gravitationally induced bias parameter as a function of z, where $b_{15} \equiv b(z = 15)$, $z = 15$ is the typical condensation scale and $b = \rho_B/\rho_{DM} = \Omega_{B_0}/\Omega_{DM_0}[\Omega_{DE_0}a^{\eta} + \Omega_{DM_0}]^{(\eta+3\omega_{DE})/\eta}$. (b) Evolution with redshift of the ratio of the gravitational coupling for DM and baryons falling on a DM halo, using the varying coupling model discussed in [11], to be compared with the simulation of [12]. $G_{(DM-DM)} = G_{(DM-B)}(\Omega_{DE_0}a^{\eta} + \Omega_{DM_0})^{-(\eta+3\omega_{DE_0})/\eta}$.

CONCLUSIONS

Observations of Cluster A586 [3] suggest evidence of departure from virialization given that A586 is very spherical and relaxed (from its mass distribution and Gyrs without mergers). The generalized Layzer-Irvine equation allows to interpret this departure as interacting DE. We therefore link the observed virialization to interaction [8, 11] with two different models, consistent with known constraints [8, refs. therein]: an interacting quintessence with constant ω_{DE} [1] and a Chaplygin gas with $\omega_{DE} = -1$ [2].

From these models, we argue that the Equivalence Principle should be violated as they impact, for example, on the overall bias parameter [8] and on Baryon/DM asymmetric collapse [using 12, 11]. This is consistent with violation that other models have reported [13].

ACKNOWLEDGMENTS

The work of MLeD is supported by FCT (Portugal),SFRH/BD/16630/2004 and hosted by J.P. Mimoso and CFTC, Lisbon University.

REFERENCES

1. L. Amendola, *Phys. Rev.* **D 62** (2000) 043511.
2. M.C. Bento, O. Bertolami and A.A. Sen, *Phys. Rev.* **D 66** (2002) 043507; A. Kamenshchik, U. Moschella and V. Pasquier, *Phys. Lett.* **B 511** (2001) 265.
3. E.S. Cypriano, G.B. Neto, L.J. Sodré and J.P. Kneib, *Ap. J.* **630** (2005) 38.
4. Z.-K. Guo, N. Ohta and S. Tsujikawa, astro-ph/0702015; L. Amendola, G.C. Campos and R. Rosenfeld, *Phys. Rev.* **D 75** (2007) 083506.
5. M.C. Bento, O. Bertolami and A.A. Sen, *Phys. Rev.* **D 70** (2004) 083519.
6. P.J.E. Peebles, 1993, *Principles of physical cosmology* Princeton, N.J.: Princeton University Press.
7. D.N. Spergel *et al.* [WMAP Collaboration], astro-ph/0603449.
8. O. Bertolami, F.G. Pedro and M. Le Delliou, *Phys. Lett.* **B** to appear, astro-ph/0703462.
9. P. Astier *et al.*, Astron. Astrophys. 447 (2006) 31
10. D.J. Eisenstein *et al.* [SDSS collaboration], Astrophys. J. **633** (2005) 560.
11. O. Bertolami, F.Gil Pedro and M.Le Delliou, astro-ph/0705.3118.
12. M. Kesden and M. Kamionkowski, *Phys. Rev. Lett.* **97** (2006) 131303; *Phys. Rev.* D 74 (2006) 083007.
13. A. Füzfa and J.M. Alimi, *Phys. Rev. Lett.* **97** (2006) 061301; *Phys. Rev.* **D 75** (2007) 123007.

New constraints on the observable inflaton potential from WMAP and SDSS

Julien Lesgourgues

LAPTH, Université de Savoie & CNRS, 9 chemin de Bellevue, BP110, F-74941 Annecy-le-Vieux Cedex, France

Abstract. We discuss the latest constraints on a Taylor-expanded scalar inflaton potential, obtained by focusing on its observable part only, and by computing the scalar/tensor primordial spectra numerically. This is in contrast with other works in which an extrapolation of the potential is applied using the slow-roll hierarchy. We find significant differences.

Keywords: Cosmological inflation
PACS: 98.80.Cq

INTRODUCTION

In order to create the coherence on scales as large as the Hubble horizon today, inflation must last about 30 – 60 e-folds after the presently observed spectrum has left the Hubble horizon. The shape of the spectrum of these observed perturbations presently gives us the only hint on the physics that can have driven inflation. The observations alone though, give no information on inflation after the creation of the observed spectrum. Many previous works have used the formalism of flow equations [1] to extrapolate beyond the observed primordial power spectrum, in order to select only those spectra that correspond to a scalar-field potential that drives inflation long enough [2, 3, 4]. This extrapolation puts strong bounds on the shape of the potentials [4], but it can never be model-independent.

In a recent work we presented new bounds on the shape of the inflaton potential within the observable window only, thereby remaining conservative about what (humble or exotic) mechanism drives inflation in the subsequent e-folds [5]. The results were obtained by matching a Taylor-expanded smooth scalar-field potential directly to the data numerically (the data being WMAP[1] 3-year data and the SDSS-LRG[2] sample). These results apply to any inflationary model such that within the observable window, one effectively has one scalar degree of freedom (one inflaton) with a smooth potential.

[1] Wilkinson Microwave Anisotropy Probe [2, 6]

[2] Sloan Digital Sky Survey - Luminous Red Galaxy, data release 4 [7]

CP957, *Particles, Strings, and Cosmology, 13th International Symposium*
edited by A. Rajantie, C. Contaldi, P. Dauncey, and H. Stoica

TABLE 1. Bayesian 68% confidence limits for ΛCDM inflationary models with a Taylor expansion of the inflaton potential at order $v = 2,3,4$ (with the primordial spectra computed numerically). The last line shows the maximum likelihood value.

Parameter	$v = 2$	$v = 3$	$v = 4$
$\Omega_b h^2$	0.022 ± 0.001	0.022 ± 0.001	0.022 ± 0.001
$\Omega_{cdm} h^2$	0.109 ± 0.004	0.109 ± 0.004	0.109 ± 0.004
θ	1.041 ± 0.004	1.041 ± 0.004	1.040 ± 0.004
τ	0.08 ± 0.03	0.09 ± 0.03	0.10 ± 0.03
$\ln\left[\frac{128\pi 10^{10} V_*^3}{3V_*'^2 m_P^6}\right]$	3.06 ± 0.06	3.07 ± 0.06	3.11 ± 0.08
$\left(\frac{V_*'}{V_*}\right)^2 m_P^2$	< 0.4	< 0.4	< 0.8
$\frac{V_*''}{V_*} m_P^2$	0.1 ± 0.5	-0.2 ± 0.6	0.4 ± 0.9
$\frac{V_*'''}{V_*}\frac{V_*'}{V_*} m_P^4$	0	8 ± 5	13 ± 11
$\frac{V_*''''}{V_*}\left(\frac{V_*'}{V_*}\right)^2 m_P^6$	0	0	200 ± 150
$-\ln \mathscr{L}_{\max}$	2688.3	2687.2	2687.2

CONSERVATIVE APPROACH

In our calculations, we evolved the universe for a given $V(\phi)$. During a stage of scalar field domination, the Friedmann equation and the equation of motion of the scalar field can be written with the usual quantities as

$$\dot\phi = -\frac{m_P^2}{4\pi} H'(\phi), \tag{1}$$

$$\left[H'(\phi)\right]^2 - \frac{12\pi}{m_P^2} H^2(\phi) = -\frac{32\pi^2}{m_P^4} V(\phi), \tag{2}$$

where $\dot{}$ denotes a time derivative and $'$ denotes a derivative with respect to ϕ. Initial conditions for the evolution can be chosen in terms of $V(\phi)$ and $\dot\phi$. By demanding that the attractor solution for $\dot\phi$ is reached just before the observable scales leave the Hubble scale, we chose to decrease the number of free parameters. In practice this implies that inflation starts *at least* a few e-folds before observable modes leave the Hubble radius. To do a Monte-Carlo simulation we used the publicly available code COSMOMC [8], which we extended with our own available module [5] to numerically calculate the primordial power spectrum (including tensors), given some potential. We assumed a spatially flat standard ΛCDM-model, with the free parameters shown in Table 1. The Taylor-expanded inflaton potential is parametrized by some convenient combinations of the derivatives at the pivot value (i.e. evaluated when the pivot scale $k_* = 0.01$ Mpc^{-1} exits the Hubble radius during inflation).

CURRENT BOUNDS

The new bounds are displayed in Table 1. The number of degrees of freedom in the potential is denoted by the parameter v, where $v = 2$ denotes an expansion up to the

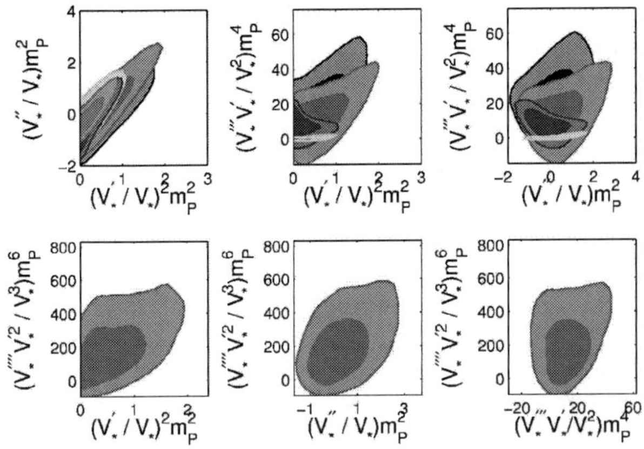

FIGURE 1. Two-dimensional 68% and 95% confidence level contours based on WMAP 3-year and the SDSS LRG spectrum, for the parameters describing the inflaton potential, obtained directly from the MCMC in the case of models $v = 2$ (magenta), $v = 3$ (blue), $v = 4$ (red), or derived from second-order formulas for models $p = 2$ (green), $p = 3$ (black).

second derivative, $v = 3$ up to V''' , etc. Central values and error bounds are obtained by marginalization over all other parameters and given at 68% confidence level. All solutions are symmetric under a sign change in both V' and V'''.

The best likelihoods of the simulations indicate that the improvement by adding the fourth derivative is negligible. As could be expected, the bounds on lower derivatives loosen when one allows higher derivatives. In Figure 1 all 2D-correlations between the potential derivatives are given, marginalized over all other parameters. In this figure results are compared with those obtained by inverting the slow-roll approximation, when probing a scalar and tensor primordial spectrum with a scalar tilt ($p = 2$) or an additional running of the tilt ($p = 3$). The covered area in parameter space is equally large in both approaches, however the overlap is not 100%, which shows that indeed information is lost when mapping from one basis to the other. This indicates that up to second order, the slow roll approximation is not completely accurate within the observed window. The primordial spectra of both simulations on the other hand, which do not depend on approximations, do fully overlap [5]. Finally, in Figure 2 the inflaton potential as bounded by present observations up to 95% confidence level is shown, allowing up to second, third and fourth derivative.

DISCUSSION

Previous works relied on an extrapolation of the potential over a number of e-folds which is an order of magnitude broader than the range probed by the data itself, which is fine if one assumes that the inflaton potential should not be too complicated. We propose here an alternative method which avoids any assumption on the inflation potential outside the observable range. In this approach, the data can be well fitted with a simple linear or

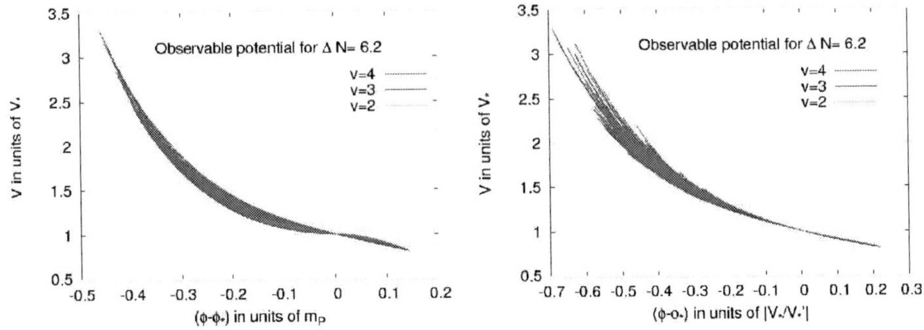

FIGURE 2. *Left:* The shape of the scalar potential at the 95%-confidence level, normalized to the value of the potentials at the pivot scale. *Right:* The same, in a normalization such that all first derivatives are equal, and focus can be laid on the higher derivatives.

quadratic potential. However, when introducing more derivatives, one finds that many potentials allowed by the data have a large tilt running, and even running of the running. The same potentials tend to satisfy slow-roll conditions only marginally on the edges of the observable range; as a consequence, our exact results differ from those obtained by inverting the second-order slow-roll approximation.

Our CosmoMC-compatible module for computing the scalar/tensor primordial spectra numerically is publicly available [5]. It can easily be generalized to other parameterizations of the potential.

ACKNOWLEDGMENTS

This work was supported by the EU 6th Framework Marie Curie Research and Training network "UniverseNet" (MRTN-CT-2006-035863), and initiated during a fruitful stay at the Galileo Galilei Institute for Theoretical Physics with partial support from INFN.

REFERENCES

1. P. J. Steinhardt, and M. S. Turner, *Phys. Rev.* **D29**, 2162–2171 (1984).
2. D. N. Spergel, et al., *Astrophys. J. Suppl.* **170**, 377 (2007), astro-ph/0603449.
3. H. Peiris, and R. Easther, *JCAP* **0607**, 002 (2006), astro-ph/0603587.
4. R. Easther, and H. Peiris, *JCAP* **0609**, 010 (2006), astro-ph/0604214.
5. J. Lesgourgues, and W. Valkenburg, *Phys. Rev.* **D75**, 123519 (2007), astro-ph/0703625.
6. L. Page, et al., *Astrophys. J. Suppl.* **170**, 335 (2007), astro-ph/0603450.
7. M. Tegmark, et al., *Phys. Rev.* **D74**, 123507 (2006), astro-ph/0608632.
8. A. Lewis, and S. Bridle, *Phys. Rev.* **D66**, 103511 (2002), astro-ph/0205436.

Propagation of a string across the cosmological singularity

Przemysław Małkiewicz and Włodzimierz Piechocki

Department of Theoretical Physics, Sołtan Institute for Nuclear Studies, Hoża 69, 00-681 Warszawa, Poland

Abstract. Our results concern the transition of a quantum string through the singularity of the compactified Milne (CM) space. We demonstrate that both classical and quantum dynamics of a string are well defined. Presently available calculations strongly suggest that the singularity of a time dependent orbifold is useless as a model of the cosmological singularity. Our results bring, to some extent, this claim into question.

Keywords: quantum string, Milne space, cosmic singularity
PACS: 98.80.Jk, 11.25.Wx, 98.80.Qc

INTRODUCTION

In the cyclic model of the evolution of the universe the cosmic singularity (CS) plays the key role because it joins each two consecutive classical phases. We propose the following basic criterion for the choice of the model of CS: reasonable model should allow for propagation of quantum p-brane (i.e., particle, string, membrane,...) from the pre-singularity to the post-singularity epoch. If CS constitutes an insurmountable obstacle for quantum elementary objects, the cyclic evolution cannot be realized.

One of the simplest models of the neighborhood of CS, inspired by string/M theory, is the compactified Milne (CM) space. It was used recently in the cyclic universe scenario [1, 2]. In what follows we apply our criterion to the evolution of a test string in CM space. The propagation is well defined both at classical and quantum levels. Our lecture is based on [3].

CLASSICAL DYNAMICS OF A STRING

The Polyakov action for a test p-brane embedded in a fixed background spacetime with metric $g_{\tilde{\mu}\tilde{\nu}}$ has the form

$$S_p = -\frac{1}{2}\mu_p \int d^{p+1}\sigma \sqrt{-\gamma}\, [\gamma^{ab}\partial_a X^{\tilde{\mu}} \partial_b X^{\tilde{\nu}} g_{\tilde{\mu}\tilde{\nu}} - (p-1)], \tag{1}$$

where μ_p is a mass per unit p-volume, $(\sigma^a) \equiv (\sigma^0, \sigma^1, \ldots, \sigma^p)$ are p-brane worldvolume coordinates, γ_{ab} is the p-brane worldvolume metric, $\gamma := det[\gamma_{ab}]$, $(X^{\tilde{\mu}}) \equiv (X^{\mu}, \Theta) \equiv (T, X^k, \Theta) \equiv (T, X^1, \ldots, X^{d-1}, \Theta)$ are the embedding functions of a p-brane, i.e. $X^{\tilde{\mu}} = X^{\tilde{\mu}}(\sigma^0, \ldots, \sigma^p)$, in $d+1$ dimensional background spacetime.

CP957, *Particles, Strings, and Cosmology, 13th International Symposium*
edited by A. Rajantie, C. Contaldi, P. Dauncey, and H. Stoica

The total Hamiltonian, H_T, corresponding to the action (1) is the following

$$H_T = \int d^p\sigma \mathcal{H}_T, \quad \mathcal{H}_T := AC + A^i C_i, \quad i = 1, \ldots, p \tag{2}$$

where $A = A(\sigma^a)$ and $A^i = A^i(\sigma^a)$ are any regular functions of p-volume coordinates,

$$C := \Pi_{\tilde{\mu}}\Pi_{\tilde{\nu}} g^{\tilde{\mu}\tilde{\nu}} + \mu_p^2\, det[\partial_a X^{\tilde{\mu}} \partial_b X^{\tilde{\nu}} g_{\tilde{\mu}\tilde{\nu}}] \approx 0, \tag{3}$$

$$C_i := \partial_i X^{\tilde{\mu}} \Pi_{\tilde{\mu}} \approx 0, \tag{4}$$

and where $\Pi_{\tilde{\mu}}$ are the canonical momenta corresponding to $X^{\tilde{\mu}}$. Equations (3) and (4) define the first-class constraints of the system.

The Hamilton equations are

$$\dot{X}^{\tilde{\mu}} \equiv \frac{\partial X^{\tilde{\mu}}}{\partial \tau} = \{X^{\tilde{\mu}}, H_T\}, \qquad \dot{\Pi}_{\tilde{\mu}} \equiv \frac{\partial \Pi_{\tilde{\mu}}}{\partial \tau} = \{\Pi_{\tilde{\mu}}, H_T\}, \qquad \tau \equiv \sigma^0, \tag{5}$$

where the Poisson bracket is defined by

$$\{\cdot,\cdot\} := \int d^p\sigma \Big(\frac{\partial \cdot}{\partial X^{\tilde{\mu}}} \frac{\partial \cdot}{\partial \Pi_{\tilde{\mu}}} - \frac{\partial \cdot}{\partial \Pi_{\tilde{\mu}}} \frac{\partial \cdot}{\partial X^{\tilde{\mu}}} \Big). \tag{6}$$

In what follows we restrict our considerations to the compactified Milne space. Its metric is defined by the line element

$$ds^2 = -dt^2 + dx^k dx_k + t^2 d\theta^2 = \eta_{\mu\nu} dx^\mu dx^\nu + t^2 d\theta^2 = g_{\tilde{\mu}\tilde{\nu}} dx^{\tilde{\mu}} dx^{\tilde{\nu}}, \tag{7}$$

where $\eta_{\mu\nu}$ is the Minkowski metric, and θ parameterizes a circle[1].

The spacetime with the metric (7) is singular at $t = 0$ and it is not a coordinate choice singularity: one of its space dimensions (denoted by θ) disappears for a moment at $t = 0$ leading to the degeneracy of the metric. It is a spacelike orbifold singularity.

In the paper we analyze the dynamics of a string in the *zero-mode* (the lowest energy state) which is *winding* around the θ-dimension. The string in such a state is defined by the conditions

$$\sigma^p := \theta \equiv \Theta \quad \text{and} \quad \partial_\theta X^\mu = 0 = \partial_\theta \Pi_\mu. \tag{8}$$

The constraint equations, (3) and (4), for a string winding around the θ-dimension reduce to the following form

$$C = \Pi_\mu(\tau)\, \Pi_\nu(\tau)\, \eta^{\mu\nu} + \breve{\mu}_1^2\, t^2(\tau) \approx 0, \qquad C_1 = 0, \tag{9}$$

where $\breve{\mu}_1 \equiv 2\pi\mu_1$, and the equations of motion (5) read

$$\dot{\Pi}_t(\tau) = -2A(\tau)\, \breve{\mu}_1^2\, T(\tau), \qquad \dot{\Pi}_k(\tau) = 0, \tag{10}$$

---

[1] Orbifolding S^1 to the segment S^1/Z_2 gives a model of spacetime in the form of two orbifold planes which collide and re-emerge at $t = 0$.

and

$$\dot{T}(\tau) = -2A(\tau)\,\Pi_t(\tau), \qquad \dot{X}^k(\tau) = 2A(\tau)\,\Pi_k(\tau), \tag{11}$$

where $A = A(\tau)$ is any regular function.

One can show [3] that the solution to (10) and (11) reads

$$X^k(t) = X_0^k + \frac{\Pi_0^k}{\breve{\mu}_1}\sinh^{-1}\Big(\frac{\breve{\mu}_1}{\sqrt{\Pi_0^k \Pi_{0k}}}\,t\Big), \qquad t \equiv T. \tag{12}$$

The solution (12) is bounded and continuous at the singularity. Thus, the classical dynamics of the zero-mode winding string is well defined in CM space.

QUANTUM DYNAMICS OF A STRING

The Hamiltonian of a string is

$$H_T = C = \Pi_\mu(\tau)\,\Pi_\nu(\tau)\,\eta^{\mu\nu} + \breve{\mu}_1^2\,t^2. \tag{13}$$

The quantum Hamiltonian corresponding to (13) reads

$$\hat{H}_T = \frac{\partial^2}{\partial t^2} - \frac{\partial^2}{\partial X^k \partial X_k} + \breve{\mu}_1^2 t^2. \tag{14}$$

According to the Dirac quantization method the physical states ψ should satisfy

$$\hat{H}_T\,\psi(t, X^k) = 0. \tag{15}$$

To solve (15), we make the substitution

$$\psi(t, X^1, \ldots, X^{d-1}) = F(t)\,G_1(X^1)\,G_2(X^2)\cdots G_{d-1}(X^{d-1}), \tag{16}$$

which turns (15) into the following set of equations

$$\frac{d^2 G_k(q_k, X_k)}{dX_k^2} + q_k^2\,G_k(q_k, X_k) = 0, \qquad k = 1, \ldots, d-1, \tag{17}$$

$$\frac{d^2 F(q,t)}{dt^2} + (\breve{\mu}_1^2 t^2 + q^2)\,F(q,t) = 0, \qquad q^2 := q_1^2 + \ldots + q_{d-1}^2, \tag{18}$$

where $q_k^2, q^2 \in R$ are the separation constants.

Two independent solutions to (17) have the form

$$G_{1k}(q_k, X_k) = \cos(q_k X^k), \qquad G_{2k}(q_k, X_k) = \sin(q_k X^k), \qquad k = 1, \ldots, d-1 \tag{19}$$

(there is no summation in $q_k X^k$ with respect to k).

Two independent solutions of (18) read

$$F_1(q,t) := \sqrt{q}\,\exp\big(-\frac{\pi}{8\breve{\mu}_1}q^2\big)\exp\big(-i\breve{\mu}_1 t^2/2\big)\,H\Big(-\frac{\breve{\mu}_1 + iq^2}{2\breve{\mu}_1}, (-1)^{1/4}\sqrt{\breve{\mu}_1}\,t\Big), \tag{20}$$

$$F_2(q,t) = \exp(-i\breve{\mu}_1 t^2/2)\, {}_1F_1\Big(\frac{\breve{\mu}_1 + iq^2}{4\breve{\mu}_1}, \frac{1}{2}, i\breve{\mu}_1 t^2\Big), \tag{21}$$

where $H(a,t)$ is the Hermite function and ${}_1F_1(a,b,t)$ denotes the Kummer confluent hypergeometric function. In what follows we consider only the neighborhood of $t = 0$, i.e., $t \in [-t_0, t_0]$. We construct a Hilbert space, $\mathcal{H}$, of our system based on the solutions (19)-(21) by making use of the mapping

$$L^2(R^{d-1}) \ni f \longrightarrow h_s \in L^2([-t_0,t_0] \times R^{d-1}) =: \tilde{\mathcal{H}}, \tag{22}$$

where

$$h_s(t, X^1, \ldots, X^{d-1}) := \int_{R^{d-1}} f(q_1, \ldots, q_{d-1})\, F_s(q,t) \prod_k \exp(-iq_k X^k)\, dq_1 \ldots dq_{d-1}, \tag{23}$$

and where $q^2 = q_1^2 + \ldots q_{d-1}^2$.
Replacing f by consecutive elements of an orthonormal basis in $L^2(R^{d-1})$ leads to an infinite countable set of vectors in $\tilde{\mathcal{H}}$. So obtained set of vectors can be rearranged into a set of independent vectors and further turned into an orthonormal basis by making use of the Gram-Schmidt procedure. One can show that the span of such an orthonormal basis, $\mathcal{F}$, is dense in $\tilde{\mathcal{H}}$. Completion of $\mathcal{F}$ in the scalar product of $\tilde{\mathcal{H}}$ defines $\mathcal{H}$.

CONCLUSIONS

Our results demonstrate that the dynamics of a *test* string winding around the compactified dimension in the zero-mode state is well defined. Our non-perturbative method of quantization leads to a non-singular quantum evolution of a string. A great challenge is examination of the dynamics of a *physical* string, i.e. a string that may change its state and modify the background spacetime during the evolution of the entire system. It is strongly suggested in [4] that, for instance, a physical string added to a time dependent singular orbifold may cause the orbifold to collapse into a large black hole, which leads finally to the creation of a big-crunch that would not be followed by a big-bang. This result suggests that the compactified Milne space is useless as a model of the cosmological singularity in the context of the cyclic universe scenario.

However, the result of [4] should be treated with caution because it was obtained by making use of semi-classical methods. Real solution of the problem requires quantization of the entire system, including string and spacetime. We plan to carry out such analyzes by making use of the loop quantum gravity formalism [5].

REFERENCES

1. P. J. Steinhardt and N. Turok, *Science* **296**, 1436 (2002).
2. P. J. Steinhardt and N. Turok, *Phys. Rev.* D **65**, 126003 (2002).
3. P. Małkiewicz and W. Piechocki, *Class. Quant. Grav.* **24**, 915 (2007).
4. G. T. Horowitz and J. Polchinski, *Phys. Rev.* D **66**, 103512 (2002).
5. T. Thiemann, *Modern Canonical Quantum General Relativity*, Cambridge University Press, Cambridge, 2007.

Effects of non-standard neutrino-electron interactions in relic neutrino decoupling

Teguayco Pinto

Instituto de Física Corpuscular (CSIC-Universitat de València)
Ed. Institutos de Investigación, E-46071 Valencia, Spain

Abstract. We consider the decoupling of neutrinos in the early Universe in presence of non-standard neutral current neutrino-electron interactions (NSI). We present the results of fully numerical and momentum-dependent calculations, including flavor neutrino oscillations. We find that the presence of neutrino-electron NSI may enhance the entropy transfer from electron-positron pairs into neutrinos instead of photons, up to a value of the effective number of neutrinos $N_{\rm eff} \simeq 3.12$ for NSI parameters within the ranges allowed by present laboratory data, which is almost three times the effect that appears for standard weak interactions. Thus non-standard neutrino-electron interactions do not essentially modify the density of relic neutrinos nor the bounds on neutrino properties from cosmological observables, such as their mass.

Keywords: Early Universe; Neutrinos; Non-equilibrium kinetics
PACS: 14.60.Pq, 98.80.Es

INTRODUCTION

In the early Universe, neutrinos were kept in thermal contact with the electromagnetic primordial plasma by rapid weak interactions with electrons and positrons. When the temperature dropped below a few MeV, these weak processes became ineffective and the process of neutrino decoupling took place, while shortly after the $e^{\pm}$ pairs began to annihilate almost entirely into photons thus producing a difference between the temperatures of the relic photons and neutrinos. However some relic interactions between $e^{\pm}$ and neutrinos exist all along the $e^{\pm}$ annihilation stage, leading to a slightly smaller increase of the comoving photon temperature and to small distortions (at the percent level) of the neutrino momentum distributions.

Presently, there exist compelling evidences for flavor neutrino oscillations from a variety of experimental data on solar, atmospheric, reactor and accelerator neutrinos (see e.g. [1]). Interestingly, non-zero neutrino masses usually come with non-standard interactions (NSI) that might violate leptonic flavor and/or break weak universality. Recent analyses [2, 3, 4] have considered the neutral current NSI in a phenomenological way, showing that they can be bounded.

We have studied the neutrino decoupling process in presence of additional interactions between neutrinos and electrons [5],a possibility already noted in [2, 3]. In this case, neutrinos could be kept in longer contact with $e^{\pm}$ and thus share a larger amount of the total entropy transfer than in the SM. Actually, if the non-standard neutrino-electron interactions were large enough, the neutrino momentum distribution would be significantly different from the standard case. In turn, this would modify the final yield of light nuclei during the epoch of Big Bang Nucleosynthesis (BBN), as well as the radiation content of

CP957, *Particles, Strings, and Cosmology, 13th International Symposium*
edited by A. Rajantie, C. Contaldi, P. Dauncey, and H. Stoica

the Universe, affecting the anisotropies of the Cosmic Microwave Background (CMB) and the power spectrum of Large Scale Structures (LSS). Our goal is to calculate how the decoupling is modified taking into account NSI with couplings which are still allowed by present laboratory data, and to discuss if cosmological observations can be used as a complementary way to bound these exotic scenarios.

Here we will follow [2, 3] and assume that new physics induces NSI only through the four fermion operators $(\bar{\nu}\nu)(\bar{f}f)$, where f is a charged lepton or quark, but not new charged lepton physics at tree level. In particular, since we are interested in the decoupling process of relic neutrinos, we consider only the NSI related to electrons which, together with the standard weak interactions, are described by the effective Lagrangian $\mathscr{L}_{\rm eff} = \mathscr{L}_{\rm SM} + \sum_{\alpha,\beta} \mathscr{L}^{\alpha\beta}_{\rm NSI}$; which contains the four-fermion terms

$$
\begin{aligned}
\mathscr{L}_{\rm SM} &= -2\sqrt{2}G_F \left\{ (\bar{\nu}_e\gamma^\mu L\nu_e)(\bar{e}\gamma_\mu Le) + \sum_{P,\alpha} g_P(\bar{\nu}_\alpha\gamma^\mu L\nu_\alpha)(\bar{e}\gamma_\mu Pe) \right\} && (1) \\
\mathscr{L}^{\alpha\beta}_{\rm NSI} &= -2\sqrt{2}G_F \sum_P \varepsilon^P_{\alpha\beta}\, (\bar{\nu}_\alpha\gamma^\mu L\nu_\beta)(\bar{e}\gamma_\mu Pe) && (2)
\end{aligned}
$$

for energies much smaller than the Z boson mass, as in our case, with G_F the Fermi constant and $P = L, R = (1 \mp \gamma_5)/2$ the chiral projectors. Greek indices label lepton flavors ($\alpha, \beta = e, \mu, \tau$) and the Z couplings are $g_L = -\frac{1}{2} + \sin^2\theta_W$ and $g_R = \sin^2\theta_W$.

Our results are shown in terms of both the impact on the abundance of ^{4}He produced during Big Bang Nucleosynthesis (ΔY_p) and the effective number of neutrino species ($N_{\rm eff}$). It is a usual way to parameterize the neutrino contribution to the radiation energy density $\rho_{\rm R} = \rho_\gamma + \rho_\nu + \rho_X$ [6, 7]

$$
N_{\rm eff} \equiv \frac{\rho_{\nu+X}}{\rho^0_\nu} \frac{\rho^0_\gamma}{\rho_\gamma}, \tag{3}
$$

where the energy densities ρ^0_γ and ρ^0_ν refer respectively to the photon plasma and to a single neutrino species in the limit of instantaneous decoupling, while ρ_γ is the actual energy content of the photon plasma and $\rho_{\nu+X}$ the total energy content of weakly interacting particles (including possible exotic contributions).

NUMERICAL RESULTS

A proper calculation of the process of non-instantaneous neutrino decoupling demands solving the momentum-dependent Boltzmann equations for the neutrino spectra, a set of integro-differential kinetic equations that are difficult to solve numerically. See [5, 8] for details. In particular, we will describe the neutrino ensemble in the usual way by generalized occupation numbers where the off-diagonal terms are non-zero in the presence of neutrino mixing or flavor-changing NSI. The kinetic equations should be numerically solved along with the equation governing the evolution of the electromagnetic temperature during the process of $e^\pm$ annihilations. This can be found from the continuity equation for the total energy density ρ.

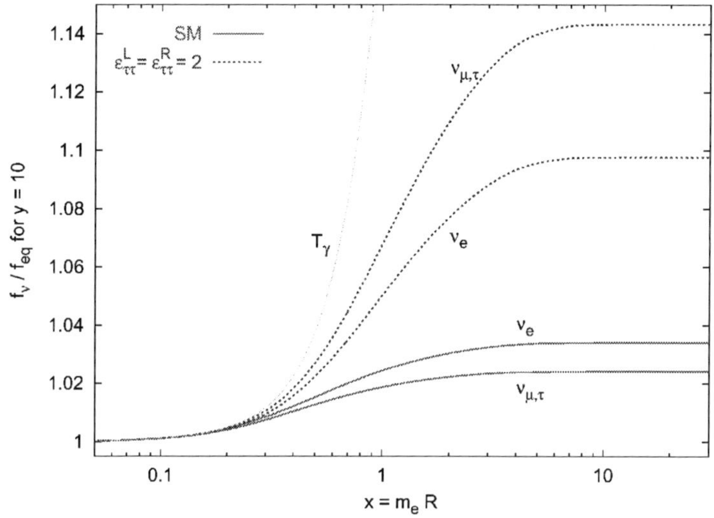

FIGURE 1. Evolution of the distortion of the ν_e and $\nu_{\mu,\tau}$ spectra for a particular comoving momentum ($y = 10$) with standard weak interactions (solid line) and with $\nu_\tau - e$ NSI (dotted line). The line labelled with T_γ corresponds to neutrinos in full thermal contact with the electromagnetic plasma [5].

We have numerically calculated the evolution of the neutrino density matrix solving a system of equations, during the full process of neutrino decoupling. To do this we define the dimensionless variables $x \equiv mR$ and $y \equiv pR$ instead of time and momenta, respectively, where m is an arbitrary mass scale which we choose as the electron mass and R is the Universe scale factor. We performed the calculation for a selection of values for the NSI parameters, representative of the regions allowed by present bounds [4].

As an example, in Fig. 1 we plot the distortion of the neutrino distribution as a function of x for a particular neutrino momentum ($y = 10$), both in the case of SM weak interactions and in presence of large NSI between tau neutrinos and $e^\pm$, corresponding to $\varepsilon^L_{\tau\tau} = \varepsilon^R_{\tau\tau} = 2$. The behavior of f_ν has been previously described e.g. in [5, 8, 9, 10]. In the case of large $\nu_\tau - e$ NSI, neutrinos are kept longer in thermal contact with $e^\pm$, leading to larger distortions for all neutrino flavors, in particular for $\nu_{\mu,\tau}$. For this case, the final values of the distribution are 9.8% for the ν_e's and 14.3% for the $\nu_{\mu,\tau}$'s larger than in the limit of neutrino decoupling before any $e^\pm$ annihilation. For comparison, the values for standard weak interactions are 4.4% for the ν_e's and 2% for the $\nu_{\mu,\tau}$'s.

Finally in Table 1 we show the results for different sets of parameters in terms of $N_{\rm eff}$, the distortions over the neutrino distribution function δ and ΔY_p. In general, we find that the presence of non-zero NSI enhances the transfer of entropy from $e^\pm$ to neutrinos, leading to values up to $N_{\rm eff} \simeq 3.12$ when all NSI parameters are close to the boundaries of the allowed regions, which is almost three times the departure of $N_{\rm eff}$ from 3 that already exists for SM weak interactions (first row in Table 1). On the other hand, for particular choices of the diagonal NSI parameters that minimize the $\nu - e$ interaction we find a reduction in the final value of the distortions or $N_{\rm eff}$ with respect to the SM value.

TABLE 1. Results for various choices of NSI parameters [5] (see text for details).

ε_{ee}^{L}	ε_{ee}^{R}	$\varepsilon_{\tau\tau}^{L}$	$\varepsilon_{\tau\tau}^{R}$	$\varepsilon_{e\tau}^{L}$	$\varepsilon_{e\tau}^{R}$	$\delta_e(\%)$	$\delta_{\mu,\tau}(\%)$	$N_{\rm eff}$	ΔY_p
0	0	0	0	0	0	0.73	0.52	3.046	2.1×10^{-4}
0	0	0	0	0.85	0.38	1.46	1.16	3.085	4.4×10^{-4}
0.12	-1.58	-0.5	0.5	-0.85	0.38	2.21	1.66	3.120	6.0×10^{-4}

CONCLUSIONS

We find that the enhancement in observables like $N_{\rm eff}$, or the ^{4}He mass yield in the BBN can be up to three times larger than the ones found in the standard case [11], for values of all NSI parameters close to the limits placed by laboratory experiments. Nevertheless, even the variation of up to 0.2% that we found in the ^{4}He abundance is yet too small (about one order of magnitude) when compared to the observational error on Y_p. Instead, a value of $N_{\rm eff} \simeq 3.12$, which is still possible within the present parameter space for NSI couplings, might even be barely detectable [12, 13].

In any case we conclude that the prediction of $N_{\rm eff} \simeq 3$ within a few $\%$ is quite robust even when taking into account neutrino-electron NSI of the four-fermion type. Thus, their existence can not modify in a significant way the bounds on neutrino properties from cosmological observables [14]. Since it is likely that future experiments may narrow further the allowed range of NSI couplings, their role in a cosmological context will be even smaller, at most a sub-leading correction to the standard prediction of the Cosmic Neutrino Background properties. Turning the argument around, however, one can conclude that a significant deviation of $N_{\rm eff}$ from 3 may require major revisions of the cosmological model, like the introduction of new relativistic relics and/or of an exotic thermal history, or both (see e.g. [15, 16]).

REFERENCES

1. M. Maltoni, T. Schwetz, M.A. Tórtola and J.W.F. Valle, New J. Phys. **6** (2004) 122 [hep-ph/0405172].
2. Z. Berezhiani and A. Rossi, Phys. Lett. B **535** (2002) 207 [hep-ph/0111137].
3. S. Davidson, C. Peña-Garay, N. Rius and A. Santamaria, JHEP **0303** (2003) 011 [hep-ph/0302093].
4. J. Barranco et al, Phys. Rev. D **73** (2006) 113001 [hep-ph/0512195].
5. G. Mangano et al, Nucl. Phys. B **756** (2006) 100 [arXiv:hep-ph/0607267].
6. V.F. Shvartsman, Pisma Zh. Eksp. Teor. Fiz. **9** (1969) 315 [JETP Lett. **9** (1969) 184].
7. G. Steigman, D.N. Schramm and J.R. Gunn, Phys. Lett. B **66** (1977) 202.
8. G. Mangano et al, Nucl. Phys. B **729** (2005) 221 [hep-ph/0506164].
9. A.D. Dolgov, S.H. Hansen and D.V. Semikoz, Nucl. Phys. B **503** (1997) 426 [hep-ph/9703315].
10. S. Esposito et al, Nucl. Phys. B **590** (2000) 539 [astro-ph/0005573].
11. G. Mangano et al, Phys. Lett. B **534** (2002) 8 [astro-ph/0111408].
12. R. Bowen et al, MNRAS **334** (2002) 760 [astro-ph/0110636].
13. S. Bashinsky and U. Seljak, Phys. Rev. D **69** (2004) 083002 [astro-ph/0310198].
14. J. Lesgourgues and S. Pastor, Phys. Rept. **429** (2006) 307 [astro-ph/0603494].
15. P.D. Serpico and G.G. Raffelt, Phys. Rev. D **70** (2004) 043526 [astro-ph/0403417].
16. A. Cuoco et al, Phys. Rev. D **71** (2005) 123501 [astro-ph/0502465].

Constraints on scale-dependent non-Gaussianity

Sarah E. Shandera

Institute of Strings, Cosmology and Astroparticle Physics, Physics Department, Columbia University, New York, NY 10027

Abstract. We review why detection of non-Gaussianity in the spectrum of primordial fluctuations would be an indication of interesting inflationary physics and discuss the observational constraints on a simple type of scale-dependent non-Gaussianity. In particular, if the amount non-Gaussianity increases during inflation then observations on scales smaller than those probed by the Cosmic Microwave Background may provide important constraints. Clusters number counts can be a useful tool in this context.

Keywords: Inflation, Structure formation
PACS: 11.25.Wx, 98.80.Bp

INTRODUCTION

Although the most recent measurements of the Cosmic Microwave Background (CMB) strongly support inflation as a mechanism to solve the horizon and flatness problems and to generate the spectrum of observed temperature fluctuations [1], these observations do not currently say much about the fundamental origin of inflation. Any theory that can produce standard slow-roll inflation with the correct normalization and scale-dependence in the power spectrum can claim success (not that this has been overwhelmingly easy to find in string theory). However, there are some observables that may prove to be more powerful discriminators. Non-Gaussianity is particularly valuable because it is sensitive to inflaton self-interactions that cannot be constrained by the power spectrum alone. In addition, a slowly and smoothly rolling scalar field will not generate observably large non-Gaussianity, so if a non-Gaussian spectrum is observed it would provide an important indication of physics different from the conventional scenario.

While standard single field, slow-roll inflation predicts only unobservable deviations from Gaussianity [2, 3], Mukhanov and others [4, 5] have investigated the case of a Lagrangian which is a general function of the inflaton and powers of its first derivative. These models are characterized by a sound speed c_s different from $c = 1$ during inflation. Terms involving powers of derivatives can generate large interaction terms without spoiling the slow-roll aspects of the potential (and in fact can allow even a steep potential to give sufficient inflation by slowing down the field [6]). The primordial bispectrum of these scenarios has been worked out in detail [7, 8]. For small sound speed, these terms can generate non-Gaussianity that is observably large. In addition, the sound speed may change during inflation, leading to scale-dependent non-Gaussianity. Interestingly, large non-Gaussianity even in these models is somewhat surprising from an effective field theory point of view [9]. However, there is some motivation for large non-Gaussianity from string theory, where an infinite sum of higher-derivative terms sometimes appears.

CP957, *Particles, Strings, and Cosmology, 13th International Symposium*
edited by A. Rajantie, C. Contaldi, P. Dauncey, and H. Stoica

Examples are relativistic brane inflation [6] (where the time derivative of the inflaton appears under a square-root) and inflation from stringy tachyon models [10].

With this motivation in mind, it is interesting to ask how scale-dependent non-Gaussianity can best be constrained. Although one should reanalyze the CMB constraints, it perhaps more useful to get a longer lever arm by looking at the effects on large-scale structure. Currently, CMB constraints extend to about $k = 0.05 h\,Mpc^{-1}$ while clusters cover a range of about $k = 0.1 - 0.3 h\,Mpc^{-1}$. If the primordial distribution is skewed, then the number of large fluctuations which develop into galaxy clusters can change significantly [11, 12, 13, 14]. Below we will briefly outline the necessary calculation and some results. This talk is based on [15].

SCALE-DEPENDENT NON-GAUSSIANITY

The ansatz for non-Gaussianity often used in the literature is

$$\zeta(\mathbf{x}) = \zeta_G(\mathbf{x}) + \frac{3}{5} f_{NL} \left(\zeta_G(\mathbf{x})^2 - \langle \zeta_G^2(\mathbf{x}) \rangle \right) \tag{1}$$

where f_{NL} is a constant, ζ is the primordial curvature and ζ_G is a Gaussian field. One arrives at an expression for the matter fluctuations δ (in Fourier space) using the Poisson equation and evolving the perturbations forward to the matter era

$$\delta(\mathbf{k}, a) = \frac{2}{5} \frac{1}{\Omega_m} \frac{1}{H_0^2} D(a) T(k) k^2 \zeta(\mathbf{k}) \tag{2}$$

Where Ω_m is the fractional density of matter today, $H_0 = 100h$ is the present value of the Hubble parameter, $D(a)$ is the linear growth function, and $T(k)$ is the transfer function.

However, Eq.(1) does not allow for a scale-dependent non-Gaussianity, and in fact the shape of its 3-point function in Fourier space is quite different from that predicted by small sound speed models. For non-Gaussianity generated by higher derivative terms, the maximum contribution comes from triangles with all three momenta equal. One can then define an f_{NL}^{eq} by evaluating the 3-point at $k_1 = k_2 = k_3$ and incorporate a scale dependence by writing

$$f_{NL}^{eq}(k) = f_{NL,\,pivot}^{eq} \left(\frac{k}{k_{pivot}} \right)^{-2\kappa} \tag{3}$$

Here the parameter κ is related to the changing sound speed.

CONSTRUCTING THE DISTRIBUTION AND COUNTING CLUSTERS

The 3-point function is only one of the moments of the probability distribution $P(\delta)$, and although in principle they can all be found from the action, in practice this is a lengthy calculation. However, one can approximate the distribution using a well-known technique (see [16] for a derivation), which leads to the Edgeworth expansion:

$$P(\nu)d\nu = \frac{d\nu}{\sqrt{2\pi}}e^{-\nu^2/2}\left[1+\sigma_R\frac{S_3(R)}{6}H_3(\nu) + \sigma_R^2\left(\frac{S_4(R)}{24}H_4(\nu)+\frac{S_3(R)^2}{72}H_6(\nu)\right)+\ldots\right] \quad (4)$$

Here $\nu = \delta_R/\sigma_R$ (σ_R is the smoothed variance), H_n are Hermite polynomials and $S_n(R)$ are the reduced cumulants of the distribution (which vanish for a Gaussian) smoothed on a scale R. If only the first few terms in the expansion are kept, this expression has a limited range of validity. In particular, it can develop negative regions. Here we include only S_3, and so our results are limited in range in both mass and red-shift.

To count the number of clusters expected for a particular inflationary theory, we use the Press-Schechter approach and compute the fraction of matter in collapsed objects, f_c. That is, we find the probability that the density fluctuation smoothed on the scale corresponding to mass M objects is above the threshold for collapse, δ_c:

$$f_c = 2P(>\delta_c, M) = 2\int_{\delta_c}^{\infty} d\delta P(\delta, M) \quad (5)$$

where the distribution Eq.(4) has been used with $S_3(R)$ calculated from the results of [8]. One can use more accurate expressions, which we do in [15]. Then with simple manipulations, one can find expressions for the number of clusters as a function of mass for fixed redshift, $\frac{dn}{dM}(M,z)$ or count all clusters above some limiting mass per red-shift interval, $\frac{dn}{dz}(M_{lim})$. Results for a negative skewness model (and Gaussian window function) are shown in Figure 1.

DISCUSSION

Of course, cluster counts are sensitive to quantities other than the primordial non-Gaussianity. To distinguish the curves in Figure 1, we must be able to determine the cluster mass and σ_8 (the amplitude of fluctuations on 8Mpc scales) to an accuracy of about 2%. Then with an all-sky survey we might hope to constrain f_{NL}^{eq} on these scales to within ± 300 (one can do better if f_{NL} constant is assumed) and, in the event large non-Gaussianity is observed, one can distinguish κ of about 0.1. In addition to being sensitive to scale-dependence, clusters counts have very different systematics from CMB measurements, and so provide an important independent constraint. Finally, one might expect even better constraints from the galaxy bispectrum, which we discuss briefly in [15] but a full analysis remains to be done.

ACKNOWLEDGMENTS

I am grateful to my collaborators in this work: Marilena LoVerde, Amber Miller and Licia Verde.

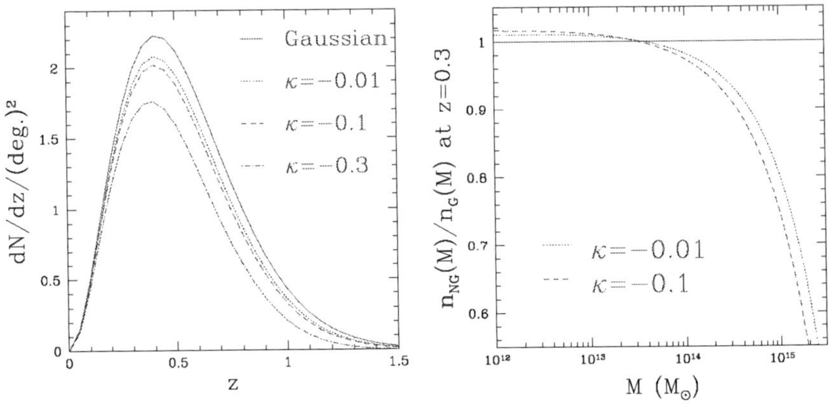

(a) (b)

FIGURE 1. Results for non-Gaussianity consistent with CMB constraints and growing with scale. The example plotted has negative skewness, so the number of clusters is more suppressed for larger $|\kappa|$. Panel (a) shows clusters with mass above 2.5×10^{14} solar masses. Panel (b) shows the ratio of expected cluster counts for the non-Gaussisn case to the Gaussian for redshift $z = 0.3$. For this redshift the Edgeworth expansion becomes unreliable around $M = 2 \times 10^{15}$ solar masses.

REFERENCES

1. D. N. Spergel, et al. (2006), astro-ph/0603449.
2. V. Acquaviva, N. Bartolo, S. Matarrese, and A. Riotto, *Nucl. Phys.* **B667**, 119–148 (2003), astro-ph/0209156.
3. J. M. Maldacena, *JHEP* **05**, 013 (2003), astro-ph/0210603.
4. C. Armendariz-Picon, T. Damour, and V. F. Mukhanov, *Phys. Lett.* **B458**, 209–218 (1999), hep-th/9904075.
5. J. Garriga, and V. F. Mukhanov, *Phys. Lett.* **B458**, 219–225 (1999), hep-th/9904176.
6. E. Silverstein, and D. Tong (2003), hep-th/0310221.
7. D. Seery, and J. E. Lidsey, *JCAP* **0506**, 003 (2005), astro-ph/0503692.
8. X. Chen, M.-x. Huang, S. Kachru, and G. Shiu (2006), hep-th/0605045.
9. P. Creminelli, *JCAP* **0310**, 003 (2003), astro-ph/0306122.
10. N. Barnaby, and J. M. Cline (2007), arXiv:0704.3426[hep-th].
11. F. Lucchin, and S. Matarrese, *Astrophys. J.* **330**, 535–544 (1988).
12. W. A. Chiu, J. P. Ostriker, and M. A. Strauss, *Astrophys. J.* **494**, 479–490 (1998), astro-ph/9708250.
13. J. Robinson, and J. E. Baker (1999), astro-ph/9905098.
14. L. Verde, R. Jimenez, M. Kamionkowski, and S. Matarrese, *Mon. Not. Roy. Astron. Soc.* **325**, 412 (2001), astro-ph/0011180.
15. S. Shandera, M. LoVerde, A. Miller, and L. Verde, *to appear* (2007).
16. F. Bernardeau, S. Colombi, E. Gaztanaga, and R. Scoccimarro, *Phys. Rept.* **367**, 1–248 (2002), astro-ph/0112551.

Inflaton Decay in Supergravity and Gravitino Problem

Fuminobu Takahashi

Deutsches Elektronen Synchrotron DESY, Notkestrasse 85, 22607 Hamburg, Germany

Abstract. We have recently shown that, if the inflaton has a nonzero vacuum expectation value, it generically couples to any matter fields that appear in the superpotential at the tree level, and to any gauge sectors through anomalies in the supergravity. Through these processes, the inflaton generically decays into the supersymmetry breaking sector, producing many gravitinos. The inflaton also directly decays into a pair of the gravitinos. Taking account of these processes, we derive constraints on both inflation models and supersymmetry breaking scenarios for avoiding overproduction of the gravitinos.

Keywords: inflation, reheating, gravitino, supergravity
PACS: 98.80.Cq,11.30.Pb,04.65.+e

INTRODUCTION

Recently there has been much progress concerning the decays of scalar fields such as moduli [1, 2, 3] and inflaton [4, 5, 6, 7, 8, 9] in a framework of the local supersymmetry (SUSY), i.e., the supergravity (SUGRA). The supersymmetric extension is one of the most promising candidates for the theory beyond SM. If SUSY exists at the TeV scale, the inflaton dynamics is quite likely described in SUGRA. In addition, since the existence of a flat direction is mediocre in SUSY models, one can find extremely flat potentials appropriate for the slow-roll inflation. Throughout this paper we consider inflation models in SUGRA. We have investigated the reheating of the universe in this framework, and found that the gravitinos are generically produced from the inflaton decay in most inflation models. In particular, Ref. [4] has first pointed out that the inflaton can directly decay into a pair of the gravitinos. Moreover, incorporating the gravitational effects, Refs. [6, 8] have shown that the inflaton generically decays into the SUSY breaking sector, which produces the gravitinos (in)directly. The gravitino production rates due to these processes depend on the inflaton parameters as well as the detailed structure of the SUSY breaking sector. Such gravitino production clearly goes beyond the simplification of the reheating that has been adopted so far, and interestingly enough, it provides severe constraints on inflation models as well as the SUSY breaking scenarios. These constraints, together with the future collider experiments and observations on CMB, should become an important guide to understand the high energy physics and the early universe. The purpose of the article is to provide both a rough sketch of the gravitino production and the constraints on the representative inflation models. For more details, the reader should refer to the original references or the review [9].

CP957, *Particles, Strings, and Cosmology, 13th International Symposium*
edited by A. Rajantie, C. Contaldi, P. Dauncey, and H. Stoica

GRAVITINO PRODUCTION

Let us first briefly review the recent development on the gravitino production from the inflaton decay. There are three gravitino production processes; (a) the gravitino pair production [4, 5, 1, 2, 3]; (b) spontaneous decay at tree level [6]; (c) anomaly-induced decay at one-loop level [8]. For the processes listed above, the gravitino production rate can be expressed as

$$\Gamma_{3/2} = \frac{x}{32\pi}\left(\frac{\langle\phi\rangle}{M_P}\right)^2 \frac{m_\phi^3}{M_P^2}, \tag{1}$$

where m_ϕ is the inflaton mass, $\langle\phi\rangle$ a vacuum expectation value (VEV) of the inflaton, and $M_P = 2.4\times10^{18}$GeV the reduced Planck mass. Here it should be noted that $\langle\phi\rangle$ is evaluated at the potential minimum after inflation. The precise value of the numerical coefficient x depends on the production processes, possible non-renormalizable couplings in the Kähler potential, and the detailed structure of the supersymmetry (SUSY) breaking sector [9]. To be concrete, let us assume the minimal Kähler potential and the dynamical SUSY breaking (DSB) [10] with a dynamical scale Λ. In the DSB scenario, the SUSY breaking field z can acquire a large mass m_z, which is assumed to be roughly equal to the dynamical scale $\Lambda \sim \sqrt{m_{3/2}M_P}$ in the following. Such a simplification does not essentially change our arguments. For a low-inflation model with $m_\phi < \Lambda$, the process (a) becomes effective, and $x = 1$. On the other hand, for the inflaton mass larger than Λ, the processes (b) and (c) become effective instead. The inflaton decays into the hidden quarks in the SUSY breaking sector via Yukawa couplings (process (b)), or into the hidden gauge sector via anomalies (process (c)). Since the hidden quarks and gauge bosons (and gauginos) are energetic when they are produced, they are expected to form jets and produce hidden hadrons through the strong gauge interactions. The gravitinos are likely generated by the decays of the hidden hadrons as well as in the cascade decay processes in jets. We denote the averaged number of the gravitinos produced per each jet as $N_{3/2}$. Then x is given by [9] [1]

$$x = \frac{N_{3/2}}{8\pi^2}\left(\frac{1}{2}N_y|Y_h^2| + N_g\alpha_h^2(T_g^{(h)} - T_r^{(h)})^2\right), \tag{2}$$

where Y_h and α_h are the Yukawa coupling and a fine structure constant of the hidden gauge group, respectively, N_y denotes a number of the final states for the process (b), N_g is a number of the generators of the gauge group, and $T_g^{(h)}$ and $T_r^{(h)}$ are the Dynkin indices of the adjoint representation and the matter fields in the representation r. Although x depends on the structure of the SUSY breaking sector, its typical magnitude is $O(10^{-3}-10^{-2})$ for $m_\phi > \Lambda$ [2]. To be explicit we take $x = 1/(8\pi^2)$ in the following analysis.

---

[1] If the Kähler potential takes a form of the sequestered type, the spontaneous decay through Yukawa couplings is suppressed [8, 9].

[2] Roughly, we expect $N_{3/2} = O(1-10^2)$, $N_g = O(1)$, $\alpha_h = O(0.1)$, and $T_g^{(h)} - T_r^{(h)} = O(1)$, while Y_h strongly depends on the SUSY breaking models. Note also that the gravitino can be produced through the Yukawa interaction in the messenger sector, if the inflaton mass is larger than the messenger scale.

Using the gravitino production rate given above, we can estimate the abundance of the gravitinos non-thermally produced by an inflaton decay:

$$
\begin{aligned}
Y_{3/2}^{(NT)} &= 2\frac{\Gamma_{3/2}}{\Gamma_\phi}\frac{3T_R}{4m_\phi}, \\
&\simeq 7\times 10^{-11} x \left(\frac{g_*}{200}\right)^{-\frac{1}{2}} \left(\frac{\langle\phi\rangle}{10^{15}\text{GeV}}\right)^2 \left(\frac{m_\phi}{10^{12}\text{GeV}}\right)^2 \left(\frac{T_R}{10^6\text{GeV}}\right)^{-1}, \qquad (3)
\end{aligned}
$$

where g_* counts the relativistic degrees of freedom, and $\Gamma_\phi \sim T_R^2/M_P$ denotes the total decay rate of the inflaton. It should be noted that the gravitino abundance is inversely proportional to the reheating temperature, which should be contrasted to the standard thermal production of the gravitinos.

CONSTRAINTS ON INFLATION MODELS

Now we would like to derive constraints on the inflation and SUSY breaking models, using the non-thermal production of the gravitinos discussed in the previous section. The abundance of the gravitinos are tightly constrained by e.g. BBN, depending on the properties of the gravitino. Using (3), therefore, we can constrain the inflaton parameters.

In Fig. 1, we show the constraints on the inflaton mass and VEV for $m_{3/2} = 1\,\text{GeV}, 1\,\text{TeV}$, and $100\,\text{TeV}$, together with typical values of the inflation models. The region above each solid line is excluded. We find that in the case of $m_{3/2} = 1\,\text{TeV}$ with the hadronic branching ratio $B_h = 1$, all the inflation models shown in the figure are excluded. For the gravitino mass lighter or heavier than the weak scale, the constraints become relaxed. The inflaton mass and its VEV depend on the inflation models. Generically speaking, for larger m_ϕ and $\langle\phi\rangle$, the constraints become severer, simply because more gravitinos are produced by the inflaton decay (see (3)). On the other hand, if the inflaton is charged under some symmetries, its VEV becomes suppressed or even forbidden especially when the symmetry is exact at the vacuum. Then the bounds can be avoided for such inflation models. This is the case of the chaotic inflation model with a discrete symmetry (note that the chaotic inflation model shown in Fig. 1 is the one without such a symmetry). For the other possible solutions [7], see Ref. [9].

The solid lines which denote the constraint are jaggy at an intermediate value of m_ϕ. This is because the dominant production channels of the gravitinos changes. In the right side, the gravitinos are produced by the spontaneous and anomaly-induced decays, while the inflaton directly decays into a pair of the gravitinos in the left side.

In Fig. 1, we have set T_R to be the highest value allowed by the cosmological constraints. As mentioned before, the abundance of the non-thermally produced gravitinos is inversely proportional to T_R, which is different from that of the thermally produced one. If T_R takes a smaller value, the constraints shown in the figure becomes severer. Thus, the bounds shown in Fig. 1 are the most conservative ones. Note that one may have to introduce couplings of the inflaton with the SM particles to realize the highest allowed reheating temperature.

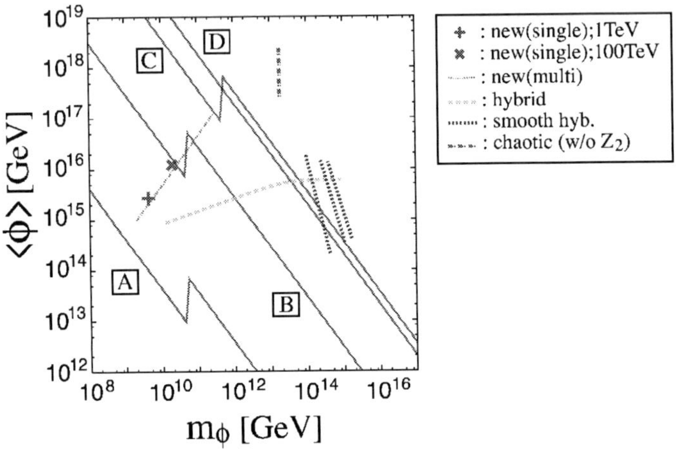

FIGURE 1. Constraints from the gravitino production by the inflaton decay, for $m_{3/2} = 1\,\text{TeV}$ with $B_h = 1$ (case A), $m_{3/2} = 1\,\text{TeV}$ with $B_h = 10^{-3}$ (case B), $m_{3/2} = 100\,\text{TeV}$ (case C), and $m_{3/2} = 1\,\text{GeV}$ (case D). The region above the solid (gray) line is excluded for each case. For $m_\phi \gtrsim \Lambda$, we have used the anomaly-induced inflaton decay into the hidden gauge/gauginos to estimate the gravitino abundance, while the gravitino pair production has been used for $m_\phi \lesssim \Lambda$. Since T_R is set to be the highest allowed value, the constraints shown in this figure are the most conservative ones.

CONCLUSION

The above studies may provide us with a breakthrough toward the full understanding of the inflationary universe. In addition to the standard analysis on the density fluctuations, the inflation models in supergravity are subject to the constraints due to the (non)-thermally produced gravitinos. Whether a consistent thermal history after inflation is realized now becomes a new guideline to sort out the inflationary zoo, and hopefully it will pin down the true model, together with data in the future collider experiments such as LHC.

REFERENCES

1. M. Endo, K. Hamaguchi and F. Takahashi, Phys. Rev. Lett. **96**, 211301 (2006); S. Nakamura and M. Yamaguchi, Phys. Lett. B **638**, 389 (2006).
2. M. Dine, R. Kitano, A. Morisse and Y. Shirman, Phys. Rev. D **73**, 123518 (2006).
3. M. Endo, K. Hamaguchi and F. Takahashi, Phys. Rev. D **74**, 023531 (2006).
4. M. Kawasaki, F. Takahashi and T. T. Yanagida, Phys. Lett. B **638**, 8 (2006); Phys. Rev. D **74**, 043519 (2006).
5. T. Asaka, S. Nakamura and M. Yamaguchi, Phys. Rev. D **74**, 023520 (2006).
6. M. Endo, M. Kawasaki, F. Takahashi and T. T. Yanagida, Phys. Lett. B **642**, 518 (2006).
7. M. Endo, K. Kadota, K. A. Olive, F. Takahashi and T. T. Yanagida, JCAP **0702**, 018 (2007).
8. M. Endo, F. Takahashi and T. T. Yanagida, arXiv:hep-ph/0701042.
9. M. Endo, F. Takahashi and T. T. Yanagida, arXiv:0706.0986 [hep-ph].
10. E. Witten, Nucl. Phys. B **188**, 513 (1981).

An AdS/CFT Correspondence that Establishes a Cosmology and Field Unification

Joseph Towe

Department of Physics, The Antelope Valley College, Lancaster, CA 93536-5426 USA

Abstract. Pure SUGRA interactions and consequent cumulative absorptions of spin-2 action are associated with a continuum of Huygens patterns, each pattern describing a history of secondary inflation events that potentially complements the isotropic shock wave of global expansion. A unique Huygens pattern is established by absorptions (phase transitions) that preserve Riemannian geometry. This unique history establishes observed structure.

Keywords: super-gravity, scale transformations, isotropic expansion, gauge invariance
PACS: 01.30.Cc, 04.50.+h, 04.65.+e, 11.25.Hf, 11.25.Mj

INTRODUCTION

Osp(1/4) pure super-gravity on M_4:

$$\mathcal{L} = \sqrt{-g}\,\mathbf{R} + \mathbf{e}\overline{\psi}_{\mu}\gamma_{\nu}\gamma^{5}\mathbf{D}_{\sigma}\psi_{\rho}\varepsilon^{\mu\nu\rho\sigma}, \tag{1}$$

and a sector of fermions are simultaneously established by AdS_7XS^4 [1]. Based upon SUGRA terms $i\overline{\psi}_{\mu}\gamma_{\nu}\gamma^{5}M_{A}\omega^{A}{}_{\sigma}\psi_{\rho}\varepsilon^{\mu\nu\rho\sigma}$ one obtains first-order interactions

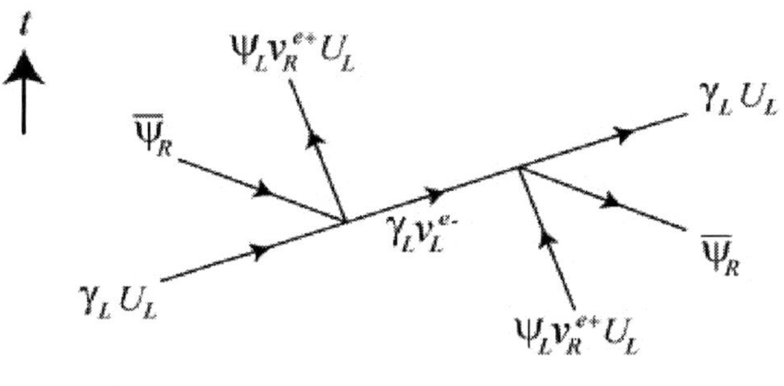

FIGURE 1. SUGRA mediated quark-lepton transitions

CP957, *Particles, Strings, and Cosmology, 13th International Symposium*
edited by A. Rajantie, C. Contaldi, P. Dauncey, and H. Stoica

between gravitons and spin-(3/2) composites of photons and quarks (or leptons).

Observe that the proposed pure SUGRA interactions produce an amplitude for similar (pure SUGRA) interactions that are characterized by absorptions of *composite*, spin-2 particles; e.g. interactions described by SUGRA vertices $\gamma_L U_L + \psi_L \nu_L^{e^-} \overline{U}_R \rightarrow \gamma_L \nu_L^{e^-} + \psi_L$ and $\gamma_L \nu_L^{e^-} + \psi_L \nu_R^{e^+} U_L \rightarrow \gamma_L U_L + \psi_L$. Such interactions clearly involve net absorptions of spin-2 action by spin (3/2) particles, which cumulatively produce increments of 4-curvature (energy-momentum), and (subsequent to inflation) increments of 4-volume.

These pervasive and isotropically distributed inflation events are regarded as complementing the isotropic shock front of global expansion with wavelets of secondary expansion, as exemplified by the Huygens pattern:

Figure 2. Secondary shock waves of universal expansion

Such inflation events are associated with SUGRA-induced phase transitions $\exp(i/\hbar)\int M_A \omega^A{}_\mu dx^\mu$, which can be written in the form $\tau = \tau_0 \exp\int \phi_\mu dx^\mu$, where $\phi_\mu = \frac{1}{2}\log f(x^\lambda)_{|\mu}$ and where $f(x^\lambda) = (i/\hbar)\int M_A \omega^A dx^\mu = (i/\hbar) M_A \omega^A{}_\mu x^\mu$ is a Finsler factor; i.e. a factor that modifies the scale of the metric tensor $\hat{g}_{\mu\nu} = f(x^\lambda) g_{\mu\nu}$, as in the theory of Weyl [2]. The string landscape is identified with the continuum of phase transitions and associated Huygens patterns (large-scale histories). This continuum is compared with the continuum of hydrogenic orbitals that is admitted by the classical theory of Weyl. Thus the advertised, unique cosmological process (from small scale to large scale) is selected in terms of the same principle of gauge invariance as that by which F. London selected the physically relevant hydrogenic orbitals [3]. Specifically, the continuum of classically admissible patterns is replaced by the unique pattern consisting of the events that preserve Riemannian geometry; i.e. the events that satisfy $(2\pi i/2h)\log M_A \omega^A{}_\mu x^\mu = 2\pi i N$; i.e. that correspond to

$$f(x^\lambda)=10^N: N=0,1,2,3,... \quad (2)$$

Each value in the sequence of scale factors (2) is associated with the proposed, collective absorptions of action $\mathcal{L}d^4x$ which are significant for the large scale. Based upon equation 2, it is concluded that each such action increment results in a mass-energy density and subsequent proliferation of mass-energy distribution which is ten times that associated with the previous state. Specifically, N=0 is associated with a single galaxy, N=1 with an isotropic configuration of galactic clusters, N=2 with an isotropic configuration of super-clusters etc.

CONFIRMATIONS OF THE PREDICTED STRUCTURE

The large-scale that is indicated by Fig. 2 is confirmed by observations. Specifically, the sector labeled 'Friedman limit' is revealed by observations of the immediate and intermediate past. The states represented by the Huygens fringe are revealed by observations of the early universe (of regions near the event horizon), where 4-distributions are in various stages of pre-inflation (ongoing inflation). The inflating 2-dimensional hyper-surfaces, which constitute the proposed secondary inflation events (2-dimensional expansions, approaching isotropic distribution, within a 3-dimensional expansion) exhibit significant radii of curvature. Such hyper-surfaces are confirmed by radio surveys:

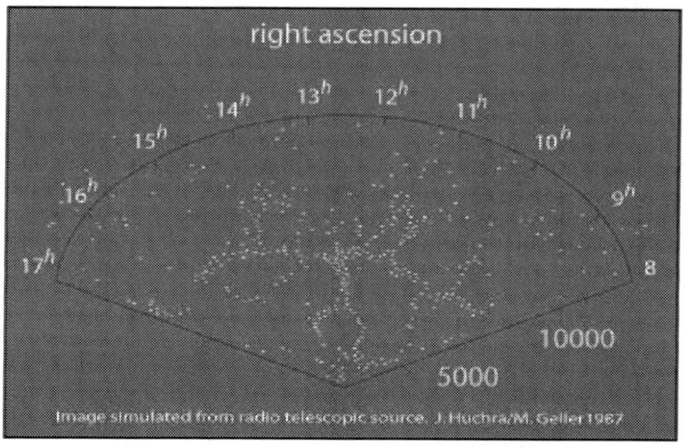

FIGURE 3. A radio-survey of regions near the event horizon.

In this well-known figure, reproduced from [4], one observes galactic clusters distributed on intersecting, semi-spherical surfaces. After surveying many such clusters, J. P. Huchra and M. Geller concluded that "galactic clusters lie on thin, sharply defined, semi-spherical shells that surround vast, empty voids". They compared these phenomena with the "bubble-like structures that constitute the foam in a kitchen sink" [4].

A quantitative feasibility argument for the proposed model is also motivated by the sequence of cosmological scale factors that is prescribed by equation 2. This theoretically generated equation is confirmed by observations indicating *that galaxies are typically separated by a distance that is about ten times the diameter of the typical galaxy; that galactic clusters are typically separated by a distance that is about ten times the diameter of the typical cluster etc.* Moreover, the scale factors that are described by (2) motivate an iterative calculation, which predicts a number of galaxies that is approximately equal to the number indicated by observation. Specifically, observations of local conditions establish that the typical galaxy is about $ct_0 = 10^5$ light years (ly) in diameter, and that five galaxies populate the typical basic cluster. Thirdly, homogeneity and isotropy permit that the number of galaxies be implemented to express approximate volume on scales through 10^9 light years (through this scale, galaxies appear to fill volumes rather than to populate the surfaces of semi-spherical shells). One therefore adopts the equation $(4/3)\pi R_1^3 = 5$, describing the volume of the typical basic cluster in terms of a number of galaxies, and calculates the radius $R_1 \cong 1.06$ galaxies. According to equation 2, $R_N = ct_N = ct_0 10^N$, implying that the radius of the typical galactic cluster in light years is about $ct_1 = ct_0 10 = 10^5(10) = 10^6$ ly. Continuing the iteration that is prescribed by equation 2, one determines the approximate radius (as a number of galaxies) of the N=2 state, the N=3 state, the N=4 state etc. Finally, one determines the radius of the N=5 state in terms of a number of galaxies: 10(diameter of N=4 state) = 10(2(8400)) = 168000 (where 8400 is a number that is obtained if the iterative calculation proceeds in detail from the beginning). Since galactic clusters populate the surfaces of semi-spheres on scales larger than 10^9 light years, the number of galaxies in the N=5 state is (summing areas of opposite semi-spherical hyper-surfaces) given by $4\pi R_5^2 = 4(3.14)(168000)^2 = 3.54 \times 10^{11}$. The radius of the N=5 state in light years is $ct_5 = ct_0 10^5 = 10^{10}$ ly, which is the approximate radius of the observable universe.

REFERENCES

1. O. Aharony, S. Gubser et.al., *ArXiv*: hep-th/9905111 v3, 1999, 48-52.
2. H. Weyl, *Space, Time and Matter*, London, 1922, Chapter 4, section 35.
3. F. London, *Z. Physik*, **42**, 1927, 375-389.
4. J. P. Huchra and M. Geller, *Semi-annual Report: Harvard-Smithsonian Observatory*, 1 July-31 December, 1987, 4 -5.

Effects of The Primordial Magnetic Field on The CMB

Dai G. Yamazaki*, Kiyotomo Ichiki†, Toshitaka Kajino*,** and Grant J. Mathews‡

*National Astronomical Observatory, Japan Mitaka, Tokyo 181-8588, Japan
†Research Center for the Early Universe, School of Science, the University of Tokyo, 7-3-1 Hongo, Bunkyo-ku 113-0033, Japan
**Department of Astronomy, Graduate School of Science, University of Tokyo 7-3-1 Hongo, Bunkyo-ku, Tokyo 113-0033, Japan
‡Center for Astrophysics, Department of Physics, University of Notre Dame, Notre Dame, IN 46556, U.S.A.

Abstract. The existence of a primordial magnetic field (PMF) would affect both the temperature and polarization anisotropies of the cosmic microwave background (CMB). It also provides a plausible explanation for the disparity between observations and theoretical fits to the CMB power spectrum. Here we report on calculations of not only the numerical power spectrum of the PMF, but also the correlations between the PMF power spectrum and the primary curvature perturbations.

Keywords: Primordial magnetic field, CMB
PACS: 98.62.En,98.70.Vc

INTRODUCTION

Magnetic fields in clusters of galaxies have been observed with a strength of $0.1-1.0\ \mu$ G. The existence of a primordial magnetic field (PMF) of order 1 nG whose field lines collapse as structure forms is one possible explanation for such magnetic fields to galactic clusters. The origin and detection of the PMF is, hence, a subject of considerable interest in modern cosmology. Moreover, a PMF could influence a variety of phenomena in the early universe [1] such as the cosmic microwave background (CMB) [2, 3, 4, 5], and the matter density field [6]. In the present work, in order to clarify the role of the PMF in CMB precisely, we must take into consideration the correlation between the CMB fluctuations induced by the PMF and those due to the primordial curvature and tensor perturbations. Also, we numerically evaluate the stochastic PMF power spectrum of the stochastic PMF and therefore avoid recourse to analytic approximations. We use the isocurvature initial condition for estimations of the CMB with the PMF effect in this paper. Throughout this letter we fix the best fit cosmological parameters of the ΛCDM + Tensor model as follows [7]: $h=0.792$, $\Omega_b h^2=0.02336$, $\Omega_m h^2=0.1189$, $n_S=0.987$, $r=0.55$, $n_T=-r/8=-0.069$, and $\tau_c=0.091$ in flat Universe models, where h denotes the Hubble parameter in units of 100 km s^{-1} Mpc^{-1}, Ω_b and Ω_m are the baryon and cold dark matter densities in critical density units, n_S is the spectral index of the primordial scalar fluctuation, r the ratio of the amplitude of tensor fluctuations to scalar potential fluctuations, n_T is the spectral index of the primordial tensor fluctuation, and τ_c is the optical depth of Compton scattering, respectively. We use natural units $c=\hbar=1$.

CP957, *Particles, Strings, and Cosmology, 13th International Symposium*
edited by A. Rajantie, C. Contaldi, P. Dauncey, and H. Stoica

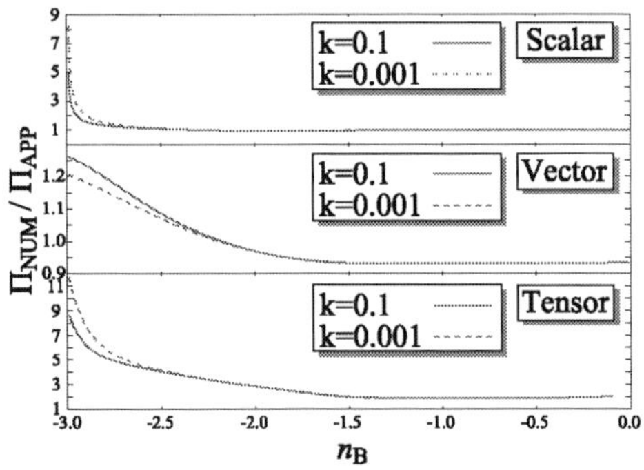

FIGURE 1. The comparison between our exact estimation and the previous approximation of the PMF power spetrum. The vertical axis is the ratio of our exact estimation(Π_{NUM}) to the previous approximation(Π_{APP}), and The horizontal axis is the power spectral index of the PMF(n_B).Red bold and Green dotted curves are $k = 0.1$ and $k = 0.001$, respectively.

PRIMORDIAL MAGNETIC FIELD

Since the trajectories of plasma particles are bent by Lorentz forces in a magnetic field, photons are indirectly influenced by the magnetic field through Thomson scattering. The energy density of the magnetic field can be treated as a first order perturbation upon a flat Friedmann-Robertson-Walker (FRW) background metric. In the linear approximation, the magnetic field evolves as a stiff source. Therefore, we can discard all back reactions from the magnetohydrodynamic (MHD) fluid onto the field itself. The conductivity of the primordial plasma is very large, so that the magnetic field is “frozen-in” [8]. Furthermore, we can neglect the electric field, i.e. $E \sim 0$, and can decouple the time evolution of the magnetic field from its spatial dependence, i.e. $\mathbf{B}(\tau,\mathbf{x}) = \mathbf{B}(\mathbf{x})/a^2$ for very large scales , where a is the scale factor. We assume that the PMF is statistically homogeneous, isotropic and random. For such a magnetic field, the power spectrum can be taken as a power-law $S(k) =< B(k)B^*(k) >\propto k^{n_B}$ [8] where n_B is the power-law spectral index of the PMF. The index n_B can be either negative or positive depending upon the physical processes of the field creation. Although possible origins of a PMF have been studied by many authors, there is no consensus how the PMF correlates with the primordial density fluctuations. In order to study the PMF effect in a more general manner, we here consider possible correlations between the PMF and the primordial density and tensor fluctuations. Therefore, we introduce a coefficient "s" to parameterize the correlation between the PMF and the primary power spectra[6, 9],

$$
\begin{aligned}
(2l+1)^2 C_l^{(X)} &= \frac{4}{\pi}\int dk k^2 [\Theta_{\rm p}^{(X)}(k)\Theta_{\rm p}^{(X)*}(k) + \Theta_{\rm PMF}^{(X)}(k)\Theta_{\rm PMF}^{(X)*}(k) \\
&\quad + s^{(X)}\{\Theta_{\rm p}^{(X)}(k)\Theta_{\rm PMF}^{(X)*}(k) + \Theta_{\rm PMF}^{(X)}(k)\Theta_{\rm p}^{(X)*}(k)\}], \qquad (1)
\end{aligned}
$$

where $X = S$ or T for scalar or tensor modes. The last term is the correlation between the primary and the PMF power spectra.

In previous analyses , an analytic approximation to the PMF power spectrum was utilized. Specifically, $(k+k')^{n_\mathrm{B}+2}-|k-k'|^{n_\mathrm{B}+2}\sim 2(n_\mathrm{B}+2)k^{n_\mathrm{B}+1}k'$ for $k'<k$, and $2(n_\mathrm{B}+2)kk'^{n_\mathrm{B}+1}$ for $k'>k$. Those approximation is better for $k'\ll k$ or $k'\gg k$, or $n_B<-2$, however worse for otherwise since neglected terms by such approximation are not able to be neglected. When previous works estimate PMF power spectra, approximated equations are integrated for $0<k'<k$. Therefore previous approximation can not calculate correctly in this case. it leads to smaller values than the numerically evaluated spectrum for n_B < -2(Fig.1). Thus, in this paper we have implemented a numerical method ([6, 10]) to evaluate the PMF power spectrum. Using this method, we are able to take quantitative account of the time evolution of the cut off scale and thereby reliably calculate the effects of the PMF.

RESULTS AND DISCUSSIONS

We have explored effects of the PMF on the CMB for the allowed PMF parameters ($B_\lambda<10\,n$G and $n_\mathrm{B}<-2.4$) deduced in our previous work [4]. In Fig. 2 we depict the CMB temperature and polarization anisotropies in the presence of the PMF taking into account the correlations. Since we obtain the temperature and polarization anisotropies of the CMB with the PMF from the isocurvature initial condition, the phase of CMB perturbation with the PMF is different $\pi/2$ from those without the PMF (by the adiabatic initial condition) and odd peaks of the scalar mode of the CMB perturbations raise up for the positive correlation between the PMF and the energy density perturbations, or suppress for the negative one. These are compared with the observed power spectra (WMAP [11, 12], ACBAR [13], CBI [14, 15], DASI [16], BOOMERANG [17], and VSA [18]). Since the models of the lower n_B give better fits to the observations than those of the higher n_Ba PMF for the lower ℓ regions of the TE mode [(2a) and (2b) of Fig.2], we see that the power spectral index of the PMF, n_B, is more effectively constrained from CMB observations for lower ℓ than those for higher ℓ. Furthermore,there is no discrepancy between observations and theories of the CMB polarization for models with the PMF for higher ℓ [(2c) and (2d) of Fig.2]. Especially, we see that the models of the lower n_B give better fits to the observations than those of the higher n_Ba PMF for the lower ℓ regions of the TE mode [(2c) of Fig.2]. This shows that the PMF is one of the most effective candidates for explaining the CMB polarizations for higher ℓ. The power of CMB emission depend on frequency, and such difference expect to be due to foreground effects (e.g. the Sunyaev-Zel'dovich effect) , on the other hands, effects of the PMF is frequency-independent since the PMF affect the primary CMB as background. Therefore the degeneracy between the PMF and others foreground effects for higher ℓ is weaker, and we can simultaneously constrain the PMF and foreground effects individually by more than two different frequencies observation data set. In summary, we have found that we can constrain more precisely the power law index n_B and amplitude B_λ of the PMF from all modes of CMB temperature and polarization anisotropies. The strong degeneracy of these parameters [3, 4], is broken by the different effects of the PMF on the CMB power spectrum for lower and higher ℓ. The scalar mode from the PMF can be a main source for lower ℓ, while the vector mode can dominate for higher ℓ in the CMB temperature anisotropies. Furthermore, we have placed a limit on the correlation parameters $s^{(X)}$, which have a close relation to the origin of the PMF, along with the other PMF parameters.

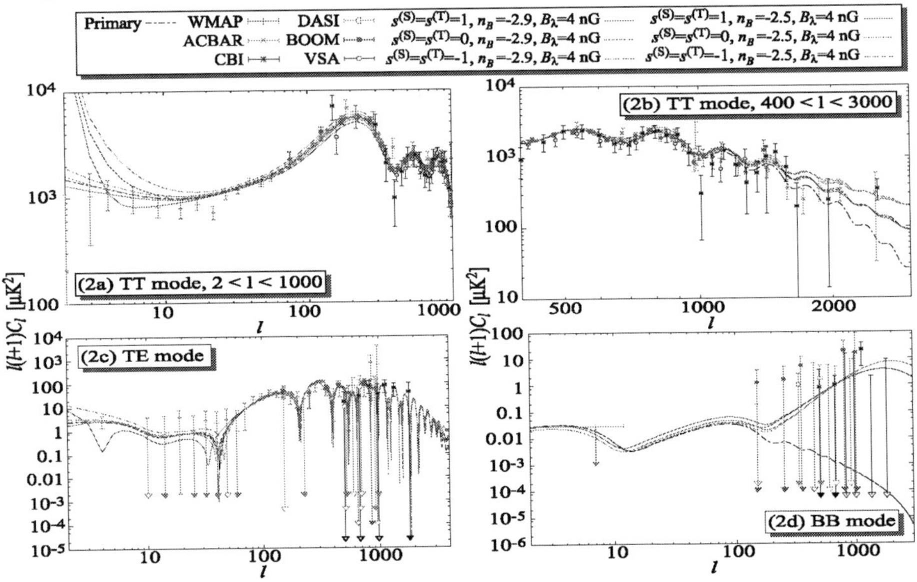

FIGURE 2. CMB temperature and polarization anisotropies from the PMF. Panels (2a), (2b), (2c), and (2d) show TT($2 < \ell < 1000$), TT($400 < \ell < 3000$), TE($2 < \ell < 4000$), and BB($2 < \ell < 3000$) modes, respectively. Curves in all Panels are theoretical lines as indicated in the legend box on the figure. Lines in (2c) are plotted in the absolute value. Downward arrows for the error bars of Panels (2c) and (2d) indicate that the data points are positive and the lower error negative.

REFERENCES

1. D. Grasso, and H. R. Rubinstein, *Phys. Rept.* **348**, 163–266 (2001).
2. D. G. Yamazaki, K. Ichiki, and T. Kajino, *Nuclear Physics A* **758**, 791–794 (2005).
3. D. G. Yamazaki, K. Ichiki, and T. Kajino, *Astrophys. J.* **625**, L1–L4 (2005).
4. D. G. Yamazaki, K. Ichiki, T. Kajino, and G. J. Mathews, *Astrophys. J.* **646**, 719 (2006).
5. D. G. Yamazaki, K. Ichiki, T. Kajino, and G. J. Mathews, *PoS(NIC-IX)*. p. 194 (2006).
6. D. G. Yamazaki, K. Ichiki, K. Umezu, and H. Hanayama, *Phys. Rev.* **D 74**, 123518 (2006).
7. D. N. Spergel, et al. (2006), `astro-ph/0603449`.
8. A. Mack, T. Kahniashvili, and A. Kosowsky, *Phys. Rev.* **D 65**, 123004 (2002).
9. M. Giovannini, *Class. Quant. Grav.* **23**, 4991–5026 (2006).
10. D. G. Yamazaki, K. Ichiki, T. Kajino, and G. J. Mathews (2007).
11. G. Hinshaw, et al. (2006), `astro-ph/0603451`.
12. L. Page, et al. (2006), `astro-ph/0603450`.
13. C.-L. Kuo, et al. (2006), `astro-ph/0611198`.
14. A. C. S. Readhead, et al., *Astrophys. J.* **609**, 498–512 (2004).
15. J. L. Sievers, et al. (2005), `astro-ph/0509203`.
16. E. M. Leitch, et al., *Astrophys. J.* **624**, 10–20 (2005).
17. W. C. Jones, et al., *Astrophys. J.* **647**, 823 (2006).
18. C. Dickinson, et al., *Mon. Not. Roy. Astron. Soc.* **353**, 732 (2004).

Distinguishing between $w < -1$ Dark Energy Models

Houri Ziaeepour

Mullard Space Science Laboratory, Holmbury St. Mary, Dorking, Surrey, RH5 6NT, UK.

Abstract. Recent data and new data analysis methods show that most probably the parameter w in the equation of state of the dark energy is smaller than -1 at low redshifts. We briefly review some of the models with such a property and without violating null energy condition. We investigate the difference between the observables and predictions of these models, and how they can be explored to single out or constrain the origin of dark energy and its properties.

Keywords: Dark energy, Dark matter, Cosmology, Early universe

Understanding the nature of dark energy is one of the biggest challenges in present cosmology and particle physics. In order to achieve this goal, the measurement of the cosmological evolution of the dark energy, parameters of the candidate models, and their origin and relation with other contents of the Universe are of extreme importance. This also means that the way we model the data and extract parameters affects our interpretation of what is the dark energy and how it evolves.

At present all the determination of dark energy parameters is based on the simplest extension of the LCDM. The energy content of the Universe is usually considered to be composed of 3 components: cold visible and dark matter, hot matter, and a dark energy component with a perfect fluid equation of state:

$$\frac{H^2(z)}{H_0^2} = \frac{\rho(z)}{\rho_0} = \Omega_m(1+z)^3 + \Omega_{hot}(1+z)^4 + \Omega_{de}(1+z)^{3\gamma(z)} \tag{1}$$

When $\gamma(z)$ does not depend on redshift, $\gamma = w+1$, and $w \equiv P/\rho$. We note that in this definition no interaction between various components is included. Although observations of the CMB anisotropy shows that the non-gravitational interaction between the hot matter - mainly CMB - and dark and baryonic matter is very small, the constraints on the non-gravitational interaction between dark components are not very strong[1].

Using 1 for fitting data from CMB, LSS, and supernovae, the estimations of recent measurements are summarized in Fig.1. Many of these estimations relay on multiple type of data to remove the degeneracy between cosmological parameters.

Therefore in this sense not all of them are independent. Nonetheless, if we assume that the difference in their estimation is statistical, the best estimation of w would be the weighted average of all the measurements. What one obtains in this way is $w = -1.056 \pm 0.023$ at 2σ level. This value as well as the individual measurements are all very close to the critical value $w = -1$ (a cosmological constant). Nevertheless, the best estimation and most of the measurements are consistent with $w \lesssim -1$. In the most simplistic view of the data we can say that the deviation from a cosmological constant is

CP957, *Particles, Strings, and Cosmology, 13th International Symposium*
edited by A. Rajantie, C. Contaldi, P. Dauncey, and H. Stoica

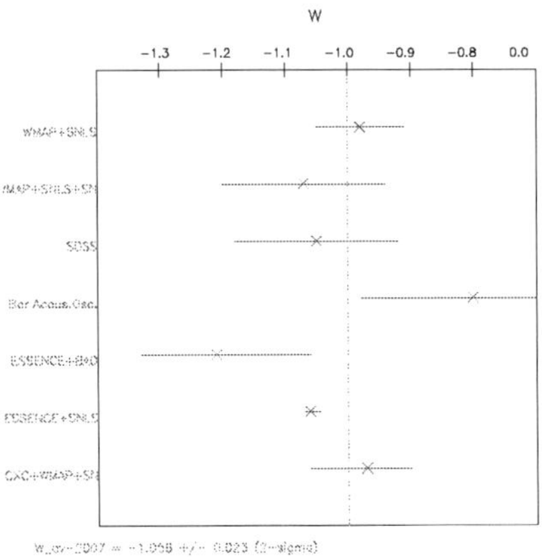

FIGURE 1. Recent estimations of w and their weighted average.

due to the errors. However, considering the best estimation mentioned above, even at 2σ level $w < -1$. This can be the evidence of a much richer physics behind the dark energy than just a putative cosmological constant.

Evidently just one number is not enough to decide about the nature of such a complex entity as the dark energy. There have been a number of attempts to estimate the redshift evolution of w. Although present data is not ideal for this purpose, the minimal conclusion is that it does not significantly vary up to $z \sim 1$[2]. Here we want also to mention very briefly the issue of the techniques usually employed to determine the equation of state of the dark energy from data. In most data sets one has to fit a large number of parameters including w or its redshift expansion parameters together. Although this method is acceptable for quantities that their exact value is not crucial for their physics such as Ω_{dm} - at least not at our present level of knowledge e- the degeneracy in fit results can be very important for parameters close to a critical value such as w. A method based on geometrical properties of $d\rho/dz$ is suggested [3] which in one hand permits to directly measure the sign of w and its redshift dependence. On the other hand, up to certain limits its results are less affected by the uncertainty in prior parameters such as H_0 and Ω_m. Application of this method to the publicly available SN type I data shows that $w < -1$ at least up to $z \sim 0.5$ and its variation, if exists, is very small.

The main argument - or rather we can call it "fear" - against $w < -1$ is that it violates the weak energy condition i.e. $P_{de} + \rho_{de} \geqslant 0$. However, we should remember that this condition is written for a perfect fluid without interaction. If the interaction between various components, in particular with the dark energy, is taken into account the form of null energy condition would be much more complex and eventually w can be less than -1 without violation of any fundamental law of physics. In this case the evolution of total density can be written as:

$$\frac{\rho(z)}{\rho_0} = \Omega_m(1+z)^3 + \Omega_{hot}(1+z)^4 + \Omega_{de}(1+z)^{3\gamma(z)} + g^2 f(\Omega_m, \Omega_{hot}, \Omega_{de}, z) \quad (2)$$

where f is an unknown function and depends on the dark sector model. It is possible to show that for a number of models in which dark matter and dark energy interact with each other or are not an ideal fluid, *the effective* w when 1 is used to analyze the data in place of 2, is smaller than -1. Here we mention a few examples of these models. The simplest example is a decaying dark matter with cosmological constant as dark energy[4]. The next one is a decaying dark matter with a very small branching ratio to a light, axion-like scalar[5, 6]. The condensation of this field plays the role of a quintessence field with an evolution very similar to a cosmological constant from very early times after formation of the meta-stable dark matter. Finally the last example we mention here is a stable dark matter interacting with a quintessence field as dark energy[7]. Many other examples can be found, but at least some of them imply a violation of equivalence principal or cosmological variation of particle masses and couplings which are strongly constrained by the non-observation of density dependent effects of dark energy at cosmological distances.

How can we distinguish between these models ? This task is specially more difficult in the situation where dark energy does not strongly vary with redshift. But even if it has some variation, it is very difficult to conclude the nature of the underlaying model just from a couple of parameters that determine its variation. We need additional observables more closely related to the field theoretical aspects of the models. At present level of our knowledge about the physics beyond the Standard Model, it is very difficult to place the dark energy field - quintessence scalar - in the zoo of particles. Nonetheless, it is possible to guess some of the possible observables. For instance, if dark matter has a non-gravitational interaction with dark energy condensate, we expect that *dark energy particles* are released from the condensate. As they are expected to be very light, they should make a hot dark background of non-SM nature. This process is very similar to the scattering from a Bose-Einstein condensate[8]. The cosmological density of this hot component and its evolution depends on the type and strength of the interaction. These are unknown, but from constraints on the clustering of the dark energy we expect that the coupling must be very small, and therefore the density of the corresponding hot matter should be small too - probably less than CMB, nonetheless the observation of such a component can significantly help to understand the origin of dark energy. On the other hand, we should also expect some anisotropy in the dark energy at large that may be observable specially in the high precision CMB data.

If the dark energy is a cosmological constant the situation is more ambiguous because we don't yet have a generally acceptable definition for a vacuum energy. If its origin is some physics at very high energy scale - Planck or superstring scale - and fixed at very early stage in the evolution of the Universe, the dark energy would be completely static, isotropic, and have no effect other than gravity up to any measurable redshift. In this case an *effective* $w < -1$ would be the signature of a non-gravitational interaction in dark matter sector, for instance slow decay of dark matter. If at least part of the remnants of the decay are SM particles, we must be able to observe them as an additional component in cosmic rays. Their energy range depends on the mass of the decaying dark matter particles and can be $E \gtrsim 10^{13}$ GeV or $E \lesssim 100$ GeV. Moreover, it must somehow correlate with the large structures. By contrast, if the remnants are all dark, then they can be only detected through their effect on the structure formation. But this can be very small and very difficult to observe.

The last case we consider here is when the dark energy is the result of the condensation

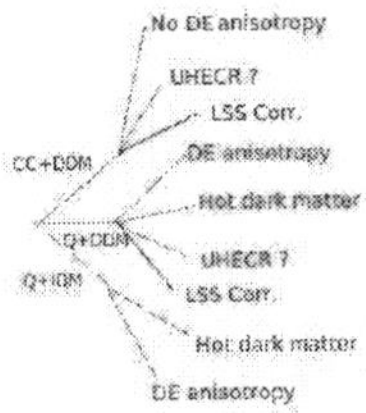

FIGURE 2. Dark energy models with effective $w < -1$ and their observables.

of a light scalar produced during the decay of a meta-stable dark matter. It has been shown that the equation of state of the condensate is very similar to a cosmological constant [5]. This model can produce a light hot dark matter - the scalar particles - an excess of visible cosmic rays, and/or dark remnants. The anisotropy of the dark energy would be very small but worth to be searched for.

Fig.2 summarizes the possible origins of a dark energy with effective $w < -1$ considered here and their observables.

Evidently, the most important question is how we can observe the tiny and weak interacting observables explained above. In one hand, this needs drastic improvement in the precision of measurements. On the other hand, more precise observations of cosmological effects means more sensitivity to the foreground phenomena that can mislead interpretation of observations. Therefore, the first step in this direction is a better understanding of the foreground. On another front, the discovery of physics beyond the Standard Model by LHC can be very important for clarifying which direction(s), both theoretical and experimental, we should investigate more closely to find the origin of the dark energy.

In conclusion, we have discussed some of the possible models for dark energy with effective $w < -1$ and without violation of the null energy condition. At present these types of dark energy are preferred by the data. Without considering any explicit implementation of these models we investigated the difference between their observables and how this can help to pin-down the underlaying model.

REFERENCES

1. B. Bochner (2007), URL `http://arXiv.org/abs/astro-ph/0702730`.
2. e. Riess, A.G., *ApJ* **656** (2007), URL `http://arXiv.org/abs/astro-ph/0611572`.
3. H. Ziaeepour, *Mod. Phys. Lett. A* **22**, 1569 (2007), URL `http://arXiv.org/abs/astro-ph/0702519`.
4. H. Ziaeepour (2000), URL `http://arXiv.org/abs/astro-ph/0002400`.
5. H. Ziaeepour, *Phys. Rev. D* **69**, 063512 (2004), URL `http://arXiv.org/abs/astro-ph/0308515`.
6. H. Ziaeepour (2006), URL `http://arXiv.org/abs/hep-ph/0603125`.
7. K. J. Das S., Corasaniti P.S., *Phys. Rev. D* **73**, 083509 (2006), URL `http://arXiv.org/abs/astro-ph/0510628`.
8. e. Band Y.B., *Phys. Rev. Lett* **84**, 5462 (2000), URL `http://arXiv.org/abs/cond-mat/0005021`.

PROGRAMME

Monday 2 July

09:00 - 09:25 Registration

Chair: Tom Kibble (Imperial)

09:25 - 09:35 *Welcome*
Donal Bradley, Head of the Physics Department

09:35 - 10:10 *Inflation and string theory landscape*
Andrei Linde (Stanford)

10:10 - 10:45 *Cosmological constraints on fundamental parameters*
Uros Seljak (Zurich)

Chair: Giuseppe Vitiello (Salerno)

11:15 - 11:50 *Neutrino oscillations - past, present, and future*
David Wark (Imperial)

11:50 - 12:25 *Neutrinos and the origin of fermion mass structure*
Graham Ross (Oxford)

12:25 - 13:00 *High-energy cosmic rays*
Giorgio Matthiae (Rome)

Chair: Kellogg Stelle (Imperial)

14:30 - 15:30 *The measure of the universe*
Stephen Hawking (Cambridge)

15:30 - 16:05 *The search for gravitational waves*
Jim Hough (Glasgow)

Chair: Renata Kallosh (Stanford)

16:35 - 17:10 *The search for tensor modes*
George Efstathiou (Cambridge)

17:10 - 17:45 *Brane inflation, an update*
Henry Tye (Cornell)

17:45 - 18:20 *Kac-Moody symmetries in supergravity and M theory*
Chris Pope (Texas A&M)

from 18:30 Reception

Tuesday 3 July

Chair: Wilfried Buchmüller (DESY)
09:00 - 09:30 *Heavy ion collisions and new forms of matter*
Larry McLerran (BNL)
09:35 - 10:10 *Gauge/gravity duality and the quark gluon plasma*
Dam Son (Washington)
10:10 - 10:45 *Black holes in high energy collisions*
Steve Giddings (Santa Barbara)

Chair: Andreas Albrect (Davis)
11:15 - 11:50 *Dark matter: Status and perspectives*
Joseph Silk (Oxford)
11:50 - 12:25 *Current status of direct dark matter searches*
Nigel Smith (RAL)
12:25 - 13:00 *Testing the LCDM paradigm on large and small scales*
Carlos Frenk (Durham)

14:30 – 18:30 Parallel sessions

Tuesday Parallel Sessions

14:30 – 16:15

Particles (Chair: Ben Allanach)

Hunting for the top partner in the littlest Higgs model with T parity at the CERN LHC
Shigeki Matsumoto

O'Raifeartaigh models with spontaneous R-symmetry breaking
Luca Ferretti

Recent results from HERA ep collisions
Francois Corriveau

Precision Standard Model tests at the Tevatron
Tommaso Dorigo

Prospects for SUSY with the first data at the LHC
Ulrich Goerlach

Strings (Chair: Michael Duff)

Stringy model of the cosmological dark energy
Irina Arefeva

Large gravitational waves in string-inflation
Axel Krause

Magnetogenesis from rotating cosmic string loops
Thorsten Battefeld

Bouncing and colliding branes
Jean-Luc Lehners

Brane annhilation and defect formation
Horace Stoica

Cosmology (Chair: Hiranya Peiris)

Propagation of a string across the cosmological singularity
Wlodzimierz Piechocki

Phenomenology of D-brane inflation with general speed of sound
Hiranya Peiris

New constraints on the observable inflaton potential from WMAP and SDSS
Julien Lesgourgues

Child universe production
Eduardo Guendelman

Inflaton decay in supergravity
Fuminobu Takahashi

16:45 – 18:30

Particles (Chair: Steven Abel)

The standard model on a domain-wall brane
Raymond Volkas

Sparticle spectra from large-volume string compactifications
Joseph Conlon

Probing the geometry of warped throats at the LHC
Bret Underwood

Standard Model and Beyond the Standard Model searches at the Tevatron
Gerard Sajot

Prospects for beyond SUSY with the first data at the LHC
Jon Butterworth

Strings (Chair: Axel Krause)

Inflation from branes
Louis Leblond

Dark matter and dark energy - fact or fantasy?
Philip Mannheim

The effective theory of codimension-two branes
Claudia de Rham

Field theoretical approach to the formation of junctions of cosmic strings
Petja Salmi

Motion in brane world models: The Bazanski approach
Magd Kahil

On Newton's law in supersymmetric braneworld models
Gonzalo Palma

Cosmology (Chair: Christopher Gordon)

The effect of SN Ia peculiar velocities on dark energy constraints
Christopher Gordon

Distinguishing between w < -1 dark energy models
Houri Ziaeepour

Suppression of Meszaros effect in coupled DE models
Silvio Bonometto

Solar system constraints on Gauss-Bonnet mediated dark energy
Stephen Davis

Dark energy-dark matter interaction from the Abell Cluster A586
Morgan Le Delliou

Wednesday 4 July

Chair: Anne Davis (Cambridge)

09:00 - 09:30 *Inflation and its cosmic probes, now and then*
Dick Bond (CITA)

09:35 - 10:10 *Testing string theory with CMB*
Renata Kallosh (Stanford)

10:10 - 10:45 *Quantum resolution of cosmological singularities using AdS/CFT*
Neil Turok (Cambridge)

Chair: David Bailin (Sussex)

11:15 - 11:50 *Search for Higgs-bosons and supersymmetry at the Tevatron*
Volker Büscher (Bonn)

11:50 - 12:25 *Metastable SUSY breaking in the cooling universe*
Vadim Kaplunovsky (Texas)

12:25 - 13:00 *Lattice QCD*
Ron Horgan (Cambridge)

13:00 - 14:30 Poster Session with Buffet Lunch

Daniel Cumberbatch	*Constraints on sub-dominant sterile neutrino dark matter from x-ray background observations*
Gianfranca De Rosa	*Light-nuclei identification with PAMELA space telescope: preliminary in-flight results*
László Gergely	*Cosmological tests of the generalized Randall-Sundrum brane-world model with late-time dark radiation*
Jinn-Ouk Gong	*Multiple scalar particle decay and perturbation generation*
Choon-Lin Ho	*Cosmological constant from gauge fields on extra dimensions*
Choon-Lin Ho	*(Quasi)-exactly solvable quasinormal modes*
Masahiro Kawasaki	*Latetime Affleck-Dine bayogenesis*
Biagio Lucini	*Persistent currents in QCD with compact spatial dimensions*
Leila Maghlaoui	*Bosons-parafermions Wess-Zumino model*
Farrin Payandeh	*The CMB anisotropies, and the choice of vacuum in QFT in de Sitter spacetime in Krein space quantization*
Joe Towe	*Large-scale structure as a consequence of SUGRA GUT Symmetry*
Yosef Verbin	*Sigma Model Q-balls and Q-Stars*

Thursday 5 July

Chair: Dick Bond (CITA)

09:00 - 09:30 *The case for an aggressive program of dark energy probes*
Andreas Albrecht (Davis)

09:35 - 10:10 *Attempts to model dark energy*
Ed Copeland (Nottingham)

10:10 - 10:45 *Everpresent Lambda*
Rafael Sorkin (Syracuse)

Chair: Jihn E. Kim (Seoul)

11:15 - 11:50 *CP violation and quark flavour experiments; current status and prospects for the LHC era*
Tatsuya Nakada (CERN and EPFL)

11:50 - 12:25 *Beyond the Standard Model at the LHC and beyond*
John Ellis (CERN)

12:25 - 13:00 *Attractors in all extended supergravities*
Sergio Ferrara (CERN)

14:15 - 16:00 Parallel sessions

16:30 - 17:30 GP Thomson Lecture
Recent performance of GaN-based blue light emitting diodes and laser diodes
Shuji Nakamura (Santa Barbara)

19:00 - 22:00 Conference Dinner
Riverbank Park Plaza Hotel

Thursday Parallel Sessions

14:15 – 16:00

Particles (Chair: Gavin Davies)

Neutralino dark matter in the NMSSM
Ana Teixeira

Signatures of pseudo-degenerate WIMP dark matter
Adam Ritz

The dark matter as a light gravitino
Gilbert Moultaka

Cryogenic dark matter search experiment status and future: CDMS to SuperCDMS
Tobias Bruch

Results from the Xenon10 dark matter search experiment at Gran Sasso laboratories
Alfredo Ferella

Strings (Chair: Steven Abel)

Almost the supersymmetric standard model from intersecting D6-branes
David Bailin

Particle creation in braneworld collisions
Paul Saffin

Gauge-Higgs unification in six dimensions from the heterotic string
Jonas Schmidt

Why did the universe start in such a low entropy state?
Laura Mersini-Houghton

Properties of nonlocal cosmological models derived from string field theory
Liudmila Joukovskaya

Cosmology (Chair: Carlo Contaldi)

Constraints on non-Gaussianity in string-inspired inflation
Sarah Shandera

Density perturbations in the universe from massive vector fields
Konstantinos Dimopoulos

Has PVLAS detected the chameleon?
Anne Davis

Cosmic string loops and gravitational lensing
Katherine Mack

Is the cold spot in the CMB a texture?
Neil Turok

Friday 6 July

09:00 – 13:00 Parallel sessions

Chair: Peter Dornan (Imperial)
14:30 - 15:05 *The LHC machine and experiments: Status and prospects*
Jim Virdee (CERN and Imperial)
15:05 - 15:40 *Baryogenesis - 40 years later*
Wilfried Buchmüller (DESY)
15:40 - 16:15 *Non-perturbative (D-instanton) effects in string vacua: New implications for particle physics*
Mirjam Cvetic (Pennsylvania)

Chair: Sergio Ferrara (CERN)
16:45 - 17:20 *Amplitudes in field theory*
Bo Feng (Imperial)
17:20 - 17:55 *Emergent quantum mechanics and emergent symmetries*
Gerard 't Hooft (Utrecht)

Friday Parallel Sessions

09:00 – 10:45

Particles (Chair: Paul Dauncey)

LFV, LHC and DM: Synergy for a (multi)TeV SUSY
Antonio Masiero

New two-loop contributions to hadronic EDMs in the MSSM
Minoru Nagai

Latest CPV results at NA48
Giuseppina Anzivino

Search for new physics Beyond the Standard Model at BaBar
Matthew Barrett

Strings (Chair: Horace Stoica)

Lorentz violation and Finsler geometry
Chris Pope

Second order perturbative calculation of quasinormal modes
Hsien-chung Kao

A new non-perturbative bosonic string configuration
Jean Alexandre

Renormalisation of N=1/2 supersymmetry
Ian Jack

Towards lattice simulation of the gauge theory duals of black holes
Toby Wiseman

Cosmology (Chair: Kazuya Koyama)

Neutrino mass constraint from CMB, BAO and SN
Kazuhide Ichikawa

Effects of non-standard neutrino-electron interactions on relic neutrino decoupling
Teguayco Pinto

The role of micropancakes in dark matter detection experiments
Alexander Shirokov

Non-perturbative effect on thermal relic abundance of dark matter
Masato Senami

Increasing the effective number of neutrinos with decaying particles
Kazunori Nakayama

11:15 – 13:00

Particles (Chair: Ben Allanach)

Flavor charges and flavor states of mixed neutrinos

Massimo Blasone

Neutrino mixing as a source of dark energy

Giuseppe Vitiello

Dirac neutrinos, leptogenesis and a vanishing Higgs @ LHC

Thomas Underwood

Latest results from MINOS

Peter Litchfield

MiniBooNE neutrino oscillation result

Morgan Wascko

Strings (Chair: Jean-Luc Lehners)

Cycling branes in warped throats

Damien Easson

Low energy supersymmetry breaking from non-geometry

Eran Palti

Properties of some CFTs with M-theory duals

Oscar Varela

Hyperkahler sigma models on cotangent bundles of Hermitian symmetric spaces using projective superspace

Masato Arai

Realistic Yukawa couplings through instantons in intersecting brane worlds

Mark Goodsell

Cosmology (Chair: Kazuya Koyama)

Flavour-dependent effects on thermal leptogenesis

Stefan Antusch

Primordial nucleosynthesis as a probe of fundamental physics parameters

Steffen Stern

Aspects of a supersymmetric Brans-Dicke theory

Riccardo Catena

Effects of a primordial magnetic field on low and high multipoles of the CMB

Dai Yamazaki

Gravitino production from late-decaying scalar field

Motoi Endo

Saturday 7 July

Salam+50 Conference

Chair: Professor Donal Bradley FRS

09:30 - 09:45	*Welcoming address* Sir Richard Sykes FRS, Rector, Imperial College
09:45 - 10:45	Physics Department Centenary Salam Lecture *Salam and the grand view of physics* Professor Gerard 't Hooft, Nobel Laureate, University of Utrecht

Chair: Professor Sir Peter Knight FRS

11:15 - 11:30	*Abdus Salam and Pakistan* Dr Maleeha Lodhi, Pakistan High Commissioner, UK
11:30 -11:35	*Greetings from K. R. Sreenivasan,* *Director of the Abdus Salam ICTP, Trieste* read by Professor Michael Duff, Imperial College
11:35 - 12:35	*The International Centre for Theoretical Physics* Professor Faheem Hussain, Lahore University of Management Sciences

Chair: Professor Christopher Isham

13:30 - 14:30	*Abdus Salam at Imperial* Professor Tom Kibble, FRS, Imperial College
14:30 - 14:40	*Obituary of Abdus Salam by Yuval Neeman* read by Professor Kellogg Stelle, Imperial College
14:40 – 14:55	*Presentation of Salam Undergraduate Prize* Mirza Ahmad, President of the Ahmadiyya Muslim Youth Association UK
14:55 - 15:40	*Recollections of the Salam family* Dr Aziza Rahman, Ahmad Salam, Umar Salam, Saif Rahman

Chair: Professor Michael Duff

16:00 - 17:00	*Salam's mantle of Britishness* Dr Gordon Fraser, Salam biographer
17:00 - 17:30	Salam Memorabilia Display

PARTICIPANTS

Steven Abel	Durham IPPP
Augusta Abrahamse	UC Davis
Andreas Albrecht	UC Davis
Jean Alexandre	Kings College London
Benjamin Allanach	University of Cambridge
Stefan Antusch	MPI Munich
Giuseppina Anzivino	University of Perugia and INFN
Masato Arai	University of Helsinki and Helsinki Inst. Phys.
Irina Aref'eva	Steklov Mathematical Institute
David Bailin	University of Sussex
Myron Bander	University of California -- Irvine
Michael Barnard	UC Davis
Matthew Barrett	Brunel University
Thorsten Battefeld	DAMTP Cambridge University
Massimo Blasone	INFN and Universita di Salerno
Andrew Blechman	University of Toronto
J. Richard Bond	CITA
Silvio Bonometto	Universita di Milano-Bicocca
Leron Borsten	Imperial College London
Salima Boudjemaa	University of Southampton
Brandon Bozek	UC Davis
Mark Brook	University of Nottingham
Tobias Bruch	I Physical Institut,RWTH-Aachen
Wilfried Buchmüller	DESY
Clare Burrage	DAMTP, Cambridge University
Volker Büscher	Universität Bonn
Jonathan Butterworth	UCL
Riccardo Catena	DESY
Claudio Caviezel	Max-Planck-Institut for Physics, Munich
Alex Chambers	Imperial College London
Sherrie Chatzkel	Olympic Medical Center
Chan-Chi Chiou	Durham University
Joseph Conlon	DAMTP, Cambridge
Carlo Contaldi	Imperial College London
Ed Copeland	Nottingham University
Francois Corriveau	IPP / McGill University
Daniel Cumberbatch	Oxford University
Mirjam Cvetič	University of Pennsylvania
Duminda Dahanayake	Imperial College London
Paul Dauncey	Imperial College London
Gavin Davies	Imperial College London
Anne Davis	DAMTP, Cambridge University

Stephen Davis	Lorentz Institute, Leiden University
Claudia de Rham	McMaster/Perimeter Institute
Gianfranca De Rosa	Universita di Napoli Federico II
Pierre-Philippe Dechant	Cavendish Laboratory, Cambridge University
Thomas Dent	ITP, University of Heidelberg
Marco Di Mauro	INFN - Salerno
Konstantinos Dimopoulos	Lancaster University
Tommaso Dorigo	INFN - Sezione di Padova
Peter Dornan	Imperial College London
Michael Duff	Imperial College London
Damien Easson	Durham University
Hajar Ebrahim Najafabadi	Imperial College London
George Efstathiou	IoA, Cambridge
John Ellis	CERN
Motoi Endo	DESY
Mboyo Esole	KU Leuven
Bo Feng	Imperial College, London
Tom Ferbel	Dept of Energy/Univ. of Rochester
Alfredo Davide Ferella	RWTH-Aachen University
Sergio Ferrara	CERN
Luca Ferretti	SISSA/ISAS
Filipe Freire	Univ Leiden/Univ Utrecht
Carlos Frenk	Durham University
Laszlo Gergely	University of Szeged
Steve Giddings	University of California, Santa Barbara
Joshi Girish	University Of Melbourne
Ulrich Goerlach	ULP IPHC
Jinn-Ouk Gong	Harish-Chandra Research Institute
Mark Goodsell	University of Durham
Christopher Gordon	University of Oxford
Steven Gratton	Institute of Astronomy, Cambridge University
Eduardo Guendelman	Ben Gurion University
Lawrence Hall	University of California, Berkeley
Lisa Hall	University of Sheffield
Stephen Hawking	University of Cambridge
Choon-Lin Ho	Department of Physics, Tamkang University
Jenny Hogan	Nature magazine
Ron Horgan	University of Cambridge
James Hough	University of Glasgow
Stephan Huber	University of Sussex
Faheem Hussain	Lahore University of Management Sciences
Kazuhide Ichikawa	University of Tokyo
Alberto Iglesias	University of California Davis
Yukinori Iwashita	Tokyo Institute of Technology
Ian Jack	University of Liverpool
Steven Johnston	Imperial College London

Liudmila Joukovskaya	DAMTP, Cambridge
Magd Elias Kahil	Modern Sciences and Arts University
Renata Kallosh	Stanford University
Hsien-chung Kao	University of Durham
Vadim Kaplunovsky	University of Texas
Masahiro Kawasaki	University of Tokyo
Tom Kibble	Imperial College London
Jihn E. Kim	Seoul National University
Simon Koers	Max-Planck-Institut für Physik
Tomi Koivisto	Helsinki Institute of Physics
Kazuya Koyama	ICG, University of Portsmouth
Axel Krause	Arnold Sommerfeld Center, LMU München
Gaetano Lambiase	Università di Salerno
Morgan Le Delliou	CFTC, Lisbon University, Portugal
Louis Leblond	Texas A&M University
Jean-Luc Lehners	DAMTP, Cambridge
Julien Lesgourgues	LAPTH (Annecy, France)
Andrei Linde	Stanford University
Peter Litchfield	University of Minnesota
Biagio Lucini	Swansea University
Christoph Luedeling	Universitaet Heidelberg
Katherine Mack	Princeton
Leila Maghlaoui	Mentouri University-Constantine
Muhammad Maher	Queen Mary, University of London
Bienvenue Malumba	Université de Kinshasa
Philip Mannheim	University of Connecticut
Antonio Masiero	Univ. of Padua and INFN, Padua
Shigeki Matsumoto	Tokoku University
Giorgio Matthiae	University of Roma Tor Vergata
Larry McLerran	Brookhaven National Laboratory
Dhagash Mehta	Imperial College London
Noppadol Mekareeya	University of Cambridge
Zeeya Merali	New Scientist
Laura Mersini-Houghton	University of North Carolina
Ian Moss	Newcastle University
Gilbert Moultaka	LPTA-Montpellier
Tuomas Multamäki	University of Turku
Minoru Nagai	University of Tokyo
Tatsuya Nakada	CERN and EPFL
Kazunori Nakayama	University of Tokyo
Horatiu Nastase	Tokyo Institute of Technology
Siew Phang Ng	University Of Toronto
Gustavo Niz	Nottingham
Eoin Ó Colgáin	Imperial College London
Gonzalo Palma	DESY, Hamburg
Eran Palti	Oxford University

Luca Panizzi	University of Trieste - Department of Physics
Farrin Payandeh	Amir Kabir U. of Technology
Paniez Paykari	Imperial College London
Iain Peddie	Aristotle University of Thessaloniki
Hiranya Peiris	University of Chicago
Lydia Philpott	Imperial College London
Wlodzimierz Piechocki	Soltan Institute for Nuclear Studies
Christopher Pope	Texas A&M University
Poulose Poulose	University of Southampton
Andrea Quadri	Dip. di Fisica - Università di Milano
Arttu Rajantie	Imperial College London
Vappu Reijonen	Helsinki Institute of Physics
Adam Ritz	University of Victoria
Ray Rivers	Imperial College London
Nikoleta Rizopoulou	Imperial College London
Graham Ross	University of Oxford
Paul Saffin	Nottingham University
Gerard Sajot	LPSC Grenoble Universite Joseph Fourier
Petja Salmi	Leiden University
Jonas Schmidt	DESY Hamburg
Kai Schmidt-Hoberg	DESY Theory
Uros Seljak	University of Zurich
Masato Senami	ICRR, University of Tokyo
Sarah Shandera	Columbia University, ISCAP
Alexander Shirokov	CITA
Arunansu Silk	CEA/Saclay
Joseph Silk	University of Oxford
Nigel Smith	STFC Rutherford Appleton Laboratory
Paul Smyth	K.U. Leuven
Dam Son	University of Washington
Rafael Sorkin	Syracuse University
Kepa Sousa	Lorentz Institute (Leiden University)
Kellogg Stelle	Imperial College London
Steffen Stern	Institute for Theoretical Physics, Heidelberg
Florea Horatiu Stoica	Imperial College London
Gerard 't Hooft	Utrecht University
Fuminobu Takahashi	DESY
Ana Teixeira	LPT-Orsay
Steven Thomas	Queen Mary, University of London
Andrew Tolley	Perimeter Institute
Joseph Towe	The Antelope Valley College
Neil Turok	University of Cambridge
Henry Tye	Cornell University
Yoshi Uchida	Imperial College London
Bret Underwood	University of Wisconsin-Madison
Thomas Underwood	University of Durham

Carsten van de Bruck	University of Sheffield
Oscar Varela	Imperial College London
Yosef Verbin	The Open University of Israel
Tejinder Virdee	CERN and Imperial College
Giuseppe Vitiello	University of Salerno
Raymond Volkas	The University of Melbourne
Daniel Waldram	Imperial College London
Kameshwar C. Wali	Syracuse University
Lingyu Wang	Imperial College London
Matthias Wapler	Perimeter Institute
David Wark	Imperial College London
Morgan Wascko	Imperial College London
David Weir	Imperial College London
Toby Wiseman	Imperial College London
Dai Yamazaki	National Astronomical Observatory of Japan
Motohiko Yoshimura	Okayama University
Andrea Zanzi	University of Bonn
Houri Ziaeepour	University College London

Author Index

A

B

C

D

E

F

G

H

I

J

K

L

M

N

O

P

R

S

T

U

V

W

Y

Z